Rescuing Reason

BOSTON STUDIES IN THE PHILOSOPHY OF SCIENCE

VOLUME 230

RESCUING REASON

A Critique of Anti-Rationalist Views of Science and Knowledge

by

ROBERT NOLA

*The University of Auckland,
New Zealand*

KLUWER ACADEMIC PUBLISHERS

DORDRECHT / BOSTON / LONDON

A C.I.P. Catalogue record for this book is available from the Library of Congress.

ISBN 1-4020-1042-7 (HB)
ISBN 1-4020-1043-5 (PB)

Published by Kluwer Academic Publishers,
P.O. Box 17, 3300 AA Dordrecht, The Netherlands.

Sold and distributed in North, Central and South America
by Kluwer Academic Publishers,
101 Philip Drive, Norwell, MA 02061, U.S.A.

In all other countries, sold and distributed
by Kluwer Academic Publishers,
P.O. Box 322, 3300 AH Dordrecht, The Netherlands.

Cover of the paperback edition:
Painting by Charles F. Goldie and Louis John Steele: The arrival of the Maoris
in New Zealand used by permission of the Auckland Art Gallery Toi o Tamaki.

Printed on acid-free paper

Dedication

To the memory of my parents,

Jean Beatrice and Bože

TABLE OF CONTENTS

ACKNOWLEDGEMENTS

Some of this book was begun while I was a Visiting Fellow at the Center for the Philosophy of Science, University of Pittsburgh, during 1995. I wish to acknowledge the stimulating research environment provided by the Center. The completion of the book was assisted by a Marsden Fund Grant.

The following offered valuable comments on parts of the book, often, but not always, heeded: Peter Anstey, Jan Crosthwaite, Gürol Irzik, Fred Kroon and Bob Solomon. I am indebted to the publisher's anonymous reviewer.

Some of this book derives from earlier papers I have published on its various subjects. Sometimes portions of them reappear here as they were originally published. But more often than not they have been totally revised for the book; in some cases their different revised sections have been split up and used in different places in the book. I would like to thank the publishers for their kind permission to use previously published material in this way.

The original sources are as follows.

Reprinted by permission of Taylor and Francis, Oslo, Norway:
'The Strong Programme for the Sociology of Science, Reflexivity and Relativism'; *Inquiry: An Interdisciplinary Journal of Philosophy,* vol. 33, 1990, pp. 273-96;
'Postmodernism, a French Cultural Chernobyl: Foucault on Power/Knowledge'; *Inquiry: An Interdisciplinary Journal of Philosophy,* vol. 37, 1994. pp. 3-43.

Reprinted by permission of Frank Cass Publishers, London:
'Knowledge, Discourse, Power and Genealogy in Foucault', *Critical Review of International Social and Political Philosophy,* **1 #2**, 1998 pp. 109-54. This also appeared in R. Nola (ed.) *Foucault* (London, Frank Cass 1988).

Reprinted by permission of Kluwer Academic Publishers, Dordrecht, The Netherlands:
'Nietzsche's Naturalism: Science and Belief', in Babette Babich and Robert S. Cohen (eds.) *Nietzsche and the Sciences II: Nietzsche, Epistemology, and Philosophy of Science,* (Dordrecht, Kluwer, 1999); pp. 91-100;
'On the Possibility of a Scientific Theory of Scientific Method', *Science & Education* **8**, 1999, pp. 427-39;
'Saving Kuhn from the Sociologists of Science', *Science & Education* **9**, 2000 pp. 77-90;
(with H. Sankey) 'A Selective Survey of Theories of Scientific Method', in R. Nola and H. Sankey (eds.) *After Popper, Kuhn and Feyerabend; Recent Issues in Theories of Scientific Method,* (Dordrecht, Kluwer, 2000), pp. 1-65.

INTRODUCTION

The theme of this book is an old one going back to the beginning of philosophy in Socrates' and Plato's encounters with the anti-rationalists of their day, the sophists. Though it cannot emulate the depth and subtlety with which these philosophers reclaimed rationality and knowledge from their detractors, in a nutshell the book investigates the following. Do knowledge and science arise from the application of canons of rationality and scientific method? Or is all our scientific knowledge caused by socio-political factors, or by our interests in the socio-political – the view of sociologists of "knowledge"? (As will be argued, the scare quotes are quite deliberately intended.) Or does it result from the interplay of relations of power – the view of Michel Foucault? Or does our knowledge arise from "the will to power" – the view of Nietzsche? It is argued here that, despite their large contemporary following, the latter fail badly to make their case against advocates of rationality and methodology in science and knowledge.

Our contemporary sophists come in a number of guises, such as relativists, sceptics, nihilists and sociologists of "knowledge" – their most recent incarnation being as postmodernists. 'Postmodernism' is a broad term said to cover a number of intellectual stances within a number of disciplines from architecture to philosophy; the term covers not only intellectual fashions but also alleged social currents and changes. As its name suggests, postmodernism is "post-", setting itself against "the modern", whatever this may be. But such a definition, which says what postmodernism is not, is unhelpful, particularly in the case of postmodernism in philosophy. What counts as modern in philosophy is not clear; but two common philosophical targets of postmodernists are advocates of rationality and method such as Descartes and Kant. Postmodernists also hope to turn out the lights first turned on by eighteenth century enlightenment philosophers. They were called 'enlightened' in large part because of the way in which they attempted to free their generation from the dogmas of religion; but the enlightened few did not make for an enlightened age.

The Latin term 'modus' (from which our term 'modern' derives) means 'measure', of which the ablative is 'modo', 'with measure', especially in respect of time; this in turn gave rise to an adverbial form meaning 'just now'. In the later Roman Empire the term 'modernus' was derived meaning 'our current time', by which was intended a contrast between the then present

Christian empire and the previous pagan empire. Granted this etymology, the term 'modern' indicates that there has been a change from a previous mode or manner of being or doing to a later mode or manner of being or doing. Thus the term 'postmodern' tautologically points to a further recent change that allegedly now characterises the very latest in fashionable modes of thought, or social trends – from Paris to New York. For our purposes we can take the postmodernists to be at least pointing to a new conception of what is generally thought to drive change in our knowledge and science. We cannot say 'growth' of science, because this might be misconstrued teleologically. The "driver" is no longer anything rational but something quite different – something social, political, cultural or psycho-social.

The contrast between the rational and political or social, one important theme of this book, is well described by Cahoone, a commentator on current postmodernist writings. He proposes a number of characterisations of postmodernism in terms of what postmodernists typically accept or reject of the philosophical tradition (Cahoone (1996), pp. 14-5). Of importance for our purposes is the contrast between "the transcendence of norms as opposed to their immanence":

> The denial of the *transcendence* of norms is crucial to postmodernism. Norms such as truth, goodness, beauty, rationality, are no longer regarded as independent of the processes they serve to govern or judge, but are rather products of and immanent in those processes. For example, where most philosophers might use the idea of justice to judge a social order, postmodernism regards that idea as itself the product of the social relations it servers to judge; that is, the idea was created at a certain time and place, to serve certain interests, and is dependent on a certain intellectual and social context, etc. ... This leads postmodernists to respond to the normative claims of others by displaying the processes of thought, writing, negotiation, and power that produced those very normative claims. (*ibid.*, p. 15).

An important part of the philosophical tradition, well exemplified from Plato to Kant, is the view that there is an independent, and even an *a priori* justifiable, realm of the normative, whether it concern norms of reason, logic, epistemology and methodology, or norms of morality, politics and the law. Here the focus will be on the first set of norms, collectively called for convenience 'norms of rationality'. Some have been critical of the status claimed for these norms from Plato to Kant and beyond; these critics have attempted to give a different account of that status while still assuming that some of these norms at least retain their authority. However postmodernists, and others for different reasons, have claimed not only that the status assigned to norms of reason is wrong, but also that the claim that the norms are authoritative in any way is to be debunked. Cahoone says as much about postmodernist claims concerning the norms of justice. And the same considerations can be extended to the norms of rationality. Nowhere has this been more evident than in the recent "science wars" in which some critics aim to dethrone the status of scientific knowledge by undermining the claims of scientific rationality on which it depends. The full import of the denial of a role for norms of rationality is a central theme of this book.

All people by nature desire to know, said Aristotle in the very first sentence of his *Metaphysics*. But was he over-optimistic? Certainly we have evolved as primates who believe, this being one main characteristic that advocates of naturalism use to distinguish us, despite our otherwise animal nature, from all other primates and the rest of the animal kingdom. But the path from mere belief to knowledge is a difficult one, as much epistemology shows. Do we all desire to *know*? Is this desire due to our very natures? Just how widespread is not only this desire, but also the capacity to exercise our abilities to know? Recent empirical research in cognitive psychology indicates that the capacity is variable and often fitful, leading some to posit the idea of the "irrational" human.[1]

Clearly many lack the desire to know; these range from the merely dull to the dangerously dogmatic. Others say we can make little headway, or none at all, along the path to knowledge; these are the sceptics. Yet others say that it is an illusion that there is any such path to be found and that we should simply be content to live with the variety of beliefs that surround us; these are the relativists and their present-day fellow travellers amongst who are many postmodernists. Yet others, including sociologists of "knowledge" and some postmodernists, say that power relations or political interests, but not anything rational, determine, or cause, our beliefs. Finally, some think that there is a path to knowledge about some matters, and strive to find it. To this end they equip themselves with theories of rationality and method to enable them to detect not only appropriate paths through the thicket of belief, but also to detect when they have reached their goal, knowledge. Aristotle's *Organon* is one of our first treatises on such path-breaking, and knowledge detection. Perhaps it is just to these optimistic rationalists with their methods and principles of reasoning that Aristotle should refer, and not humankind as a whole with its burgeoning number of postmodernists who are often relativistic, sceptical or nihilistic about claims to scientific knowledge.

There have always been naysayers to both the possible fulfilment of the Aristotelian desire for knowledge and the idea that there may be methods for certifying some beliefs as knowledge. Plato showed us how to deal with the naysayers to method and knowledge of his own time in a number of his dialogues. Thus in the *Theaetetus* he wrestles with the relativism of the sophists, especially that of Protagoras, in order to draw out unacceptable conclusions for the doctrine and to show that it is self-refuting.[2] And in the *Gorgias* he wrestles with advocates of rhetoric and power who would give us merely conviction and not knowledge. While criticising his opponents, Plato looked to what intellectual equipment we would need if we were to attain any knowledge, thereby developing one of our earliest theories of knowledge[3];

aspects of Plato's position are discussed in section 2.1. Of course, there will be those who maintain that Plato was singularly unsuccessful in refuting the views of his opponents and that either the task has to begun anew or victory is to go to his opponents. The view taken here is that Plato makes some important points that his opponents must take seriously or be in intellectual peril.

As the history of human intellectual frailty shows us, Plato's encounter with the sophists of his day has to be replayed anew for every generation unless dogmatism, scepticism, relativism, or whatever new intellectual fashions, such as postmodernism or the doctrines that inform much of cultural studies, get their grip upon us. As a result much of this book might appear negative in character; but even discovering that not-p is the case can be of value in any critique of intellectual fashions. Parts II, III and IV each deal with the doctrines of three contemporary heirs to the views of those very sophists Plato encountered. The three are chosen for scrutiny because of their salience in our current intellectual scene and their wide influence. Part II deals with sociologists of knowledge, from Marx and Mannheim to advocates of the "Strong Programme" for the sociology of scientific knowledge, and the claims they make against any autonomous theory concerning the rationality of science. Part III concerns Foucault's archaeology, genealogy and his power/knowledge doctrine. Part IV deals with Nietzsche's influential views, said to be the for-runner of much of current postmodernism. What will be discussed is his view that our entire conceptual scheme is false, but it is projected on to the world in order that we might survive. The projectionist or constructivist account of reality is currently influential amongst postmodernists and others, and it often has a source in Nietzsche.

In contrast, Part I is more positive in setting out some of the familiar grounds for thinking that there are methods for obtaining knowledge by reviewing some theories of scientific rationality. Though there is some defence of this position, a more detailed defence is not possible within the scope of this book. As familiar as these ideas may be, a secure grasp of them is needed as one sails, like Ulysses, the oceans of belief trying to resist the siren calls of sociologists, cultural studies enthusiasts, Foucault and Nietzsche, to mention just a few of those who would woo us into the whirlpools of irrationality. As we sail by we can glimpse the havoc these sirens have wrought on many a captured university discipline – English, Education, Sociology, History of Science, to mention a few.

Perhaps the most self-conscious way we have of breaking paths to knowledge is by means of science. True, the optimism of Francis Bacon that science would cure all our ills did not take into account that science might produce some ills of its own, such as the nightmare of nuclear warfare, pollution or global warming. Whether it is science that is to blame for this or, as is more likely in many cases, the use of science within processes of production dominated by profit maximisation that are not subject to public and democratic control, is not my theme. Rather my focus is upon the methods of science and the idea of a *pure inquiry* into what the world is like. Whatever other connotations the word 'science' might have, the focus here is on science in which there is a product, knowledge (or certified belief), and a process for getting some, i.e., methods of inquiry. This is reflected in the etymology of the Latin noun 'scientia' that derives from the verb 'scire', 'to know'. Scientific inquiry has been spectacularly successful in producing pure knowledge of the world. But how? By using principles of rationality and method, say a host of philosophers from Plato and Aristotle down to philosophers of science as diverse in their views as to how that success has come about, for example, Popper, Lakatos, Hempel, the Bayesians, and many others. But there are some naysayers, even amongst contemporary philosophers of science, for example, Kuhn and Feyerabend.

But are they really the naysayers they are often thought to be? Part I of this book takes up this theme. Chapter 1 begins by setting out what are some of the core ideas behind what might be called our "critical tradition", part of which includes the methods of science and their claim to rationality. A number of detractors from this tradition are briefly reviewed, especially sociologically inclined writers on science. Kuhn and Feyerabend have also been taken to have rejected the critical tradition. However their position is somewhat complex. They have not bucked the tradition totally; rather it will be argued that they are critics of particular versions of the tradition, while maintaining an important place in it. The case for Feyerabend is much less clear than the case to be made for the later Kuhn who revised many aspects of his earlier, more widely known, views.

Chapter 2 raises the question 'why bother with knowledge rather than belief, or true belief?' Plato, who first drew the knowledge/belief distinction, provides us with an answer. And so do others who insist that the two questions '*Quid facti?*' (viz., questions about factual matters concerning our beliefs), and '*Quid juris?*' (viz., questions about what *right* we have to our beliefs, or justufucation for them) have importantly different answers. Once the significance of the distinction is recognised then it is important to develop a theory of the normative nature of knowledge. Two contrasting theories, amongst the many that are available, are set out to emphasise the normative and critical character of knowledge as opposed to mere belief. Those who would downplay the whole idea of knowledge overlook its normative and critical aspects. As will be seen, this is a common feature of the writers discussed in Parts II, III and IV. They ignore the significant difference between knowledge and belief, and simply stay with the latter, because they are either indifferent or oblivious to the normative and critical aspect of knowledge which mere belief lacks (but which is possessed by *rational* belief).

The writers discussed in Parts II, III and IV adopt some version of naturalism. They also turn out to be extreme nominalists or particularists when it comes to matters to do with ontology (i.e., what exists). However their nominalism is rarely argued for; rather it is just asserted. Their naturalist stance is shared here, but not their nominalism. All naturalists must face a problem about the status of norms of any sort. Are the norms of reason, logic and methodology, as well as ethics, super-natural items, or can they be located within a naturalistic worldview? In chapter 3 some of the difficulties about the status of norms of rationality are set out; and a proposal is made about how a quite objective account of norms can still be held while giving them a place within naturalism. As will be seen, the writers discussed in Parts II, III and IV, while recognising that there is a problem of normativity within naturalism, either become sceptical about the status and role of norms in adjudicating between beliefs, or they become quite subjectivist or relativist about norms, or they become norm nihilists denying that any exist. Whichever position they adopt, they drop norms out of the picture altogether, and then look for other naturalistic grounds on the basis of which beliefs are acquired. As a result, the much-vaunted rationality of science is undermined. Chapter 3 provides an important launch pad for a critique of the writers discussed subsequently, all of who have abandoned, in one way or another, a role for the norms of rationality.

One of Plato's more striking images is of his relativist rival Protagoras whose head would pop up from the ground every once in a while to reassert his account of relativism – and then to disappear again before any Socratic interrogation could take place (*Theaetetus* 171D). Protagoras is represented as a denizen of an underworld of philosophical connections. This underworld of connections still thrives amongst the heirs to the views of Plato's opponents. Each makes their contribution towards nihilism, scepticism, relativism or postmodernism while at the same time downplaying, or completely rejecting, any connection with the traditional concerns of epistemologists and methodologists. There are in fact too many present-day connections to deal with in one book. But a few are chosen for scrutiny given their current significance. There is no European connection today that is not sustained by its USA connections. In fact the USA connections have vastly outgrown their European origins, and have spawned home-grown connections that in turn have become as globalised as other aspects of USA culture. None of these connections feature in this book, though it is hoped that what is said of the European connections will carry over to some of the globalised USA versions.

Only one French Connection is discussed here in any detail; this is Foucault whose influence has been large in sociological, cultural, and science studies. Part III is a critical evaluation of aspects of his views on the "archaeology" of knowledge (chapter 8), and of his account of genealogy and the accompanying doctrine that "knowledge" is part of a nexus of power relations (chapter 9). Only passing mention will be made of other French connections to theories of science and knowledge that deserve fuller treatment on their own, such as Lyotard or Latour.[4] A fuller summary of the topics to be covered concerning Foucault is given in the 'Introduction to Part III'.

Though there are also many important German connections, only one is explored in Part IV – the *éminence grise* of anti-modernism, Nietzsche. His doctrines have deeply influenced a number of postmodernist writers, including Foucault who has adapted much of Nietzsche's account of truth and power for his own purposes. Chapter 10 discusses the way Nietzsche envisages that his "will to power" operates to produce our conceptual scheme of objects, our notion of truth and our moral beliefs (as set out in his *The Genealogy of Morals*). In fact Nietzsche holds that our entire set of beliefs is a projection that we make, and that what items we think there exist are merely fictional constructs of our own making. The agents of construction are sometimes individuals, but they can also be groups through which the "will to power" operates as an important drive. Mannheim is in part right when he claims that Nietzsche was one of the two founders of the sociology of "knowledge", the other being Marx. But unlike Marx, Nietzsche's explanations of our beliefs are psychological, or psychosocial. The chapter will also show how explanations in terms of "power" replace those commonly given in terms of rationality, thereby debunking rational explanations.

Part II is entitled 'The Poverty of the Sociology of Knowledge'. It deals with the different inroads that many sociologists of science have allegedly made against the rationality of science. Chapter 4 discusses two further German connections. The first is Karl Marx who attempted to show that much of science was conditioned by the prevailing forces and relations of production, especially those of capitalism. But Marx's thesis trades on a highly ambiguous claim that pervades much sociological theorising that will be disentangled in the course of Part II. The other German connection is Mannheim who, in contrast, produced a restricted thesis about the way in which the social can impinge on the scientific.

The remaining chapters 5, 6 and 7 of Part II focus on the most thoroughgoing account of the sociology of scientific knowledge due to two leading members of the University of Edinburgh School of sociology, David Bloor and Barry Barnes. This is the "Edinburgh Connection". They have developed and defended what they call the 'Strong Programme' in the Sociology of Scientific Knowledge, with its core Causality Tenet. This is the view that "knowledge" in science (and elsewhere) is mainly caused by social, political and social factors (they use the term 'knowledge' but 'belief' will be the preferred term here). This view is to be found in many other writers, but it gets its best expression and elaboration within the Strong Programme. The various tenets of the Programme are set out in Chapters 5 and 6 and critically evaluated as rivals to rational explanations of scientific knowledge and belief.

An Austrian Connection plays a role at this point, though a shadowy one – Wittgenstein. There are many different interpretations of his often obscure texts. Here we are not concerned with Wittgensteinian hermeneutics directly, but with the interpretation that advocates of the Strong Programme say arises from his writings. They have adopted what they call a 'finitist' view about the meaning of terms within science, and they locate the source of this in one understanding of Wittgenstein's views on rule following. Such finitism is merely one of the many ways in which social factors can impinge not only on science, in this case the very meaning of its terms, but also on the very norms of science.

This raises an important ambiguity within the Strong Programme. Does it deny that there is a role for the norms of rationality and methodology of science in explaining scientific belief, as is clearly the case on one understanding of its central Causality Tenet? Or is there a role for these norms in such explanations, but these norms must be construed very differently from the way in which they have been traditionally understood? It is here that advocates of the Strong Programme appeal to the notoriously difficult Wittgensteinian notion of rule-following and form(s) of life, giving them a particular gloss in which rule-following (and thus the norms of rationality) are understood to be ultimately social. They also claim that logical relations are really social relations of constraint. The critical issue here is whether advocates of the Strong Programme deny that there is such a thing as scientific rationality; or whether they claim there is such a thing, but that a quite different account of its status and nature needs to be given in terms of their understanding of Wittgenstein.

Many of the ideas examined in Parts II, II and IV can be found in other writers on science and society in a modified or adapted form. The doctrines discussed in these three Parts have been chosen because of the seminal role they play in current critiques of science and knowledge, for example, in the "science wars". But they also share a common form. For a given body of belief, Marxists will try to explain why it is held in terms of its connection with the prevailing forces and relations of production. Sociologists such as Mannheim will look to its connection with the existing social conditions. Advocates of the Strong Programme will look to the range of social, political and cultural factors that prevail as an explanation. Foucault's "power/knowledge" doctrine tells us that our bodies of belief, as well as much else, are bound up with social relations of power. For Nietzsche it is the particular operation of the "will to power" which is efficacious in producing any of our bodies of belief, whether moral or non-moral. For all these theorists, even though they might have an in-house dispute as to whether it is politics, economics, socio-cultural factors, power or the "will to power" which causes belief, they all agree that the cause is not anything rational. Or if it is something rational, then rationality has to be reconstrued as something social, or as something aligned with power, which lacks independent authority. Once such items are accepted as causes of belief, then rational explanations are to be debunked by exposing them as ideological because they mask what is allegedly really at work in belief formation. While it can be agreed that *some* bodies of belief might be held by *some* people at *some* time in the way they specify, it is not the case that *all* bodies of belief for *all* people at *all* time are held in the specified way. The stakes are high for the viability of a critical tradition in which rationality plays a role in science and knowledge.

It is no news to most in the academic community that there is a deep division between two contrasting positions often loosely labelled as 'modernist' or 'rationalist' versus 'post-modernist', 'relativist' or whatever, especially as these terms are applied in philosophy and to accounts of science. And it is no news to most that the dispute that rages over the positions so labelled has "hotted up" with books such as *Higher Superstition* (Gross and Levitt (1994)), with the hoax perpetrated on the editors of *Social Text* who accepted for publication a spoof by the mathematical physicist Alan Sokal,[5] and with the host of articles by writers battling it out along the front of the "science wars". No attempt will be made here to give an account of what these labels, and others, might mean; nor will their associated doctrines be discussed. Nor will the "science wars" be directly discussed (see Brown (2001)), though many of the philosophical issues on which they depend will be. An occasional gesture will be made, as in the above paragraphs, in using these terms to paint with broad brush strokes the fault lines that lie across our current intellectual landscape. However in singling out for critical evaluation aspects of the work of sociologists of scientific knowledge, Foucault, Nietzsche and, in passing, other postmodernists and constructivists, the nature of the fault lines will become clearer.

Given that much of science has come under attack, what do the scientists themselves say? Usually they are too busy doing science to become preoccupied with reflections about the status of their own activity. But when they do reflect, they tend to be drawn more to one or other of the philosophical accounts of scientific belief than to any of the sociological accounts. Some scientists who have taken up the cudgel on behalf of science include the already mentioned Sokal, and Gross and Levitt; but also there are Wolpert, Perutz and Weinberg, to mention just a few.[6] Whatever difficulties scientists might have had in recognising themselves in philosophical accounts of science, they have had far greater difficulties in the case of sociological and other such accounts. The absence of self-recognition by scientists in sociological studies is the reason one sociologist of science gives for rejecting the approach to the nature of scientific belief of many of her profession discussed in this book (see Segerstråle (1993)). In fact the difficulties have led to outrage about the misunderstandings and distortions that scientists have perceived in sociological accounts. Are scientists right to be outraged? Much turns on the theory one has of the scientific enterprise, or one's image of science. In the absence of a viable theory, outrage is an understandable, but inadequate, response.

Is my account of the views espoused by the writers discussed in Parts II to IV correct? As an interpreter of their position I can only hope that I have their views as correct as I can get them. But the question is also asked quizzically since these writers often adopt views about which it is hard not to be incredulous. And they often express themselves in ways that invites parody. Another antipodean philosopher, viewing the detritus of the outside intellectual world washing up onto his own shores, asked the question: 'What is wrong with our thoughts?' He replied: 'What is needed in order to answer it, but what we have as yet scarcely the faintest glimmerings of, is a *nosology* of human thought. (A nosology is a classification of diseases.)' (Stove (1991), p. 187). This book is in part a contribution to such a nosology.

NOTES

[1] For an account of empirical research into human reasoning capacities that reveals its inadequacy in many cases, see Stich 'Could Man be an Irrational Animal? Some Notes on the Epistemology of Rationality' reprinted in Kornblith (1994), pp. 337-57; see also Samuels, Stich and Tremoulet (1999).
[2] See *Theaetetus* 152A-186E. Invaluable is the commentary by Burnyeat (1976) with its interpretation of Plato's argument that Protagorean relativism is self-refuting. There is a very useful application of relativism to postmodernists and others engaged on the opposite side of the "science wars" in Paul Boghossian 'What the Sokal Hoax Ought to Teach Us' in Koertge (ed.) (1998), pp. 23-31.
[3] The relevant dialogues are those such as *Meno* and *Theaetetus* that focus on the distinction between belief and knowledge. Plato also developed a theory about timeless abstracta, the Forms or Ideas, and how we might have knowledge of these; this is not the epistemological theory intended here.
[4] For a critique of Lyotard's brand of postmodernism, see Nola and Irzik (2003). For a broader purview of the many French connections the reader should consult Sokal's and

Bricmont's *Intellectual Impostures* (1997). There one can follow up many of the pronouncements on science and knowledge by at least Derrida, Lacan, Kristeva, Irigary, Baudrillard, Deleuze, Guattari and Latour.

[5]   Sokal, (1996a) and (1996b). For an account by Sokal of his hoax, see his 'What the *Social Text* Affair Does and Does Not Prove' in Koertge (ed.) (1998), pp. 9-22. Sokal's hoax (Sokal (1996a)) appeared in a special edition of *Social Text*, Spring 1996, which had the title 'Science Wars'. See also Ross (ed.) (1996) which is a reprint, with the same title, of the papers of the special edition of *Social Text* but which omits the Sokal hoax paper. Sokal's papers just cited, along with many others for and against the hoax, are collected in Editors of *Lingua Franca* (2000).

[6]   See Wolpert (1993) who in chapter 6 is enamoured of neither philosophers nor sociologists of science, and who has entered into public controversy with the latter. The same can be said of Weinberg's many writings, in particular Weinberg (1993) chapter VII 'Against Philosophy' which is against both philosophers and sociologists of science. For one philosopher's response to Weinberg see Salmon (1998) chapter 26 'Dreams of a Famous Physicist: An Apology for Philosophy of Science'. For a recent collection of his writings on these matters see Weinberg (2001). Max Perutz's angry review of the historian Geison's book *The Private Science of Louis Pasteur* entitled 'The Pioneer Defended' can be found in the *New York Review of Books* December 21, 1995, pp. 54-8; it is followed by an exchange of letters between Perutz and Geison in the same journal April 4, 1996, pp. 68-9. The works just cited are merely a few of the many that have been engaged on both sides of the "science wars".

PART I

KNOWLEDGE, SCIENCE AND THE

EPISTEMOLOGICAL ENTERPRISE

SYNOPSIS OF PART I

Part I sets out, and briefly defends, the idea that there are norms of rationality to be found in logic, epistemology and methodology, and that these norms, central to our critical tradition, apply to the sciences. In Chapter 1, sections 1.1 and 1.2, the very idea that there is such a thing as a critical tradition for assessing beliefs in science, and elsewhere, is introduced. Some normative methodological principles that are characteristic of the critical tradition are also sketched, along with ways in which these principles might be justified, or legitimated. This is contrasted in section 1.3 with a sampling of some of the claims made by present-day leading naysayers concerning our critical tradition. The task of critically evaluating some of these claims is left to the remaining parts of the book.

It is often alleged that Kuhn and Feyerabend are to be included amongst the naysayers to rationality and method in science. In the case of Kuhn it is argued in section 1.4 that, while his earlier work of 1962 *The Structure of Scientific Revolutions* seems to undercut the whole idea of the rational comparison of competing deep theories or paradigms, Kuhn later denied that he was committed to either irrationalism and relativism. There is some doubt as to whether the 1962 Kuhn can escape these charges. But as he developed his ideas on theory choice from the 1970 second edition of his book, his thinking on these matters underwent a change. In fact it changed so much that the 1980s Kuhn mounted a defence of the rationality of his theory of scientific values and attacked those postmodernists and sociologists of science who would co-opt him to their cause of deflating the pretensions of rationalist accounts of theory change. The case of Feyerabend, discussed in section 1.5, also shows that, despite the impression he tried to create (or is thought to have created), there is much in his position that supports rationalism against irrationalism (though any attempt at Feyerabend interpretation is not without its difficulties). In particular, it will be argued that the methodological nihilism of "anything goes" in methodology commonly attributed to Feyerabend is not a principle that he endorsed.

Chapter 2 focuses on the following question: given that we can entertain an indefinitely large number of beliefs (which we can express through language), why do we bother with knowledge and not stay with mere belief? In the light of the nihilism, scepticism or relativism about knowledge to be found in the theorists investigated in Parts II to IV, this is an important matter to resolve. Of the many theories of knowledge that are currently available only two are discussed to any extent here. The first is a traditional foundationalist account of knowledge in terms of justified true belief; this is set out, along with some of its difficulties, in section 2.2. It provides one ideal model of scientific knowledge that sharply contrasts with any attempt to play down the rational elements within science. Such a model, which might be advocated by what is called a "Cartesian rationalist", stands in stark contrast with models provided by sociologists of science, so much so that each would take the other to have given a discreditable account of science. The second theory of knowledge discussed in section 2.3 is reliabilism. A version of reliabilism is favoured, not least because of the way in which it sits more happily with scientific naturalism than does the foundationalist model just mentioned. Moreover the three theories investigated in Parts II, III and IV need some reliabilist account of knowledge if they are to preserve both their naturalism and any account of knowledge that gets beyond mere belief. As we will see, all three theories adopt failed versions of naturalism that are incapable of resurrecting any notion of knowledge. The final section 2.4 discusses some of the ways in which social aspects can enter into our concept of knowledge. But none of this supports the distinctive claims, made by the theorists discussed in Parts II to IV, about how social factors come to replace the standard rationalist account of how scientific knowledge is to be critically evaluated.

If one is a naturalist then an urgent question arises, viz., what is the status of norms of reason and method employed in a naturalistic account of scientific knowledge? Advocates of the three anti-rationalist theories discussed in Parts II to IV do feel the force of this question; but since they can find no place for norms of reason and method within naturalism, they conclude: 'so much the worse for these norms'. This has been commonly said of the norms of ethics when placed in a naturalist setting; either there are no ethical norms, or one ought to be sceptical of them, or they are merely expressive or imperatival and do not have truth values – or if they do, they are all false. The same can be said of the norms of reason and of scientific method; either there are no such norms, or one should adopt a sceptical view of them, or they are merely expressive of attitudes. Understood this way, they can lack authority or any warrant. This issue is taken up in chapter 3 where it is argued that in reconciling norms of reason and method with naturalism, the norms do not have to come off second best. On the contrary, there can be, within naturalism, an account of norms that allows that they are cognitively significant, objective and have truth values. The force and authority of methodological norms is thus restored.

Chapter 3 addresses these matters. It begins with Quine's initial attempt to relocate norms of method as a chapter of psychology, and his later resurrection of them as a branch of the technology of reasoning. Though norms are finally retained, it is unclear what status they have with respect to the naturalism Quine also espouses. The idea of naturalism is explored in section 3.2 while section 3.3 sets out some further examples of norms of method to give substance to the following more abstract discussion of the status of methodological norms. Section 3.4 investigates at least four ways in which the normative and the natural might be related to one another. Two of these accounts of norms are singled out for attention. The first is anti-objectivist naturalism (see sections 3.4.1 and 3.5.1) in which nihilist and sceptical attitudes are adopted towards norms. Also included under this heading is the view that normativity arises from our *expressing acceptance* of a set of norms. The second position, defended in the remaining sections, is an objective naturalism in which norms can be "defined", in a sense to be explained, in terms of the non-normative, or the natural.

In order to provide such a "definition", section 3.6 argues that the practice of scientists at least shows that there is such a thing as a "folk scientific rationality" exhibited in the quite particular choices scientists make about their theories and hypotheses in "playing the game of science". And they makes these choices even though there may be no overall theory of scientific rationality that they endorse, or even apply, in making the choice. Granted such a folk scientific rationality, it is shown in section 3.7 how the normative concepts that occur in the norms of method (and logic) can be "defined" using the device of the Ramsey sentence, as modified by David Lewis. Finally it is argued in section 3.8 that the normative in epistemology and methodology supervenes on the non-normative. In this way norms remain objective yet are locatable within naturalism.

The particular way in which norm nihilism and/or scepticism are manifested amongst the three philosophical positions of Parts II to IV will be examined separately in each Part. Once again a mistaken conception of naturalism come to the fore. Part I provides the basis for one kind of diagnosis (there are others) of some of the recent intellectual ills due to postmodernism and other "anti-rationalistic" approaches to science and knowledge. Some remedial medicine is, however, available; and scattered throughout Parts II to IV can be found the prophylactic of more detailed argument.

CHAPTER 1

THE CRITICAL TRADITION AND SOME OF ITS DISCONTENTS

One of the cornerstones of much philosophy is the idea that it has a critical tradition, and that this tradition has informed our general knowledge, science, mathematics and even philosophy itself. Also the tradition has not remained static and given once and for all, but has grown and developed alongside these subjects. The earliest aspects of the critical tradition, as they arise in Western philosophy (but also elsewhere), can be found in the Presocratic philosophers, Plato and Aristotle. Since then, the tradition has grown and developed in the course of the history of philosophy and science, sometimes strongly, sometimes fitfully. How some contemporary philosophers of science have understood aspects of this tradition is outlined in sections 1.1 and 1.2. Some of the detractors from the tradition are discussed in section 1.3. They include some sociologists of scientific knowledge and postmodernists, with other leading detractors being left for discussion in Parts II, III and IV. Some philosophers of science, such as Kuhn and Feyerabend, have also been thought to be renegades from the critical tradition. However as the final two sections 1.4 and 1.5 argue, the critical tradition can be understood broadly enough to contain Kuhn's position, and a good deal of Feyerabend's position, without it being undermined by either.

1.1   ON THE VERY IDEA OF A CRITICAL TRADITION

1.1.1 Critical and Non-Critical Traditions

One useful way to orient oneself in the debate about the rationality of science is to take a cue from Karl Popper's idea of different traditions of belief for understanding the world as set out in his 'Towards a Rational Theory of Tradition' (Popper (1963), chapter 4). People have attempted to understand the world through traditions of belief and practice ranging from myth and religion to science. Thus using human agency as a model and appealing to super-human beings, one Ancient Greek tradition claims that lightning in a thunderstorm is to be explained by the displeasure of the God Zeus, or a storm at sea by the anger of the God Poseidon. Again, the Homeric tradition held that dreams were objective visions of the supernatural in which other-worldly things and persons paid the dreamer a visit. In contrast the rival tradition of the atomists held that, for example, storms at sea were a complex interaction of the surface of the water with the movement of the atmosphere and its different levels of temperature. A rival tradition involving Xenophanes, Heraclitus and Aristotle rejected the supernatural view of dream contents. Aristotle criticised the theories of his predecessors and claimed that dreams were a 'replay of previous waking experience, sometimes bizarrely scrambled as a result of physiological disturbance' (Gallop (1990), p. 19), while normal sensory function is suspended in sleep.

Why is there a change from one tradition of belief to another? What is the cause of such change? Why bother criticising alternative views and proposing new ones, as Aristotle commonly did? Why not remain content with the old beliefs, or be content with a plurality of old and new beliefs in some area even if they conflict? That a later tradition provides easier understanding than an earlier tradition may not suffice as an adequate explanation; often it is around the other way. Moreover, what is so great about 'easier understanding' so that it (or any other) criterion is to be adopted as the arbiter between different traditions?

Popper's suggestion is that alongside the stories that we tell in our various traditions, there emerged a second-order tradition – a tradition of critically discussing the first-order traditions, whether of myth, science or whatever. The second-order tradition would ask us to try to tell a better first-order story – in some sense of 'better' to be spelled out in the second-order tradition. It would also bid us introduce new requirements on explanation, and adopt explanatory mechanisms other than, for example, mythical explanations modelled on human agency. It would also bid us to adopt a critical attitude to the first-order stories and look for ways of challenging both the storyteller and the story by inventing other stories and then comparing the two stories according to various criteria. This much is evident in Aristotle's work on dreams and divination, in which he reviews, as he does in writings on many other topics, the views of his predecessors, criticises them and then proposes his own account. In this way, says Popper, science emerged as a first-order tradition competing with other traditions such as those of myths – but with this difference: the scientific tradition is accompanied by a second-order tradition in which first order traditions, mythical, scientific or whatever, can be critically evaluated. (Popper does not discuss whether or not pre-scientific mythical traditions have their own second-order critical tradition. There is no reason to suppose that myths are always to be treated in a non-critical fashion and that there may not be, in some cases, a second-order tradition for the evaluation of rival mythical traditions. However for Popper what demarcates science from myth is that the first-order tradition has a second-order *rational* critical tradition applied to it – and this makes the first order tradition scientific.)

What can be said of the second-order tradition? We need not, as Popper does, identify it with his falsificationist account of scientific method. The second-order tradition has a long history, starting even before Aristotle's first self-conscious attempt to codify the canons of critical inquiry in science in his *Organon*. Amongst other important contributors to logical, epistemological and methodological aspects of the critical tradition we can selectively list: philosophers and scientists such as Descartes, Bacon, Galileo, Newton, Whewell and Mill; inductivists from Hume to Reichenbach and Carnap; the growing legion of Bayesians who, on various interpretations of the probability calculus, devise Bayesian theories of confirmation and decision; the anti-inductivists and anti-Bayesians such as Popper, Lakatos and Kuhn; the host of statisticians who have developed principles of statistical inference and experimental techniques to accompany them, such as double- and triple-blind experiments. Our second-order critical tradition has not been static; it was not given once and for all. Instead it has historically evolved, developing novel ways in which scientific reasoning can be carried out, such as the reforms suggested in R. A. Fisher's statistical analysis of experimental design (Fisher (1926)).

We are familiar with many first-order traditions. For example, few societies lack a theory about dreams. Our earliest western records begin with theories found in Homer and criticised in Aristotle; and the twentieth century begins with rival theories such as those of Freud and Jung and ends with theories quite opposed to the psychoanalytic tradition such as those of Francis Crick and Allan Hobson.[1] Again there are theories of motion from Aristotle to Einstein. And so on. Perhaps less familiar are the theories of our second-order tradition mentioned in the previous paragraph. Popper suggests two such theories, his own critical and revolutionary model in which theories replace one another through falsification, and a cumulative model in which science arises from the steady accretion of observational facts. True to the critical tradition of which he is a member, Popper proceeds to criticise the second and argue for the first. Just as there is a long history of theories within the first-order tradition many of which rival or complement one another, so there is a long history of theories of scientific method, or of scientific rationality, in the second-order tradition some of which rival one another, and others of which complement one another.

1.1.2 Values, Rules and Principles of the Critical Tradition

In general we can say that the critical tradition embodies a number of theories of method. These contain principles of method, each principle containing either a value (goal, aim or prized end) that our first order theories ought to realise, or rules which are categorical imperatives telling us what to do concerning theory choice, or hypothetical imperatives telling us what to do in order to realise some value. Sometimes the principles of the second-order tradition may be explicitly set out and followed (as in the case of Descartes or

Newton); or they may be merely implicit in the practice of scientists as they make choices concerning their theories in the first-order tradition. Amongst the purely epistemic values to be found in the critical tradition that we might wish our theories to exemplify are the following: simplicity; truth, or verisimilitude; fitting all the observable facts; high testability; precise predictions; explanatory depth; increased understanding; high confirmation on available evidence; ability to solve problems rival theories cannot; and so on. Methodologists often suggest their own prized list, such as Quine's list of theoretical 'virtues': conservatism, modesty, simplicity, generality and refutability (Quine and Ullian (1978), chapter VI). Kuhn also provides his own list, to be discussed further in section 1.4: simplicity, plausibility, internal and external consistency, accuracy, scope, fruitfulness and social utility. Importantly, as these methodologists emphasise, not all these values may be realised in any one theory. Further some subset of values may be either compatible in that a theory realises all of them, or they may be incompatible in that not all the values can be equally realised in any theory. Methodologists can also differ over the values they advocate. Thus non-realists will reject truth and verisimilitude; anti-inductivists will reject high probability; others such as Lakatos and Feyerabend downplay the value of consistency; and so on. Controversies over which are our prized values will not be entered into here.

Alongside each value are rules that ought to be followed if the values are to be realised. Thus Popper in *The Logic of Scientific Discovery* (Popper (1959), section 11) emphasises the fact that throughout the history of the sciences theories have been overthrown and replaced by others. For such revisability to occur a theory must be testable, which Popper argues is the same as being falsifiable. So to ensure revisability, Popper adopts falsifiability as a supreme value for our theories and proposes a number of conventionally adopted rules which, if followed, would realise this value. He proposes that we should accept (provisionally) theories only if they have passed severe tests; and we should reject theories (as candidates for the truth even though we may continue to work on them) either if they fail to pass a severe test or are superseded by another theory which is even more testable. He also proposes that we should not stop testing theories unless we fall into a dogmatic attitude of acceptance without continued testing. Finally he proposes a supreme rule that says that all the rules of scientific method must be such that they do not protect a theory against falsification. One such rule is the anti-*ad hoc* rule: do not adopt *ad hoc* saving hypotheses unless they increase the degree of falsification overall.[2] Another rule bids us not treat theories as sets of implicit definitions (which are then untestable), or to change the meanings of words to protect against falsification.

In a completely different style are Newton's four 'Rules of Reasoning in Philosophy' contained in Book III 'The System of the World' of his *Principia*. The first two rules bid us to be parsimonious in the postulation of causes; the final two rules tell us what inductive inferences we are permitted to make. In addition the fourth rule suggests a demarcation criterion between genuine science based on induction, and mere speculation in science that is not so based. There are also methods for judging causal connections from the more simple text-book methods of J. S. Mill to more complex causal modelling to be extracted from correlations (Glymour *et. al.* (1987); Pearl (2000)). These methods are important throughout all the sciences. They are particularly significant for those sociological opponents of the critical tradition who wish to claim that socio-cultural factors *cause* scientific beliefs. Without an account of causal methodology, then the very claim central to the sociology of scientific "knowledge", viz., that some particular socio-cultural circumstance *'gives rise to'*, *'influences'*, *'determines'* etc, some particular scientific beliefs, cannot be justifiably made. This criticism of sociological theories of scientific "knowledge" is explored in chapter 5.

In what follows we will take a *principle* of method to be either a value, or a rule, or a combination of a rule and a value, such as: 'if one wishes to realise some value v then rule r ought to be followed (rather than some other rule r*)'. Any theory of method will comprise one or more such principles. Sometimes principles of method will be elliptically expressed as when a rule is mentioned but no value. However the value is either implicit in the methodology, or is a presupposition of it, and can be readily provided. At the other extreme a principle may be merely a value with no rule suggested for achieving the value. All that is required is that theories of the first order tradition exemplify the value; whether there is any effective means for realising the value remains an unanswered question. More commonly, principles of method are to be found fully expressed as a rule paired with a value that the rule realises with a high degree of reliability.[3]

As an example, consider the controversial principle of method, Inference to the Best Explanation (IBE). The rule part of IBE bids us to adopt that hypothesis from a bunch of hypotheses that explains best some given set of facts. The value part of IBE, not explicitly mentioned, is truth. Thus IBE captures the idea that the best of a rival bunch of explainers of some facts is the true hypothesis. This formulation is brief leaving a number of matters unspecified; for example, the bunch of hypotheses must meet minimally satisfactory criteria for being good explainers in the first place. Further, like any principle of method, we need a guarantee that following IBE will always (or with high probability) realise its value, truth. That the rule, as formulated, is unreliable in this respect is argued by Laudan (see Laudan (1984), chapter 5). How we guarantee the reliability of rules is not a matter discussed so far;

but it emerges in section 1.2 concerning the ways in which we might justify, or 'legitimate', any methodological principle. Other versions of IBE focus not on the best explanation but the best theory, or the most likely cause. Yet other versions recommend that the value to be reliably realised is not truth but *reasonable belief* in the truth, or increased verisimilitude. It is not my purpose here to enter into the controversy over how IBE is best formulated or what status the formulations have.[4] The purpose is much more limited in merely showing that one important methodological principle, in its various formulations, has many advocates and fits the general schema for methodological principles set out above.

Those who adopt a more historical approach to the critical tradition see the principles of method evolving and changing with the sciences of the first-order tradition. But then one needs a justification for adopting one principle at one time and then dropping it at another. The need for a justification also arises where there are two or more rival theories of method available. This raises the further normative question concerning what principles *ought* to be adopted within the critical tradition, or which principles are "better" than some other principles ("better" for what purposes also being spelled out within the methodology). Positing some third-order tradition to resolve disputes within the second-order tradition would invite an unsatisfactory regress of justifications of never-ending ascending orders of critical traditions. So in what way can we justify the adoption of some theory of method in the second-order critical tradition? Let us dub this 'the legitimation problem'; this is the topic of the section 1.2.

1.1.3 Social Aspects of the Critical Tradition

That science is a social activity is a commonplace. But just what, and how much, of science is social? The sociologists of knowledge with which we will be dealing see social factors everywhere. Two sociologists consequently think their views are widely confirmed, and also widely adopted; so they feel they can drop the word 'social' from the title of their book.[5] Much of this book is devoted to finding the limits of the involvement of the social in scientific knowledge. Here we will address only its involvement in our second-order critical tradition. Is such criticism only something which a solitary individual can carry out? Or does criticism require two or more people to be actively engaged? That is, is there something social or collective about there being a second-order critical tradition? On the face of it a critical tradition requires both a critic and the person criticised; since this involves two, it is minimally social. But could an individual be both protagonist and antagonist for the same set of beliefs? There is a model of such a critic in the later work of Wittgenstein who often conducts an internal dialogue with himself as both supporter and critic of a particular claim. Sociologists of science often criticise epistemology and methodology of science as being too individualistic and for ignoring their social dimensions. But many of the same sociologists of science are also avid supporters of the views of the later Wittgenstein whose own critical practice is often entirely individualistic, even allowing for his own internal dialogues. What has not been noticed is that even an opponent of the sociology of knowledge such as Karl Popper argued that there is something importantly social about the second-order critical tradition and that it cannot merely be an individualist enterprise.[6]

Popper says of the critical tradition that there are '*social aspects of knowledge*, or rather of scientific knowledge' (Popper (1962), p. 217). As far back as his 1934 *The Logic of Scientific Discovery* (Popper (1959), section 8) Popper had argued that all the sciences require *intersubjective* testing, and this is social in character. The objectivity of science does not arise from the attempts of an isolated individual to be impartial in, say, their theorising, experimenting or following the canons of method. Rather:

> ... objectivity is closely bound up with the *social aspect of scientific method*, with the fact that science and scientific objectivity do not (and cannot) result from the attempts of an individual scientist to be 'objective', but from the co-operation of many scientists. Scientific objectivity can be described as the inter-subjectivity of scientific method. (Popper (1962), p. 217)

Popper suggests that there are two aspects to 'the public character of scientific method' (*ibid.*, p. 218). The first is that when scientists put forward any idea then, even if they think it is beyond criticism, it is at least in the public domain and thus open to the (freely offered) criticism of others. This much is hinted at by Popper in his remarks, mentioned above, about the emergence of a second-order critical tradition in which a person who holds a particular set of beliefs is challenged by another to tell a better story, in some sense of 'better'. Here two or more must play a role in the dialectic of criticism that evolves.

The second public aspect of scientific method is the intersubjective character of observation (as opposed to experience), and of experimentation in which several must collaborate. In addition there are publicly agreed test conditions for any publicly expressed theory. This not only guarantees the objectivity of tests but also the objectivity of the theory in the sense that there is public agreement about what it says and thus what would count as a test. This is not to say that there may not be idiosyncratic and cranky views within science; but these do 'not seriously disturb the working of the various *social institutions* which have been designed to further scientific objectivity and criticism; for instance the laboratories, the scientific periodicals, the congresses' (*ibid.*, p. 218).

In support of his views Popper offers two putative examples of science for consideration. The first is that of a clairvoyant who produces a book full of hitherto unknown revolutionary science that results from a dream process, or from automatic writing. Subsequently scientists, using their usual methods, vindicate the contents of the book. Did the clairvoyant produce a scientific book? Suppose the contents of the book are true, and were believed by the clairvoyant; further, suppose that the book even contains justifications for the claims made that are correct. Then there is still something missing. First, even if what the clairvoyant says is true we do not think that the processes whereby the clairvoyant's beliefs were acquired are reliable for the truth; the processes used by the clairvoyant might hit on the truth occasionally but they do not reliably take us to the truth (see section 2.3 on reliabilist theories of knowledge). Further, neither the clairvoyant nor anyone else actually carried out any experiments, or proposed any theories along with their test conditions. Nor did the clairvoyant, or anyone else, actually carry our any of the reasonings, or apply any of the principles of scientific method to obtain specific results – and this despite the fact that the book contains lots of such justificatory claims. Nor did the clairvoyant subject the contents of the book to public scrutiny. On these grounds we can dismiss the claim that the book is scientific in the same way that claims to knowledge can be dismissed because the evidence condition is not properly satisfied; the alleged knower has not worked out the reasons for his or herself and has not subject them to external check (see section 2.1 on the problem of Plato's tether). A person may accidentally and miraculously hit on the truth without having knowledge; they may even mention reasons without actually acquiring beliefs on the basis of the use of such reasons.

Popper's second case is that of Robinson Crusoe, who on his island without the help of any Man Friday, proposes theories, makes observations, constructs and uses laboratories, carries out experiments, tests theories, applies principles of scientific method, and writes up all his results in the form of standard scientific papers. Has Crusoe indulged in science? Could Crusoe, even by extending the style of Wittgenstein as both protagonist and antagonist for his own views, have been able to produce some science? What Popper says is missing is that, even though Crusoe might be right in his science, 'there is nobody but himself to check his results', nobody to correct his prejudices due to his mental history, nobody to act co-operatively as a devil's advocate for the opposite of his ideas, and nobody to whom he has to clearly communicate his results (ibid., p. 219). Suppose that Crusoe indulges in some astronomy. Each astronomer who uses an optical telescope has a personal equation, the idiosyncratic reaction time of each astronomer who make observations with optical telescopes and then, as quickly as possible, notes the time of the observation using a clock. The personal equation was discovered in our public science by noting the discrepancy between various observers. Here the public character of science makes the discovery of observational discrepancies highly probable; in contrast the discovery would be somewhat miraculous on the private Crusoe model.

Though Popper does not put his point this way, we can say the following. In acquiring scientific knowledge we need to use methods and have procedures in place that are reliable for the goals contained in those methods. In particular any principle of scientific method must be highly reliable in yielding theories that exemplify some value (such as truth, or explanatory power, etc). The principles themselves may have independent grounds of test which in turn would establish their reliability. But this may not suffice to show that their application in any given situation is correct. We might have to appeal to others to go over our results, or repeat our experiments or reasoning, to ensure that we have applied the methods correctly. To the extent that this must be done, the reliable application of methods and processes cannot turn on the processes of belief formation of separate individuals only; the co-operation of many is required to make the belief-forming processes reliable. That is, the reliability of the processes and procedures employed by an individual depends as much on the actions of others as a public check, as upon the individual himself or herself. (Social aspects of reliability are discussed in sections 2.3 and 2.4.)

To this end Popper talks of submitting one's results to scientific periodicals with their peer review processes, or to congresses and conferences which allow for open criticism, or even merely making one's views public thereby opening them to criticism. It is such critics that Crusoe lacks. Suppose Crusoe's results (like those of the clairvoyant) are correct and (even better than the clairvoyant) that Crusoe has employed scientific methods and processes to the best of his ability. Then what is lacking is the extra reliability of the social processes of criticism and scrutiny by others that puts the final stamp of 'scientific' on his results. In the absence of such a public critical process Crusoe's results are less than scientific because of the absence of the kind of reliability conferred by the joint action of others. The question concerning Crusoe the scientist is not whether he is logically possible (he is); rather it is whether in a completely individualistic context we have scientific knowledge and scientific method as the certifier of knowledge. (More is said on social aspects of reliability in section 2.4.3.)

Such public checkability arises in mathematics. The very long proofs of the "four colour" theorem, or of Fermat's Last Theorem, left much room for doubt about the results because of two problems; those of surveying the whole proof, and of checking each step in the proof. The proof may be too long for any single individual to find the time to survey it from beginning to end. And any individual may fail to note that, in a long proof, some step is not correct, a fault which would undermine the justification condition for knowledge. Checking by others can remedy this oversight. In the case of the recent proof of Fermat's Last Theorem some errors were publicly uncovered, and the proof was only finally accepted as mathematical knowledge once sufficient public checking had taken place by the relevant mathematical community.[7] The same could be said of the results of empirical theories; they also require the assent of the relevant scientific community such that each, in applying the same procedures and methods, has obtained the same results. Public checkability becomes an important part of Popperian methodology on other grounds. His methodology can be thought of as processes of detecting and eliminating any error in our theories. The claim 'there are no errors here' is a negative existential claim. These can be hard to show true; but they can be readily shown false, as when one comes across some error in the process of checking. Here public checkability plays an important role in ensuring the detection of error.

Popper's point about the personal equation of astronomers who make systematic errors in their observations is of a different kind. Since much science turns on experiment and observation, peculiar characteristics of observers and experimenters need to be taken into account. In collective public science it is possible to turn the very observers and experimenters themselves into objects of study; in Crusoe's individual science this is not possible. Even though Crusoe might have hit upon his own personal equation, the lack of check upon himself as an observer detracts from the scientific character of his overall work. Though Popper does not mention this, there is also another sense in which science is co-operative and not individualistic. That there is a division of labour within science is well expressed by Newton's aphorism about his own discoveries, viz., that only by standing on the shoulders of giants was he able to see so far. Most of the enterprise of science cannot be carried out by one individual but, like the building of medieval cathedrals, requires the co-operation of the many over time and place.

None of the above remarks on public checkability should give comfort to social constructivists. The following constructivist claim is not endorsed by the above, viz., 'that some result is correct' merely means 'that the community gives its assent to the result'. One can ask the following two questions. (a) Does the community give its assent because the result is correct (as shown by its best methodological procedures)? Or, (b) is the result correct because the community gives its assent?[8] Only the most rabid constructivist would endorse (b). (a) supports objectivity. There is a fact of the matter as to the votes for candidates in some election; this is determined by the ticks or crosses on each voting slip. Given a large number of votes to count, a reliable procedure is set in place involving a number of counters and scrutineers who ensure, to the best of their ability, that the final outcome of the vote has been reliably arrived at. Similarly, there can still be some fact of the matter that a scientific investigation is trying to determine. But given the methods and procedures that have to be applied, the reliability of the outcome may well depend on social processes of checking. Granted these points, a role can be found for social processes in the application of scientific method. As Popper puts it:

> ... what we call 'scientific objectivity' is not a product of the individual scientist's impartiality, but a product of the social or public character of scientific method; and the individual scientist's impartiality is, so far as it exists, not the source but rather the result of this socially or institutionally organised objectivity of science. (*ibid.*, p. 220)

More will be said about the social character of knowledge[9] in section 2.4. But this should suffice to show that while one can agree that scientific methodology can be social in some respect, one need not be committed to more radical involvements of the social in science of the sort set out and criticised in Part II.

1.2   SOLVING THE LEGITIMATION PROBLEM[10]

1.2.1   A Priori Approaches to Legitimation

There are many proposals for legitimating principles of method that take us to the heart of epistemology, some features of which will be followed up later in this, and the next, chapter. What will be provided here is a brief thumbnail sketch of two broad approaches to legitimation that have been taken: the *a priori* and the empirical. There are several *a priori* approaches, one of which is Kantian. A Kantian approach to the legitimation problem would be transcendental; it would attempt to argue from the bare possibility of science to one of its presuppositions, viz., that there must be some appropriate and correct theory of method. Given that the bare possibility has been realised (because we in fact do have some successful science), then we would be able to conclude that there is, within the second-order tradition, some legitimate theory of method. But it is hard to see how any substantive principle of method of the sort mentioned in the previous section could be established in this way from the bare possibility of science. So such a Kantian approach, even though it is *a priori* in character, is rather empty.

A second *a priori* approach would be to show that just as there are *a priori* grounds for legitimating deductive inferences, so there are *a priori* grounds for legitimating the non-deductive inferences used in science. Since science uses both deductive and non-deductive patters of inference, then its accompanying second-order tradition would be legitimated. However, since Hume there has been an outstanding problem about the justification, or legitimation, of induction. Many philosophers, from Kant to Russell, have taken an *a priori* approach to solve this problem. These cannot be reviewed here;[11] but there appears to be no satisfactory *a priori* justification of inductive reasoning. (This leaves open the possibility that there might be non-*a priori* justifications available). The best prospects for a solution to Hume's and other problems associated with induction (e.g., Goodman's 'new riddle') lies with their subsumption within Bayesian modes of inference.

A final quasi-*a priori* approach to be mentioned here is that of the early Popper of the *Logic of Scientific Discovery* (see chapter II) in which a method for legitimating principles of method is proposed, each containing a rule and a value as illustrated in the previous section. Popper argues that principles of method do not have an *a priori* justification of the sort given for the principles of deductive logic. Nor do they have what he calls a naturalistic justification in which they are to be tested against experience as are ordinary scientific claims. Rather the rules and values are adopted as *conventions* which are still open to examination: 'It is only from the consequences of my definition of empirical science, and from the methodological decisions which depend on this definition, that the scientist will be able to see how far it conforms to his intuitive idea of the goal of his endeavours' (Popper (1959), p. 55).

For scientific theories Popper adopts a hypothetico-deductive method in which consequences are drawn from test hypotheses applied in some situation, the consequences being tested against observations or evidence gathered from experiment. The hypotheses stand or fall according to the Popperian rules governing such testing. How are Popper's principles of method to be tested? In much the same way. Principles of method become the test hypotheses in which the hypothetico-deductive method is used at a higher or meta-level. Popper regards his conventionally adopted methodological principles as open to test by comparing the consequences that can be drawn from them with one's intuitive idea of the goal of scientific endeavours. Though much more needs to be said of this, what Popper in effect suggests is that the hypothetico-deductive method he recommends for first-order theories also be applied, when suitably reconstrued, to decide between principles of method, i.e., the rival principles of our second order critical tradition. Thus hypothetico-deductivism reappears in meta-method as a way of testing, and thus of legitimating, methodological principles themselves.

Even though hypothetico-deductivism is at the heart of Popper's methodology, it need not be tied to that methodology with its associated rules of method. It can also be accompanied by a theory of confirmation that is unlike Popper's theory. Such recent developments of hypothetico-deductivism provide one of the more fertile approaches in methodology (see Kuipers (2000)).

1.2.2 Empirical Approaches to Legitimation

The later Popper adopted a different 'quasi-empirical' approach to the justification of his methodology (Popper (1974), pp. 976-87). Popper drew up a list of what he called great or heroic science (call these the 'good sciences'), a few of which are exemplified by the scientific work of Galileo, Kepler, Newton, Einstein and Bohr (the list can be extend further). Then he drew up a list of non-sciences or pseudo-sciences or unacceptable sciences (call these the 'bad' sciences). They are exemplified by the work of Marx, Freud and Adler (and the list can be extended). These two lists then comprise the empirical test base against which Popper's demarcation criterion for science is to be tested, using the hypothetico-deductive schema as a meta-criterion for testing the demarcation proposal. Suppose in such a schema that Popper's definition of science, or his demarcation criterion, is the hypothesis under test. Then the demarcation criterion will pass its test if it captures all the sciences on the good list and none on the bad list. It will be inadequate (and so falsified) if it leaves out some of the good sciences even if it captures none of the bad; and it will definitely be falsified if it captures some of the sciences on the bad list. In this way methodological principles of science can be legitimated.

Where do the two lists come from? They are Popper's own pre-analytical value judgements about the good and the bad in science. The idea of such a list can be made more precise. First, instead of merely listing the names of scientists one could instead list some of their more characteristic hypotheses or theories. Second, one could conduct a sociological survey of other scientists working in the field to determine their collective judgement about which of the two lists some theory or hypothesis might be placed. Such a modification was proposed by Lakatos. A test basis is to be found in the value judgements the 'scientific élite' make about the scientificness or otherwise of particular hypotheses or theories, or moves in the "game of science". In this way a less arbitrary, fallible and more empirically well-founded list of good and bad theories and/or hypotheses than Popper's can be drawn up.

Concerning such particular judgements and more general methodological principles, Lakatos says:

> While there has been little agreement concerning a *universal* criterion of the scientific character of theories, there has been considerable agreement over the last two centuries concerning *single* achievements. While there has been no *general* agreement concerning a theory of scientific rationality, there has been considerable agreement concerning whether a particular single step in the game was scientific or crankish, or whether a particular gambit was played correctly or not (Lakatos (1978), p. 124).

Let us agree that there has been dispute over what is our more general theory of scientific rationality, or our demarcation criterion, or whatever one would like to call our best theory about the nature and methodology of science. If, as Lakatos says, there has been considerable agreement amongst members of the scientific élite over single achievements, or particular gambits, in science, then we can use these judgements about particular cases as a test basis. The scientific élite will tell us not only what is good or bad about accepting a particular hypothesis, but they will also tell us which are the other good or bad moves or gambits concerning theory change in the game of science.

Not only does Lakatos propose a different test basis from that of Popper, but he also adopts a different meta-methodological criterion for deciding between differing conceptions of scientific method. In order to legitimate theories of method, his meta-criterion bids us select that theory of method which, as he puts it, yields a 'reconstruction of a growing bulk of value-impregnated history [of science] as rational' (Lakatos (1978), p. 152). Given his preferred test basis and his own meta-criterion, Lakatos gave us grounds for rejecting aspects of other theories of method, including Popper's, while retaining his own meta-methodological criterion (based in his theory of scientific research programmes).[12] Other methodologists have developed Lakatos' programme of determining which methodology is best vindicated, or legitimised.[13]

Lakatos makes it quite clear that not all beliefs within science can be explained in terms of adherence to principles of rationality and method. Let us call these explanations 'internalist', i.e., those explanations of a person's scientific beliefs by means of the principles of method used to acquire the beliefs. There will always be a residue of belief not explainable 'internally' but explained 'externally', i.e., by appeal to matters other than principles of method. It is here that the sociology of scientific knowledge might be able to step in and, in a limited way, fill the gap by offering alternative explanations. According to Lakatos, each theory of method will draw the internal/external division differently and, depending on what can be explained internally, will leave more, or less, of scientific belief to be explained externally. According to Lakatos, theories of method ought, as far as possible, to maximise internalist explanations of scientific belief – thereby maximising the rationality of science. This is something he alleges his own methodology of scientific research programmes does. Thus there is a clear criterion for improvement in theories of method; they increase the bulk of scientific belief to be explained internally (rather than externally) by means of methodological principles, leaving any 'irrationalist' residues to the sociologists to explain.

Yet another empirical approach is that of Larry Laudan's *normative naturalism*. Principles of scientific method are taken to be hypothetical imperatives of the form 'if one's value is v then one ought to follow rule r to achieve v', where 'v' stands for any epistemic or cognitive value for science and 'r' stands for a rule which, if followed, reliably achieves the value. Given the plethora of such hypothetical imperative principles, which should be adopted? Laudan recommends that we turn each hypothetical imperative into a hypothetical assertion about the regularity with which r realises v. Thus we arrive at an empirical law-like claim that says: always (or with a high degree of probability) following r realises v. Such a claim is open test against the historical record of science. Our current and past science is a vast repository of successful and unsuccessful strategies in which in some particular episode in science, some rule r is, or is not, regularly linked to some goal v. It is the task of the programme of normative naturalism to test each such law-like claim against the history of these strategies to determine which are the most reliable principles of scientific method. Like any other scientific hypothesis if they fail their test, they can be modified or replaced by better hypotheses. In the way it builds up its methodological principles, normative naturalism reflects the very procedures of science itself.

What method of test does Laudan's normative naturalism employ? He appeals to a meta-inductive principle in which we infer, from the successful employment of means to realise a given end in some given scientific situation, to the subsequent use of the same means to achieve similar ends in other scientific situations. Thus normative naturalism makes clear its presupposition of induction, not at the level of scientific theories where induction may not be directly employed, but at the level of assessing principles of method. In fact normative naturalism must rely on a broader notion of statistical sampling if it is to take the vast repository of successful employment of methodological means-ends strategies in the history of the sciences as its test basis. Whether or not it presupposes too much in the way of statistical methodology to be a viable independent approach remains to be seen.[14] In any case statistical theories themselves comprise an important independent domain of principles of methodology to be employed in the sciences which lie at the heart of our critical tradition, but which are often overlooked.

Statistical methodology finds a ready place in a quite different account of scientific method, viz., Bayesianism, currently the most widely adopted theory of method. Prominent are the subjective Bayesians, i.e., those whose account of scientific reasoning and method depends on an interpretation of the probability calculus as a rational degree of subjective belief. Central to the Bayesian approach is Bayes' Theorem which follows from the axioms of probability and says in its simplest form, where 'H' is a hypothesis and 'E' some evidence and 'p' stands for probability understood personalistically as rational degree of subjective belief: $p(H, E) = p(E, H) \times p(H)/p(E)$. Also central to the Bayesian approach is a principle of updating old probabilities to new probabilities according to: $p_{new}(H) = p_{old}(H, E)$ (this being the simplest of the updating principles.) Degrees of belief are said to be rational or coherent when they are so distributed over hypotheses and evidence that they obey the probability calculus, including the above principles. Why are they rational, or coherent? It is argued that for any person whose distribution of degrees of belief does not obey the probability calculus, a 'Dutch Book' can be made against them; that is, if a person bets on those probabilities they will either never win or always lose. Thus in playing a game of betting with nature on hypotheses and evidence, scientists will lose, or at least will never win, if they adopt a distribution for their degrees of belief other than that which accords with the probability calculus. This consideration, along with some others, provides a basis for the legitimation of claims about the rationality of a Bayesian account of scientific inference that places Bayesianism within our second-order critical tradition.[15]

The above will suffice as an incomplete list of both the kinds of theory that can be found in the second-order critical tradition and the kinds of legitimation provided for these theories. What the list reveals is that talk of *the* critical tradition can be misleading. It has undergone stages of evolution and there has been rivalry about which principles ought to prevail. However there is large amount of agreement about particular principles of method. For example, the anti-Bayesian Popper is in agreement with Bayesians, with Lakatos, and with a number of other methodologists from Huygens to Whewell, about the role played by true novel predictions of a new theory which were not known given the background of old theories; these can boost the degree of support for a theory quite markedly above that provided by predictions that are not novel. There is also a large measure of agreement about the outcomes of different meta-methodologies. These need not always rival one another; in many cases meta-methodologies can give complementary justifications for the same aspects of theories of method.

Given the absence of a single preferred theory of method, some might be willing to adopt the tolerant stance of admitting that there is a plurality of critical traditions. Others might argue that perhaps there has been a patchwork of several such traditions only some of which are satisfactory and some of which are inconsistent with one another. And some historical periods may have been so unenlightened that not even a patch of this tradition has prevailed. Importantly there have been intellectual trends, now currently resurgent, in which even the patchwork second-order tradition has been abandoned. It is alleged that there is nothing privileged in talk of a *second-order* tradition. All traditions are first-order, which is tantamount to abandoning any distinction in order between traditions. In this way any theory within the critical tradition is downgraded to merely one of many traditions of belief that compete with one another for our assent. Critics of the very idea of Popper's second-order tradition argue that there is nothing privileged about the critical tradition; its postulation is merely another manifestation of the wide range of beliefs within our culture. Adopting some critical tradition is merely an honorific way of talking about one of our 'forms of life', to adopt Wittgenstein's phrase. All talk of legitimating the principles of our 'critical tradition' simply serves to mask this fact. Thus the critical tradition, and any legitimacy it might be thought to have, is dethroned.

Who are the dethroners? And why? Many are to be found amongst sociologists of scientific knowledge – though not all sociologists of science have joined the throng of dethroners. Few sociologists ever discuss principles of method, even though, as will be seen, they need them in their very theorising, and often employ them uncritically. Some sociologists enlist the assistance of philosophers of science such as Thomas Kuhn and Paul Feyerabend, greeting them as soul mates who have undermined the pretensions of philosophy to underpin any critical tradition. But matters are not that clear cut. As will be seen, both of these philosophers distance themselves from many of the claims made within the sociology of scientific knowledge, which are nihilistic, sceptical or relativist about any second-order critical tradition. Before examining the cases of Kuhn and Feyerabend, let us turn to a sampling of those who would dethrone the critical tradition, often for sociological reasons.

1.3   SOME DETHRONERS OF THE CRITICAL TRADITION.

1.3.1   Some Sociological Connections

The dethroners are many and varied. One of the more serious attacks on the critical tradition comes from those who work in sociological studies of science. A leader in the field, David Bloor, adapts a remark of Wittgenstein when he entitles a chapter in one of his books 'The Heirs of the Subject that Used to Be Called Philosophy', making it clear who these heirs are: 'they belong to the family of activities called the sociology of knowledge' (Bloor (1983), Chapter 9 and p. 183). Not any sociology of knowledge would count as an heir, but the Strong Programme is claimed to be. Whether Wittgenstein would agree is hard to determine (see section 7.5). Here advocates of the Strong Programme latch on to the pessimistic side of Wittgenstein who held that not only was philosophy a sick discipline, but also the times in which he lived as well as our present – the two sicknesses not being unrelated according to von Wright who has an illuminating paper on Wittgenstein's views of his times.[16] The later Wittgenstein is famous for having held that our "form of life" is deeply embedded in social structures; the most important of these are language games, which link linguistic items with our activities. But the very language-games themselves can go awry along with our form of life, including the philosophy that limns the very language games. Though he was not a reformer of our linguistic practices, Wittgenstein claimed that the aim of philosophy was to expose the distortions of our language games.

It is unclear what follows from such claims. Do they allow that, once the distortions are exposed, a language game can be changed and a distortion-free game then be introduced? Or are the distorted games to be abandoned and not replaced? Wittgenstein says very little about what alternative language games might be like, or how possible they are. But for some the easy prospect of alternative language games is readily taken on board; and this quickly leads to some version of relativism with respect to language games. If a language game can be changed through philosophical therapy, does there still remain a role for philosophy? Or does philosophy in some sense 'come to an end'? The idea that there is an end to philosophy in the sense that it has a terminus, and not a positive goal to achieve, is an important theme in recent philosophy. So, what would be the prospects for a philosophy that gives a prominent role to a critical tradition concerning knowledge, method and rationality that, one must presume, is also distorted by faulty language games? It too would have to undergo therapy to cure its philosophical ills. What, then, is left? Following Bloor's construal of Wittgenstein's position, the critical tradition will have come to an end and will have been replaced by one of its heirs ... the sociology of knowledge!? Von Wright paints a picture of Wittgenstein's deep disquiet about our society dominated by science and technology, and about the decay of its culture. Is the sociology of knowledge from Marx to its most recent incarnations also part of the scientism of our age that Wittgenstein would also despise? It is hard not to draw that conclusion, despite what sociologists of knowledge might say about their being the Wittgensteinian heirs to philosophy. More will be said on this issue in section 7.5.

Wittgenstein is one philosopher who, on a certain interpretation, has been taken to advocate replacing the critical tradition within philosophy by sociology; Thomas Kuhn is another. The editors of a collection of papers on the sociology of science, Thomas Brante, Steve Fuller and William Lynch, tell us in their 'Preface' that their studies have led them to abandon the distinction between the very idea of a 'calm' core of the content of science around which swirls the contingencies of controversy. Rather 'controversy goes to the very core of science, which makes any neat boundary between "science" and "society" always elusive and temporary. What STS [science and technology studies] has failed to face is the way in which this finding politicises science as well as our picture of science' (Brante *et. al.* (1993), p. viii.) For the editors, it was Thomas Kuhn in his *Structure of Scientific Revolutions* who

> pointed the way toward the integrated study of history, philosophy, and the sociology of science (inclusive of technology). ... It alerted STS practitioners to the mystified ways in which philosophers talked about science, which made the production of knowledge seem qualitatively different from other social practices. In the wake of STS research, philosophical words such as *truth, rationality, objectivity*, and even *method* are increasingly placed in scare quotes when referring to science – not only by STS practitioners, but also by scientists themselves and the public at large.'
> (Brante *et. al., ibid.*, p. ix.)

Perhaps all is not well in the STS camp because the editors continue with the confession: 'the field [of STS] has yet to articulate aspirations that go beyond this deflation of philosophical pretensions'. But what is clear is that, despite some serious limitations they allege concerning his book, Kuhn's work has set the scene for their new discipline in which the spectre of a critical tradition in philosophy with its talismanic words, *truth, rationality, objectivity*, and *method*, can be finally laid to rest between scare quotes. Though many sociologists of science have co-opted Kuhn to their cause, we will see in the next section that Kuhn himself was not a willing fellow-traveller.

The idea that scientific knowledge is politicised has been around for a while. Thus Shapin and Schaffer tell us at the beginning of the final chapter of their book entitled 'The Polity of Science': 'Solutions to the problem of knowledge are solutions to the problem of social order' (Shapin and Schaffer (1985), p. 332). And they end the chapter with the equally political claim: 'As we come to recognise the conventional and artefactual status of our forms of knowing, we put ourselves in a position to realise that it is ourselves and not reality that is responsible for what we know. Knowledge, as much as the state, is the product of human actions.' (*ibid.*, p. 344). The passages in between merely spell out the science-politics nexus they discern. One of their central discoveries is expressed succinctly: they have attempted to show that 'the solution to the problem of knowledge is political' (*ibid.*, p. 342). If this is right, then the legitimation problem as formulated in the previous section, not to mention the whole career of epistemology and methodology from Ancient Greek times to our own, has simply been mad-dog "barking up the wrong tree". As will become evident in chapter 2, talk of 'the problem of knowledge' by Shapin and Schaffer confuses, on the one hand, the conditions and *processes* by which we come to know (which on one construal may include human activity in some political context), and, on the other hand, the *product* of that activity, knowledge, one condition of which is that what is known is also true.

Andrew Pickering emphasises the notion that science is a social construct when he endorses the views of Brante *et. al.*, saying of SSK, the sociology of scientific knowledge: 'SSK insisted that science was interestingly and constitutively social all the way to its technical core: scientific knowledge itself had to be understood as a social product' (Pickering (ed.) (1992), p. 1). If by 'social' is meant that two or more people interacting in some way have in fact been involved in the production of scientific knowledge, then most would not demur. This would be plain historical fact. Nor would most demur if the involvement was stronger in the sense that if two or more people had not interacted in various ways then we would not have had the scientific knowledge we currently possess; the previous interaction of the many is, in some sense, necessary. But that the many might be involved in no way tells us what certifications, based in reason, evidence and/or justification, are required to make something a piece of knowledge rather than mere commonly held belief amongst the many. The distinction between sociality and epistemic certifications for knowledge is made in section 1.1.3 in the discussion of Popper's view that there are both social and non-social aspects to scientific method. The point can be made again in an important lesson from Plato on constructivism in knowledge.

Plato's *Meno* contains a dialogue between Socrates and a Slave Boy who has had posed to him the following geometrical question: 'what is the length of the side of a square double the area of a given square?'. Through Socrates' asking questions and eliciting answers from the Boy, he come to see that two of his answers are wrong and that a suggestion of Socrates', viz., the hypotenuse of the given square, is the right answer. The dialectic of question and answer is social in that there is a discursive interaction of a logical character between two people; without that interaction the Slave Boy would never have hit upon the answer. But more than this, the Boy comes to see, by going through again by himself the reasoning process that Socrates initially led him through, that he has not merely obtained a true belief; he has acquired a piece of knowledge. (The knowledge/belief distinction is discussed further in section 2.1.) So, in one sense one can agree with Pickering about the social character of much knowledge through dialogues of discovery involving two or more persons. But what Pickering and others of a social constructivist persuasion consistently leave out is an account of the certification, provided by reason, evidence and/or justification, that makes knowledge. That this is not to be omitted is made quite clear in the *Meno*. To acquire knowledge might require the efforts of the many, and thus in one sense be "socially constructed"; but this tells us nothing about the necessary additional requirement of certification for knowledge. This is important for the traditional rationalist account of knowledge, but it is not part of the constructivist conception of "knowledge".

The social character of knowledge leads Pickering to claim that science is constitutively social. But what aspect of science? Experiments? Two or more are often involved in experimentation. Reasoning? Two or more, like Socrates and the Slave Boy, might be needed to provide the correct body of reasoning for some science. Producing knowledge? Two or more can be involved in the task of evidence gathering and justification-making that leads to knowledge. The product, propositional knowledge? Since Pickering's topic is SSK, where 'K' stands for (propositional) knowledge, then what the producers of such knowledge must provide is the certification (or warrant, both of these being normative) that distinguishes mere belief from knowledge. But nothing is said about such certification; all emphasis is on the social, which, we can happily admit, plays its role in the *production* of knowledge. But this has little to do with the *certification* which turns belief into knowledge; and it is this, rather than the social, which can appropriately be said to be 'constitutive' (to employ Pickering's undefined term). (The matter of certification is discussed in the first sections of Chapter 2 through talk of 'Plato's tether'. Also in section 2.4 there is a brief discussion of whether there is also a social condition to be placed on knowledge and what this might mean.)

One more British Connection to SSK will be mentioned mainly for his adherence to scepticism and relativism. Like many other sociologists, Harry Collins wishes to replace the rationally based decision-making about theories employing the purely epistemic criteria beloved by philosophers in the critical tradition, by a process of negotiation within specific scientific communities in which personal, professional career and socio-political interests are the "drivers". Collins also advocates what he calls the *Empirical Programme of Relativism* which 'rests on the prescription: treat descriptive language as though it were about imaginary objects' (Collins (1985), p. 16). This is a surprising injunction that no one ought to adopt. We can only suppose for Collins that it results from a relativism in which 'cultures differ in their perception of the world' because these 'perceptions … cannot be fully explained by reference to what the world is really like' (*loc. cit.*). Since the real world underdetermines our perceptions, or our descriptions, of the world then it seems as if these can run riot – in fact run riot to the extent that there is scepticism about whether our perceptions and descriptions latch onto the world at all! For all we know they could be about the merely imaginary. So, Collins says, let them be about the imaginary! Even granting the oddness of much of the theory of perception on which this relies, it is a fallacy to argue from the claim that it is *possible* that our perceptions and/or descriptions be about the imaginary, to the claim that they are, or to an injunction to treat them as if they are so. For Collins the only constrains on perceptions and descriptions would appear to be social, and nothing to do with the world, which drops out of the picture all together. It is the luck of our social negotiations, and nothing to do with reality, that keeps us out of the way of the proverbial oncoming bus, thereby keeping us alive to continue to use 'descriptive language as if it were about imaginary objects'.

Collins is also a sceptic about the results of experimentation because of the 'experimenters regress'. We get correct results in an experiment when the experimental apparatus is functioning properly; but we can only tell when the experimental apparatus is functioning well when we get correct results. Since evidential matters are alleged not to be able to break this circle but are part of it, the only way out left to us is to appeal to social matters, such as the negotiations carried out by members of the relevant scientific community which decide whether the experiment was correct or not. This concoction of scepticism and relativism is common amongst advocates of SSK; but it has its critics. Opponents such as Franklin[17] argue that the relativism based on underdetermination is faulty. Moreover scepticism is not justified by the experimenter's regress since there is no regress; it can be broken on epistemic grounds, as Franklin shows. There are ways of checking the correct functioning of the apparatus independently of taking its results to be correct.

1.3.2 A Case of Sociological Truth-Phobia: Shapin

While most philosophers are lovers of truth, sociologists have a phobia about truth. The neglect, even disparagement, of truth is one of the great divides between philosophers and SSK practitioners. For those who wish to review the nosology of truth phobia in much current thought, the catalogue of sociologists and postmodernists gathered in Rosenau (1992) in a chapter entitled 'A Theory of Truth and the Terrorism of Truth' is a good starting point for their pathological examinations. Here we will restrict ourselves to the views of one sociologist of science, Steve Shapin. Sometimes sociologists are right to set truth aside. If one is a sociologist investigating the epidemiology of belief, say, about the existence or non-existence of (the Christian) God, across a community, then one need not take an interest in whether the proposition *that God exists* is true or false; we can set matters of truth or falsity within religion aside. But the epidemiologist does have an interested in some truths such as whether, of some person A that they are interviewing, *it is true that A believes that God exists*, or *it is not true that A believes that God exists*. There are alternative locutions here; the word 'believes' can be replaced by phrases like 'accepts true' and the same point holds. The important point is that epidemiologists had better not have such a bad dose of truth-phobia that they cannot investigate the truth or falsity of the following claims; *A believes (accepts true) that God exists*, or *A does not believe (accept true) that God exists*. Otherwise their very sociological investigation is in jeopardy.

In general, for the epidemiologist of belief, *that p* is true, or false, is not the focus of their interest. But what is of interest to them is the intensional proposition *that A believes that p*. The epidemiologist had better get right the truths about who believes what if they are to get their statistics right.

Importantly, advocates of SSK fail to notice the simple logical point, well known since Frege, that from the claim *A believes that p* nothing follows about the truth or falsity of the content of the belief, *that p*. In the claim *A believes that p* the context of 'p' is said to be *intensional*; that is, the whole claim is not a truth function of its partial content expressed by *that p*. This point makes it possible for epidemiologists of belief to investigate claims like *x believes that p*, where 'x' ranges over the members of some community, free of any concern as to whether the contained claim, *that p*, is true or false. The research world of sociologists of belief is made safe by the simple logical point about the intensionality of 'believes'; they can quite rightly and happily ignore the matter of whether *that p* is true or false.

The above elementary point is perhaps over-laboured, but there are sociologists of scientific knowledge who have difficulties with it, Shapin being a particularly good example. For those with finely cultivated sensibilities about truth and related notions, the title of his book *The Social History of Truth*, jars. Truth does not have a history; but what we *accept* as true, or what we *believe*, does have a history. Shapin is not unaware of the difference and says of the 'distinction between what is "true" and what is merely taken to be so, by some people at some time':

> Indeed, there is a special community of language-users called 'academic philosophers' who insist very vigorously on such a distinction. The body of locally credible knowledge – what is taken to be true – cannot be the same as 'truth', since truth is one and what people have taken to be true is known to be many'. (Shapin (1994), p. 3)

Academic philosophers could agree with the general thrust of this (but they could quibble about the use of scare quotes to some unclear purpose and raise an eyebrow about what the iteration of epistemic notions comes to in the phrase 'credible knowledge' – but then relax if all that this means is 'what is taken to be true', i.e., 'belief'.). Shapin recommends that the more restrictive notion of *truth* beloved of academic philosophers be set aside and replaced by the more liberal notion of *accepted truth* (*ibid.*, pp. 4-5). (Sometimes Shapin talks of *accepted truth*, on others *accepted belief*; though the last locution is odd, any difference between these will not be insisted upon here.) But what is the point of this liberality if it engenders intellectual confusion? Shapin tells us that he wants 'to preserve from the restrictive sensibility the loose equation between truth, knowledge and the facts of the matter' (*ibid.*, p. 4). But such an equation, however loose, ruins important distinctions without which the whole field of epistemology is rendered useless. For epistemologists, since knowledge implies truth but not conversely, there is no equivalence, no matter how loose. But there is the well-known equivalence between 'that p is true' and 'that p'. Talk of the fact that p will only enter into this if one is some kind of correspondence truth theorist, or one holds that there are truthmakers such as facts. So there is no equivalence, however loose, between the three notions, truth, knowledge and facts. Hence the therapeutic role of the first three sections of the next chapter for those sociologists who fail to take care of important distinctions in epistemology.

Shapin also tells us: 'There are groups of people dedicated to the disinterested understanding of cultural variation in belief, and for them the restrictive sensibility lacks both value and legitimacy. For historians, cultural anthropologists, and sociologists of knowledge, the treatment of truth as accepted belief counts as a maxim of method, and rightly so' (*loc. cit.*). The elevation of the confusion of *truth* with *accepted belief* to the status of a maxim of method is truth-phobia in one of its most virulent forms. If it is a maxim of method, then the epidemiologist mentioned earlier has the following problem when he/she interviews A (whom we may suppose is an atheist). We can agree that the truth value of the content of what A believes, viz., *that God does not exist,* is not up for discussion (though it does have a truth value and is not to be downgraded to merely a belief that A accepts). But what does the epidemiologist put in their survey?

What they ought to enter is the truth they have uncovered in their survey: *it is true that A believes that God does not exist* (which is simply equivalent to *that A believes that God does not exist*). But on Shapin's faulty maxim, the epidemiologist's investigation cannot culminate in any truth, but only in what is *accepted* as true (presumably accepted by the epidemiologist). At best, following the maxim leads us to: *it is an accepted belief (by the epidemiologist) that A believes that God does not exist.* We never get from the epidemiologist a true account of what A believes (or anyone else that is surveyed). What we get at best is what *the epidemiologist accepts as their belief* about what the surveyed believe, and not a real story about the variation of belief across a community. There is the world of difference between the claim 'that p is true', and the claim 'person A accepts that p as one of their beliefs'. But it is this difference Shapin wants to obliterate with his faulty maxim. For those who have their truth sensibilities intact, the point made against Shapin is one aspect of the many objections Plato made to Protagoras about his truth relativism (see Burnyeat (1976)).

The last cited remark of Shapin ends with the emphasis, 'and rightly so'. But this is merely a form of words in which a person expresses their endorsement of a claim. And this we can do using the equivalent expression 'and this is true', the word 'this' picking up as its reference the methodological maxim M 'truth is to be treated as accepted belief'. So let us apply M to the very words of Shapin's endorsement 'and rightly so'. This is, in common language, equivalent to 'and this is true'. But by Shapin's maxim we have to replace this by 'and this is accepted belief'. Who does the accepting? Let us suppose it is Shapin (or perhaps one of the dedicated historians, cultural anthropologists, and sociologists of knowledge Shapin mentions). But again there is the world of difference between saying of some maxim M 'M is true', and what Shapin's maxim bids us replace this by, viz., 'Shapin accepts maxim M as a belief of his'. In sum, always replacing the restrictive *truth* by the liberal *accepted belief* leads to disaster.

Why adopt a disastrous maxim when a fact of simple logic about the intensional context of 'p' in 'A believes that p' already guarantees the independence of the truth values of *A accepts the belief that p*, and *that p* itself? Evaluating the truth of the contained *that p* is a superfluous task over which no sociologists of belief should exercise themselves; it is irrelevant to their proper task of evaluating the truth of *x believes that p*, for the many xs of the community under investigation. As a prophylactic to the truth-phobia with which we have been dealing, here is a counter-maxim: *truth* and *accepted belief*, or *accepted as true*, are to be treated as distinct (in the light of the logical independence of one from the other). Only through the counter-maxim can epidemiologists of belief have viable theories. The counter-maxim relieves sufferers from truth-phobia and enables sociologists of belief to become healthy members of the critical tradition.[18]

1.3.3   A French Connection: Latour

This and the next sub-section, concerns two French connections, Latour and Lyotard, who are dethroners of philosophy within science and science studies. Latour's books are a veritable mine of material against the critical tradition in science, only two of which will be mentioned here. The first half of his book *The Pastuerization of France*, is a social study of aspects of Pasteur's science in France. The second half is devoted to an account of what Latour calls 'irreduction' in the sciences. Though oddly named, this is Latour's account of the autonomy of social studies of science freed from any philosophical doctrine within the critical tradition – with comments in passing on what such a philosophy might be like and why it should be abandoned. Latour begins with a diagnosis of what many people find puzzling about empirical studies of science. They are puzzled because they expect the studies to yield more than they do, and so conclude that something important has escaped the attention of those doing the studies. What is this missing *je ne sais quoi*? Latour tells us: 'I quickly unearthed what appeared to me to be a fundamental presupposition of those who reject "social" explanations of science. This is the assumption that force is different in kind to reason; right can never be reduced to might' (Latour (1988), p. 153). In so far as social studies of science are deemed to be deficient because they leave out the important ingredient of reason, they are said to be 'reductionist' (in Latour's idiosyncratic use of the term to characterise his opponents). In contrast Latour calls his own position 'irreductionist' since it investigates 'how knowledge and power would look if no distinction were made between force and reason' (*loc. cit.*).

Latour's exposition of his irreductionism proceeds by a series of numbered aphorisms with more discursive interludes interpolated between them. Thus aphorism 2.4.3 tells us: 'We cannot distinguish those moments when we have might and when we are right' (*ibid.*, p. 183). So it is not merely that we can simply suspend the force/reason distinction within science studies and then see what happens; we cannot even distinguish between force and reason (sometimes for the odd reason that the two are the same). As aphorism 1.1.2 tells us in Nietzschean fashion: 'There are only trials of strength, of weakness. Or more simply, there are only trials' (*ibid.*, p. 158). As samples of the titles of two interludes we find: 'Interlude VI: In Which the Author, Losing His Temper, Claims that Reducers are Traitors' (*ibid.*, p.208); and 'Interlude VII: In Which we Learn Why This Précis Says Nothing Favorable About Epistemology' (*ibid.*, p.215).

In the last-mentioned Interlude, Latour recognises that 'we would like to be able to escape from politics' (*loc. cit.*) and find refuge in reason and epistemology. But Latour claims there is no such refuge from politics, power or violence:

> ... this faith [in epistemology] has become the main obstacle that stands in the way of understanding the principle of irreduction. Its only function is passionately to deny that there *are* only trials of strength. "Be instant, in season, out of season," to say that "there is something in addition, there is also reason." This cry of the faithful conceals the violence that it perpetrates, the violence of *forcing* this division.
>
> All of which is to say that this précis, which prepares the way for the analysis of science and technology, is not epistemology, not at all. (ibid., p. 216)

Members of the critical tradition will take Latour's remarks as their own critics. What they express vigorously is the dethroning of the critical tradition and its usurpation by a sociology of power and force in science studies and in theories of scientific "knowledge". Even allowing that much of the above might be Gallic insouciance, it is also an irrationalist *cri de coeur* which, from the stance of the critical tradition, has done much damage in science studies. Even within science studies a few, such as Bloor, are now distancing themselves from Latour's position.[19] But not because there is anything worthwhile to be found in the epistemology that Latour attacks! The connection between power and knowledge that Latour alleges, is discussed more fully in Part III on Foucault who also adopts related views.

In 'Appendix 1' to *Science in Action,* Latour gives us seven 'Rules of Method' that are used in his book. Consider only the third of these. It continues the theme of right and might, but in a confused way: '*Rule 3* Since the settlement of a controversy is the *cause* of Nature's representation, not its consequence, we can never use this consequence, Nature, to explain how and why a controversy has been settled' (Latour (1987), p. 258). As Sokal[20] points out, there is an important ambiguity in the switch from talk in the first clause of 'Nature's representation' (which we can suppose are things like the theories we develop to represent how nature is) and, in the second clause, talk of just 'Nature' with no mention of 'representations'. If we stay with talk of 'Nature's representations', or better 'theory of Nature that scientists accept', all the way through Rule 3, then we have the following claim: 'since the settlement of controversy is the [more strictly 'a'] cause of scientists generally accepting some theory of Nature, not its consequence, we can never use this consequence, scientists' acceptance of the theory, to explain the settlement of controversy'. There is something here with which we can agree: settling controversy within a scientific community *can* lead to the overall acceptance of a given theory. But consider the converse, viz., acceptance of a theory by a community leads to the settlement of controversy. This also seems possible, contrary to Latour; and it does not follow from the claim of which it is the converse.

Now consider Rule 3 with talk of 'Nature' *simpliciter* all the way through. Then we have the following bizarre claim: 'Since the settlement of a controversy is the cause of Nature, not its consequence, we can never use this consequence, Nature, to explain how and why a controversy has been settled'. The first clause expresses the grossly idealist claim that Nature itself is a consequence of the settlement of controversy within a scientific community. And this must include all of Nature even before humans ever came on the scene. The second clause is also faulty because Nature can enter into explanations as to why controversy is settled. Thus one part of actually existing Nature, viz., that there is no South Pacific Ocean continent *Terra Australis Incognita*, is a component, along with what Cook's voyages in that region showed, in any explanation about why controversy in Europe about its claimed existence was settled.

Latour's Rule 3, like his other Rules, needs to be subject to careful analysis before being accepted. On one reading the settlement of controversy has enormous world-creating powers. Though one may well wonder *about what* is there some controversy before settlement is reached; it is only at settlement, and not before, that the relevant bit of the world is created! But on another reading it is an inflated way of saying that some explanations as to why we accept theories depend, *in part*, on the agreements we make. Rule 3 as it stands is to be rejected; otherwise, in all the confusion it engenders, Latour's irreductionist replacement of right by might gets its grip.

1.3.4 Yet Another French Connection: Lyotard

The French postmodernist, Jean-François Lyotard, also proclaims the dethronement of the critical tradition when he speaks of 'the end of all metanarratives'. What does this mean? Lyotard tells us: 'Science has always been in conflict with narratives. Judged by the yardstick of science, the majority of them prove to be fables' (Lyotard, (1984), p. xxiii). But if 'narrative' means the telling of a story which can be either true or false (with appropriate attention to the telling which may be engaging, amusing, well expressed, and so on for other criteria for a *good* story), then Lyotard's remark is misleading. Science, too, can have its narratives, as when Hooke described (narrated) in his *Micrographia* (1665) the world he was first to observe under the microscope, or Jane Goodall did ground-breaking scientific work in her descriptions (narratives) of the behaviour of the chimpanzees she carefully observed in Gombe Park, Kenya.

Lyotard continues saying: 'But to the extent that science does not restrict itself to stating useful regularities and seeks the truth, it is obliged to legitimate the rules of its own game' (*loc. cit.*). But wrong again. Suppose science were to restrict itself to useful regularities only, then what are they useful for? One use for observed regularities is to make predictions about the next unobserved case. In the light of this, suppose one were to limit oneself only to talk of such 'useful regularities'. Then one would not avoid the problems of the 'legitimation' of inductive inference, or its justification, of which inference to the next case is an example.

What of seeking the truth? This, as will be seen, has its "grand narrative" which Lyotard invites us to reject (though it is often unclear whether, when the grand narrative is debunked, it is truths that are to be rejected, or just some *theory* (or all theories) of what truth is). Though Lyotard does not say anything like this, his remarks might be taken to concern the use of patterns of inference, such as Inference to the Best Explanation, to establish the truth (or verisimilitude) of laws, or of theories. As mentioned in the previous section, principles of method embodying such patterns of inference stand in need of justification. This is hardly news to anyone in the philosophy of science who has investigated the patterns of argument used to establish truths about the existence of unobservables, or our theories of them. What might be novel is Lyotard's talk of the legitimation of *rules of the game of science* in this context. This reveals one way in which Wittgenstein has had a strong influence upon Lyotard. Though Lyotard does not say as much, we might see in his talk of 'rules of its own game', a link to an idea expressed in the previous sections 1.1 and 1.2, viz., that there are principles of method, such as Inference to the Best Explanation, within our critical tradition that stand in need of justification. Looked at in this light, Lyotard's remarks bring him close to the idea of principles of method and meta-method that stand in need of legitimation.

Lyotard then continues:

It [science] then produces a discourse of legitimation with respect to its own status, a discourse called philosophy. I will use the term *modern* to designate any science that legitimates itself with reference to a metadiscourse of this kind making an explicit appeal to some grand narrative, such as dialectics of the Spirit, the hermeneutics of meaning, the emancipation of the rational or working subject, or the creation of wealth. (*loc. cit.*)

Once again we can interpret Lyotard's remarks about 'grand narratives' and 'metadiscourse' as having a bearing on much that has been said in the previous two sections. There we pointed out that our critical tradition will involve both meta-methods to be used to justify (legitimate) the use of various principles of method (understood as rules and/or values); and these principles are in turn used to criticise the claims of the sciences themselves. However none of the four 'grand narratives' that Lyotard mentions play any role in this account of our second order critical tradition. Talk of 'the dialectics of the Spirit' has little current standing in Anglo-American philosophy (but might be a Hegelian hang-on in France). Talk of 'the hermeneutics of meaning' is obscure. Relating this to principles of method, few have looked to a theory of meaning to justify them. (As will be seen in the next section Kuhn is an exception but his claims have little to do with a 'hermeneutics of meaning'.) The final two 'grand narratives' have nothing to do with the *epistemic* justification of principles in our critical tradition. But they might have a bearing on the quite different *social* legitimation of science base in the hopes, realised or unrealised, that humankind has had of science in enabling us to create greater freedom or wealth for ourselves.

Focusing on epistemic rather than social legitimation, Lyotard's talk of 'grand narratives' shows no familiarity with attempts by those in the critical tradition to legitimate science and its methods. He nowhere comes to terms with the issues raised by Popper, Lakatos, Laudan or the Bayesians concerning the justification of methodological principles (as set out in the previous section). So our (plausible) attempt to view Lyotard in this way yields no critical evaluation of our second order critical tradition. Nor does Lyotard makes clear why, or what aspect of, truth is allegedly another grand narrative (*ibid.*, p. xxiv). Talk of truth in this context could mean that one of the aims of science is to uncover true observational claims, laws and theories. But does this constitute a 'grand narrative'? Talk of truth could also mean that at the meta-level we seek viable theories of truth (another sort of 'grand narrative'), or viable methodological principles the use of which guarantees we achieve the goal of truth. But no theory of truth or principles of method (like Inference to the Best Explanation) are mentioned. Once again no case can be found in Lyotard either for against such meta-considerations of epistemic legitimation which are part of our second order critical tradition. This is significant for the next important claim he goes on to make.

The punch line of Lyotard's discussion is: 'I define *postmodern* as incredulity toward metanarratives' (ibid., p. xxiv). But this is merely asserted and not argued. Merely defining 'postmodernism' gives us no reason as to why we *ought* to be incredulous toward metanarratives, and so be a postmodernist. However if one is a postmodernist then, on this definition, no meta-story, or justificatory meta-method is to be believed. This has some drastic consequences. Meta-methods were meant to justify the rules and principles of scientific method, but now we are to be incredulous about any meta-method; so it follows that we are also to be incredulous towards principles of scientific method. And if no principles of method are to be believed, then no science got by the use of these methodological principles is to be believed either. (This may be an unintended consequence of the postmodernist position, but it is a consequence – unless they can give an account of what authority and legitimation our principles of method and logic have.) For postmodernists it follows that there is a positive disbelief in any attempt to justify not only the claims made in meta-methodology and the very principles of method that might be used to justify claims in science, but also the claims in the various sciences themselves. Conclusion: scepticism about science, its methods and meta-methods. Lyotard goes on to speak of the 'obsolescence of the metanarrative apparatus of legitimation' and of its correspondence to 'the crisis of metaphysical philosophy' (*loc. cit.*). But it is hard to see that this is a genuine crisis in philosophy rather than a crisis in Lyotard's own version of philosophy. The upshot of all of this is that the very idea of something like Popper's critical tradition is simply one of the objects of postmodernist incredulity (though no argument is given for this important conclusion). We are also to remain incredulous about truth and/or theories of truth since both of these are alleged to be some kind of meta-narrative. Lyotard's postmodernism suffers from a serious attack of truth-phobia, which undermines even his own claims presumably advanced as truths.

The above merely sets out Lyotard's position and discusses no argument he might give for it; in fact little argument can be found in Lyotard's book, thereby creating incredulity all round. What one does get is an account, somewhat along Wittgensteinian lines, of language games and of the pragmatics of language and what little legitimation these might provide: 'The philosopher at least can console himself with the thought that the formal and pragmatic analysis of certain philosophical and ethico-political discourses of legitimation, which underlies the report [i.e., Lyotard's book], will subsequently see the light of day' (*ibid.*, p. xxv). So perhaps not all is lost in the abyss of incredulity. Significantly Lyotard adds that his book 'will have served to introduce that analysis from a somewhat sociologizing slant' (*loc. cit.*). Thus for Lyotard the legitimation narratives of epistemology will get replaced by what legitimation, if any, the social analysis of language games can provide. Once more the critical tradition is usurped by sociology.[21]

This section merely samples the top of an iceberg of anti-philosophy stances within current studies of science dominated by sociology of knowledge, postmodernism, various kinds of truth phobia, and so on. But it shows that the stakes have become high in the opposition between the critical tradition within philosophy and its various detractors concerning our understanding of knowledge and science. The criticisms contained in the above do not entail that there is not a viable subject called the 'sociology of science' in which various aspects of science can be explored from a sociological perspective. What is criticised is a different beast, the sociological study of the very knowledge content of science itself. This point is explored more fully in Part II. We turn now to two philosophers of science who have often been thought to have prepared the ground for the dethroning of the critical tradition in science studies, with or without a usurping sociology of knowledge.

1.4   KUHN AS DETHRONER OF THE CRITICAL TRADITION?

Some sociologists of science cite Kuhn and Feyerabend as two philosophers of science who have assisted in the dethroning of the rationality of science thereby voiding the critical tradition of any content. But have they? The situation is not that simple for either philosopher, though both might be said to have played a role in getting us to rethink what the critical tradition might be like. Consider first the case of Kuhn who throughout his career has adopted many positions on the question about the rationality of science. Kuhn can appear to be an advocate of rationality in science. For those who overlook it,[22] Kuhn says in *The Structure of Scientific Revolutions*:

> ... there is another set of commitments without which no man [*sic*] is a scientist. The scientist must, for example, be concerned to understand the world and to extend the precision and scope with which it has been ordered. That commitment must, in turn, lead him to scrutinize ... some aspect of nature in great empirical detail. And, if that scrutiny displays pockets of apparent disorder, then these must challenge him ... to a further articulation of his theories. Undoubtedly there are still other rules like these, ones which have held for scientists at all times.' (Kuhn (1970), p. 42).

What Kuhn here calls 'rules' he latter calls 'values'; here they include the somewhat unspecific *understanding* the world, and the more specific *precision* and *scope* (especially in respect of the amount of order disclosed by our theories). The significance of Kuhn's remark is that our theories ought to exemplify these three values (which can be understood more generically and which can come in degrees); and we can choose between theories on the basis of the extent to which they exemplify these values. Importantly these values are said to hold for all scientists at all times, and not merely within some paradigm. There is also a tinge of necessity when Kuhn says that no person can be a scientist *unless* they hold these values (and some other values only hinted at). This puts Kuhn squarely in the camp of those advocates of the critical tradition who place epistemic value on the understanding our theories provide, and their precision and scope – these values being universal and necessary for anything to be a science. But at other places in his book Kuhn can appear to be anti-rationalist and relativist about such values.

The notion of incommensurability looms large in Kuhn's philosophy of science. In this section I wish to focus on just one aspect, viz., methodological incommensurability. This is the view that there are no shared, objective standards, rules or values whereby either theories or paradigms can be rationally evaluated. Hence there may be alternative theories or paradigms which are incommensurable in the sense that there are no standards whereby we can rationally chose between them (on whatever grounds). Was Kuhn ever an advocate of methodological incommensurability? There are some grounds for this claim in his earlier writings. However the later Kuhn made his position more specific and the charge of methodological incommensurability is hard to sustain. What will be argued is that the later Kuhn moved much closer to traditional preoccupations of methodologists of science who have tried to develop theories of scientific rationality. In so doing he left behind many who saw in the earlier Kuhn a fellow supporter of sociological and/or postmodernist approaches to theory change in science. The later Kuhn's rejection of their position will be discussed in the final sub-section.

1.4.1 Kuhn Phase I: Methodological Incommensurability in The Structure of Scientific Revolutions

Those who view Kuhn as holding either an irrationalist, or anti-methodology stance, or endorsing a paradigm-relative account of scientific method, can find passages in the first edition of Kuhn's 1962 book, *The Structure of Scientific Revolutions*, that support methodological incommensurability. Using a political metaphor to describe scientific revolutions Kuhn says of scientists working in different paradigms that '... because they acknowledge no supra-institutional framework for the adjudication of revolutionary difference, the parties to a revolutionary conflict must finally resort to the techniques of mass persuasion, often including force' (Kuhn (1970), p. 93). Continuing the political metaphor, there is also a suggestion that if there are methods for the evaluation of theories *within* a paradigm or throughout the course of normal science, they do not carry over to the evaluation of rival paradigms:

> Like the choice between competing political institutions, that between competing paradigms proves to be a choice between incompatible modes of community life. Because it has that character, the choice is not and cannot be determined merely by the evaluative procedures characteristic of normal science, for these depend in part upon a particular paradigm, and that paradigm is at issue. When paradigms enter, as they must, into a debate about paradigm choice, their role is necessarily circular. Each group uses its own paradigm to argue in that paradigm's defence. (*ibid.*, p. 94)

Later he speaks of paradigms '... as the source of the methods ... accepted by any mature scientific community at any given time' (*ibid.*, p. 103). Since the notion of a scientific community is fundamental and is sometimes used to individuate the notion of a paradigm, then these passages strongly suggest that even if methodological principles hold within a paradigm, there are no paradigm transcendent principles available.

Kuhn's position is more starkly presented when he continues saying:

> As in political revolutions, so in paradigm choice – there is no standard higher than the assent of the relevant community. To discover how scientific revolutions are affected, we shall therefore have to examine not only the impact of nature and of logic, but also the techniques of persuasive argumentation effective within the quite special groups that constitute the community of scientists. (*ibid.*, p. 94)

Kuhn does not spell out what he means by the 'impact of nature and logic' in theory choice. Perhaps we may see a role for logic in the deduction of test consequences from a theory, and a role for nature in saying 'yes' or 'no' to the truth of the consequences. What is clear is that logic and nature are not denied a role in theory choice and can be persuasive in getting community-wide assent to a particular point of view. However 'persuasive argumentation' can be understood in a much broader sense which allows a place for rhetorical argumentation in scientific reasoning. By linking 'persuasive argumentation' to the context of political revolutions, the assent of a community might well be achieved by means that go beyond the bounds of an acceptable rationality. And many have read Kuhn in this way in the light of his above-mentioned remarks about 'mass persuasion' and 'force'.

Other passages in Kuhn are not always so clear. Thus he tells us: 'In learning a paradigm the scientist acquires theory, methods, and standards together, usually in an inextricable mixture.' (*ibid.*, p. 109) If this is a point about a pupil's induction into a science, then the acquisition of standards (which we may suppose include principles of theory choice), along with theories and methods peculiar to particular sciences, is unproblematic. But then he goes on to say, in separating issues to do with the posing of problems and their solution from matters of theory choice:

> Therefore, when paradigms change, there are usually significant shifts in the criteria determining the legitimacy both of problems and of proposed solutions.
> That observation ... provides our first explicit indication of why the choice between competing paradigms regularly raises questions that cannot be resolved by the criteria of normal science. To that extent ... that two scientific schools disagree about what is a problem and what is a solution, they will inevitably talk through each other when debating the relative merits of their respective paradigms. ... Like the issue of competing standards, that question of values can be answered only in terms of criteria that lie outside of normal science altogether, and it is that recourse to external criteria that most obviously makes paradigm debates revolutionary.' (*ibid.*, pp. 109-10)

This passage suggests that Kuhn has in mind an (unsatisfactory) argument from the premise that there are always disagreements about what counts as problems to be posed and their solution from within the perspective of any pair of paradigms, to the conclusion that there are no standards, values or criteria for choice between paradigms. As will be seen shortly, elsewhere Kuhn downplays the generality of the premise of this argument (he modifies the claim that there are *always* incomparable problems), thereby making possible inter-paradigm choice. But the passage also makes clear that if there are values concerning theory choice outside normal science then paradigm choice become 'revolutionary' (though perhaps we should not read this as 'irrational').

Kuhn often expresses methodological incommensurability in terms of his talk of 'world changes' and 'conversion' when he mentions a:

> ... third and most fundamental aspect of the incommensurability of competing paradigms. In a sense that I am unable to explicate further, the proponents of competing paradigms practice their trades in different worlds. ... [This] is why, before they can hope to communicate fully, one group or the other must experience the conversion that we have been calling a paradigm shift. Just because it is a transition between incommensurables, the transition between competing paradigms cannot be made a step at a time, forced by logic and neutral experience. Like the gestalt switch, it must occur all at once (though not necessarily at an instant) or not at all. (*ibid.*, p. 150)

Once again not only logic and experience can bring about theory change (providing rational grounds), but also conversions which are akin to gestalt switches (which are definitely non-rational grounds). Of course both logic and experience, on the one hand, and gestalt switch conversions, on the other, can, for two different people, bring about the same change in theory; but obviously the grounds for the change are significantly different. Kuhn continues declaring that scientists are human in that they '... cannot always admit their errors, even when confronted with strict proof. I would argue, rather, that in these matters neither proof nor error is at issue. The transfer of allegiance from paradigm to paradigm is a conversion experience that cannot be forced' (*ibid.*, p. 151). The problem of 'elderly holdouts' is an example of an allegiance which would not yield to a forced conversion. But there still might be resistance to 'conversion' on other grounds, such as belief that a paradigm in difficulties will, with a bit of ingenuity, eventually overcome its problems.

However the 'conversion' view is tempered when Kuhn admits that '... the question of the nature of scientific argument has no single or uniform answer' (*ibid.*, p. 152). In making this concession Kuhn allows for the rational comparison of paradigms, one of which is their problem-solving powers: 'Probably the single most prevalent claim advanced by the proponents of a new paradigm is that they can solve the problems that have led the old one to a crisis. When it can legitimately be made, this claim is often the most effective one possible' (*ibid.*, p. 153). Thus there may be conditions under which some competing paradigms can be compared for their problem-posing and problem-solving abilities. Contrary to what has been indicated a few paragraphs above, it is not generally true for Kuhn that paradigms are incomparable in respect of their ability to pose and solve problems; some pairs may be fully comparable while other pairs might not be. However we are still left with talk of 'conversion' from one paradigm to another. While this might be a correct factual description of how some have changed paradigms, Kuhn still leaves in limbo the "ought" and the rationality of the canons of principles of method that might be used in bringing about the change.

Many of the above claims made by Kuhn do show that Lakatos is correct when he says that the theory choice involved in paradigm change is a matter of 'mob psychology' (Lakatos and Musgrave (1970), p. 178). And Kuhn does say in reply to those critics who claim that he advocated "...'irrationality', 'mob rule' and 'relativism'" that he needs 'to eliminate misunderstandings for which my own past rhetoric is doubtless partially responsible" (*ibid.*, pp. 259-60). But in eliminating some of the misunderstandings of his 1962 book, both in his 'Reflections on my Critics' (see his (1970a), but presented much earlier in 1965) and in the 'Postscript-1969' to the 1970 second edition of his book, Kuhn in fact articulates new ideas only dimly discernible, if at all, in the original version of the book.

1.4.2 Kuhn Phase II: Values and a Model for Theory Choice

In a retrospective interview published in Italian in 1991 Kuhn said: 'I'm not sure I would use the term paradigm in such a wide sense anymore' (Borradori (1994), p. 166). Alert readers of the 'Postscript – 1969' would have already noticed that Kuhn effectively abandoned the notion of a paradigm in favour of the more articulated notion of a disciplinary matrix. A disciplinary matrix has four main features: symbolic generalisations (SG), or the formalisable components of the disciplinary matrix such as laws or principles; models (M), such as the model of a gas as inelastic billiard balls, etc; values (V); exemplars (E) such as the text-book examples of concrete problem solutions typical of the disciplinary matrix which students master in the course of their education. The focus here will be upon values, just one of the elements of a disciplinary matrix.

Contrary to the impression that might be gathered from Kuhn Phase I, values '... are more widely shared among different communities than either symbolic generalisations or models, and they do much to provide a sense of community to natural scientists as a whole' (Kuhn (1970), p. 184). This is an important claim to note. In the 'Postscript' one of Kuhn's concerns is to individuate the notion of a paradigm. How is this to be done? Kuhn recognises the circularity in the claim: 'A paradigm is what the members of a scientific community share, *and*, conversely, a scientific community consists of men who share a paradigm'. (*ibid.*, p. 176) Hence the importance of the sociological aspect of Kuhn's account of science in which sociological criteria are used to individuate scientific communities. Whether or not Kuhn's sociological mode of individuation is successful and whether it can be carried out without appeal to a paradigm or the contents of a paradigm, is not a matter that need concern us here.[23] Granted independently individuated scientific communities, Kuhn then sets out to individuate paradigms, rather than the other way around: 'Scientific communities can and should be isolated without prior recourse to paradigms; the latter can then be discovered by scrutinising the behaviour of a given community's members' (loc. cit.). Thus it follows that if values are widely shared among different communities, as alleged by Kuhn (as cited at the beginning of this paragraph), then values can be shared across paradigms, i.e., across disciplinary matrices.

The point is important. Suppose we are given two successive disciplinary matrices, $DM_1 = <SG_1, M_1, V_1, E_1>$ and $DM_2 = <SG_2, M_2, V_2, E_2>$. Then it could quite regularly be the case that the set of values V_1 *is the same as* the set of values V_2 and that there is not the disparity of values across paradigms, or disciplinary matrices (DMs). If DMs are to differ, then they will have to differ either in their SGs, or their Ms or their Es. In fact it is quite likely that they do differ in these three respects since different SGs will apply in different Ms and thereby generate different Es. And all this can take place while the Vs remain invariant.

What does Kuhn allege that scientists value? First they value predictions. But not any old predictions. They should be quantitative rather than qualitative; and they should be accurate, or accurate to within some desired degree. They also value theories which 'first and foremost, permit puzzle-formation and solution' (ibid., p. 185) Here earlier Kuhnian qualms about the incomparability of successive paradigms with respect to their problem posing and solving have been set aside in proposing this value. Scientists also value simplicity, internal consistency and plausibility (i.e., external consistency with other currently deployed theories). Kuhn also allows that there can be other values, citing social utility as an example.

In a 1977 paper 'Objectivity, Value Judgement, and Theory Choice' (Kuhn (1977), chapter 13) he re-endorses these values and adds the following. Theories ought to have broad scope; that is, they ought to apply to new areas beyond those they were designed to explain. And theories ought to be fruitful; that is, they ought to introduce scientists to hitherto unknown facts or to unknown areas of research. Also simplicity is mentioned again, but more in the sense of '... bringing order to phenomena that in its absence would be individually isolated and, as a set, confused' (Kuhn (1977), p. 322). But a much better name than 'simplicity' for this value would be 'unity'.

Kuhn's list is a fairly standard list of values that nearly all methodologists would like to see our theories exemplify; there is nothing radical in it. Nor does Kuhn regard the list as 'exhaustive' (*ibid.*, p. 321). There are other values, but the ones Kuhn lists are central to science. Importantly, in contrast to his earlier more radical Phase I, Phase II Kuhn does not restrict the application of these values to only *within* paradigms or disciplinary matrices; being inter-paradigmatic, they transcend them. Kuhn also notes that many methodologists express similar claims in terms of rules rather than values, such as: 'choose, or accept, the simplest theory', and so on. Kuhn prefers to express his criteria as values, saying that '... the criteria of choice ... function not as rules, which determine choice, but as values, which influence it' (*ibid.*, p. 331). However there is no reason why rules without complex antecedents which fully determine their application, may not fully 'determine' choice, but may merely 'influence' it. This aside, the distinction Kuhn draws between influencing and determining is important for other reasons.

The first is that for Kuhn different scientists can give the very same values different weightings. And it is the differing subjective, or idiosyncratic, weightings that will also influence choice as well as the value which is weighted. More will be said shortly about Kuhn's model of subjectively weighted shared values. The second reason is that Kuhn declares that his set of values are imprecise, or ambiguous in application. But this is not a strong ground for declaring that values influence rather than determine choices. It is always open to scientists to make their values more precise. Thus Kuhn's value of simplicity is notoriously vague; however we have already noted one way in which it can be made more precise by introducing 'unity' for one cluster of characteristics that Kuhn lists under 'simplicity'. Further precision can be made by qualifying the kind of simplicity such as in simplicity of equations employed (for which there is a sharp criterion), or simplicity in the number of *ad hoc* assumptions introduced, and so on. Despite what Kuhn says, some of his values are themselves quite precise. Thus accuracy can be specified to any required degree; and consistency, either internal or external, can be expressed fairly sharply, as Kuhn himself admits.[24]

A further ground on which Kuhn alleges that values influence rather than determine theory choice is that the values taken together can lead to different choices. Kuhn admitted that this was the case with the corresponding rule versions. Thus the rules 'accept theories with minimum degree of accuracy of predictions 99% (say)' and 'accept theories which have simple equations' can pull us in different directions. But the same inconsistency can also apply to these criteria expressed as values. It is important to note the possibility of inconsistent sets of values which can pull us in different directions in our evaluation of theories. But note that it is a mere possibility; there is no argument that always, or necessarily, our values form an inconsistent set when applied to some theory. On the contrary there is reason to suppose that in a good many cases the values we endorse can all be consistently realised together. Given that sets of values can be consistently realised then we can lessen the need to always resolve the inconsistency by differentially weighting the values; they can all be applied with the same weighting.[25]

Let us now focus on Kuhn's list of values. It omits several important values that have been endorsed by other methodologists. Thus inductivists and Bayesians put store on high degree of support of hypotheses by evidence. Constructive empiricists put high value on theories which are empirically adequate. Realists wish to go further and value not only this but also truth, or increased verisimilitude, about non-observational claims. Yet others value high explanatory power. And so on. Kuhn rejects the idea that our theories do approximate to the truth about what is "really there" (Kuhn (1970), p. 206); so we ought not to view him as endorsing the realists' value of truth. But we can view him as adopting the value of empirical adequacy espoused by constructive empiricists such as van Fraassen (i.e., truth at the observational level, this being an extension of the idea of accurate predictions). Finally some methodologists would downplay some of the values Kuhn endorses, such as external consistency or social utility (a value that Kuhn himself later questioned).

Even though the strong anti-inductivist strain in Kuhn takes him away from notions of high degree of support, he at least requires a notion of degree of support in adopting the values of scope and fruitfulness. For what does the successful fruitfulness of a theory do but increase its degree of probability? There is a simple proof of this using Bayes' Theorem in the form: $p(T, E\&K)$ $= p(E\&K, T) \times p(T, K)/p(E, K)$, where T is some theory, E is new or novel evidence and K is background old evidence. For simplicity also suppose that both old and new evidence are deductive consequences of theory T so that $p(E\&K, T) = 1$. Since E is new or novel evidence with respect to the background evidence K, then $p(E, K)$ will be low. Given that E is unexpected with respect to K, it is easily seen that T is given a confirmatory boost by E with respect to K. Results such as these have led some commentators, such as Earman and Salmon,[26] to argue that all that is valuable in Kuhn's theory of method can be incorporated into a Bayesian framework. Given these considerations, Kuhn cannot easily omit the value of high probability on evidence; he is in fact committed to it given some of the values he does endorse. Further, given his endorsement of high predictive accuracy, Kuhn is also committed to some account of verisimilitude as applied to sequences of predictive consequences (though, as has been seen, Kuhn is not necessarily committed to verisimilitude at the theoretical level).

The position of Kuhn on methodology after the first edition of *Structure* yields the following picture of a model of weighted values. (1) There is a set of values which can vary over time, and can vary from methodologist to methodologist. A sub-set of these values could comprise a cluster of central values which hold for most sciences and throughout much of the history of science. For Kuhn's own model, the values he mentions fall into this sub-set with few, if any, falling outside the sub-set but within the larger set. (Kuhn's liberality with respect to admissible values in his work of the 1970s is not always endorsed in his 1980s work; in fact the values that have been so far mentioned seem to be the only ones he later deems to be central to science.) (2) These values are used to guide and inform theory choice across the sciences and within the history of any one science, including its alleged 'paradigm' changes. That is, the cluster of values transcend all the sciences and any 'paradigms' within each science. (3) The model may be either descriptive or normative. Kuhn does not make it clear whether his model is to be understood as a description of how scientists do in fact make their choices, or whether it is to be understood normatively in that it tells us how we ought to make choices. If the latter then there is a need for a justification of the norms it embodies. (4) Kuhn also says that the values may be imprecise and be applied by different scientists in different ways. But as has been argued, Kuhn underestimates the extent to which either the values are themselves precise or can be made precise. (5) Different scientists do, as a matter of fact, give different weightings to each of the values. Again we have seen that while this is a possibility it is by no means always, or necessarily, the case. All values may be consistently realisable without the need for weighting.

However if we grant Kuhn the points made in (4) and (5), there emerges two aspects in theory choice – an objective aspect in shared values and a subjective aspect in idiosyncratic weightings of values (and interpretation of the value where this might arise). In the light of this, Kuhn claims that there is no general 'algorithm' for theory choice – though there is hardly any methodologist who has required that methodological principles should be algorithmic. This allows that different scientists can reach different conclusions about what theory they should choose. First, they might not share the same values; but where they do share the same values they might interpret them differently or give them different weightings. Shared values (with the same interpretation) and shared weightings of these values will be sufficient for sameness of judgement within a community of scientists. However this might not be necessary; it might be possible for scientists to make the same theory choices yet to have adopted different values and/or have given them different weightings. Thus there is the possibility that consensus might be a serendipitous outcome despite lack of shared values and different weightings. However it is more likely that, where values and weightings are not shared, different theory choices will be made, and there is no consensus.

Whether scientists do or do not make theory choices according to Kuhn's model is a factual question to answer. But what does the model say about what we *ought* to do, and what is its normative/rational basis? In particular why, if T_1 exemplifies some Kuhnian value(s) while T_2 does not, should we adopt T_1 rather than T_2? Kuhn's answer to the last meta-methodological question is often disappointingly social and/or intuitionistic in character. In his 1977 paper Kuhn refers us to his earlier book saying: 'In the absence of criteria able to dictate the choice of each individual, I argued, we do well to trust the collective judgement of scientists trained in this way. "What better criterion could there be", I asked rhetorically, "than the decision of the scientific group?"' (Kuhn (1977), pp. 320-1). As to why we ought to follow the model, Kuhn makes a convenient is-ought leap when he says in reply to a query from Feyerabend: 'scientists behave in the following ways; those modes of behaviour have (here theory enters) the following essential functions; in the absence of an alternative mode *that would serve similar functions*, scientists should behave essentially as they do if their concern is to improve scientific knowledge' (Lakatos and Musgrave (eds.) (1970), p. 237). The argument is not entirely clear, but it appears to be inductive: in the past certain modes of behaviour (e.g., adopting Kuhn's model of theory choice) have improved scientific knowledge; so in the future one ought to adopt the same modes of behaviour if one wants to improve scientific knowledge. What is surprising about this line of argument is its dependence on meta-induction from the past success of the use of Kuhn's model to its future success. However Kuhn later offers other reasons for adopting his model.

1.4.3 Kuhn's Meta-Methodological Justification of his Values

In the last phase of his work on method dating from the 1980s, Kuhn attempted to give a meta-methodological justification for his values and his model. His views arise in a paper entitled 'Rationality and Theory Choice' which was presented as part of a 1983 symposium on 'The Philosophy of Carl G. Hempel' with Salmon and Hempel. In that paper Kuhn addresses a point that Hempel had made in an earlier paper (Hempel (1983)) about Kuhn's position, viz., that Kuhnian values are goals at which science aims, and not means to some goal such as puzzle-solving. Both Kuhn and Hempel take puzzle-solving, accuracy, simplicity, etc, to be values (ends) rather than rules (means) – a position that has been scrupulously adopted in this section. Given that theories are judged by the values they exemplify, Kuhn takes up a further point that Hempel makes, viz., that rationality in science is achieved through adopting those theories that satisfy these values *better*. Hempel thinks that this criterion of rational justification is near-trivial. However Kuhn turns Hempel's near-triviality into a virtue by proposing that the criterion is analytic, thereby adopting as his meta-methodology a theory of analyticity concerning the term 'science'. In developing his views in this paper Kuhn tells us that he is 'venturing into what is for me new territory' (Kuhn (1983), p.

565; or Kuhn (2000), p. 211). So we can take it that the meta-methodological justification developed here is not one that Kuhn had had in mind before.

Kuhn's account of analyticity is based on what he calls 'local holism'. This is the view that the terms of any science cannot be learned singly but must be learned in a cluster; the terms have associated with them generalisations that must be mastered in the learning process, and the cluster of terms form contrasts with one another that can only be grasped as a whole. If 'learning' is understood as 'understanding the meaning', then analyticity becomes an important part of the doctrine of 'local holism' for the central terms of each sufficiently broad scientific theory. In Kuhn's view the doctrine applies not only to specific theories such as Newtonian Mechanics with its terms 'mass' and 'force' which must be learned together holistically. It also applies to quite broad notions signified by the terms 'art', 'medicine', 'law', 'philosophy', 'theology', and so on; the central terms associated with these notions must also be learned holistically. Importantly 'science' is another such broad notion to be learned holistically since 'science' is in part to be understood in contrast to these terms.

Kuhn recognises that not every science we adopt should possess every value since the values are not necessary and sufficient conditions for theory choice; rather they form a cluster associated with the local holism of the term 'science'. But what he does insist upon is that claims such as 'the science X is *less* accurate than the non-science Y' is a violation of local holism in that 'statements of that sort place the person who makes them outside of his or her language community' (Kuhn (1983), p. 569; or Kuhn (2000), p. 214). For Kuhn, Y's being more accurate than X is just one of the things that the local holism of the words 'accuracy' and 'science' makes Y scientific; Y cannot be non-scientific. For Kuhn, Hempel's near-triviality is not breached because a convention has been violated; nor is a tautology negated. Rather 'what is being set aside is the empirically derived taxonomy of disciplines' (*loc. cit.*) that are associated with terms like 'science'. Like many claims based on an appeal to analyticity, meaning or taxonomic principles, one might feel that the later Kuhn has indulged in theft over honest toil. However in linking his model of weighted values to the alleged local holism of the term 'science', Kuhn comes as close as any to adopting the meta-methodological stance that his theory of method has an analytic justification for its rationality. In this respect Kuhn's position has close affinities with that of Strawson ((1952), Chapter 9, part II) who tried to justify the rationality of induction in much the same way.

The position just indicated was one which Kuhn continued to adopt in his quite late papers. Thus in his 'Afterwords' to the collection of papers on his work edited by Horwich he says of values and theory choice:

Accuracy, precision, scope, simplicity, fruitfulness consistency, and so on, simply *are* the criteria which puzzle solvers must weigh in deciding whether or not a given puzzle about the match between phenomena and belief has been solved. Except that they need not all be satisfied at once, they are the 'defining" characteristics of the solved puzzle. ... To select a law or theory which exemplified them less fully than an existing competitor would be self-defeating, and self-defeating action is the surest index of irrationality. Deployed by trained practitioners, these criteria, whose rejection would be irrational, are the basis for the evaluation of work done during periods of lexical stability, and they are basic also to the response mechanisms that, at times of stress, produce speciation and lexical change. As the developmental process continues, the examples from which practitioners learn to recognise accuracy, scope, simplicity, and so on, change both within and between fields. But the criteria that these examples illustrate are themselves necessarily permanent, for abandoning them would be abandoning science together with the knowledge which scientific development brings. (Horwich (1993), p. 338)

This is a striking passage not least for the values it re-endorses, including the value of accuracy which had been endorsed in his 1962 book). It also gives a central role to them in saying what science is and what comprises rationality; and they apply not only within a 'paradigm', now identified with lexical stability, but also across 'paradigms' in which there is lexical change. The position of the later Kuhn in which epistemic values are linked to rationality is a far cry from the earlier Kuhn in which these values were barely discernibly at work in paradigm choice. It might be no exaggeration to say that Kuhn's views of paradigms and theory choice underwent a 'paradigm change' in the course of Kuhn's exploration of these issues.

1.4.4 Kuhn's Rejection of the Sociology of Scientific Knowledge

The striking doctrines of the earlier, but not the later, Kuhn have been taken over by sociologists of scientific knowledge in their co-option of Kuhn as a precursor to their own radical programme for the study of science. As noted in the previous section, Brante, Fuller and Lynch tell us: 'The publication of Thomas Kuhn's *The Structure of Scientific Revolutions* in 1962 pointed the way toward the integrated study of history, philosophy and the sociology of science (including technology) known today as science and technology studies (STS)'. And in turn these studies have shown that 'philosophical words such as *truth, rationality, objectivity*, and even *method* are increasingly placed in scare quotes when referring to science' (Brante *et. al.* (1993), p. ix.). However what is evident is that the sociologists who find Kuhn's 1962 book path-breaking do not follow the later Kuhn through his next two phases, as indicated above. This final sub-section addresses the later Kuhn's attempt to distance himself from the very sociologists who embrace the earlier Kuhn.

During the more than thirty years after he published *Structure* Kuhn worked at modifications of its views, projecting even another book which remained unpublished at the time of his death. The new book, he said, would return to the philosophical problems *Structure* bequeathed to him by the incomplete account he had given of them in his earlier work. The list of philosophical problems include: 'rationality, relativism and, most particularly realism and truth ... [and a notion of] incommensurability ... [which] is far from being the threat to rational evaluation of truth claims that it has frequently seemed' (Kuhn (1991), p. 3; or Kuhn (2000), p. 91).

Surprisingly for a book whose historical examples were largely culled from physics and chemistry, *The Structure of Scientific Revolutions* has been taken up by social scientists and lauded by sociologists of scientific knowledge as one of the works which inaugurated the new sociological approach. What was Kuhn's attitude to the burgeoning literature of sociological studies of science? Though he says that there is a lot to learn from such studies, he expresses his view of its underlying philosophical stance with uncharacteristic harshness. His projected book was to provide an account of incommensurability consistent with the notion of rational evaluation of truth as a corrective to the sociological stance:

> ... incommensurability is far from being the threat to rational evaluation of truth claims that it has frequently seemed. Rather, it's what is needed, within a developmental perspective, to restore some badly needed bite to the whole notion of cognitive evaluation. It is needed, that is, to defend notions like truth and knowledge from, for example, the excesses of post-modernist movements like the strong program' (Kuhn (1991), pp. 3-4; or Kuhn (2000), p. 91).

Even though Kuhn would not have adopted a realist correspondence theory of truth, his unfinished project was going to conserve, and address, some of the traditional concerns of scientific method, the idea of 'a rational evaluation of truth', and provide criteria for theory evaluation with 'bite', all (surprisingly in the light of Kuhn Phase I) alongside a notion of incommensurability.

In a 1991 address in which Kuhn distances himself from 'people who often call themselves Kuhnians' (Kuhn (1992), p. 3; or Kuhn (2000), p. 106), he said: 'I am among those who have found the claims of the strong program absurd: an example of deconstruction gone mad.' (*op. cit.*, p. 9; or p. 110). He begins by agreeing with advocates of the Strong Programme:

> Interest, politics, power and authority undoubtedly do play a significant role in scientific life and its development. But the form taken by studies of 'negotiation' has, as I've indicated, made it hard to see what else may play a role as well. Indeed, the most extreme form of the movement, called by its proponents 'the strong program', has been widely understood as claiming that power and interest are all there are. Nature itself, whatever that may be, has seemed to have no part in the development of beliefs about it. Talk of evidence, of the rationality of claims drawn from it, and of the truth or probability of those claims has been seen as simply the rhetoric behind which the victorious party cloaks its power. What passes for scientific knowledge becomes, then, simply the belief of the winners. (*op. cit.*, pp. 8-9, or p. 110)

With these remarks in mind, David Bloor, in an obituary notice for Kuhn, thought it was still possible to co-opt him to the sociologists' cause. In relation to Kuhn's remarks about the role of nature Bloor says: 'How nature is to be described is not predetermined, or independent of the traditional resources that are bough to bear on it. That is why, despite some uncharacteristically ill-focused remarks of his own to the contrary, Kuhn is properly called a 'social constructivist' (Bloor (1997a), p. 124). Setting aside the obscure matter as to whether Kuhn was a 'social constructivist', his model of consensus through individually weighted shared values makes theory choice at best weakly 'social constructed'. What Kuhn also requires is 'cognitive evaluation with rational bite'.

Kuhn clearly agrees with the sociologists' claim that interests, power and politics do play a significant role in *scientific life* and its development. This is almost a banal truth with which no one would disagree. The extent to which interests of power, money and politics support, say, modern scientifically based agribusinesses, or timber-felling companies, over those who wish to retain elements of biodiversity in the countryside is an empirical matter to decide – and one in which it is very clear who is the winner in a number of cases. But it is a very different matter to allege that power, politics, money, interests and authority influence not only *scientific life* (as Kuhn says) but also the *acceptance* of *scientific hypotheses* or *observations* in scientific investigations. Concerning the adjudication of theories, Kuhn would apply the 'cognitive bite' of his model of weighted values – and this says nothing of power and politics amongst its list of values.

This highlights Kuhn's worry. If the power politics of negotiation is the only, or dominant, factor in determining what we are to accept as scientific facts, or to accept as hypotheses, or accept as what evidence supports what hypothesis, then Kuhn can see no role for nature as that which enters into our very experimental interventions. Nor is there even a partial role for nature as that to which our theories and observations are a response, or that which our theories are about. Nature appears to be irrelevant while society is paramount. For Kuhn the Strong Programme so marginalizes the role of nature in our theory choices, or in our accounts of what science is about, that it seems to drop out of the picture all together. In sum, despite some of the more radical claims of the first edition of *Structure* which subsequently underwent refinement, Kuhn clearly sees himself as a member of the long tradition which still has a use for the notions of truth, knowledge and rationality in theory choice and does not jettison them – despite Kuhn's somewhat erratic earlier encounter with these issues that, on his own admission, might have given a wrong impression to some.[27]

1.5 THE ANARCHIST FEYERABEND AS DETHRONER OF THE CRITICAL TRADITION?

For a person who is famous for alleging that the only universal principle of rationality is 'anything goes', or giving his books titles such as *Against*

Method, or *Farewell to Reason*, it might come as a surprise to some to find that Feyerabend, in his autobiography completed just before his death, makes the following claim on behalf of rationality: 'I never "denigrated reason", whatever that is, only some petrified and tyrannical version of it' (Feyerabend (1995), p. 134). Or, 'science is not "irrational"; every single step can be accounted for (and is now being accounted for by historians ...). These steps, however, taken together, rarely form an overarching pattern that agrees with universal principles, and the cases that do support such principles are no more fundamental than the rest' (*ibid.*, 91). Inspecting his earlier career, one will find that Feyerabend even proposed some principles of method, such as the Principle of Proliferation, the aim of which is 'maximum testability of our knowledge' and its associated rule is '*Invent, and elaborate theories which are inconsistent with the accepted point of view, even if the latter should happen to be highly confirmed and generally accepted*' (Feyerabend (1981), p. 105).[28]

1.5.1 Feyerabend's Defeasible Principles of Method

Feyerabend opposed the following view of scientific method (call it 'Rationalism' with capital 'R') espoused by Popper and Lakatos (in fact his criticism of Rationalist methodology is almost entirely narrowly focused upon the principles proposed by the 'Popperian school' and hardly any others):

(I) There is a universal principle (or small unified set of principles) of scientific method/rationality, R, such that for all moves in the game of science as it has historically been played out in all the sciences, the move is an instance of R, and R rationally justifies the move.

His opposing position can be easily expressed by shifting the order of the quantifiers in (I) from 'there exists – all' to 'all – there exists', and then adding a qualification about the nature of the principles. Call this doctrine 'rationalism' with a little 'r'.

(IIa) For all moves in the game of science as it has been historically played out in all the sciences, there is some principle (or set of principles) of scientific method/rationality, R, such that the move is an instance of R, and R rationally justifies the move.[29]

This leaves open two extreme possibilities: that a different rule is needed for each of the moves, or the remote possibility that there is still one universal rule which covers all the moves. Feyerabend's position is more particularist and is close to the first alternative. There might be rules which cover a few moves, but the rules are so contextually limited that they will not apply to a great number of other moves and other rules will have to be invoked. Feyerabend also has an account of the nature of these rules:

(IIb) Rules of method are highly sensitive to their context of use and outside these contexts have no application; each rule has a quite restricted domain of application and is defeasible, so that other rules (similarly restricted to their own domain and equally defeasible) will have to apply outside that rule's domain. (Feyerabend often refers to these as rules of thumb.)

Sometimes Feyerabend adopts quite traditional cognitive values for science, such as the rather Popperian Proliferation Principle 'maximum testability of knowledge'. But Feyerabend is also a social critic of science who asks 'what is so great about science with such aims?', and then argues that for some of the moves in the game of science we would have been better off if they had never occurred. For Feyerabend the whole game of science may not be worth playing because science might make a monster of us (Feyerabend (1975), pp. 174-5). There are better things in life than science, such as acting in plays, singing opera or being a Dadaist, and such choices are highly contextual.

For Feyerabend there are a number of broad aspects of science to consider, each of which has its respective values, and these in turn are to be associated with contextual and defeasible rules. By shifting focus from one aspect of science to another, Feyerabend is able to abandon contextual and defeasible methodological rules which are allegedly designed to promote epistemic progress in science, in favour of other goals, for example, dialectical, humanitarian, aesthetic and moral goals, which have little to do with scientific method as standardly understood. In respect of these other goals Feyerabend asks 'What is so great about science?' (Feyerabend (1978), p. 73), and then he answers by saying 'science is one ideology among many' and can be a threat to the democratic life (*ibid.*, p. 106). In what follows let us focus on the methodological principles that are alleged to pertain to science.

In taking the position he does Feyerabend is not beyond the pale of rationality; but he is beyond the pale of Rationality. There is much textual support, only a little of which can be cited here, for the view that Feyerabend is not a *Rationalist* but a *rationalist*. His opponents are Rationalists who advocate a small unique set of universal rules to be applied to all sciences at all times. On occasions he refers to such Rationalising methodologists as 'idealists', or as 'rationalists' (Feyerabend's little 'r' must be read with big 'R' connotations), thereby creating the impression that he must be an irrationalist. But such talk masks Feyerabend's real position.

Direct support for claims (IIa) and (IIb) comes from Feyerabend's response to critics of the 1975 *Against Method,* and in his restatement of his position in Part I of his 1978 *Science in a Free Society.* Much of Part I of this book has been incorporated into the third 1993 edition of *Against Method,* thereby supporting the view that Feyerabend's 1978 position was one he could still endorse in 1993.[30] Feyerabend contrasts two methodological positions: *naive anarchism*, with which his own position should not be confused (presumably Feyerabend is a sophisticated anarchist); and *idealism* in which there are absolute rules – but they are conditional with complex antecedents which spell out the conditions of their application within the context of some universal Rationalist methodology. Of naive anarchism he says:

> A naive anarchist says (a) that both absolute rules and context dependent rules have their limits and infers (b) that all rules and standards are worthless and should be given up. Most reviewers regarded me as a naive anarchist in this sense overlooking the many passages where I show how certain procedures *aided* scientists in their research. For in my studies of Galileo, of Brownian motion, of the Presocratics I not only try to show the *failures* of familiar standards, I also try to show what not so familiar procedures did actually *succeed*. I agree with (a) but I do not agree with (b). I argue that all rules have their limits and that there is no comprehensive 'rationality', I do not argue that we should proceed without rules and standards. I also argue for a contextual account but again the contextual rules are not to *replace* the absolute rules, they are to *supplement* them. (Feyerabend (1978), p. 32; or Feyerabend (1993), p. 231)

Thesis (b) marks the crucial difference between naive anarchism and Feyerabend's own position; moreover the denial of (b) shows that Feyerabend cannot be an arch irrationalist. There *are* rules worth adopting, but not those commonly advocated. Oddly enough, universal rules are not to be replaced but are to be supplemented. What this means is unclear; but it might be understood in the following way. Consider as an example of a universal rule Popper's 'do not adopt *ad hoc* hypotheses (if you wish to advance knowledge)'. For Feyerabend this is not to be understood as a universal ban which applies to all sciences and in all circumstances come what may. Sometimes adopting *ad hoc* hypotheses will realise our aim (advancing knowledge) better than not adopting them. Feyerabend seems to suggest that alongside this universal rule are other rules (equally open to supplementation one supposes) about its application. The task of these other rules will be to tell us about the occasions when we should, or should not, adopt this Popperian rule. What Feyerabend needs is the notion of a defeasible rule, but defeasibility is not something he ever discusses. If rules are defeasible, then universalising 'idealists' and Rationalists will not be able to apply rules regardless of their situation. Viewed in this light the passage cite above supports the position outlined in (IIa) and (IIb), as do other similar passages in his writings.[31] The passage does not support the view that Feyerabend rejects our critical tradition; rather he proposes a revised view of it in which methodological principles are not universal but defeasible.

In the passages surrounding the last quotation Feyerabend attempts to distinguish between a modified idealism, in which universal rules are said to be conditional and have antecedents which specify the conditions of their application, and his own view in which rules are contextual and defeasible. Presumably the difference is that for Rationalist 'idealists' the conditions of application of the rules are spelled out in the fully specific antecedents of conditional rules. But in Feyerabend's view such conditions cannot be fully set out in some antecedent in advance of all possible applications of the hypothetical rule; at best such conditions are open-ended and never fully specifiable. So Feyerabend opts for a notion of a rule which is not conditional in form but categorical, and is best understood as contextual and defeasible. So even if rules appear to be universal, as in 'do not adopt *ad hoc* hypotheses', there will always be vagueness and imprecision concerning their application. There is also no mention of the conditions under which they can be employed or a time limit imposed on their application; presumably this task is to be left to supplementary rules.

Does Feyerabend adopt as his one and only methodological principle *'Anything goes'* (Feyerabend (1975), p. 28)? No:

> As for the slogan 'anything goes', which certain critics have attributed to me and then attacked: the slogan is not mine and it was not meant to summarise the case studies of *Against Method* ... (Feyerabend (1987), p. 283)
> 'anything goes' is not a 'principle' I defend, it is a 'principle' forced upon a rationalist who loves principles but who also takes science seriously'. (*ibid.*, p. 284)

Once again Feyerabend uses the term 'rationalist' to name his opponents; but from this it does not follow that his own position is 'irrationalist'. Instead his view is that one cannot have both the complexities of the actual history of science and a universal methodology of the sort loved by those he variously dubs as 'Rationalists' or 'Idealists'.

As he explains in several passages of *Science in a Free Society* (Feyerabend (1978) pp. 32, pp. 39-40 and p. 188): *'But anything goes' does not express any conviction of mine, it is a jocular summary of the predicament of the rationalist'* (*ibid.*, p. 188).[32] The joke has backfired, and has been costly in misleading many about Feyerabend's real position. If the Rationalists within the critical tradition want universal rules of method then given that, according to Feyerabend, all such rules have counterexamples outside their context of application, the only universal rule left "will be empty, useless and pretty ridiculous – but it will be a 'principle'. It will be the 'principle' 'anything goes'". (*loc. cit.*) But this is hardly convincing since the Rationalist need not take up Feyerabend's invitation to adopt 'anything goes'. Any Rationalist will see that from 'anything goes' it follows (by instantiation) that every universal rule of method also 'goes'; but if the Rationalist also accepts Feyerabend's claim that these have counterexamples and are to be rejected as universally applicable then, by *Modus Tollens*, the Rationalists can infer that 'anything goes' cannot be an acceptable rule of Rationalist method – even if jocularly imposed by Feyerabend on serious universalising

Rationalists. 'Anything goes' does not mean what it appears to say; it is not even a principle of method that Feyerabend endorses.

1.5.2 Feyerabend's Attempted Justification of Defeasible Rules of Method

Given that one can adopt defeasible principles and remain a rationalist about method (but not a Rationalist), Feyerabend does not appear to be the opponent of theories of scientific method that he is often made out to be, or says that he is. Granted such Feyerabendian principles, what does he say about their justification? There are two approaches. The first would be to take some principle that Feyerabend advocates (such as the Principle of Proliferation or rules of counter-induction) or some principle he criticises (such as the alleged principle of consistency or Popper's anti-*ad hoc* rule) and attempt to evaluate them either on logico-epistemological grounds or on the historical record of the decision context of various scientists. But the latter would take us into a long excursion through episodes in the history of science, and the former has to some extent been carried out.[33] Instead we will look at Feyerabend's meta-methodological considerations in justification for his views on scientific methodologies.

If we are to attribute to Feyerabend a meta-methodology concerning his defeasible rules, then it veers between that of a Protagorean relativist and that of a dialectical interactionist in which principles and practice inform one another. In setting out his position he adopts the same Popperian notion of a *tradition* as discussed in section 1.1; Feyerabend uses this to apply not only to mythical and scientific systems of belief, but also to traditions including those of religion, the theatre, music, poetry, and so on. He also speaks of the rational tradition; but unlike Popper he does not privilege it by claiming some special 'second-order' status for it. For Feyerabend all traditions, including the critical or rational tradition, are on an equal par. In resisting the idea that there is a special status to be conferred upon the rules that comprise the tradition, Feyerabend adopts a Protagorean relativism about traditions. About traditions he makes three claims:

> (i)*Traditions are neither good nor bad, they simply are.* ... rationality is not an arbiter of traditions, it is itself a tradition or an aspect of a tradition. ...
> (ii) *A tradition assumes desirable or undesirable properties only when compared with some tradition.* ...
> (iii) *(i) and (ii) imply a relativism of precisely the kind that seems to have been defended by Protagoras.* (Feyerabend (1978), pp. 27-8; or Feyerabend (1993), pp. 225-6).

For Feyerabend there still remains a rational tradition; and it has contextual defeasible rules which can be used to evaluate claims in other traditions (which in this context we can take to include not only the principles of any scientific methodology but also any meta-methodology which attempts to justify any such principles). But given his Protagorean relativism about traditions, all such evaluations are from *within* some tradition and are directed at some other tradition. There is no absolute tradition, encapsulated in some meta-methodology, which stands *outside* all other traditions and from which we can evaluate them. In this sense no tradition is an absolute arbiter of any other. Putting it another way, there is no privileged or *a priori* story to be told about the justification of our principles of method. In this respect Feyerabend is close to the postmodernism of Lyotard (mentioned in section 1.3.4) with its 'incredulity toward metanarratives'. What does this mean for the contextual defeasible rules of method that Feyerabend endorses? Perhaps it means that such rules of method have no further justification other than that they are what we have simply adopted as part of *our* critical tradition. Their truth, validity or correctness is at best relative to our tradition; there is no further meta-methodological account of their status to be given by appealing to some absolute or privileged tradition of Rationality. It is this relativism that has led some to claim that Feyerabend, even if he admits there are defeasible rules of method, is at heart an irrationalist. But this too strong a claim. Rather Feyerabend is akin to Wittgenstein in this respect in claiming that these are our practices, and there is nothing behind them that could act as a justification of them. This Wittgensteinian interpretation is discussed further in section 7.4 on Wittgenstein's claims about a lack of further justification for our practices.

If Feyerabend really adopts a Protagorean relativism about rules of the sort Plato describes in the *Theaetetus*, then at best we can say that there are rules R-relative-to-tradition-T, and rules R*-relative-to-tradition-T*, and not merely rules R and R*. But the former pairs do not, while the latter pairs might, come into some logical relation with one another. Such a version of relativism undercuts the very possibility of rules ever being assessed with respect to one another. But this is often something Feyerabend requires we do. This suggests that Feyerabend is not really a relativist but a pluralist about rules and the traditions they embody (and pluralism need not entail any relativism). The running together of these two notions is evident in the following passage: 'Protagorean relativism is *reasonable* because it pays attention to the pluralism of traditions and values' (Feyerabend (1978), p. 28). What is still excluded by this stance is any attempt to invoke meta-methodology to give an *a priori,* or even an empirical, justification of his defeasible rules of method. But pluralism does make possible the critical 'rubbing together' of different traditions, something that Feyerabend would endorse given his principle of proliferation (as applied to different methodologies). And it does make possible the following more dialectical view of the interaction between traditions, rules and practices. It is this critical 'rubbing together' of different traditions through their 'dialectical interaction' that allows us to say that there remains a whiff of meta-methodological evaluation for Feyerabend.

How does Feyerabend's dialectical interactionism differ from the above kind of relativism commonly attributed to him? There are remnants of the positions of the later Popper and Lakatos, with their appeal to the intuitions of a scientific élite, in Feyerabend's talk of the interaction between reason and practice, of which he distinguishes three aspects. First, he acknowledges that reason can be an independent authority which guides our practices in science – a position he dubs 'idealistic'. But also 'reason receives both its content and its authority from practice' (*ibid.*, p 24) – a position he dubs 'naturalism'. Though he does not say how reason gets its authority, his position is one in which a strong role is given to intuitions about good practice. This is reminiscent of the later Popper, Lakatos, and even the meta-method of reflective equilibrium advocated by Nelson Goodman. (In the meta-method of reflective equilibrium, particular cases and general rules are bought into relation with one another, in a process which may result in some particular cases being dropped, or in some rules being modified or dropped). But both naturalism and idealism have their difficulties, says Feyerabend. Idealists have a problem in that too ideal a view of rationality might cease to have any application in our world. And naturalists have a problem in that their practices can decline because they fail to be responsive to new situations and need to critically re-evaluate their practice. He then canvases the suggestion 'that reason and practice are not two different kinds of entity but *parts of a single dialectical process*' (*ibid.*, p. 25).

But Feyerabend finds that the talk of reason and practice being separate 'parts' of a single process draws a misleading distinction; so he concludes: '*What is called 'reason' and 'practice' are therefore two different types of practice*' (*ibid.*, p. 26). The difference between the two *types* of 'practice' is that one is formal, abstract and simple, while the other is non-formal, particular and submerged in complexity. Feyerabend recognises that the conflict between these two types of practices (or 'agencies' as he goes on to call them) recapitulates 'all the "problems of rationality" that have provided philosophers with intellectual ... nourishment ever since the "Rise of Rationalism in the West"' (*ibid.*, pp. 26-7). As true as this may be at some level of abstraction, Feyerabend's shift to a position of dialectical interactionism with its additional plea for a Principle of Proliferation with respect to interacting traditions (including traditions about theories of method), does have the characteristics of an appeal to some meta-methodological theory; so this might take Feyerabend out of the postmodernist camp.[34]

For Feyerabend there is always the dialectical task of bringing our practices (in science and elsewhere) into line with our theories of those practices (i.e., our theories of method); and conversely, of bring our theories into line with our practices. In this respect Feyerabend can be viewed as an advocate of the methodology of reflective equilibrium advocated by Goodman and others in which particular practices and general principles are bought into accord with one another. Looked at this way, we can resist the temptation to go relativist by viewing the activity of bringing the rules of 'reason' and the particularity of 'practice' into accord with one another as yet just one more activity on a par with any other activity. Rather this activity is a substantive one that is part of assessing the principles of our second-order critical tradition. In this respect Feyerabend adopts a somewhat different position to that of Wittgenstein (as suggested above) in that our practices do admit of criticism through "dialectical interactionism" and are not beyond criticism (though it is clear that even this is just one more practice, though a meta-practice). In adopting this approach one needs to be both a methodologist and an historian of science in bringing together, in "dialectical interaction" or "reflective equilibrium", both methodology and practice.

In conclusion, can Kuhn and Feyerabend be viewed as philosophers of science who stand outside the critical tradition and attempt to overturn it? Not so. Both have recast in some ways our conception of what that tradition ought to be like by correcting over-simple accounts of the role of rationality in the growth of scientific knowledge. As different as their views might be from others within the tradition, there is no strong case for claiming that they have abandoned rationality altogether. In the case of Kuhn this is quite clear in his later writings. The case of Feyerabend is harder to determine. But on one interpretation he does say that there are defeasible principles of method, and there is always the task of bring practice and principles together to prevent both the decline of practice through lack of criticism, and our principles becoming too abstract and irrelevant. And these two aspects of his work, despite its other maverick aspects, are common to those who do view themselves as members of the critical tradition.

NOTES

[1] See for example Hobson (1988), especially the review of a number of theories about the function of REM sleep and dreaming in chapter 15. In the same chapter there is a discussion of a 1983 paper by Crick and Mitchison in which dreaming is an attempt to eliminate unwanted information.

[2] Many of Popper's rules for the "game of science" can be found in Popper (1959) Chapters II and IV.

[3] In Laudan et. al. (1986) there is an extensive list of over two hundred claims about the methods governing scientific change culled from the writings of five recent methodologists, Popper, Kuhn, Feyerabend, Lakatos and Laudan. Though not always set out as principles with

an explicit rule and value, with further research this deficiency could be remedied. Given the plethora of principles, it is part of Laudan's programme of *normative naturalism* to provide means for testing the principles; see Laudan (1996), Part Four. The account of methodological principles given in the text is consistent with that presented in Laudan (1996) chapter 7.

[4]   For recent defences of IBE against the criticisms of writers such as Fine, Laudan and van Fraassen (and also for reference to their criticisms), see Psillos (1999) chapter 4 and 5 and Niiniluoto (1999), chapter 6.4.

[5]   See Latour and Woolgar (1986) who tell us why, in the first edition title of their book *Laboratory Life: The Social Construction of Scientific Facts*, they feel it now safe to drop the word 'social' from the sub-title. See p. 7 and the part of the 'Postscript to the Second Edition' ironically entitled 'The Demise of the "Social"', p. 281.

[6]   An exception is Bloor (1997), pp. 108-10, who does recognise that Popper proposed a social, and not an individualist, account of scientific method, even though the principles of his method appear as if they are binding only on individuals who act in a solitary fashion. Popper maintains that criticism is the heart of his methodology and that there are principles governing the manner in which the criticism is to be carried out. Though the principles do not wear their social character on their face, criticism could not take place unless the many participate in it. For a quite different account of argumentation within a social context see Goldman (1999), chapter 5 'Argumentation'.

[7]   There is an illuminating paper, Tymoczko (1979), on the use of computers to prove the Four Colour Theorem. Proofs, says Tymoczko, have to be convincing, surveyable and formalisable; but the proof of the theorem is certainly not surveyable by any single mathematician. Moreover the use of computers introduces an empirical element into the proof since mathematicians do rely on the correct working of computers in their acceptance of the proof. Would it be too fanciful to imagine the work of computers being replaced by legions of mathematicians producing the required computations, and then their results being checked for their reliability by other legions of mathematicians? Perhaps the power of such computers outstrips the ability of all of humanity to be involved in the computation. In addition, we depend on the empirical reliability of the working of computers when we come to believe, and even know, their outputs. Turning to scientific method, there are also procedures which have to be surveyable and checkable (in ways that may or may not involve computers) and this may often require the co-operation of the many in order that the scientific result become an acceptable piece of scientific knowledge.

[8]   The parallel with Plato's argument in the *Euthyphro* is deliberate. The definition of 'good' in terms of 'what God commands' invites similar questions raised by Socrates. Does God command x because x is good? Or is x good because God commands x?

[9]   The position of Popper outlined above is also discussed in other works, such as the theses, especially numbers twelve and thirteen, of his 'The Logic of the Social Sciences' in Adorno *et. al.* (1976). Popper's view gets independent support from other treatments of the social aspect of knowledge. Thus Goldman (1999) has several relevant chapters, particularly chapter 5 'Argumentation' which deals with arguments in a social context. And in Schmitt (ed.) (1994) there are a number of papers which, while not adopting radical accounts of the involvement of the social in epistemology, do offer important considerations about how the involvement of the social can yield, in the context of a reliabilist theory of knowledge, belief-forming processes which are more reliable; see the papers in Schmitt (ed.) by Alston (chapter 2), Kornblith (chapter 5) and Kitcher (chapter 6).

[10]   This section draws on material in Nola and Sankey 'A Selective Survey of Theories of Scientific Method' (Nola and Sankey (eds.) (2000), pp 1-65), which discusses not only the second order theories of method and their meta-methodological presuppositions mentioned below, but also a number of others.

[11] For an excellent overview of attempts to solve Hume's problem about induction, including *a priori* and other approaches, see Salmon (1967) chapter II.

[12] In Lakatos (1978) there are reprinted two papers which set out most clearly Lakatos' formulation of the legitimation problem for theories of method and ways of legitimating them. The papers are: chapter 2 'History of Science and its Rational Reconstructions' which shows us how to compare methodologies of inductivism, conventionalism, falsificationism and Lakatos' own methodology of scientific research programmes; and chapter 3 'Popper on Demarcation and Falsification'.

[13] Lakatos' project of evaluating a number of theories of method, including his own, in the light of various episodes in the history of the sciences is best illustrated in the collection of Howson (ed.) (1976).

[14] For more on Laudan's normative naturalism see the collection Laudan (1996) Part Four. For a discussion of some of the difficulties of the meta-methodological test procedures of normative naturalism see section 11 of Nola and Sankey 'A Selective Survey of Theories of Scientific Method' in Nola and Sankey (eds.) (2000); see also Nola (1999).

[15] See Earman (1992) or Howson and Urbach (1993) for more on subjectivist accounts of probability, the probability calculus and its theorems, along with the many forms of Bayes' Theorem and the range of 'Dutch Book' arguments which underpin the rationality of adopting subjectivist Bayesianism.

[16] For an excellent brief account of Wittgenstein's connected views about the decline of both philosophy and of our times see Von Wright's essay 'Wittgenstein in Relation to his Times' in von Wright (1982), pp. 203-16.

[17] For a criticism of the use of underdetermination by advocates of SSK to bolster their relativism, see Laudan (1996), Part I chapters 2 and 3, and Brown (1989), pp. 50-56. On the so-called 'experimenter's regress' see Alan Franklin's criticisms in 'Avoiding the Experimenter's Regress' in Koertge (ed.) (1998), pp. 151-65. Franklin's researches into modern physics, and his re-researching of the same material presented by advocates of SSK, are a useful corrective to the excesses of scepticism, relativism, constructivism and irrationality found in their empirical case studies which purport to bolster their philosophical doctrines, in particular the idea that all theory acceptance is based in negotiation.

[18] Goldman (1999), chapter 1.2, is entitled 'Veriphobia', from which I acknowledge the origin of my term 'truth-phobia'; and it makes related points against Shapin. Goldman also has an excellent argument as to why in logic we cannot replace truth by accepted belief, that is, always replace the restrictive notion by the liberal notion. We would destroy the very notions of validity and invalidity as standardly employed in reasoning. But perhaps sociologists of science would not be moved by this disaster.

[19] See Bloor's (1999) 'Anti-Latour'. This is followed by a reply by Latour (pp. 113-129), and a response by Bloor (pp. 131-6) in which he says that Latour has conceded much to him. Part of their falling out has to do with in-house issues concerning underdetermination, the tenets of the Strong Programme (in particular the Symmetry Tenet) and whether 'progress', whatever that might be for sociologists, has been made in science studies. Of course that science itself might be said to have 'progressed' is a modernist myth not to be countenanced; meanwhile progressive leaps and bounds can be made in science studies itself!

[20] This rule is also criticised on pp.13-14 in Sokal's 'What the Social Text Affair Does and Does not Prove' reprinted in Koertge (ed.) (1998). The remarks in the text reiterate the excellent points that Sokal makes.

[21] For a fuller evaluation of Lyotard's postmodernism see Nola and Irzik (2002).

[22] I overlooked it until David Stump pointed it out to me; I thank David for the reference.

[23] My view is that it is not successful for reasons set out in Musgrave (1980); contrary to Kuhn's project, the notion of a paradigm is needed to individuate scientific communities.

24   Kuhn himself admits that the value of consistency can be unequivocal when he says: 'The consistency criterion, by itself, … spoke unequivocally for the geocentric tradition' (Kuhn (1977), p. 323). For more on the way in which Kuhn underestimates the precision of his values see Laudan (1984), pp. 88-92.

25   The above point that value sets need not always be inconsistent is forcibly made in Laudan (1984), pp. 92-4.

26   See Earman (1992), chapter 8 and Salmon (1990a).

27   While the later Kuhn returned to the fold of the critical tradition (telling us that he was never really out of it), he persisted in being an anti-realist. In fact his latter position became even more extremely antirealist in which not only were there no natural kinds, but there were no objects existing independently of us. The lexicon we adopt makes not only kinds but also objects. This is well set out in Ghins (2003, forthcoming). However some recent work on Kuhn does show that he is not the irrationalist he has commonly been taken to be, and that despite some disclaimers, he had a great deal in common with so-called "positivists" such as Carnap. This is effectively argued in Irzik and Grünberg (1995). Irzik and Grünberg (1998) also sets out a neo-Kantian background to the later Kuhn.

28   A list of methodological principles that Feyerabend at one time endorsed is given in Preston (1997) section 7.5. The Principle of Proliferation is criticised in Laudan (1996), pp. 105-10, and in Achinstein's 'Proliferation: Is it a Good Thing?' in Preston et. al. (eds.) (2000), pp. 37-46.

29   Feyerabend often talks of 'rules' or 'standards' rather than 'principles', the terminology adopted in this book. The difference in terminology between that adopted here and by other writers should cause no problems.

30   Of the 1978 Science in a Free Society pp. 16-31, pp. 31-7 and pp. 252-67 become, respectively, sections 17, 18 and 20 of the 1993 third edition of Against Method (though there are small changes here and there). Feyerabend's note to the third edition on p. xiv about the changes hardly indicates their extent and their source. All citations in the text from the 1978 book can readily be relocated in the 1993 third edition.

31   See, for example, Feyerabend (1975) p. 32 and chapter 15; Feyerabend (1978) pp. 98-9 and pp. 163-4.

32   In the 'Preface' to the revised 1988 edition of Against Method Feyerabend again points out: "'anything goes' is not a 'principle' I hold" (Feyerabend (1993), p. vii). Rather he thinks that Rationalists must hold this principle. He explains that he cannot endorse this claim because: "I do not think that 'principles' can be used and fruitfully discussed outside the concrete research situation they are supposed to affect". This is hardly clear; however it might be taken to be a reference to the contextual character of principles of method.

33   Earlier criticisms of Feyerabend's position appeared in reviews of Against Method to which Feyerabend replied; many of the replies are collected in Feyerabend (1978) Part Three. For two recent evaluations of Feyerabend's views on particular principles see Laudan (1996) chapter 5, and Preston (1997), chapter 7.

34   It is possible to make a case for the late Feyerabend falling more within the postmodernist camp than indicated here. See Preston (2000) entitled 'Science as Supermarket: 'Postmodern' Themes in Paul Feyerabend's Later Philosophy of Science'. But it would be wrong to say that Feyerabend's position was always within the ambit of postmodernism, not least because of his claim that there are (defeasible) principles of method, and because they can be critically evaluated in accordance with his idea of dialectical interactionism (which is hardly akin to Lyotard's 'incredulity towards metanarratives').

CHAPTER 2

THE PROBLEM OF KNOWLEDGE

As has been seen in the previous chapter (section 1.3), many of those who attempt to dethrone the critical tradition do so by rejecting the very idea of knowledge and of scientific knowledge based in method. Further attempts by sociologists of scientific knowledge, Foucault and Nietzsche along these lines are critically reviewed in Parts II to IV. In contrast, in this chapter and the next the task is more positive. They sketch some of the expectations we have concerning theories of knowledge and method, and how we might defend certain conceptions of these against attacks mounted upon them. This chapter outlines some of the issues that arise in epistemology in an attempt to spell out a satisfactory conception of knowledge, and its impact on various conceptions of scientific knowledge. As will be seen subsequently, some of the dethroners of our critical tradition have given up on the goal of knowledge and have either embraced scepticism about our ever knowing anything, or have fallen into an easy relativism.

The first section asks: 'Why bother with knowledge? Why don't we just stay with belief?' That something needs to be added to mere belief to obtain knowledge is a core doctrine within philosophy. It is even part of the core view of subjective Bayesianism, currently the dominant position in methodology. Bayesianism downplays the notion of knowledge and makes the notion of belief do much of the work of that older, grander notion. But in eschewing knowledge Bayesians do not simply adopt the notion of mere belief. Rather their interest is focused on the idea of *rational* degree of belief (i.e., one's degrees of belief must be in accord with the probability calculus). The dethroners of our critical tradition have as much difficulty with the notion of Bayesian rationality as they do with the idea of knowledge or principles of method. Since their doctrines focus on mere belief, they shun not only traditional conceptions of knowledge but also Bayesianism with its central notion of *rational* belief. Here our focus will be on knowledge, and on why it was originally suggested by Plato, and others, that there has to be more to knowledge than mere belief (or more correctly true belief). What this extra is will be called here 'the problem of Plato's tether' in honour of Plato who insisted, against the sophists and other 'postmodernist' critics of his day, that there was something else to be considered.

Often the solution to "Plato's tethering problem" has been thought to lie in some theory of justification of our beliefs, thus leading to a classical foundationalist account of knowledge as justified true belief. One of the results that can be truly attributed to twentieth century 'analytic' philosophy is that such a theory is a non-starter.[1] However many traditional accounts of knowledge, particularly scientific knowledge, have commonly been presented within the framework of such an epistemological theory. Since sociologists of science and many postmodernists reject this theory of knowledge (largely because it is foundational), then they also reject any account of the sciences that is cast within this theory of knowledge. Most philosophers would agree with the position adopted by these critics of this classical theory, but not for their reasons. In section 2.2 this theory is outlined and criticised along the lines of the Hellenistic Philosopher Agrippa who produced some of the earliest objections to it.

In rejecting the classical justificationist theory some dethroners of knowledge and method all too readily assume that *all* other theories of knowledge are to be rejected. But this does not follow. In section 2.3 the reliabilist theory of knowledge is outlined, its solution to Plato's tether problem being that for knowledge our beliefs must arise from belief-forming processes which are reliable for their truth. This is not a foundationalist theory of the sort just outlined with a commitment to certainties about immediate experience; but it does have different foundational aspects. And one of its virtues is that it can be made to fit within a naturalist framework. As will be seen, sociologists of knowledge, Foucault and Nietzsche also wish to view the formation of our beliefs within a naturalist framework. But from this they infer that there can be no role for the normative aspects of knowledge emphasised by the classical justificationist theory. It will be argued that this inference, important for their anti-rationalist stance, is faulty; and the reliabilist theory sketched will enable us to see why. In section 2.4 some aspects of a social epistemology are discussed which, in part, build on reliabilism. It is important to distinguish viable aspects of social epistemology from both empirical sociology of science and the non-viable sociology of scientific knowledge (see Part II) that attempts to show that there are causal links between socio-political circumstance and scientific belief (not to be confused with knowledge). Knowledge is a normative and critical notion; the normative character of knowledge and principles of method is explored in chapter 3.

2.1   KNOWLEDGE – WHY BOTHER? THE PROBLEM OF PLATO'S TETHER.

If there is one thing that distinguishes us humans from the animals it is our ability to have, and express, a potentially infinite number of beliefs. We have so evolved as animals that our brains enable us to represent, more or less accurately, features of the world in which we must survive. But our brains are also quite fertile and can produce beliefs that do not represent reality at all. And they produce beliefs that go well beyond those bits of reality we can check by simple observation thereby leaving us in a quandary as to whether they are true or not. Such is the case with our most primitive beliefs about spirits, gods, souls, witches, and the like, that populate the myths and stories of religious world-views, both early and recent. Our most self-conscious enterprise for getting belief, namely science, also makes claims about what we cannot observe, from sub-atomic particles to tectonic plates and neutron stars. But even here we have made erroneous postulations about unobservables, such as the alleged existence of celestial spheres, epicycles and deferents postulated in pre-Copernican astronomy and even the astronomy of Copernicus himself, the electromagnetic ether of nineteenth century theories of light, and so on.

Given our proneness to belief, and even our all to easy disposition to credulousness, there ought to be a subject that studies the epidemiology of belief in a population of believers spread over a historic-geographic region. We find such a study embryonically in history from Herodotus onwards, and in our current sciences of social psychology, sociology and anthropology. Daniel Sperber's book *Explaining Culture: a Naturalist Approach* outlines such a study under the description 'the epidemiology of representations' (Sperber (1996), chapters 3 and 4). Its focus is not so much on the truth value of the contained belief but on, first, describing the epidemiology of belief, and secondly, and even more importantly, providing an explanation, within a naturalist (or scientific) framework, of that epidemiology. As will be seen, the theorists discussed in this book, sociologists of science, Foucault and Nietzsche, are epidemiologists of certain kinds of belief; and they do attempt to find naturalistic explanations of belief. The serious objections raised here against their projects have to do much more with the explanations they offer of the epidemiology they detect than the epidemiology itself.

2.1.1 Why Bother With a Distinction Between Knowledge and Belief?

So far, there has been no talk of knowledge, only belief. Given our credulousness some more discerning believers wish to impose a 'filter' on our beliefs. We might use a filter to ensure that we take on board only those beliefs that make us feel comfortable, secure, happy, or whatever. Or beliefs might be selected for the functional role they play. Thus Hume marvelled at the fact of the 'easiness with which the many are governed by the few' and offered the following early explanation in terms of the ideological function of belief:

> When we enquire by what means this wonder is affected, we shall find, that, as FORCE is always on the side of the governed, the governors have nothing to support them but opinion. It is therefore, on opinion only that government is founded; and

this maxim extends to the most despotic and most military governments, as well as to the most free and most popular. (Hume (1985), p. 32)

Here Hume points to the functional and ideological role that belief can play in maintaining those who govern in power. This in turn requires some selection process whereby beliefs that play such a role come to the fore while those that do not are eliminated.

Other cognitive filters can be deliberately applied by believers to themselves as when they attempt to take on board only those beliefs that can be shown to be true, reject those shown to be false, and to suspend belief in the remainder. Such a filter would be that of a Cartesian pure inquirer. The Cartesian filter does not produce beliefs that play some functional role; nor is its aim non-cognitive goals such as producing certain pleasing states of mind, or satisfying political interests. Such a Cartesian filter also promotes scepticism, in that it invites suspension of belief where it can establish neither truth nor falsity. Scepticism has played a positive role in promoting mental hygiene by submitting our beliefs to critical examination. The origin of the word 'scepticism' is instructive here. It comes from the Ancient Greek 'to look about', but later came to mean 'considered examination' or 'inquiry'. Given the high standards enquiry is expected to meet, many beliefs fail to past muster. Hence the modern meaning of scepticism in which very few, sometimes no, beliefs survive the filter of critical inquiry – those that do being given the honorific title 'knowledge'.

What are beliefs? For naturalists they are at least mental states that are capable of representing the world. In Frank Ramsey's apt phrase, beliefs are also maps we steer by.[2] We would not knowingly steer our lives by false beliefs, this not being conducive (on the whole[3]) to our general well being, or even survival. That is, there is adaptive benefit in having a cognitive apparatus that yields true beliefs in most circumstances of perception and thought, and a serious maladaptive disadvantage if our cognitive apparatus did not. We can also say that beliefs are what we take to be true. As G. E. Moore pointed out, there is some deep kind of contradiction in the first person claim involving saying while disbelieving: 'I believe that p, but p is false'.[4]

Why would we want more than beliefs that are true? The first philosopher to make much of the distinction between belief or opinion (*doxa*) and true belief on the one hand, and knowledge (*episteme*) on the other, was Plato. In his very first discussion of the distinction in the *Meno*, Plato shows that he is acutely aware of the 'why bother with knowledge?' question. He produces an argument, which shows that for all practical purposes there is no difference.

Socrates: If someone knows the way to Larissa, or anywhere else you like, then when he goes there and takes others with him he will be a good and capable guide, you would agree?
Meno: Of course.

Socrates: But if a man judges correctly which is the road, though he has never been there and doesn't know it, will he not also guide others aright?
Meno: Yes, he will.
Socrates: And as long as he has a correct opinion on the points about which others have knowledge, he will be just as good a guide, believing the truth but not knowing it.
Meno: Just as good.
Socrates: Therefore true opinion is as good a guide as knowledge for the purpose of acting rightly. That is what we left out just now in our discussion of the nature of virtue, when we said that knowledge is the only guide to right action. There was also, it seems, true opinion.
Meno: It seems so.
Socrates: So right opinion is something no less useful than knowledge.
Meno: Except that the man with knowledge will always be successful, and the man with right opinion only sometimes.
Socrates: What? Will he not always be successful so long as he has the right opinion?
Meno: That must be so, I suppose. (Plato, Meno 97A-98A; Guthrie (1956), p. 153)

The upshot is that a map of true beliefs is a much better map to steer by than a map of false beliefs; the treacherous reefs of false belief, or no belief, are either mismarked on the map, or simply unmarked. But a map of true beliefs is just as good a map to steer by as a map that represents knowledge (which we may suppose is at least true belief with some further condition such as the beliefs are filtered by properly conducted inquiry). So why bother with the extra effort to obtain knowledge when true belief can do just as well? Plato's worry, as we will see in the next sub-section, is that true beliefs can arise quite accidentally. What we want are true beliefs that do not arrive in our minds by luck or accident.

The 'why bother?' question also arises in Bernard Williams' account of a Cartesian project of pure enquiry in Chapter 2 of his book *Descartes: The Project of Pure Enquiry*. Pure enquirers will have the high aim of possessing beliefs that track the truth and only the truth. But since we are all prone to error our beliefs can often be false. How can we, as pure enquirers, guard against this? One way might be to collect a lot of beliefs and then check them for their truth or falsity. Not only might this be a laborious task, it fails to get to the bottom of the problem as to why we are prone to falsity. What we need to ensure is that the very processes whereby we get our beliefs in the first place are also processes that are reliable for producing *true* beliefs. That is, whatever means (such as perception) or whatever methods (of inference, or of science, etc) a pure enquirer employs to get beliefs in the first place, there is an onus on the enquirer to ensure that the very processes they use are also reliable for the truth. The answer to the 'why bother?' question is that in order to overcome our proneness to error we need to ensure that, as pure enquirers, we employ reliable methods for producing truths in the first place.

Papineau ((1993), chapter 5.2) turns this answer to the 'why bother?' question into a definition of knowledge, viz., person A knows that p iff p is true, A believes that p, and A has belief-forming processes which produce p and which are reliable for the truth. This leads us to the reliabilist conception of knowledge, to be discussed further in section 2.3. Such a conception of knowledge can be found in Ramsey: 'I have always said that a belief was knowledge if it was (i) true, (ii) certain, (iii) obtained by a reliable process.' (Ramsey (1990), p. 110). This can also be seen, albeit dimly, in Plato's problem of the *tether* that marks the difference between knowledge and belief.

2.1.2 Knowledge and the Problem of Plato's Tether

In the *Meno* Plato continues from the previous quotation, using the metaphor of the statues of Daedalus that are so life-like they appear to move.

> Socrates: If you have one of his works untethered, it is not worth much: it gives you the slip like a runaway slave. But a tethered specimen is very valuable, for they are magnificent creations. And that, I may say, has a bearing on the matter of true opinions. True opinions are a fine thing and do all sorts of good so long as they stay in their place, but they will not stay long. They run away Meno: In that case, I wonder why knowledge should be so much more prized than right opinion, and indeed how there is any difference between them.
> Socrates: Shall I tell you the reason for you surprise, or do you know it?
> Meno: No, tell me.
> Socrates: It is because you have not observed the statues of Daedalus. Perhaps you do not have them in your country.
> Meno: What makes you say that?
> Socrates: They too, if no one ties them down, run away and escape. If tied, they stay where they are put.
> Meno: What of it?
> Socrates: If you have one of his works untethered, it is not worth much; it gives you the slip like a runaway slave. But a tethered specimen is very valuable, for they are magnificent creations. And that, I may say, has a bearing on the matter of true opinions. True opinions are a fine thing and they do all sorts of good so long as they stay in their place; but they will not stay long. They run away from a man's mind; so they are not worth much until you tether them by working out the reason (*aitias logismos*). That process, my dear Meno, is recollection, as we agreed earlier. Once they are tied down, they become knowledge, and are stable. That is why knowledge is something more valuable than right opinion. What distinguishes one from the other is the tether. (Meno 97D-98A; Guthrie (1956), pp. 153-4)

Plato bequeathed to epistemology the problem of the tether which, when properly understood, marks the difference between mere true belief and knowledge. Knowledge is *tethered* true belief; untethered true belief can slip away or run from our minds unlike knowledge that is tethered, stays in the mind or is stable and does not run or slip away. What kind of tether produces stability in our beliefs akin to physically tied down statues? With a little retrospective hindsight one might say that those very belief-forming processes that are also reliable for the truth yield the requisite stability for belief in the mind. But other parts of Plato's account detract from this suggestion.[5]

Tethering is said to be 'working out the reasons' or '*aitias logismos*', 'aitia' being cause or more generally explanation or reasons, and 'logismos' being calculation; the whole expression could also be rendered as 'reasoning out the explanation', or even 'giving a chain of proof'.[6] Such knowledge is clearly to be distinguished from dogmatism, and it challenges belief founded on authority or tradition. It also makes clear that it is the enquirer himself or herself who must do the work here. This makes Plato's theory of knowledge internalist rather than externalist (under which heading reliabilist theories of knowledge are usually classified). An *internalist* theory of knowledge is one in which *all* the justifying factors that make true belief knowledge must be cognitively accessible to any enquirer; in contrast, an *externalist* theory of knowledge allows that these factors not be cognitively accessible to, or even be beyond the ken of, an enquirer. Both theories come in a weaker and a stronger form. A stronger form of internalism is one in which the enquirer actually have cognitive access to all the justifying factors, and employ them, when they know; the weaker version only requires that all the factors be accessible, but not all actually accessed. The latter come close to a weaker form of externalism that may only require that the justifying factors not actually be accessed in the course of knowing, even though they are readily accessible and not beyond an enquirer's ken. The internalism of Plato's account comes to the fore when the elements that do the tethering, viz., the *working out* of the reasons or explanation, must be at least weakly cognitively accessible as the enquirer does the 'working out'.

Such an account of 'working out the explanation' is given in Socrates' encounter with a Slave Boy in which he is asked to solve a geometrical problem, viz., what is the length of the side of a square whose area is double the area of a given square with side, say, 2 metres? Through a process of Socratic question-answer, the Boy is led to the correct answer, viz., the hypotenuse of the given square. But the Boy only has a true opinion, and not knowledge, because he has yet to grasp fully the tether of reasons implicit in Socrates' question/answer process. As Socrates puts it: 'At present these opinions, being newly aroused, have a dream-like quality. But if the same questions are put to him on many occasions and in different ways, you can see that in the end he will have a knowledge on the subject as accurate as anybody's.' (*Meno* 85C; Guthrie (1956), p. 138). There is also talk of recollection (*anamnesis*) in which true belief becomes tethered. But again knowledge need be recollection only in the weak sense of 'working out the reasons' for the belief (in the course of recollection). We are not obliged to accept a further part of Plato's theory of knowledge as recollection, viz., that the knowledge is already allegedly implanted innately, or in some *a priori* fashion, in our minds.

There is an important issue about the syntax of 'know' that tells us something about the wide range of the possible 'objects" of knowledge, an issue that arises in different guises in Plato's epistemology. In most of this book knowledge and belief will be assumed to have propositional content. Thus we will speak of a person knowing, or believing, *that* p (where 'p' expresses a proposition). However there are other uses of 'know' not of this kind. The original sense of the English 'know' (not unrelated etymologically to the word 'cunning'), as well as the Greek 'episteme', had to do with 'know *how to*', i.e., a skill or ability. Thus one is said to know how to play the trumpet, differentiate second order equations, speak Chinese, tame a wild horse, and so on. Other constructions include 'know *why,* (for example) the heart pumps blood', or know *how* the heart pumps blood'. Such knowledge is of different explanations that can be given concerning the heart's action. The construction 'know *what* x is' has to do with definition or nature of some x. And so on for other constructions such as 'know *whether'*, 'know *who'*, etc. Finally there is knowledge by acquaintance where the word 'know' takes a direct object. Thus one can know, say, the Pope, know the way home, know Bach's preludes and fugues, know mathematics, know French cooking, etc. All these different uses of 'know' are important in science, since science involves skills and abilities, offers explanations, provides definitions, etc. The issues raised by the wide variety of "objects" that 'know' can take will emerge in different ways throughout the book (see especially 8.1 concerning Foucault on objects of knowledge). For the time being the focus will be on knowledge and belief where the object is propositional.

There are a variety of theories about knowledge expressed by the construction 'knows that p'. Understanding the tethering relation as one of justification, yields the common internalist account of knowledge as justified true belief. This time-honoured theory has been challenged by two main objections, one new and due to Gettier, and one quite old and due to Agrippa.[7] In the next section we will follow up an epistemology based on the structure of justified true belief that we might call 'Cartesian', leaving it open as to whether the philosopher Descartes actually held such a theory. And we will explore Agrippa's objections to such a theory. Its significance is two-fold; first, it is a foundationalist theory and, second, it is a picture of knowledge to which much philosophy of science is said to subscribe. Many infer that if foundationalist theories of knowledge fail then not only does the traditional epistemological framework for science allegedly collapse, but also the way is open to relativist, nihilist, postmodern or some other quite non-traditional theory of knowledge and science that eschews any rationalist account of Plato's tether. But this is a wholly bad inference to make. There is still a wide range of intellectually respectable theories of knowledge available. Only one of these, externalist reliabilism, will be explored in section 2.3.

2.2 AGRIPPA'S PROBLEM FOR KNOWLEDGE AS JUSTIFIED TRUE BELIEF

In his *Outlines of Pyrrhonism*, Sextus Empiricus mentions Agrippa's[8] 'Five Modes Leading to the Suspension of Belief' which are intended to show that if dogmatists are forced to defend their beliefs then, in any dialectic of argument, one falls into either *circular reasoning, infinite regress* or reliance upon *unsupported assumptions*. Two other modes are important in any epistemology, but they do not arise in the dialectic discussed below. The mode of *discrepancy* arises between people who disagree over some subject; this even includes disagreements between philosophical theories of knowledge and ordinary belief, as Hume was acutely aware in the case of his own epistemology. The mode of *relativism*, as the name suggests, invites the response between disputants: 'that might be true for you and your lot but it is not true for me and my lot'. Here disagreement is avoided because even though 'p' and 'not-p' are contradictory, the relativised versions 'p is true for x' but 'not-p is true for y' are consistent. No attempt will be made to examine these ancient sources. However it will be shown how aspects of the first three modes resurface in recent accounts of one form of foundationalism in epistemology.[9]

The theory of knowledge to be investigated here is one in which Plato's tether is understood to be a sufficiently strong justification for the truth of some claim: person A knows that p iff (i) p is true [the truth condition]; (ii) A believes that p [the belief condition]; and (iii) A has a sufficient justification that p [the justification condition]. We will suppose, in common with a vast range of epistemological theories, that the truth and belief conditions while necessary for knowledge are not sufficient[10]; so we will discuss only how the justification condition might be set out and whether it can provide a third necessary condition which, in conjunction with the other two, yields a set of sufficient conditions. Without the justification condition then, concerning Agrippa's dialectic, p would appear to be an unsupported dogmatic assertion of person A. Later we will return to the issue of whether there are examples of propositions p which are known in the sense that they are true beliefs but stand in need of no justification thereby avoiding Agrippa's dialectic. Finally we will for convenience restrict ourselves to cases of typical ps that might arise in the context of ordinary and scientific knowledge, leaving logical, mathematical, moral, metaphysical and many other types of examples of p out of consideration (though in many cases the same Agrippan problems can arise for them).

Finally, we need a name for the model to be discussed. We will call it a '"Cartesian" model', leaving it open as to whether the real Descartes actually adopted it. Though it has some features in common with the real Descartes' epistemology, in some respects it is different. Moreover, as some have argued, Agrippa's challenge to knowledge employed here is somewhat different from Descartes' sceptical challenge.[11]

2.2.1 Agrippa's Dialectic and the Justification Condition

What is the structure of justification on the "Cartesian" model? There are two components, evidence e and a relation of logical support (either deductive or inductive/probabilistic) holding between e and p such that e gives strong support to p. Evidence e can be of different kinds. There is the evidence presented to the jurors in a court of law, the evidence provided by symptoms and other medical examinations for a person's medical condition, evidence provided by what we hear and observe about the verbal and non-verbal behaviour of others, the evidence provided by scientists based on observations or experimental results, and so on. Evidence e need not always be directly observational. We obtain evidence about events on, say, the stock market by reading newspapers. Newton obtained evidence for his Law of Universal Gravitation by citing some of Kepler's laws of planetary motion and Galileo's laws of free fall close to the Earth. The evidence may even be of a highly theoretical character as when one theory is cited as evidence for an even more general theory. In all cases what is of concern is the strength of the support given by e to p.

That e strongly supports p is necessary, but not sufficient, for the justification condition. Clearly, if e fails to strongly support p then no person could use e as the basis of a knowledge claim that p. A further necessary condition is that the knower A must be able to make the justificatory inference from e to p (Plato's tether requires that knowers work out the reason themselves). Thus the justification requires that knower A at least make an inference that is correct and sound. The correctness requirement is that there must be some correct principle R of which the inference from e to p is an instance (thus the correct R underpins the claim that e gives *strong* support to p). The soundness requirement is that the evidential premises e must be true; false evidential claims cannot be used to make knowledge claims. But must A know the evidence e? If we require this, then the definition of knowledge comes up against one of Agrippa's dialectical ploys, viz., circularity, in that the term 'know' is employed in its own very definition. We will return to this point below. But A must at least have some cognitive attitude to e, such as belief. In what follows let us suppose that A at least believes that e, and that e is true, and that e does strongly support p (as an instance of R).

Consider the principle R used in the justification. Grant that A, as well as believing p (this is already supposed in the belief condition for knowledge), also believes e as well. It may well be the case that the penny has not dropped for A that e really does support p by means of principle R; A might not have clicked to the fact that e really is positively relevant evidence for p. Thus as well as requiring that A believe both p and e, we must add that e be operative in A's mind in leading to the conclusion that p, and that A is not lead to p by some other pathway. But more must be operative in A's mind than this. A might get from the true e to the true p not by the correct rule R but by a totally fallacious process of inference. So we need to add that A's reasoning from e to p is by means of a correct rule R.

Finally, must we require that A *know* that R is a correct principle of inference, and know how to apply it in the case of e and p? It would not be enough that there merely exist a correct principle R that justifies the inference from e to p; somehow A must have a cognitive attitude to R and use R in the inference from e to p. To merely require that A believe that R would still be too weak a claim on which to build knowledge that p. It might be possible that A believes that e supports p and that R licences the inference, but A holds R for quite bad reasons. At this point advocates of the "Cartesian" model might appeal to what they call the "natural light of reason" to do some philosophical work in their epistemology. We may be unclear about when our "natural light" is and is not shining, or it might shine only fitfully. But if knowledge of rule R is required, then the light of reason must shine quite brightly for A. Since the full structure of A's reasoning in using R, its correctness and its application in the case of e and p, must be something that is present in A's consciousness when A 'works out the tether', then this definition will be strongly internalist.

In the above we noted that the Agrippan dialectic has struck again through the circularity in the requirement that A know that R; the very term 'know' that we are trying to define recurs in the definition. We might try to remedy this by pointing out that what A is required to know is a principle of inference and how to apply it in the case of e and p. We could argue that this is a special kind of *a priori* knowledge which A possesses and which is somehow in A's consciousness and even immediately present to A as A reasons from e to p. But this might be too much to ask of the light of natural reason in order to rescue us from the grip of Agrippa's dialectic. The theory is stuck with circularity, viz., a piece of knowledge about reasoning.

Not only does A have to know that R; it seems, as already noted, that A must also know that e. To merely have a true belief that e is fatal to the theory of knowledge under consideration; knowledge would then be merely underpinned by true beliefs. But the whole point about the investigation into knowledge was to find out about how it differed from true belief. Thus a second circularity: A must know that e. But there is a way of avoiding this circularity; simply reapply the definition of knowledge over again to 'A knows that e'. Such a definition will appeal to some evidence, say f, which supports e. But now an infinite regress threatens because the same considerations which applied in respect of e as evidence for p now apply to f as evidence for e, and so on, indefinitely. Plato was aware of such a regress in the final sections of his *Theaetetus*. Agrippa also ensnares us in a similar regress within his overall dialectical strategy; we are now caught in either circularity or infinite regress.

2.2.2 Responses to Agrippa's Problem within Epistemology

One response to the regress is that of scepticism. On the basis of the argument just given a person might well conclude that we cannot know anything (except perhaps this very claim, evidence for which is the argument above). How does the kind of theory we are exploring deal with this regress and at the same time ward off scepticism? It supposes that the regress eventually comes to a stopping point in what might be called 'foundations for knowledge'. So far the definition of knowledge has been formulated for cases of inferential knowledge, knowledge that is obtained by inference from other knowledge claims. The inferential structure of knowledge is what makes the regress possible. But if the regress is to stop, not any stopping point will do. The stopping point, the *foundations* for knowledge, must be in beliefs that are not arrived at by inference. But now another aspect of Agrippa's dialectic threatens to takes hold, viz., unsupported belief.

The theory we are exploring also requires that the non-inferential beliefs at which we stop not be open to the possibility of correction (i.e., they are incorrigible, or indubitable, or certain). That is, there are beliefs, such as q, which are such that it is *not* logically possible that A believe that q and yet q is false. Are there any such beliefs? The belief, discussed by Descartes, that I exist, is a good candidate. But this in itself can hardly establish other knowledge claims. Other candidates might be beliefs not about observable states of affairs such as 'I see light' (to employ an example of Descartes in the second of his *Meditations*), but rather about one's current experiential states such as 'it seem to me (or it appears to me) that I see light'.

Much ink was spilt in twentieth century philosophy exploring whether such beliefs about one's immediate experience were beyond correction, beyond doubt, or could be established with any degree of certainty. The philosophical jury has now returned and its verdict is negative.[12] Agrippa's dialectic strikes once more; there are, as far as this theory of knowledge is concerned, unsupported beliefs. Whether such unsupported beliefs are a challenge to the very existence of knowledge, or that the theory of knowledge can get by and live with Agrippan unsupported belief is an open question (this is the way out taken by Popper (1959), chapter V who decides to live with the infinite regress). What is certain, though, is the following. Setting aside Agrippa's dialectical qualms, suppose we do have unproblematic beliefs about our experience, such as: 'it seem to me (or it appears to me) that I see light'. Then we are each trapped on the island of reports of our own experience and we cannot reach out to the mainland of knowledge of other persons and the external world. For, from the premise 'it seems to me (or it appears to me) that I see light' it does not logically follow 'I see light'. We may not even be able to avail ourselves of other principles of inference (not free from doubt in some quarters), such as inference to the best explanation, in which the best explanation of my seeming to see light is that there is light. Inference to the best explanation is, however, a mode of non-deductive inference to which even Descartes helped himself in the final *Meditation* in arguments for an external world and against the view that our world is really a dream.

The analysis of the proposed definition of knowledge in terms of justified true belief appears to require any knower to have full conscious awareness of the entire noetic structure of any knowledge claim. Much of the structure is provided by the rules of inference, such as R, which take A from evidence e to the belief that p; A must have full epistemic access to this structure of reasoning. But to avoid the regress requires epistemic access to more than this. Knower A must also have full epistemic access to the very foundation of incorrigible beliefs upon which the inferential structure of A's knowledge that p rests. It might appear that few have knowledge in this sense. Even scientific knowledge would need to be set out in this way so that it, too, reveals its full noetic structure to individual consciousness. But in fact it is never set out in this way.

On a weak account of internalist knowledge, some of the justificatory factors can remain merely cognitively accessible to an enquirer without actually being accessed in the course of coming to know. A strong internalism would require that all the justificatory factors actually be cognitively accessed by the enquirer in order that they know. In contrast, an externalist account of knowledge allows that some of the justificatory factors not be cognitively accessed in the course of knowing, or not be accessible at all and thus beyond a person's ken yet they have knowledge. On a strong internalist understanding of the "Cartesian" model of knowledge set out above, A needs to have full epistemic access to any rules R and their application in the structure of the justification, and have full epistemic access to the foundational beliefs on which that structure rests. Weak internalism and externalism relax this strong requirement on knowledge, and in particular on the knowledge requirement for rules R. All that might be required is that

A's reasoning be in conformity with R without A using R or even knowing either R or how to apply it.

On the face of it, the "Cartesian" model of knowledge appears to have a minimal externalist element in that A's belief must at least be true, whatever other accompanying access to the noetic structure of the knowledge claim there may be. That is, A's belief that p must track the truth. So far nothing has been said about truth. But it is open to the internalist to adopt a theory of truth which may be non-epistemic (such as the classical correspondence theory), or which is epistemic (e.g., they might adopt a coherence or warranted assertability theory of truth.) Whatever the case, there must be some feature of the noetic structure supporting A's knowing that p that guarantees the truth of p; however it is not easy to ensure this as it might be possible that A's noetic structure has this feature yet p be false.

Can this "Cartesian" conception of knowledge be made less fully internalist? One could drop the requirement that A know the rule of inference R upon which A reasons from e to p. In its place we could merely require that A's inference from e to p be one which is correct under some rule or other which need not be accessed by A. That is, A relies on the correctness of the inference without knowing the full inferential structure involved. On this assumption one of the circularity objections to this "Cartesian" model of knowledge would lose its force. As will be seen in the next section, this is a feature that the reliabilist theory of knowledge takes on board. Could the "Cartesian" model employ a modicum of externalist reliability? There is no reason why a mixed theory of knowledge could not employ both internalist and externalist features; they do not exclude one another.

The "Cartesian" model of knowledge requires, in order for a person to have cognitive knowledge, that they have, accessible to them, a full, or reasonably full, account of the noetic structure underpinning their knowledge claim. Even though such a structure of knowledge is not available in science, perhaps rational reconstructions of what is accepted in science could be made to conform to this "Cartesian" model. Has such a rational construction ever been carried out? Such a "Cartesian" programme underlies one aspect of Descartes' view of science as it does Hume's attempts to found knowledge of the world on sense impressions. With the emergence of the techniques of modern logic, the programme, according to Quine, reached its culmination in Carnap's 1928 book whose English title is *The Logical Structure of the World*. [13] Despite to the efforts of Carnap, Russell and others, the failure of these attempts at reconstruction is one of the well-established outcomes of twentieth century philosophy. It has led many epistemologists to look for different approaches to the theory of knowledge.

2.2.3 Does the "Cartesian" Model Discredit Science?

If the "Cartesian" model is applied to scientific knowledge, either in its modest weak internalist or immodest strong internalist form, we can see why its advocates would take sociological studies of science to discredit science.[14] Since the "Cartesian" model is strongly internalist, the noetic structure of any knowledge claim must be accessible to each individual consciousness. Moreover, in accordance with the real Descartes' ethics of belief,[15] the only things that *should* incline the mind to assent are those claims that have an appropriate and accessible noetic structure. For the "Cartesian" model, the inquiring scientific mind is a pure mind which, in order to admit some belief, ought only to be persuaded by evidence and reason. If anything else were to impinge upon the processes which lead to the formation of the belief that p, then the norms of the ethics of belief would require us not only not to give our assent to p, but to positively reject p. A parallel to the requirements of an ethic of belief can be found in sociologists of science such as Merton who proposes sociological norms for the ethos of science (see section 4.3).

The ethics of belief aside, many have found the foundationalism of the "Cartesian" model unacceptable. It turns on a strong version of a foundation for knowledge in reports of immediate indubitable experience (and not mediated objects or facts), that many who would endorse other versions of foundationalism would reject. Others have also contrasted it with claims about the theory-ladenness of observation and evidence in both extreme and moderate versions of theory-ladenness – advocates of the enterprise of the sociology of scientific knowledge (SSK) usually taking the extreme position. But they also have other criticisms of the "Cartesian" model. They allege, on empirical and other grounds, that many of our beliefs, even in science, are not the result of such a pure inquiry; rather our beliefs are caused by socio-political factors and interests of which the "Cartesian" model of science makes no mention (see chapter 5).

Advocates of SSK would attack even the alleged purity of the very inferential rules which are central to the "Cartesian" model claiming that they, too, are grist to the sociologists programme and are not to be grounded by some "Cartesian" account of the "natural light of reason", or more generally to be given any kind of *a priori* underpinning. Advocates of SSK and the "Cartesian" model differ profoundly over what are admissible causes of belief in science. Each accuses the other of discrediting real science. Advocates of the "Cartesian" model are critical of the admission of social factors as the cause of belief in that this undermines the normativity of the rationality of science, while advocates of SSK view the "Cartesian" model as hopelessly unreal, irrelevant and unsustainable for actual science. The advocates of SSK go even further and claim that the "Cartesian" model of scientific knowledge gives not only a misleading epistemological account of science but, worse, it also gives an ideological picture of science since its misleading idealisations obscure the real social processes which allegedly shape science. In their view the true discreditor of real science is the "Cartesian" model; it ignores the important social dimension of science that allegedly penetrates to its very content.

Even if the "Cartesian" model of knowledge outlined above were to be abandoned, it does not logically follow that advocates of SSK would be free from other versions of foundationalism, or from other non-foundational epistemological theories, to pursue their sociological account of scientific knowledge. There are a wide variety of rationally based theories of knowledge to consider, such as Bayesian theories, coherence theories, theories which seek a fourth condition, naturalistic theories, other varieties of foundationalism (that do not depend on reports of indubitable experience), etc.[16] These rival the justificationism of the "Cartesian" model just described in their attempt to give a better account of what has actually happened in science concerning theory construction and choice. In the next section we explore the reliabilist theory of knowledge; in the next chapter we will give an account of the epistemic authority of norms of method.

2.3 RELIABILISM AND THE DEFINITION OF KNOWLEDGE

2.3.1 Varieties of Reliabilism

Plato's condition of a tether requires that knowledge be a true belief which is 'tied down', or 'secured', in the mind. Within reliabilist theories of knowledge a number of different 'tethers' have been suggested. Broadly, all externalist naturalised epistemologies treat a knower as a physical and biological system which interacts with the world in a range of ways; if the biological system of an organism has developed sufficiently to have cognitive states, then these states, including the epistemic states of knowing, will also be treated as part of the world/organism interactions. In accordance with this are causal theories of knowledge in which what tethers a belief is the right kind of causal connection between the belief and the state of affairs that make the belief true.[17] One particular proposal along these lines is that knowledge arises from an 'information flow' from the world to the believer: 'A knows that p = A's belief that p is caused (or causally sustained) by the information that p' (Dretske (1981), p. 86). Such causal definitions obviously apply to perceptual knowledge, and do have a "foundational" aspect. But clearly such knowledge is not foundational in the sense of being founded on incorrigible or indubitable reports either of experience or of everyday objects.

Other versions of reliabilism require that our beliefs counterfactually track the true. Thus to the belief and truth conditions is added a counterfactual tracking condition (in which appeal might be made to some principles of method M – a point of interest in the philosophy of science): if p were to be false (and an inquirer A were to employ method M to discover whether or not p) then A would not believe (*via* M) that p, and if p were to be true (and an inquirer A were to employ method M to discover whether or not p) then an inquirer would believe (*via* M) that p. (A case is made for this in Nozick (1981), chapter 3 (I)). Yet others propose that there be a law-like connection between the world and our beliefs, as illustrated by the workings of a thermometer. Just as there is a law-like connection between the ambient temperature and the reading on a thermometer, so there is a law-like connection between the states of the world and the beliefs we hold of it (Armstrong (1973), chapters 12 and 13).

There are many good detectors of particular states of the world. As well as thermometers, iron is a detector of moisture in that it rusts in the presence of moisture, litmus is a good detector of acid in that it turns red in the presence of an acid, and so on. We humans differ from iron and litmus in that we have mental as well as physical states (but for a naturalist these are all on a par). These mental states are good detectors of certain facts about the world (such as our perceptual states); and humans are sometimes reliable users of their cognitive apparatus (such as that for inference making). Given our cognitive abilities we form beliefs about the world when confronted with bits of it; and on the reliabilist theory, the more reliable the belief-forming process the more likely we are to have knowledge of the world. On Armstrong's view the reliability is to be founded in law-like connections between bits of the world and our cognitive states such as belief. Since the belief/fact connections are law-like and not accidental then the regularity here underpins the reliability of the belief-forming process, thereby yielding knowledge.

Such externalist accounts of knowledge readily fall within the scope of *naturalised* epistemology with their talk of belief states being *caused* by, or *tracking*, or being linked in a *law-like* way to, states of the world, or facts. Does this make epistemology akin to a science that studies the linkages between belief states of a cogniser and external states of the world? Traditionally understood, epistemology, and in particular definitions of knowledge, are *normative* and are not merely another science understood as "factual" and thereby devoid of normativity. The theory of knowledge cannot merely be about beliefs and their formation; that would make it a "chapter" in psychology or sociology. What is missing is a requirement that the linkage between beliefs and facts in the world be "correct", "appropriate" or "right". Importantly the aim of knowledge is to acquire true beliefs which are also *justified*, or *warranted*, or *well* grounded, or *rational*, the italicised words indicating epistemology's strong and irreducible commitment to the normativity of knowledge. As we will see in Parts II to IV, there are sociologists of science and philosophers, such as Foucault or Nietzsche, who are strongly eliminative with respect to the normativity of the notion of knowledge. This has the consequence that they focus only on the natural processes (usually involving power, or political or psycho-social relations) that bring about belief and ignore any further considerations that might pertain to knowledge. However there are those philosophers who are not eliminativist and who claim that we can both naturalise knowledge while maintaining its normative character. Such is the programme of, amongst others, Alvin Goldman who attempts to define knowledge while observing the constraint that the normative concepts of epistemology (such as knowledge, justification, warrant, etc) be defined in terms that are not normative (and thus non-epistemic given the involvement Goldman finds of the normative in our epistemic notions).[18] In the next chapter we will return to the matter of "defining" the norms of epistemology and methodology in non-normative terms that differs from the approach of Goldman; it will not use the standard "necessary and sufficient" conditions model of Goldman. Here we will focus on his reliabilist conception of knowledge that meets the above constraint on admissible definitions.

2.3.2 A Reliabilist Definition of Knowledge

The core idea of a *reliabilist* theory of knowledge is the following: person A knows that p = (1) p is true, (2) A believes that p and (3) A's belief that p is produced by a belief-forming process (BFP) which is reliable for the truth of the belief that p. More will be said shortly about what BFPs are, and when they are reliable. There are at least three points to highlight about this definition of knowledge. It is BFPs which are reliable that cash out Plato's metaphor of a "tether" for beliefs. What talk of a tether rules out are true beliefs which are acquired by purely accidental means. Thus to use Plato's example, a person who tossed a die to find out which of, say, six roads went to Larissa, might hit upon the correct road; but they would be very lucky indeed. Coin tossing even when the outcome is correct does not, on either a justificationist or a reliabilist account, provide any justification for the belief that it is, say, that the third road that gets one to Larissa. Here talk of a "tether" is metaphorical, but suggestive. The tether for the belief could be a justification; but it could also be the reliability of the BFP. What both theories of knowledge rule out are cases of the lucky acquisition of beliefs. And this, as Plato recognised, is a necessary requirement for any viable theory of knowledge. Second, the definitional constraint required by naturalists is met in that the normative notion of knowledge is defined using only non-normative notions, especially in clause (3). How normativity is still retained under such a naturalist conception of knowledge is discussed further in the next chapter.

Third, there is no epistemological requirement that A know, or believe, that A's belief that p arises from a reliable process. But there is an ontological requirement, viz., that the BFP actually be reliable. This makes reliabilism an externalist rather than internalist theory of knowledge. This does not necessarily mean that the reliabilist account is always at odds with the justificationist account. Rather it could be viewed as a more general account of justificationism in which the strong internalist requirement that a person have a further justification of the rules of justification they use, is abandoned. In its place is a weaker more internalist requirement that a person can rely on the correctness of the justifications they employ. The way in which reliabilists and justificationists might clash over this issue will be addressed briefly at the end of section 2.3.4.

When is a BFP reliable? Reliability comes in degrees. Minimally, a BFP is reliable when it produces more true beliefs than false beliefs amongst the beliefs it produces; otherwise it is not reliable. Maximally a BFP is reliable just when it produces only true beliefs and no false beliefs. For cases in between we may say that a BFP is reliable if its 'truth ratio' (i.e., frequency of true beliefs in the beliefs the BFP produces) is some percentage greater than 50%. For knowledge (close to) 100% reliability ought to be required. In cases of less than (close to) 100% reliability, we can make a simple link to the notion of a degree of justification: the degree to which a BFP is reliable is the degree to which a belief is justified. Though there is some vagueness here this is all to the good; it reflects an original vagueness in the very notion that is under analysis, viz., the concept of knowledge.

The above specifies a notion of reliability in this world in terms of the empirically determined frequency of true beliefs. But what about the use of BFPs in counterfactual situations? In particular can there be BFPs that are reliable in a range of, or all, possible worlds? Here we may talk about the counterfactual *robustness* of BFPs across possible worlds including this one. Shortly we will argue that deductive inferences are BFPs which exhibit not only 100% reliability in producing true beliefs as conclusions (given true beliefs as premises) in our actual world, but they also exhibit the same high reliability in all other possible worlds.

Though it is desirable to have BFPs that have a high truth ratio, we might also want more than this. Suppose a BFP produces very few beliefs, but when it does it usually produces true beliefs. Such a BFP would have a high truth ratio and thus be reliable. But it might not be of much use to us as believers because of its conservativeness; it produces very few beliefs. What we want are BFPs that are powerful in that not only do they produce mainly (or only) true beliefs, i.e., their truth ratio is high, but they also produce lots of beliefs. On the whole our perceptual and inferential BFPs are powerful. A visual BFP which is conservative would be quite unhelpful; what we want is what we actually have, viz., a powerful visual BFP. Power is also important when we consider principles of scientific method used to adjudicate between a large number of scientific and other hypotheses. We want them to be not only reliable but also powerful in that they produce and sustain a wide variety of beliefs in any science; they are not conservative in that they produce few beliefs (though reliably when they do).[19]

BFPs that are little used raise a problem for the truth ratio formulation of reliability that is based on the frequency of true beliefs in a sequence of beliefs produced by the BFP. Instead of talk of frequency it might be more useful to talk of the *propensity*, or the *chance*, of some BFP to produce truths. In many examples of BFPs it will turn out to be more useful to talk of the propensity that some BFP has, in the situation in which it is used, of producing truths, rather than its truth ratio (understood as a frequency), as an account of reliability. Such a change would also overcome another difficulty, viz., an untypical runs of true beliefs within the class of beliefs actually produced by a BFP over some period time thereby giving an untypical frequency, and thus untypical truth ratio for the BFP.

2.3.3 Examples of Belief-Forming Processes

What are examples of BFPs? A BFP may be some habitual way in which we have formed beliefs; or it may be some innate cognitive mechanism at work in us; or it might be some disposition that we have built up through training or learning; and so on. Perception is one kind of process that yields, for us humans, beliefs about the everyday objects that surround us. (This BFP will be, in part, a hard-wired cognitive mechanism, but also one which, if we are to take into account a limited amount of the "theory-ladenness" of observation, is open to the dispositions we have to describe our experiences in certain ways based on the language and theories we have acquired). That is, we have visual BFPs, auditory BFPs, and so on, which give us beliefs about what we see, hear, etc. But note that perception is not globally reliable even though it is reliable over a wide range of circumstances. Thus the belief 'it is an elephant' is reliably formed when a perceiver is close to an elephant but not, say, when viewing it at a distance on a plain through dust during the shimmering mid-day heat. Modifying our claim we could say that a perceptual BFP might be reliable *in particular circumstance C* just in case *BFP in C* has a high chance of producing truths. But the same BFP in other circumstances C* might have a lower chance of producing truths.

Our visual system is reliable in forming beliefs about what is in our visual field over a wide range of circumstances, but it is not so good at night, in mists, at distances, in viewing by candle light, when we are old, and so on. In particular we might need to consider what are the chances, when we attentively view something with our unaided eyes at a distance through a mist before dawn, that our belief that there is a elephant over there is true. And we might wish to compare this with the chance, when we view something with our unaided eyes 10 metres in front of us on a clear day, that the same belief is true. Here talk of the propensity, or chance, that some *BFP* has of producing truths comes into its own when compared with the truth-ratio frequency view. And we can envisage the propensity of some BFP to produce truths altering from circumstance C to circumstance C*. Such considerations about perceptual knowledge reveal an important difference between reliabilist theories of knowledge and the theory outlined in the previous section. The reliabilist theory does not involve us in any regress of reasons; nor does it require a foundation for knowledge in indubitable, or incorrigible, reports of experience. Instead it talks of the chance that some *BFP in C* has of producing true beliefs about our perceptual situation.

Another BFP is memory; but this can be unreliable and may in fact be a lot less reliable than immediate perception. Unreliable BFPs are legion such as: wishful thinking; *déjà vu* experiences; consulting the Delphic Oracle; forming beliefs on the basis of inadequate evidence, or on hunches, or the toss of a die, or on what one's guru says; and so on. We also employ principles of inference, both deductive and non-deductive, to draw out conclusions which we then believe. Our use of such principles provides a further kind of BFP. Consider valid forms of deductive inference (such as *Modus Ponens*). In a valid argument with true premises we can be 100% guaranteed to get true conclusions; but when at least one premise is false we do not have such a 100% guarantee. Thus deductive inferences are not absolutely guaranteed to be reliable generators of true conclusions from any premises. But they have 100% *conditional* reliability in the following sense: if the statements used as "input" into the premises of a valid argument are true, then the "output" conclusion has 100% chance of being true. Thus valid deductive arguments, used as BFPs, have a very high truth ratio in that they are 100% *conditionally reliable*. Moreover such a BFP is quite robust in being equally conditionally reliable in all possible worlds in which it is used. Using the notion of conditional reliability we can make a better case for the reliability of memory. Clearly our memories will be unreliable if the "input" information is false (as it sometimes is). But if the "input" is true, then our memories can have a high "output" of truths for true "inputs", and so be *conditionally* reliable rather than reliable *simpliciter*.

The reliability of valid inferences as BFPs has the following interesting feature. We can form beliefs as a result of deduction using, say, *Modus Ponens*, without knowing either what BFP we are employing (i.e., that it is *Modus Ponens* that we are using) or that the BFP is reliable (i.e., we may have never studied first year logic and have not shown that *Modus Ponens* is valid). Yet we can all get knowledge by using conditionally reliable patterns of inference. This aspect of externalist reliabilist theories of knowledge captures an important intuition about knowledge not so easily captured by internalist theories that, especially in their strong versions, require us to have knowledge of the inference pattern and its validity.

Deductive inferences are not only 100% conditionally reliable in this world; they are also conditionally reliable in a host of possible worlds, i.e., they have counterfactual robustness in that in each world (in a range of possible worlds), they exhibit the same conditional reliability. Such reliability and robustness is a boon to inquirers no matter what world they take themselves to be in. But do these features also carry over to the use of non-deductive principles of reasoning (such as induction), and the use principles of scientific method, as BFPs? That is, are inductive inferences that have a high degree of reliability in this actual world also reliable in possible worlds close to ours? Were an enquirer to use induction as a BFP in one of these close-to-us possible worlds, would it still be reliable? This is not a question we can answer here,[20] but any answer would be relevant not only to induction but also to any principle of scientific method.

For example, given the range of possible worlds we do not know which one we inhabit; but in our processes of inquiry we hope to eliminate some, using principles of inference and scientific method. Thus assuming that special relativity is correct we now take ourselves to inhabit a world in which the speed of light c is 300,000k/sec (approximately, in any given frame of reference). We do not live in a world in which it has no upper bound (as we once previously thought), or in a world in which it is twice that value, and so on. However there still remains many possible worlds consistent with this law. What we want of our principles of reason and method is that they exhibit a counterfactual robustness in applying across a number of possible worlds, including our own, in which inquiry, such as that into the speed of light, can be carried out. Our principles of reason and method must have, as well as (conditional) reliability, robustness of the sort we attribute to "laws of nature" in that both must apply in a range of counterfactual situations. If we use principles of reasoning and method which lack such counterfactual robustness in that they are only reliable in our actual world and in no other, then that we have any science at all would be a matter of shear epistemic luck. In other worlds even very close to ours, the very principles of reasoning and method that we employ would be of no avail and inquirers in these close possible worlds who used these principles would fail. Using our principles would leave them bereft of any knowledge of their world.

2.3.4 *Some Refinements of the Reliabilist Definition.*

Because most reliabilists are naturalists it does not follow that they need adopt what might be called 'the doctrine of natural reasoning'. We may all agree that, as result of the processes of evolution, we have naturally acquired the capacity to reason. But it does not follow that we have thereby acquired only valid or correct processes of reasoning. As many psychological studies have shown, we are not very good natural reasoners.[21] We reason tolerably well using *Modus Ponens* but we are not very good at *Modus Tollens* and commit related fallacies such as "affirming the consequent", and so on. When it comes to probabilistic reasoning it appears that we do quite badly. Reliabilism does not embrace this doctrine. In fact reliabilism can take on board the lessons of an ethics of belief in requiring that we ought to become reliable reasoners. After all, why should we knowingly use a BFP with a poor truth ratio when we could do better? For example why should we continue to rely on our guru, or use the gamblers' fallacy, or attempt delicate observations when not fully sober, and so on, when we can come to see that there are better BFPs to use? But we may also unknowingly use unreliable BFPs thereby undercutting the possibility of knowledge. What we need to allow for is the possibility that we can scrutinise most of our BFPs for their reliability and reform or discard them if their reliability is low.

The above considerations highlight one matter to do with normativity within reliabilist conceptions of knowledge. We can be reflective about our BFPs, and we can be successful in eliminating those that are not highly reliable and adding those that we have not so far employed but which are highly reliable. Thus we can be reflective and stop being wishful thinkers, relying on *déjà vu* experiences as a guide to the past, failing to take a wide enough selection of evidence in forming beliefs, giving in to the gambler's fallacy during a bad night at a casino, and so on. And through learning we can equip ourselves with new, more, reliable BFPs, such as previously unknown modes of inference learned while attending a course in statistics or logic. We can even correct our visual BFPs as when we learn how different colours look to us in different conditions of lighting, for example viewing the coloured fabrics in the artificial lighting of a store. We can also learn to correct fairly robust visual illusions such as the Müller-Lyer illusion; we do not always naively believe that the lines are of unequal length but know, despite appearances, that they are equal in length. However the point made here about being reflective should not be taken too far unless it become a condition on knowledge that A have checked the reliability of a BFP before using it. It is not required that A know something about the reliability of any BFP that A employs. Such a strong condition would undercut much of the externalism of reliabilist knowledge and place it within the camp of internalist theories of knowledge. This externalist feature is consistent with our having ensured that we have equipped ourselves with reliable BFPs.

This point relates to a counterexample Goldman discusses to the reliabilist theory under consideration (Goldman (1986), pp. 51-2). Suppose A goes to a class conducted by B who provides a number of algorithms for solving arithmetical problems (such as long division, taking roots, etc). Unfortunately B is a fraud and A is rather gullible in that A takes B to be a mathematical guru. And even more unfortunately, all but one of the algorithms A learns from B is defective in that they are not reliable processes for reaching results in arithmetic. Suppose A learns the algorithms, applies each and believes their results. Then in all but one case A will have used an unreliable BFP leading to a false belief. But one of the algorithms A employs is correct; then, as A has used a reliable method, A does acquire a belief that is true. And on the reliabilist theory of knowledge it follows that A knows the arithmetical result because A has used a reliable algorithmic method. But is it correct that A knows in this case? No. A has accidentally picked up the one correct algorithm amongst the many defective algorithms. A has picked up these algorithms because of A's misplaced faith in B. We can agree that simply accepting something on faith, or because our guru has told us, is not a reliable method for acquiring correct algorithms.

Here we need to distinguish between two levels of reliability. The first level pertains to the one correct algorithm that A uses. This algorithm is a reliable BFP for those who use it to produce mathematical results. The second level pertains to the method employed to get correct algorithms in the first place. Neither being rather gullible or putting ones faith in the claims of a person one takes to be a guru are reliable methods for acquiring beliefs; any beliefs so obtained would not count as knowledge on a reliabilist approach. To overcome this difficulty would it be satisfactory to require that A have a *justified belief* in the correctness of any algorithm that they use? Such a requirement would suit some conceptions of knowledge, especially those that admit such strong internalist requirements as A's having a justification. But it would not be in accord with the spirit of a reliabilist theory of knowledge that eschews such internalist requirements. So is there any other way of formulating a further requirement which is within the spirit if reliabilism? There is.

The third clause of the reliabilist definition of 'A knows that p' can be modified to distinguish between different levels of BFP: (3) A's belief that p is produced by a first level BFP which is reliable for the truth. Now add to the definition a further requirement about how first level BFPs are to be obtained such as: (4) first level BFPs are to be acquired by second level processes which are themselves reliable. That is, we appeal to second level processes that have as their task the job of determining what sort of first level BFPs are to be made available. The sort of first level BFPs that are to be made available are those that are reliable in that they have a high truth ratio. But importantly, the second level processes must themselves be reliable in producing first level BFPs which are also reliable, or more weakly they produce first level BFPs which are at least more reliable than those being replaced. That is, we require meta-reliability of second level BFPs. It remains to define such second level meta-reliability, again within the spirit of the reliabilist conception of knowledge. This can be done in the following way: a second level process is meta-reliable iff it produces mainly (or only) reliable first level BFPs out of all the first level BFPs it produces, i.e., the second order BFP has a high reliability ratio; or more weakly, it produces first level BFPs which are more reliable than the BFPs being replaced.

What might such second level processes be? We have ruled out being a gullible collector of algorithms promoted by someone we take to be an arithmetical guru. But we have already mentioned subjecting our BFPs to thoughtful reflection and examination; this is a process which, when properly conducted, can have a high degree of reliability in producing a range of first level BFPs which are themselves reliable (from examining the inference patterns we employ to examining our perceptual processes in unusual circumstances such as that of lighting, or the situation presented in the Müller-Lyer illusion). Thus second level processes can be wide-ranging, from learning to getting a hearing aid to assist one's auditory BFP, or to begin wearing glasses. What is significant about the appeal to second level processes which are meta-reliable is that person A does not have to know, or have a justified belief, that the second level process is meta-reliable; all that is required is that the second level process *be* meta-reliable. And this is the case even when, through learning, we acquire new reliable BFPs. We do not have to have a justified belief, or know, that the (second order) learning is meta-reliable; rather, if the learning is of the appropriate sort, it will be meta-reliable.

Does a regress threaten in which our second level processes might also be unreliable in much the same way as the lower level processes of acquiring BFPs which gave rise to the initial problem in the first place were unreliable? This might well be the case. But the idea that there is some sort of on-going hierarchy of ordered reliable processes is illusory; there is no such strict hierarchy. When one examines what meta-reliable processes there are, they usually involve being sufficiently reflective and critical about the kinds of BFP one possesses. And here, unless one is severely deluded, one will employ methods of reflecting and criticism that are in fact reliable. In other cases, such as unreliable eyesight, it might simply be a matter of being told by one's optician that one needs to wear glasses. And most opticians are reliable most of the time. More complex cases would include, for example, attending a certified course on statistics to improve one's patterns of reasoning. And most such courses are reliable most of the time. There are institutional arrangements for ensuring that teachers are reliable and not like the mathematical guru envisaged earlier. Here one hits rock-bottom cases of brute reliability beyond which it is hard to go in ensuring that one does have a well-functioning BFP. Importantly there is no on-going hierarchy of meta-reliable processes. If there were no reliability at some point, there would be no knowledge.

Much more needs to be said about the reliabilist model of knowledge sketched here (and it has been said both in its further development and defence against objections in the several works cited). For our purposes the above account will suffice. The core idea of knowledge is that of a true belief that is produced by a BFP that is reliable for the truth, this core idea also being extended to the meta-reliability of processes which produce our BFPs. However there remains one problem that must be aired, viz., the difference between justificationism and reliabilism and the possible superiority of justificationism over reliabilism in respect of the need we have to offer reasons for our beliefs.

2.3.5 Internalist Justification Versus Externalist Reliability

Strongly internalist justificationism and strongly externalist reliabilism are polar opposites in one respect. At one extreme, strong internalist justificationism requires the light of natural reason to be shining brightly, so brightly that the knower has complete justification of everything in the noetic structure of their knowledge that p, including all the rules of inference employed and all the necessary evidence. At the other extreme strong externalist reliabilism allows the light of reason to be turned off. All that is required is that the BFP yield a belief that is true. It is not required that the person know that their BFP is in fact reliable. They are not required to supply any reasons backing their knowledge claim. At one extreme, justificationists appear to require total justification; at the other extreme, reliabilists appear to require none. More problematic for the reliabilist are cases in which a person is a reliable producer of beliefs but at the same time they doubt that they are reliable; and their doubt undermines even the justifications that they might appeal to, such as the reliability of their BFP.

Consider the example of an expert fossil prospector who looks at an object on the ground and declares: 'this is the upper forelimb of an antelope'.[22] Such expertise fits quite well the case of reliabilism with respect to what we can perceive. Fossil prospectors have been trained to recognise the different kinds of fossils and do so with considerable ease and skill. In much the same way, a trained botanist can classify correctly all the vegetation in a forest. Trained orchestral conductors can classify all the sounds they hear in an orchestra when they correctly say: 'such-and-such an instrument playing at such-and-such a register'. Such an ability is merely an extension of the training we all receive that enables us to form beliefs about the everyday objects we directly perceive. Fossil-prospectors, botanists and conductors are highly reliable detectors of certain kinds of object; and on the reliabilist theory it can be said, say, that A *knows* that the fossil is of the upper forelimb of an antelope. A's knowledge claim is not made on the basis of reasons but on their reliability as a detector of kinds of fossil. But this does no necessarily exclude the possibility that A give reasons for the claim if asked. A could point to the features of the fossil and say why it was so classified. But in some cases A might not be able to point to such features; A may only be able to say that it just looks the way the fossils of the upper forelimb of antelopes look. (It is said that some chicken sexers cannot say what it is about the look of a chicken that leads them to classify a bird as one sex rather than another.) Here A replies in much the same way as we do when we are pushed to say why we know, on the basis of looking, that the colour of the grass is green. Our response might be that this is just the way the grass looks to us.

At this point a tension noted by Brandom[23] come to the fore. It might be possible that fossil prospector, A, remains a reliable detector of fossils but A comes to doubt the he is reliable. That is, A does not take himself to be good at fossil classification; A becomes nervous about his knowledge ability. Then it would be *cognitively irresponsible* (as Brandom puts it) for A, and others, to claim that A knows that the fossil is of the upper forelimb of an antelope, or even to believe this. Given A's doubts, A should only believe, or know, on the basis of a check with other non-doubting fossil prospectors, or on the basis of a proper chemical and morphological analysis of the fossil. In the case of the latter there is an appeal to reasons of the sort found in justificationism. But such an appeal is not only available to the justificationist. As we have noted, it is also open to the reliabilist to say that the fossil prospector, when pressed, can give reasons for their belief. However the pathway whereby they got their belief was not one based on reasons and evidence; it was based on their being a reliable detector of states of affairs involving fossils.

If we take this line then we can abandon the extremes of totally internalist justificationism, and of totally externalist reliabilism, for a modified mixture of the two in which cognitive access to reasons and reasoning is not always explicitly required or explicitly ruled out. There may be cases where reasons and reasoning are simply not available to a knower. This is a difficulty for the radical justificationist if they do not allow not only the weaker *possible* access to reasons, but they also rule out the stronger *reliance* on reasons to which they do not have any cognitive access at all. Such mere reliance upon that to which they do not have cognitive access is not a problem for the reliabilist. But it does become a problem when person A becomes nervous; A has a BFP which is reliable in that the belief, that p, that it delivers is true – but A also doubts the output, viz., doubts that p.

What then of the charge of cognitive irresponsibility in the case of such nervousness? To become cognitively responsible the nervous, but reliable, belief former would have to give up the grounds for their doubt in the beliefs they have. But this need not always require them to become justificationist in that they must always explicitly provide reasons for their belief. Instead what they might do is to inspect the very BFP that gives them their beliefs. In particular they need to look at the mechanisms at work (e.g., the fossil prospector gets an eye test), or they examine the truth ratio of the BFP, or they check the meta-reliability of the processes that gave rise to the BFP in the first place, and so on. This is different advice from that which one would give a justificationist. If they have doubts then one could get them to check their evidence, check their reasoning, and in cases of extreme Cartesian doubt to throw out anything which is infected by fallibility. The advice to the reliabilist is different; it is to get a warrant of fitness for a BFP and its reliability. Just as justificationists can monitor their BFPs, which use a narrow range of evidential and reasoning processes, so can a reliabilist monitor the much wider class of BFPs that they employ in getting knowledge. The upshot is that there is nothing unintelligible about having beliefs for which no justification has been given, or can be given, yet the beliefs arise from a BFP that is reliable for the truth. But this does not entail that the nervous reliabilist has no remedy for their condition.[24]

Much more can be, and has been, said about the dialectic between internalist justificationism and externalist reliabilism – and other theories of knowledge not even mentioned. The task of this chapter so far has been to show that even if the "Cartesian" model of the previous section is unacceptable, there are still other theories of knowledge to consider. Knowledge is not to be jettisoned. In particular, reliabilism provides an account of knowledge which is consonant with naturalism and which preserves its normative and critical role. If one is a naturalist, one does not have to abandon epistemology or its norms.

2.4   SOME SOCIAL ASPECTS OF KNOWLEDGE

So far the focus on knowledge has been on an attempt to define 'A knows that P' where 'A' is an individual person and the predicate 'knows that p' is a property or state of A's mind (albeit one with a noetic structure that various conceptions of knowledge attempt to spell out). Such an individualistic conception of epistemology has been at the forefront of philosophical analyses of knowledge. One reason for this is the onus that philosophers have thought ought to be placed on each individual to "clean up" their epistemic act and to exercise their proper autonomy as an inquirer. But not all analyses of knowledge have been of this character; it has long been recognised that there may be important social aspects of knowledge to consider. Just as there is one project in epistemology modelled on that of an individual who uses his or her own cognitive skills to acquire knowledge, so there is another project in epistemology which investigates the ways in which, for both social groups and individuals, features due to their own social organisation, as well as their individual cognitive skills, also lead to the acquisition of knowledge.

It is important to distinguish (a) conceptual analyses or definitions of knowledge into which social factors must enter, from (b) the empirical claims of a sociology of knowledge which tell us about the way in which socio-political interests and/or circumstance might influence belief or knowledge claims, or how the production, circulation, distribution and maintenance of beliefs or knowledge claims occurs in a given society. In Part II there will be much unravelling of the potent mixture of (a) and (b) that has occurred in the discipline that goes under the name 'sociology of knowledge', especially when it is claimed that social factors determine truth or knowledge as distinct from belief. And it is important to keep (a) distinct from what is often called 'social constructivism' in respect of ontology or of knowledge, a matter that we cannot go into here.[25]

Here we will focus only on the ways in which social factors might enter into the definition of 'A knows that p'. Talk of the social must at least mean that reference must be made to two or more persons in the analysis or definition. Such a reference can arise in two ways. Firstly, examining the first relata in 'A knows that p', the individual person 'A' can be replaced by a social group or a collective; thus 'A' could stand for astrophysicists, alchemists, anti-Jacobins, adulterers, Albanians, antipodeans, etc. Secondly, the social would arise if it turns out that the analysis of the predicate expression 'knows that p' makes reference to others besides A. Thus the beliefs or knowledge of others might be involved in A's justificatory process, or the beliefs, knowledge or actions of others might be involved in the reliability of A's BFP. How this might occur will be sketched below.

This section examines four ways in which claims have been made for a social epistemology over and above claims for individual epistemology. The first requires only a brief mention, viz., Popper's theory of "objective knowledge". He dismisses the individualist account as a "subjective" view of knowledge (Popper (1972) chapters 3 and 4), and in his account prescinds from any individual A and the contents of A's mind (this being understood as a 'world 2' entity in Popper's tripartite ontology). He focuses on objective knowledge implicit in claims like 'it is known that p' in which there is no knowing subject. Instead there is an inquiring community which investigates 'world 3' objective contents which include at least theories, puzzles, problem situations and arguments. We will not elaborate on Popper's theory of objective knowledge further here. However we noted in section 1.1.3, that

Popper argues for a social aspect to the application of scientific method; this can be set within the context of a theory of social epistemology.

2.4.1 Group Belief and Knowledge

A second attempt to move away from an "individualist" epistemology is through a consideration of collective or group knowledge rather than knowledge held by individuals. Group belief and knowledge can stand in a number of different relationships to the beliefs or knowledge of each individual in the group. Thus a class of children may share some common knowledge when each pupil in the class knows, say, that 7 is a prime number but 12 is not prime. But when we speak of what astrophysicists currently know, does it follow that for some p, if one astrophysicists knows that p, then they all do? Not necessarily. Each may share a core of knowledge sufficient for each to be a member of the group of astrophysicists; but each my have specialist astrophysicist knowledge that few, or no, other astrophysicists share. Social epistemology would gain some autonomy from individual epistemology if it could be shown that there is no obvious "reduction" of what the group knows to what each individual knows, or there is no "summing up" of what each individual knows to obtain collective knowledge. A case can be made for such autonomy. Consider the case of belief. What we need to show is that for a group to believe that p it is neither necessary nor sufficient for all (or most) individuals to believe that p.[26]

  It is not sufficient; if each individual in a group believes that p it does not follow that the group believes that p. Each member of a group, say a football team, might believe some fact not relevant to their group activity, say that Chopin composed mainly for the piano; each person's believing this has nothing to do with what the team believes (for example, that they lost last week's match to the Rovers). This even holds when each member of the group believes that other individuals also have the same belief about Chopin. What is important is that they do not have the belief *qua* group. Another striking counterexample is suggested in Gilbert ((1989), p. 273) that concerns two groups that happen to have exactly the same members. Suppose a college has both a Food Committee and a Library Committee, and that exactly the same people are members of both committees. Suppose that each member of the Food Committee comes to believe that the College food has too much starch. Suppose also that the Food Committee also expresses the same belief and records it in its minutes. This is now a collective belief (even if there might be a minority of dissenters); the Food Committee believes that College food has too much starch. However the Library Committee never discusses this matter or even expresses this belief in its minutes or anywhere else, despite the fact that it may well be a belief held by each member of the

Library Committee and each member knows that the other members hold this belief.

It is not necessary; a group can believe that p yet it does not follow that each individual in the group believes that p. Such is the case where a group, such as a cabinet in Westminster-style governments, is bound by the doctrine of collective responsibility. Individuals may express the dissenting view that x is the wrong course of action behind the "closed doors" of a cabinet meeting and even retain their dissenting views; yet the cabinet expresses its belief that x is the right course of action. The case of coextensive committees makes this point more forcibly. Suppose the Budget Committee and the Library Committee of a College have the same members. After careful investigation the Budget Committee believes that the Library has adequate funding. But, again, after careful consideration the Library Committee believes that the Library does not have adequate funding (for libraries can always spend more). On pain of contradiction no individual member believes both that the Library has adequate funding and that it does not; and in fact most, or all, might believe one rather than the other of the rival propositions.

If the above is correct then group belief is not always just a matter of the "summing up" of the beliefs of individuals; nor can they be "reduced" to individual beliefs. So far only belief has been considered; but the same applies to group knowledge. Assuming some justified true belief account (or a reliabilist account), then just as in the individual case a person has to hold their belief on the basis of sufficient reasons (or have got it *via* a reliable BFP), so the group has to add to their group belief what are the sufficient reasons they have adduced as a group for their belief (or the belief has to have been obtained by a reliable social process). In the light of the above point about group and individual belief and knowledge, there are features of social epistemology that have to be investigated independently of individual epistemology. What is it about group belief that makes the difference? Gilbert's exhaustive study suggests the following: '*A group G believes that p* if and only if the members of the G jointly accept that p'. And then she offers two accounts of 'joint acceptance' one of which is: 'members of G individually have openly and intentionally expressed their willingness *to accept that p together with the other members of G as a body*' (Gilbert (1989), p. 306) The definition attempts to capture the idea of a group commitment. While more needs to be said about common knowledge and other expressions in the definition, it is clear that it does not follow from this definition that each individual member of G believes that p.

2.4.2 A Social Condition on Knowledge

There is yet a third way in which social factors can enter into epistemology. On the individualistic conception, what A knows depends entirely on what justifications A can muster, if one adopts a justificationist model of knowledge, or on what BFPs A employs, if one adopts a reliabilist model. There is no mention of what another person knows or has as evidence. In fact it is initially hard to see, given the model of ourselves as individual inquirers, how what any one else knows, or has as evidence for, could enter into the very definition of what it is for individual A to know that p. But in response to some difficulties raised by some counterexamples to definitions of 'A knows that p', some epistemologists have suggested that whether or not individual A knows may depend not on A alone but on what others know, or as is said, what A's epistemic community knows. Such an appeal to others would introduce a social aspect into the account of knowledge in a quite novel way. But it does not do this in a way that gives succour to relativism in epistemology, or to a social constructivist epistemology. Here we will not endeavour to enter into the intricate debates about such a social condition that have been in the literature since the 1960's (usually overlooked by sociologists of "knowledge"). Rather it will be used to indicate one line along which some have sought a solution by appeal to sociality.

The following story is told as a test case for some definitions of knowledge. A newspaper reporter is a witness to the assassination of a famous civil right leader. His report is included in a "stop flash" item in one edition of the newspaper. Person A reads the newspaper; and since the paper is generally reliable, then A knows that the civil rights leader has been assassinated. However in order to prevent a riot it is decided secretly by civic and governmental powers to issue a denial and merely claim (until the time is appropriate) that the civil rights leader is recovering in hospital. The conspiracy is successful in that no one else outside the few conspirators is aware of the deception. A bulletin is issued about the survival of the civil rights leader and this is published in the next edition of the same newspaper. Members of the community, B, C, D, etc (or perhaps the whole country) read both editions and now do not know what to believe. However A only reads the first edition. Does A still know? Or should A have taken advantage of further relevant information supplied by the reliable newspaper in its second edition? If A had kept up with the news, then A would have given up the claim to know and be in the same quandary as the other members of A's epistemic community.

There are a number of ways of responding to the claim 'A knows that the civil rights leader has been assassinated' in the light of this story. The one focused upon here is that given by Ernest Sosa who introduces a social requirement on knowledge. He agrees with a doubt about whether A knows, given the story as told so far. But he further embellishes the story, adding about the second edition of the newspaper:

But what if only two or three people get a chance to read the next edition before it is recalled by the newspaper? Should we now say that out of millions who read the first story and mourn the loved leader not one knows of his death? I suppose we would be inclined to say that the fake edition and the few deceived by it make no difference concerning what everybody knows. It seems plausible to conclude that knowledge has a further "social aspect", that it cannot depend on one's missing or blinking what is generally known. (Sosa (1991), p. 27)

An inquirer such as A might miss relevant evidence in two ways. One way is due to lack of attention (blinking); A shirks A's epistemic duty and does not pay attention to all the relevant evidence that A could have, such as keeping up with the news. In contrast B, C and D attend to a wider class of evidence including the evidence that A possesses. A second way is that A's cognitive apparatus fails to work as well as it should, and so A misses relevant important evidence. Either way, A works off a narrower evidential base while B, C and D work off a wider evidential base that includes A's evidence.

Now in many cases A's missing evidence that others, B, C, and D, in A's epistemic community possess can count against any claim that A knows. Where the entire country reads the second edition of the paper and A does not, then the appropriate condition for A would be not that A knows but that A ought to be in the same quandary as every one else as to what to believe. However A, and most of the rest of the nation, cannot always be blamed for missing relevant evidence of the sort possessed only by very few others within A's immediate community. And in the addition that Sosa makes to the story told, this is the case. A, and the rest of the nation, do not cease to know because B, C and D obtain further evidence that others do not possesses when they, and only they, read the second edition. So a social condition on knowing can enter in two ways. The first, makes A somewhat epistemically irresponsible in not gathering evidence that is readily available in A's community; but the second does not make A epistemically irresponsible in not noticing the evidence that few in A's community possess.

Such examples can be readily constructed in the case of evidence for scientific claims, rather than the fictitious case being considered. In order to exclude such a possibility we need to add something extra to the clauses that define 'A knows that p'. The extra would be a clause to the effect that A does not miss, or fail to attend to, relevant evidence that others in A's epistemic community possess and on the basis of which they would, or would not, make a knowledge claim. Here the appeal is to what can be reasonably expected of the inquirers in a given epistemic community in respect of the availability evidence and what could be reasonably expected of an inquirer to have taken on board as evidence in making their knowledge claims. In this limited sense Sosa's definition of knowledge is social.[27]

One important matter left unaddressed by the above social condition upon knowledge is how A's epistemic community is to be individuated. It cannot include everyone; if it did then it would include the evidence the conspirators have about their deception. Another matter is that the repair suggested by the extra clause cannot simply be the bare existential claim that there be some true propositions (such as those about the deception) such that if A were to have these true proposition as evidence then A would not know. Rather the true proposition has to be held as evidence by someone; we would not count as evidence truths that are such that no one has or will possess as evidence. To show this we need to embellish the story further. Suppose that the conspirators successfully replace the dead civil rights leader by a double and no one else can detect the swap.[28] If the conspirators themselves then disappear then there are some true propositions which if A has these as evidence then A would not know; however the true propositions will never become available as evidence. What is important in the story is that many have evidence from the newspaper that there has been no assassination. And it is the evidence that they possess that leads them to withdraw their knowledge claim – as would have A if A had been attentive enough to get the evidence.

Whatever the outcome of further discussion of the appeal to one's epistemic community, here is one way in which the social is claimed to be involved in the very definition of knowledge. Epistemology has not overlooked the social in this respect. But the kind of involvement of the social with the epistemic adumbrated here may not be sort that advocates of SSK would find attractive for, or a useful adjunct to, their account of how the social is implicated in scientific knowledge.

2.4.3   Reliabilist Theories of Knowledge and Social Epistemology

The fourth and final approach to social epistemology to be mentioned here arises naturally out of the reliabilist theory of knowledge. Social features can enter naturally into the reliability which is exhibited by the BFPs that produce A's beliefs in ways that depends on the actions and beliefs of people other than A. It might be a social fact about a community that a given BFP is *taken* to be reliable, and there may well be a relativity of what is *taken* to be reliable from community to community (for example one community may take their Delphic oracle to be reliable, while others may take the tossing of a coin while chanting to be reliable). But what is of interest here is whether in fact a given BFP is reliable, and whether or not sociality plays a role in its being reliable. Again, it is a commonplace that much of science is a social process, including its organisation, funding, reward structure, production and transmission of its findings, and so on. But again what is of concern is whether social matters contribute to the reliability of some BFP, which then produce scientific knowledge (where the very propositional content of the knowledge concerns some scientific observation, hypothesis, theory, etc).

What are examples of social BFPs that may have a high degree of reliability? Criminal law courts (in societies where there is a recognised division of powers) with their judges and adversarial system of defending and prosecuting lawyers do yield an empirically determinable degree of high (but not 100%) reliability as to the truth concerning guilt or innocence of those on trial. And a hierarchy of appeal courts enhances this reliability. We have already met another kind social enhancer of reliability in section 1.1.3. This is Popper's claim that the objectivity of science is bound up with a social aspect of scientific method that arises from the co-operation of many scientists. Principles of scientific method can be applied by individuals; and their correct application can have a degree of reliability that does not depend on social factors but on the correctness of the principles themselves. But what an individual comes to believe on the basis of their sole application of a principle of method may be less reliable than the further application of the principle by others who act as a check upon the principle's correct application. Social aspects of reliability can arise in other ways, such as in the public character and checkability of scientific observations, or the openness of each scientist's theoretical and experimental work to the free criticism of other scientists. Other social aspects of reliability turn on social institutions of science such as the co-operative work of many in a laboratory, the conferences in which scientists present their work, the peer review process of scientific journals, and so on.

Important for Popper is the open expression of criticism by others (and any social conditions which might be a prerequisite for this), in contrast to the exercise of political power, or the role of commercial interests which might impair such free criticism thereby making the institutions of science less reliable social factors in the overall BFPs that produce knowledge in science. Not only do the above social processes provide a check on the correctness of the scientific claims being advanced by both individual scientists and communities of scientists; they also act as a filter for those claims which are erroneous but whose errors have so far escaped detection by individuals. It is an empirical sociological matter to determine how reliable such social processes are, or how variable their reliability might be over time and across societies. It is clear that, though not 100%, their reliability is high. Such empirical determinations provide one role for the sociology of science.

Let us turn to a different kind of BFP. How reliable are the beliefs we acquire from the say-so of experts in some area of science? Not only is the lay public dependent on the say-so of scientific experts; few scientists know much about the sciences outside their own speciality, so they rely on what experts in other fields say. Each of us might find it hard to distinguish, by ourselves, the expert from the layperson or charlatan; so we need to defer to someone or something that can determine expertise. Here much could turn on the social processes that determine, first, that a person is a scientist and second that the person is an expert in a given field; that is, we need to turn to the social institutions that produce and maintain scientific expertise, such as universities, research institutes, scientific societies such as the Royal Society, and so on. Here we can invoke the idea of levels of reliability of the section 2.3.4. We want scientific experts whose opinions are reliable in that their claims about work in their field of expertise have a high true ratio. But we also need to have institutions that are meta-reliable in the production and maintenance of expertise. That is, the institutions confer the label 'expert' on those whose opinions in a given field have a high truth ratio and withhold it from those who do not have a high true ratio; and they do this quite reliably. Some might claim that this will take us only so far since the clash of expert opinion is not uncommon. We may also have to look to the interests that a person might have in getting their say-so adopted, from the political to the commercial and financial. But here we need to distinguish between reliability for the truth with reliability for getting one's say-so accepted. These are quite different; the social epistemology under discussion here is concerned only with the former.

A not unrelated issue is the testimony of other scientists (or the testimony of others in general); for much of science relies on testimony to the truth. Thus astronomers rely on the testimony of other astronomers, from ancient times onwards, about the times and positions of heavenly bodies. In fact in general scientists rely on the observational testimony of other scientists across all the sciences. Testimony as knowledge is one aspect of social epistemology with a long history.[29] But it is not uncontroversial. There are those who would downplay an independent role for testimony as a kind of knowledge. First, they would claim that testimony can be both correct and incorrect, and sincerely or insincerely given, these pairs of distinctions cutting across one another. Second, they would argue that when a person attests to the claim that p, then any other person should check this out either using independent grounds for assessing p, or providing a check on the reliability of the attestor. But as was noted for a reliabilist theory of knowledge, such checks on the reliability of the attesters are not necessary; this is only an important requirement for justificationist theories of knowledge that are internalist. Rather for the reliabilist it is only required that the attestor *be* reliable for the truth for another person B to take A's say-so to be knowledge for B. Looked at this way, testimony becomes one basic category of knowledge on a par with, but independent from, other basic categories such as perceptual experience, memory and inference. Moreover there is a social aspect to the general reliability of testimony.

There is a pervasive need for all of us, as social beings, to share information with one another. The eighteenth century Scottish advocate of testimony, Thomas Reid, spoke of 'The wise and beneficent Author of Nature' who implanted in us two propensities, one to speak the truth and the other to believe what others tell us'.[30] This places an overly strong emphasis on the idea that we have an individualist cognitive propensity for truth telling and for belief in what others say; in the case of the latter belief, it ignores the social factors that might also be involved in inculcating this capacity in the first place and then in maintaining it in operation. Reid, of course, recognised that these propensities could be overridden; but on the whole they operate without being overridden. Importantly Reid's view of testimony can be taken out of its theological and individualist context and relocated within reliabilist epistemology. At the first level we get knowledge from the say-so of others only if they are reliable attestors (that is, the attestor has a high truth ratio). And at the second level there are social processes at work which are meta-reliable for certifying who is, or is not, a reliable attestor; for example their are social sanctions for known liars or deceivers just as there are social processes of approval for truth-tellers. That is, there is meta-reliability that makes most of us reliable sharers of truths most of the time.

We may ask whether social processes for the transmission of belief (such as schools, the family and the church) are reliable? Certainly for any belief they may be highly reliable in transmitting the belief to a new generation of learners. For example, the social institutions for learning managed to equally well transmit through cultures, and over time, the pre-Copernican belief that the Sun orbits the Earth; and they have equally as well transmitted the Copernican belief that the Earth orbits the Sun (once it was sufficiently accepted in the community of astronomers). But this merely concerns the reliability of social processes in transmitting some input belief as an output, regardless of its truth value; it is not reliable in the sense of producing beliefs that are true. However we might say that social processes of transmission have a high *conditional* reliability in that if a given input is true, then the output will be true. This suggests one way in which the account of the reliability of testimony might be modified; the *conditional* reliability of attestors might be greater than their reliability *simpliciter*.

The sections above canvass several ways in the philosophical literature in which the social is plausibly claimed to be involved in the epistemic. There are several others.[31] For one example consider Brandom's' account of knowledge which, in its general tenor, is reliabilist. In his resolution of some of the problems that confront reliabilism, Brandom requires that if any *other* person claims that A knows that p then they attribute to A a commitment to the belief that p and an entitlement to that commitment. Brandom also adds that the *other* also undertakes the same commitment. He adds: 'Doing this is adopting a complex, essentially *socially* articulated stance or position in the game of giving and asking of reasons' (Brandom (2000), p. 119). Such a view is not opposed to reliabilism in that the business of getting entitlement to beliefs turns on the reliability of the belief forming process; but it does appeal to a layer of social involvement in its requirement of giving reasons. Brandom's theory is a complex one which cannot be discussed further here. But the way in which social factors enter into Brandom's theory, like many of the theories already mentioned, is not one that is likely to impress many advocates of SSK. What they advocate is a quite different involvement of the social with the epistemic that would find few advocates in philosophical circles. Thus in his review of Goldman (1999) Kusch makes several objections, one of which is the complaint that 'Goldman never addresses its [SSK's] most central contention about knowledge, that is, that the primary knower is the group rather than the individual' (Kusch (2001), p. 188). In section 2.4.1 the distinction between group and individual belief was drawn. But it was not claimed that group knowledge was primary. What this might mean is discussed next, drawing on Bloor's claims in this respect.

2.4.4 Distinguishing Social Epistemology from the Sociology of "Knowledge"

The above discussion indicates some of the ways in which the reliability of BFPs can be enhanced by social processes arising from the cognitive activities of others in conjunction with each individual knower. And it has been limited to just one kind of knowledge, viz., the definition of 'A knows that p'. It has ignored other constructions the term 'know' might enter into, such as 'A knows how to...'. Some sociological studies of science put emphasis on the skills of scientists in experimenting, observing, arguing, etc, the buzzword here being 'practice'. But it would not be news to most that the acquisition of skills is social and that their exercise may only be possible in the context of others exercising their skills. In this respect the know-how of scientists in research teams may be no different from the know-how of performers in, say, a theatre company or a symphony orchestra. Overall, what the four examples of the involvement of sociality in knowledge show is that there is a viable subject area, social epistemology, that is independent of the empirical subject known as 'the sociology of knowledge'. This latter subject is an empirical (scientific) investigation into what social factors, if any, determine or cause beliefs (whether scientific or not). The different kinds of theory proposed in this area, from the Strong Programme to its weaker more plausible rivals, are discussed in Part II.

But the above account of some of the ways in which sociality enters into the very conditions of the definition of 'A knows that p' (alongside other factors) might be regarded by some as too conservative. Other more radical positions will be examined in Parts II, III and VI, such as the claim that all scientific knowledge is socially caused, or that knowledge *is* power, and so on. The views examined there are strongly imperialistic in that social factors are entirely dominant either in the very definition of knowledge itself, or in any explanation of what A knows or believes. Though there might be other factors involved, the social ones are alleged to have such a dominant and determining role that our epistemological activities are thoroughly social, from top to bottom.

This more radical position might be characterised by 'the inversion thesis': there is no individual knowledge without prior community knowledge. The account of social epistemology presented so far might be judged to be too conservative in that it takes as its basic model the individualist schema 'A knows that p' but revises it in two ways: first, by expanding from individual knowledge to group knowledge, and, second, by adding social factors only at the level of the justifications, or the reliability of BFPs, which form beliefs in individuals, or in groups. A more radical approach would be the following inverted picture. What is primary is not individual knowledge but community-wide knowledge out of which is distilled the knowledge possessed by each member of the community. Such an inverted conception of knowledge appears to be endorsed by the sociologist of scientific knowledge, David Bloor who says:

> Instead of defining it [knowledge] as true belief – or perhaps, justified true belief – knowledge for the sociologist is whatever people take to be knowledge. It consists of those beliefs which people confidently hold to and live by. In particular the sociologist will be concerned with beliefs which are taken for granted or institutionalised or invested with authority by groups of people. Of course knowledge must be distinguished from mere belief. This can be done by reserving the word 'knowledge' for what is collectively endorsed, leaving the individual and idiosyncratic to count as mere belief (Bloor (1991), p. 5)

This is not entirely clear since it treats belief and knowledge as if they are on a par. What is clear is that group knowledge (belief) is fundamental while individual knowledge (belief) is at best derivative. The fundamental idea here is that of a group G which "knows" that p. But unfortunately this is said to be no different from what G collectively endorses or takes to be knowledge or, as we may say, believes. To collapse the two notions is to ignore the distinction (deliberately made at the end of section 2.4.1) between group belief and group knowledge.

By-passing this epistemological solecism, what now of the individualistic A "knows" that p? It would seem that A "knows" that p only if A is in conformity with A's group G in endorsing p. This formulation denies that there is any individual knowledge distinct from group "knowledge". At best there is individual belief, which is such that where A deviates from A's group in not endorsing the group's belief that p but some other claim q, then A cannot "know" in the sense introduced above, but merely believes, that q. Incredible as it seems, those who dissent from the community can never be knowers. Thus those who introduce novel ideas in science, and know that they have done so (in some sense of know as set out in sections 2.2 or 2.3) cannot, for sociologists of science, be said to "know" in their use of the term (which involves the inversion thesis). Further, nothing is said about Plato's tether that would distinguish even collective knowledge from collective belief, let alone individual knowledge from individual belief. This example of the 'inversion thesis' is central to the sociology of "knowledge". The involvement of this thesis in SSK will get further discussion in chapter 5.

NOTES

[1]   Though there is a suggestion of a theory of knowledge as justified true belief in the *Meno*, in the much later *Theaetetus* Plato refines what he means by this and raises difficulties for any justification understood as an account (*logos*) of the true belief where the account can be interpreted as empirical foundations for knowledge (*Theaetetus* 201C – 210D). The justified-true-belief theory of knowledge is no longer accepted largely because of Gettier's counterexamples to it, first published in 1963. The literature that has flowed from them is reviewed in Shope (1983), particularly in chapter 1. This is not considered here.

[2]   In his 1929 paper 'General Propositions and Causality' (Ramsey (1990), p. 146) Ramsey says; A belief ... is a map of neighbouring space by which we steer'. Ramsey's view is usefully developed in Armstrong (1973), especially chapter 1. I follow both here in the general understanding of what a belief is.

[3]   On some occasions a fortuitous concatenation of false beliefs can lead to survival whereas true beliefs might not. Thus a series of false beliefs might lead one to miss an aeroplane flight, which crashes, whereas the true beliefs would not have been conducive to survival. On the whole, such concatenations of false beliefs are less reliable guides to survival than true beliefs.

[4]   Following the standard usage of logicians, letters such as 'p', 'q', etc, will be used to stand for propositions, statements or sentences (no distinction need be drawn between these here). For the various places where Moore discusses his 'paradox' and responses by commentators, see Hintikka (1962), chapter 4.5, especially p. 64.

[5]   Some see in the passages cited above from the *Meno* an embryonic account of a reliabilist theory of knowledge, e.g., Armstrong (1973), p. 159. A case can be made for theories of knowledge from Aristotle to Descartes being externalist and having elements of reliabilism about them. And even Descartes can be viewed as a reliabilist, albeit a supernatural reliabilist, since God is not a deceiver. In the history of epistemology only in the last few centuries have internalist accounts of knowledge been in the ascendancy.

[6]   Standard commentaries on *Meno* 98A such as Bluck (1961), pp. 412-3, suggest a range of possible ways in which *aitias logismos* can be understood. See also Nehamas, 'Meno's Paradox and Socrates as Teacher' in Day (ed.) (1994), especially Part V, pp. 240-5.

[7]   For an account of Gettier's objection to knowledge as justified true belief, see Gettier (1963) and the many theories of knowledge that have been developed in response to it discussed in Shope (1983). For both Gettier's and Agrippa's objections to epistemology see Fogelin (1994), Part I and Part II, and M. Williams (2001), chapter 5.

[8]   Agrippa probably flourished during the first century AD; none of his writings have survived. Our main source is through Sextus Empiricus who wrote about Agrippa's doctrines during the following century.

[9]   For a brief discussion of Agrippa's problems for knowledge and the way it has resurfaced in modern epistemology see Fogelin (1994), chapter 6. See also Armstrong (1973) chapter 11

for another display of how Agrippa's problem can resurface unknowingly in discussions of the infinite regress of reasons that arises for the justification or evidence condition for knowledge. Aspects of Agrippa's problem also lie behind Popper's discussion of what he calls 'Fries' Trilemma' in chapter V of Popper (1959). See also M. Williams (2001), chapter 5.

[10]   That a true belief that p is not knowledge that p is the import of Plato's *Meno*, as we have seen, and *Theaetetus* 187A-201C. Most epistemology textbooks reiterate these points.

[11]   Some of the differences are noted in M. Williams (2001), particularly chapters 13 and 16.

[12]   Of the many critiques of such foundationalism in incorrigible belief see Armstrong (1968), chapter 6 section X), and in a different vein Sellars (1963) chapters 1 to 4 and his criticism of what he calls 'the myth of the given'.

[13]   A succinct presentation of this programme in epistemology is given in 'Epistemology Naturalized' in Quine (1969), pp. 71-9.

[14]   My views in this and subsection have been influenced by David Papineau 'Does the Sociology of Science Discredit Science?' in Nola (ed.) (1988).

[15]   Descartes' ethics of belief is set out in the fourth of his *Meditations*. For an excellent account of the requirements of an ethic of belief in Descartes and Locke, and Hume's opposed view, see Wolterstorff (1996).

[16]   For a range of theories which make a response to the regress problem for knowledge and which rival the foundationalist model set out in the text, see the diagram in Armstrong (1973), p. 160.

[17]   In a 1967 paper entitled 'A Causal Theory of Knowing' (reprinted in Goldman (1992), chapter 4) we find a version of the causal theory set out as follows: 'A knows that p if and only if the fact that p is causally connected in an "appropriate" way with A's believing p' (*ibid.*, p. 80). In the paper more is said about causal chains and when they are appropriate.

[18]   See the 1979 paper 'What is Justified Belief?' reprinted in Goldman (1992) as chapter 6. The notion Goldman defines is justified belief rather than knowledge, though the kind of justification so defined is a necessary condition for knowledge. The attempt to define 'justifies' in non-normative terms is set out in *ibid.*, section I, pp. 107-12. Much of this section is indebted to Goldman's account of reliabilism set out in his 1979 paper, and further developed in Goldman (1986), Part I, especially chapters 1 and 3. The paper 'Epistemic Folkways and Scientific Knowledge' (Goldman (1992), chapter 9) develops reliabilism in the direction of a theory about beliefs obtained through the exercise of intellectual virtues.

[19]   The power of our BFPs is discussed in Goldman (1986) section 1.4 and chapter 6.

[20]   See Lewis (1986a) section 2.5, and the discussion of induction in deceptive and undeceptive worlds. Note that reliabilism can be employed to give a justification for induction; see, for example, the arguments in Papineau (1993), chapter 5 especially sections 5.9 to 5.15.

[21]   One classic work which shows how unreliable we are as reasoners, especially in the case of probabilistic inference, see Kahneman, Slovic and Tversky (eds.) (1982). For a recent survey of literature in the area see Samuels, Stich and Tremoulet (1999).

[22]   The example is taken from Leaky (1981), p. 39 where he describes the extraordinary skill of those in his various expeditions, such as Kamoya Kimeu, who could visually sight fossils (as distinct from other rocks) from a considerable distance and then say from what part of what animal they came.

[23]   See 'Insights and Blindspots of Reliabilism' in Brandom (2000), chapter 3, especially part II. This section (and others) is indebted to Brandom's discussion of reliabilism, some of the difficulties it faces and their resolution.

[24]   The response given above concerning the nervous reliabilist is, hopefully, consistent with that found in Brandom (2000), chapter 3, part II, and Papineau (1993), chapter 5.4. Brandom does endorse the kind of response given in the text only for cases of local nervousness of some believers but not for global nervousness for all believers. For his solution to the problems

raised by this, which ultimately appeal to a modified reliabilism, see Brandom (2000), chapter 3 parts III to VI.

25   For philosophical works which attempt to clear up the non-objectivism and relativism that arises out of so much talk about social constructivism, see at least Devitt (1991) chapter 13, Hacking(1999), Kukla (2000) and Searle (1995).

26   The following, along with related points, has been influenced by the papers, collected in Schmitt (ed.) (1994), by Gilbert 'Remarks on Collective Belief' (pp. 235-256) and Schmitt 'The Justification of Group Belief' (pp. 257-87). Though their examples are developed for the case of belief, they apply equally well to knowledge. A much more thoroughgoing analysis of these issues can be found in Gilbert (1989), chapter 5.

27   The debate about a social condition on knowing is an old one within epistemology that predates more recent attempts to found a social epistemology. Sosa's social account of knowledge appeared first in a 1964 paper and is further developed in a 1974 paper (from which the quotation above is taken). These papers are reprinted as chapters 1 and 2 of Sosa (1991) from which citations are made. The assassination example cited by Sosa is given in Harman (1968), section VIII, and is used in support of the view that definitions of knowledge do need some social condition. A not unrelated role for social aspects of knowing can be found in Shope (1983), especially in chapter 7 section 5 'Rationality, Science and Social Aspects of Knowledge' pp. 222-235. Shope takes on board Sosa's points about knowledge being relative to epistemic communities to solve, in a different way, some of the objections made to common definitions of knowledge based upon some well known counterexample, such as the one cited.

28   Communist leaders, it is often alleged, were replaced by doubles on many occasions. There is also a film by Kurosawa, *Kagemusha*, which is based on an historical incident in which a warlord who has been killed is replaced by a double so that the dead leader's regime may continue in its confrontation against its enemies.

29   For a brief historical sketch of the claims and counterclaims concerning social epistemology, especially in respect of testimony, from Plato, Aristotle, Aquinas, Reid, Hume and others, see the introductory essay to Schmitt (ed.) (1994), pp. 1-27 and the bibliography under 'Historical Sources' pp. 289-91. See also Coady (1992).

30   This view of Reid is mentioned in Sosa (1991), p. 220, in a chapter 12 entitled 'Testimony and Coherence'.

31   The collected papers in Schmitt (ed.) (1994) with its extensive bibliography is one source of recent philosophical work on the social in epistemology. Another recent work is Goldman (1999) on testimony, the sociality of science, law and education, and so on. However for radical advocates of SSK these works would be regarded as adopting a rather conservative approach to the whole matter.

CHAPTER 3

NATURALISM AND NORMS OF REASON AND METHOD

The previous chapter dealt with the problem of how we are to distinguish knowledge from mere belief. In all epistemologies knowledge, along with justification, rationality, etc, are normative, evaluative notions that carry with them the idea that our critical canons have been applied to our beliefs. As we will see, a number of the theorists to be discussed in Parts II to IV deny that there is anything to the norms of knowledge and rationality other than what is found in local practices embedded in local cultural and political contexts. And the norms vary dependently on their varying context. Even though they understand such local contexts naturalistically, part of their blindness about the norms of epistemology and methodology is their belief that if there are such norms then they must transcend the natural – from which they conclude 'so much the worse for norms'. Even though there are "naturalistic" norms embodied in local practices, there is no role for the more robust norms of epistemology and methodology. So, before turning to these theorists in Part II to IV, it is necessary to investigate the ways in which the normative principles of method get their force and authority.

Section 3.1 introduces Quine's influential account of naturalised knowledge and norms. But it is unclear what status Quine assigns to the norms of method within naturalism. He initially claims that all such norms can be completely naturalised as a chapter of psychology (and perhaps other sciences), but later claims that there might be a more independent account of norms as a chapter of the technology of engineering. The 'chapter' metaphor Quine uses helps obscure the ways in which norms can be understood within naturalism. One of the central tasks of epistemology is critical; it provides norms whereby we are to assess our beliefs. It would be a defect of any theory of naturalised epistemology if it were to abandon the critical role that epistemology traditionally plays. Quine comes close to obscuring the critical role that our norms play in adjudicating our beliefs; but then he rescues his position by giving norms an independent role. As we will see the sociologists of science, Foucault and Nietzsche discussed in Parts II to IV have no residual Quinean qualms about jettisoning the norms of epistemology.

Section 3.2 gives a brief account of what is meant by naturalism while section 3.3 discusses a number of examples of norms of scientific method to be borne in mind in later more abstract discussions. Section 3.4 sets out four logically possible positions concerning the natural and the normative. Section

3.5 focuses on one of these positions – the view that norms are not to be understood objectively, thereby giving rise to positions such as scepticism or expressivism about norms. The final three sections show in what way both naturalism and objectivism about norms can be retained. In section 3.6 an account of "folk scientific rationality" is given which captures the particular practices of scientists through the particular normative judgements they have made about particular episodes in the history of the developing sciences. Using this as a basis, it is shown in section 3.7 how to "define" normative methodological notions in terms of the non-normative using the device of the Ramsey sentence. In section 3.8 it is argued that the normative in epistemology and methodology supervenes on the non-normative or descriptive. Both the "definition" of norms and their supervenience show how to locate the normative within naturalism without compromising either, in particular the objectivity of norms.

Reconciling the normative with naturalism is an important issue for this book. The dethroners of our critical tradition reject the idea that norms of reason can have any independent force and authority. For them there are just two alternatives: either such norms are transcendent or they are immanent in that they are simply more human belief that is subject to the vagaries of society, culture and history. For some the rejection of transcendence in favour of immanence is a defining feature of postmodernism.[1] With the completion of this chapter the central stance to be adopted towards the theories discussed in Parts II to IV will have been set out.

3.1   QUINE'S NATURALIZED EPISTEMOLOGY

Central to epistemology is the view that it is a normative, evaluative and critical discipline which tells us what beliefs may be rationally or irrationally held, what beliefs are justified or warranted, what are the proper and improper methods of inquiry, and so on. The central evaluative terms are 'knowledge', 'rational', justified', 'proper', etc. Some of these terms enter into many a definition of 'knowledge' (though as already noted they might be excluded from some naturalistic definitions of 'knowledge'). Many philosophers, from Descartes onwards (if not before) have presented epistemology as a "first philosophy" underpinning all other branches of philosophy. However some have declared that there is no such thing as a first philosophy; there are just the various sciences of which epistemology is merely a small part. Such are the claims of Quine who used the phrase 'naturalized epistemology' to refer to a theory of knowledge which does not aspire to be a "first philosophy", is not to be founded in some *a priori* fashion, and is not separate from, or above and beyond, the science but is "continuous" with science.

3.1.1 Epistemology as a Chapter of Psychology

The deflation of epistemology's pretensions is expressed in an often-cited passage from Quine's 'Epistemology Naturalized':

> Epistemology ... simply falls into place as a chapter of psychology and hence of natural science. It studies a natural phenomenon, viz., a physical human subject. This human subject is accorded a certain experimentally controlled input ... and in the fullness of time the subject delivers as output a description of the three-dimensional external world and its history. The relation between meagre input and the torrential output is a relation that we are prompted to study for somewhat the same reasons that always prompted epistemology; namely, in order to see how evidence relates to theory, and in what ways one's theory of nature transcends any available evidence. (Quine (1969), pp. 82-3)

Consider epistemology as narrowly concerned with the evidence relation only. Initially we might have naively assumed that the relation was, in part, a normative or evaluative relation of justification. But what has happened to it in Quine's remark? Talk of epistemology being a *chapter* of psychology is too metaphorical and lacks precision. But it is clear that any normative evidential relation between observation and theory has been replaced – by what? It has not been replaced by non-normative notions of logic such as implies, or degree of evidential support, etc, since these are not obviously notions investigated in any chapter of psychology.

What Quine offers is a descriptive account (perhaps referring to causal or law-like claims within several sciences from theories of perception to linguistics) of the chain of links between sensory input and linguistic outputs. Do norms play a role here, but lead a life in heavy disguise through definition in terms of the predicates of the psychological sciences? Or are they identical to, or supervenient upon, the properties denoted by such predicates? Or are the norms deemed to be non-existent because they have been eliminated from our talk, like our earlier talk of phlogiston or Zeus' thunderbolts? It is unclear, but elimination seems to be the most likely account of the above passage. Whatever the case, we lack an account (if there is one) of the special features of those causal links between sensory input and linguistic output that characterise the normativity of justified, or warranted, or rational belief in theory, and the special characteristics which produce unjustified, unwarranted or irrational belief. So far no difference has been indicated. Nor should we rely on a 'natural rationality' based on some innate propensity to reason; empirical investigation has shown how bad we are as natural reasoners.

Suppose some epistemologists take Quine's first sentence to heart and abandon their Philosophy Department for their university's Psychology Department. Despite the fact that there is much work being done on perception, cognitive functioning and language, they might well find that no one in psychology is doing any research on their particular epistemological notion, viz., the evidence relation. More likely they would find that while psychologists *use* the evidence relation in their statistically based research, they do not *mention* it, or even discuss it. Our epistemologists might wander even further from their home Department – to Sociology. Despite some problematic encounters with sociologists of knowledge, things turn out not as badly as they might have initially thought. There are some sociologists who have read Quine and they agree with him that epistemology should be replaced by psychology. And they show that in subsequent writings Quine has added biology and the theory of evolution to psychology as the sciences that collectively replace epistemology. But they also argue that Quine's project is still too narrowly conceived and that sociology should be added to the list of sciences. And their colleagues down the corridor, anthropologists, ethnomethodologists, and the like, want to add their subjects too.

Our wandering epistemologists encounter a special sub-group of sociologists of science, the sociologists of *scientific knowledge*, who tell him that they anticipated Quine's rejection of epistemology as a first philosophy. Long ago they abandoned epistemology totally – and without the help of Quine! The sociologists' interests are in the socio-political causes of belief and not any evidential relationship that might justify such belief – talk of the evidence relation and justification being just more in the way of belief that is grist to the sociologists' mill. Thus not only sociology, anthropology and psychology replace any normative theory of epistemology, but also political studies and economics as well. They are followers of Shapin and Schaffer ((1985), p. 332) who say: 'Solutions to the problem of knowledge are solutions to the problem of social order'. Our wandering epistemologists now have an intimation that there is some vast self-reflexive paradox involved in all this navel-gazing of the various sciences upon themselves, and they begin to think that perhaps the idea of epistemology as a 'first philosophy' somewhere outside the fray of the sciences was not such a bad idea after all. By being eliminativist with respect to epistemology and not replacing it by anything see what you get – sociologists of scientific knowledge and the postmodernists!

Our epistemologists protest, hoping to eliminate the involvement of sociology in epistemology, by claiming that Quine did not intend his project in naturalized epistemology so broadly; also Quine's physicalistic naturalism excludes an autonomous role for sociology. But the sociologists respond saying that though Quine's talk about the stimulation of our nerve endings is basic to his replacement of epistemology by neuro-psychology, it is not as basic as his notion of an observation sentence – and this is essentially social. Quine has proposed several definitions of an observation sentence all of which involve reference to a community; an observation sentence is what a community would assent to given the same stimulus in the presence of community-wide collateral information (Quine (1992), chapter 1 section 2).

Then the sociologists point out to the epistemologists a quotation they had not noticed before from Quine's *The Pursuit of Truth*:

... I have been treating rather of the testing of a theory after it has been thought up, this being where truth conditions and empirical content lie; so I have passed over the thinking up, which is where the normative considerations come in. ... But when I cite prediction as the checkpoints of science, I do not see that as normative. I see it as defining a particular language game, in Wittgenstein's phrase: the game of science, in contrast to other good language games such as fiction and poetry. (Quine (1992), p. 20)

The epistemologists are dismayed, not only because normativity is dismissed from the context of justification and surprisingly relegated to the context of discovery ('thinking up'), but also because there is an appeal to Wittgensteinian language games as characterising the rules which are said to govern the game of science. The epistemologists also know that sociologists of science find social factors everywhere in science because science is alleged to be governed by social rules embedded in Wittgensteinian forms of life. And they are all too aware of Putnam's erstwhile dictum that the Wittgensteinian 'fondness for the expression "form of life" appears to be directly proportional to its degree of preposterousness in a given context' (Putnam (1975), p. 149).

But in looking further through Quine's 1990 *Pursuit of Truth* our epistemologists, to their surprise, discovers the following on the first page:

Within this baffling tangle of relations between our sensory stimulation and our scientific theory of the world, there is a segment that we can gratefully separate out and clarify without pursuing neurology, psychology, psycholinguistics, genetics, or history. It is the part where theory is tested by prediction. It is the relationship of evidential support .... (Quine (1992), p. 1)

The baffling tangle extends even to Quine interpretation! But at least it looks as if the evidential relationship is not to be naturalized after all. Then it dawns on our epistemologists that there is a paper to be written here – and they rush back to the Philosophy Dept to write it up. Just as Quine once wrote a paper called 'Three Grades of Modal Involvement' so there ought to be a paper called 'Grades of Naturalistic Involvement' – or should it be 'Grades of Normative Involvement' with the order of the grades reversed? At one end there would be 100% naturalistic involvement with zero grade normative involvement; one example of this would be mad-dog naturalism of the sort one finds in social studies of science. At the other end would be zero naturalistic involvement and 100% normativity plus lot of other things naturalists resist. The latter looks like the despised "first philosophy" which is normative, autonomous from science and *a priori* – just the sort of thing Descartes and others advocated in one way or another. In between would be various admixtures, such as Quine's position that allows for normativity and naturalism but no *a priori* independent "first philosophy".

3.1.2 Epistemology as a Branch of the Technology of Engineering

Such an admixture seems to be the right way to proceed through the thicket of Quinean hermeneutics, a position supported by Quine's reply, in the Quine-Schilpp volume, to a critic of his views on epistemic values:

> Naturalization of epistemology does not jettison the normative and settle for the indiscriminate description of ongoing procedures. For me normative epistemology is a branch of engineering. It is the technology of truth seeking, or, in a more cautiously epistemic term, prediction. Like any technology, it makes free use of whatever scientific findings may suit its purpose. It draws on mathematics in computing standard deviation and probable error and in scouting the gambler's fallacy. It draws upon experimental psychology in exposing perceptual illusions, and upon cognitive psychology in scouting wishful thinking. It draws upon neurology and physics, in a general way, in discounting testimony from occult or parapsychological sources. There is no question here of ultimate value, as in morality; it is a matter of efficacy for an ulterior end, truth or prediction. The normative here, as elsewhere in engineering, becomes descriptive when the terminal parameter has been expressed. (Hahn and Schilpp (1986), pp. 664-5)

Can Quine have been reading Foucault when he talks of 'technologies of truth?' Perhaps not. But what is surprising is that the metaphor of normative epistemology being a *chapter of psychology* has been replaced by talk of it being a *branch of engineering technology*. Quine is quite right to point out the extent to which the claims of science are tested using mathematics and statistics, an important part of our "technology" of truth seeking. But is there anything left for philosophers to contribute? Normative epistemology emerges in its own right as a theory about means-ends where the ends are particular epistemic goals, such as truth, or prediction (or more strictly true prediction). But note that Quine's remark hardly endorses pragmatism, if pragmatism is to characterised by the advocacy of a plurality of values (in which truth might or might not be included). There are a host of other epistemic ends that Quine does not mention such as fruitfulness, or degree of openness to test, and so on. However the general form of Quine's norms will be that of the hypothetical imperatives of the sort: if you want some value v (e.g., truth) then you ought to follow rule r'.

Nor is Quine clear in the final sentence about how our means-ends norms become merely descriptive once the 'terminal parameter' (the goal, or value?) has been specified. Thus consider the valueless categorical 'follow rule r' (e.g., Newton's rule 'postulate as few independent causes as possible'). Filling in the terminal parameter with a value still gives a non-descriptive claim such as the hypothetical imperative just mentioned 'if you want v then follow r'.[2] Perhaps Quine is merely making the sociological observation that many do adopt truth as the goal of their epistemic activities; or they adopt the sociological means-ends claim corresponding to the mentioned hypothetical imperative; 'following r always (mostly, or n% of the time (for n > 50%)) leads to the realisation of v'. But then the matter of normativity within naturalism reasserts itself; such sociological claims have no normative force but their associated hypothetical imperatives do. And it is an account of these imperatival norms that we seek.

However what Quine's talk of a 'terminal parameter' and the absence of an 'ultimate value' does bring out is a view of norms that has been noted in a number of philosophers already. This is the view that we can pretty much as we wish choose any terminal parameter as a value, goal or end. It all depends on what we want or desire of our scientific theories. These can be various, with no particular value the dominant one (with truth playing a big role for some, or only a minor or no role for many pragmatists). What matters is the efficacy of the means used to reach the desired goal. It is here that an *ought* tell us what to do. It recommends means which are the best, or the most efficient, or the most reliable, or those which expend the least energy, in arriving at the goal. That is, the norms of method are strictly instrumentalities which tell us, for any chosen goal or value, v, what means, m, we ought to use to realise v, and in what way and under what conditions, C (e.g., of least expenditure of energy). But what is clear is that, at some level, truth does appear, even if just to tell us that the associated assertoric sentence is true, viz., m does in fact get us to goal v under conditions C.

Quine's fullest account of the norms of methodology comes in an earlier book, *The Web of Belief*, where we find recommended the following five virtues for hypotheses: conservatism (in coming into conflict with the smallest number of other accepted hypotheses or beliefs); modesty (in keeping down the number of independent assumptions); simplicity; generality; refutability (Quine and Ullian (1978), chapter VI). These virtues, some of which are reminiscent of Kuhn's own list of virtues, are illustrated (albeit sketchily) as norms that apply to episodes in the history of science. In the later *Pursuit of Truth* Quine pairs down these virtues to two: (a) the maximisation of simplicity and (b) the minimisation of mutilation, i.e., of what is to be jettisoned in the light of contradictions or refutations (Quine (1992), pp. 14-5).

Intriguingly in a section entitled 'Norms and Aims' (*ibid.*, pp. 19-21) Quine emphasises the role of norms within naturalized epistemology when he speaks of 'normative naturalised epistemology'. But now he draws a distinction between theoretical epistemology, which is to be naturalised within psychology, and normative epistemology which is no longer to be naturalized into a chapter of any science, contrary to the view he advocated earlier: 'Insofar as theoretical epistemology gets naturalized into a chapter of theoretical science, so normative epistemology gets naturalized into a chapter of engineering: the technology of anticipating sensory stimulation' (*ibid.*, p. 19). This is an important shift in what is to be naturalized to what; but it says nothing about the status or the authority of norms, even when "naturalized" to the technology of engineering rather than psychology. In this context Quine relocates the role of the technology of prediction within his methodology. The function of prediction as a goal, aim or purpose is played down in favour of a new role as the "checkpoint" of hypotheses, depending on whether the prediction is true or false. But here Quine has paired-down his conception of methodology to the point where it does not allow that there can also be other norms which provide a "checkpoint" role for not just predictions but also hypotheses (such as explanatory power, or making a novel prediction). Importantly, much more needs to be said about the norms that accompany the hypothetico-deductive methodology that Quine assumes here, and about the link between the confirmation or refutation of hypotheses and the truth or falsity of test predictions.

Commentators have spilled much ink on what Quine means by 'naturalized epistemology' and the role that the norms of reason play within it, one of the best being Foley (1994). The upshot seems to be that there are norms that are not to be reduced to, or eliminated in favour of, sciences such as physics or psychology. Rather they are to be understood in terms of the unclear metaphor of *engineering* and *technology* that suggests that the norms of methodology are purely instrumental. However the metaphors are to be understood, Quine is not to be thought of as a nihilist about method, a position he recognised in 'Epistemology Naturalised': 'The dislodging of epistemology from its old status of first philosophy loosed a wave, we saw, of epistemological nihilism. This mood is reflected somewhat in the tendency of Polányi, Kuhn and the late Russell Hanson to belittle the role of evidence and to accentuate cultural relativism' (Quine (1969), p. 87). Quine is clearly alive to the epistemic and normative nihilism that can come to accompany naturalism, as the discussion of the theorists in Parts II to IV will show. However Quine must have in mind the Kuhn of the 1962 *The Structure of Scientific Revolutions* that, if taken as the sole account the Kuhn gave of methodology, would support Quine's remark. But in section 1.4 it was argued that this is a limited view of Kuhn's changing views on scientific method. The same could be said of any narrow view that focuses on only Quine's earlier remarks on method in which norms are to be naturalized into a chapter of psychology. Both Quine and Kuhn changed their views on the status of the norms of method, neither reaching an entirely satisfactory account. But neither can be said, in the long run, to be nihilists about method.

3.2   VARIETIES OF NATURALISM

What is naturalism? Setting aside the terminological disagreements about the meaning of 'naturalism', I will simply declare it to be the doctrine that the only items that exist are those in the spatio-temporal system. Amongst the items can be included familiar everyday objects and their properties. And the items must include those objects, properties, relations, states of affairs or facts, events and processes that are postulated in the empirical sciences. Some of these items are causally linked with others within the spatio-temporal system (the causal relations being either "sure fire" or probabilistic). Such a broadly construed scientific naturalism is characterised more by what it excludes rather than includes. It eschews supernatural items, mind/body dualism and abstracta such as numbers, Platonic universals, possibilia and/or possible worlds, fictional characters, independently existing ethical values and norms, and so on. (However there is an issue, passed over here, about what the sciences might *require*, such as mathematics and its ontology, as opposed to what the sciences *postulate* themselves.)

What is to be done with such items? 'Locate or eliminate them'[3] says the programme of discovering a comprehensive metaphysical picture of the world based in naturalism. By locating them is meant finding a place for them within naturalism. By eliminating them is meant showing that in the final metaphysical accounting of the world, such items have no place within naturalism; they are non-existent in much the same way as Zeus or Santa Claus are non-existent.

The above characterises what might be called *metaphysical (scientific) naturalism*. In contrast *methodological naturalism* is the view that our methods of inquiry are just those of the empirical sciences; sometimes the further claim is added that there are no substantive non-empirical or *a priori* principles of method. Three problems face methodological naturalism. The first is to show that no methodological principles other than those of the sciences are needed, and to show that the methods of the sciences are themselves empirical and not non-empirical (the empirical programmes of legitimation mentioned in section 1.2.2, such as normative naturalism, attempt to establish this). The second is to show that the principles of logic and mathematics themselves are empirical; or if they are understood as non-empirical, this exception presents no special problems for methodological (or metaphysical) naturalism within the empirical sciences. Finally, principles of method are often expressed as imperatives and are not descriptive, their non-descriptive character making them neither empirical nor *a priori*, nor obviously amenable to any location within naturalism. Much of the rest of this chapter will be concerned with the last problem.

Reverting to metaphysical naturalism, the locationist strategy for non-natural items comes in a number of forms. Thus those who reject dualism of the mental and the material in favour of the monistic doctrine of naturalism will try to locate, in some fashion, a place for the mental in the natural world, while they are quite happy to simply eliminate other kinds of entities such as those of mythical religion (Zeus) or of popular make-believe (Santa Claus). In attempting to locate the mental within naturalism at least three location relations have been suggested by naturalists. The first is that the mental is said to be *reduced* to the natural, in particular reduced to our brains (including our nervous systems). The second is that each mental item is said to be *identical* to some brain item (employing either type-type identity or token-token identity). Finally, and now more commonly, the mental is said to be *supervenient* upon our physical brains (in one or other of the several senses of 'supervenient' that are available). Which of these relations hold (or some other), are not matters that will be dealt with here. But if any of these can be shown to hold, then the mental will have been located within the natural and not eliminated.

The metaphysical programme of naturalism can look outward to see what allegedly non-natural items can be either located within its framework or eliminated altogether. Or it could look inward to see whether within naturalism, as so far described, there are further location or elimination tasks to be performed. Thus in looking inward, does naturalism take the spatio-temporal system itself as an independently existing item, or is it to be located within a more narrowly conceived naturalism in which the alleged space-time container is just the spatial relations between objects and the temporal relation between events? Again is the flux of time, as described by McTaggart's A-series of past, present and future, an independent feature of the world, or is it to be located within the temporal order of the B-series of earlier and later temporal relations?

Further, what of the doctrine of physicalism (or earlier doctrines of materialism)? This is an even more narrow version of naturalism in which the only items to be countenanced are those of a sufficiently mature physics (whatever this might be!). All the other items of all the other sciences are either reducible to, or identical to, or supervenient upon, the items of physics; or they are to be eliminated. If physics is too remote a science for location then, for any science, the science next down in the hierarchy of sciences might serve as the locus of location. Thus John Searle (Searle (1995), especially chapter 1)) describes a programme for the location of all social facts within an ontology of people, their collective intentions and the functions they impose on physical objects (from the functions we impose on certain materials so that they are, say, screwdrivers, to the very words we utter when we agitate the air in audibly detectable ways, such uttering 'I do' in the right circumstances that make us married). But this is not relocation within physics but relocation within a suitably broad psychology that allows for collective intentionality. Of course this broad psychology may itself be relocated within biochemistry and our evolutionary history, and these in turn may be further relocated. In this way social facts find a location within naturalism, and not as independent items in their own right.

Even physicalism might overdo its ontology and some relocation or elimination is in order. Thus (average) density is nothing but mass per volume, or solidity is nothing but the strength of the lattice relations between the atoms and molecules of a substance, and so on. Density and solidity are not independent items but are supervenient on other items within physicalism. But the celestial spheres of pre-Copernican astronomy, or the electromagnetic ether of the 19th century wave theory of light, were once thought not to be merely locatable within physicalism but an important part of what was to be counted as the physical; they now find no place within physicalism and have been eliminated. There are interesting stories to be told here that are central to current debates in philosophy of mind and science, but this is not the place to explore locationist versus eliminativist issues, or issues to do with incommensurability, within even the historical evolution of the very science that constitutes physicalism.

There are those who identify naturalism with an austere physicalism. In contrast there are those who advocate a thesis about the disunity of science in which no reductive, identity or supervenience relations are alleged to hold between the various entities postulated in the various sciences at each level within the hierarchy of sciences. Some advocates of the disunity of science, or of a "metaphysical pluralism" about the various "levels" of the sciences including the laws and ontology of each science, will want to deny any supervenience of the non-physical on the physical; but in doing so they might also want to distinguish their view from any strong account of emergence at higher levels.[4] Suppose we were to accept that there is a range of autonomous, or irreducible or non-supervenient sciences such as psychology, sociology, economics, etc. Then naturalism could be defined as the concatenation of all such sciences. Such a broad, but loose, linking of the sciences provides a picture of a more expansive naturalism which contrasts with the more austere naturalism depicted by physicalism. Such a broad naturalism would be resisted by advocates of austere naturalism, in part because it would postulate entities that would have no realisation in a physical basis.

We will say nothing further about what a mature naturalism might be like, whether it allows for the broad conjunction of all the *sui generis* sciences that contain items that go well beyond physicalism, or it admits only the ontology of physicalism. As has been noted, there are two possible clean-up tasks of elimination or location to be performed with respect to naturalism. The first we have canvassed briefly; it looks inward and attempts to pare naturalism down so that it approaches the austere ontology of physicalism. For the purposes of this chapter there is no need to pursue this issue. We may take (scientific) naturalism to be either what a final physics postulates, or what is postulated in the concatenation of all the extant independent sciences from physics to economics. The second clean-up task looks outward and focuses on those entities alleged to exist outside naturalism. There might be further clean-up task for naturalists in paring down these entities until one approaches an ontology close to that of naturalism. Broadly, naturalism is to be contrasted with a supernaturalism that admits Gods, minds and other entities of rival idealist philosophies; and it is to be contrasted with those ontologies that admit abstracta such as independently existing universals, or mathematical items such as sets or numbers. In fact an outward looking naturalism is confronted with a host of items that it needs to "clean-up".

Only one "clean-up" task will be addressed here; this concerns the status of norms within naturalism, in particular the norms of reason, logic, epistemology and scientific methodology (but sometimes the norms of ethics where an illustration from ethics might be helpful). Are these norms to be located within naturalism as a Quinean chapter of psychology, *via* some relation of reduction, identity, supervenience or whatever? Or are they to be eliminated? Or do they lead a *sui generis* existence thereby undermining the pretentions of naturalism to be a comprehensive metaphysical picture of the world? Locating these questions on a grander canvass the question asks: what place, if any, do norms and values (of reasoning and/or ethics) have in a scientific naturalist picture of the world? The position adopted here will be that the normative in methodology and epistemology is supervenient on the non-normative, or descriptive, and that, in a sense of 'definition' yet to be defined, the normative can be defined in terms of the descriptive. Here talk of 'the descriptive' has been deliberately used rather than 'naturalistic'. The import of this will become clearer in the final two sections of this chapter.

3.3 SOME NORMS OF SCIENCE AND EPISTEMOLOGY

In what follows it will be useful to have in front of us some examples of norms from logic, epistemology and methodology.[5] Our practices of valid or correct inference making can be regarded a norm-governed. Thus, corresponding to the law of logic *Modus Ponens*, there is a rule of permission which says: from premises A, and if A then B, one is *permitted* to infer B. In contrast a rule against committing the *Fallacy of Affirming the Consequent* tells us: from premises B, and if A then B, *do not* infer A (or one *ought not* to infer A). Here one may appropriately speak of the norms of reasoning (both deductive and inductive) and their imperatival character. In epistemology we also speak of our beliefs being 'justified', 'right', 'evident', 'warranted', 'having good reason', and so on. All of these terms are normative in that they set standards that we hope our inferentially acquired beliefs will reach. In particular, 'knowledge' has been commonly defined in terms of one or more of these epistemically normative terms; so 'knowledge' too is normative.

Examples of the *values* we wish our theories to exemplify have already been given in section 1.1.2, and in sections on Kuhn (1.4.2) and Quine (3.1.2). These values can also be formulated as *rules*. Thus the Kuhnian value of fruitfulness requires us to aim for theories which introduce new facts or areas of research beyond those for the theory was originally constructed; this can be expressed in rule form as: 'adopt that theory which is more fruitful than its rivals'. Such a rule has the form of a categorical imperative, further examples of which are the following. (a) Accept a new theory only if it can explain all the (empirical) successes of its predecessors. (b) On human subjects use only double blind experimental techniques. (c) Quine's maxim of minimum mutilation (conservativeness): when revising a theory in the light of contradictory evidence then chose that theory which makes the minimum of changes. And so on. Some of these appear to be simple rules of thumb. But behind many of them lies a more elaborate theory that yields a much more carefully formulated set of rules. Thus for the last Quinean rule there is now an elaborate, if somewhat contested, theory of belief revision with its several axioms about how we are to add and remove beliefs from a given belief set.

There are also hypothetical imperatives, or instrumental principles of reason, which link a rule r with a value v such as the following. (d) If you want theories with high probability (v) then accept those that make surprising novel true predictions over those that predict what is already known (r). (e) Inference to the Best Explanation: given a set of exclusive and exhaustive explanatory hypotheses and some facts, then if you want the true theory about those facts (v) pick the best explainer of those facts (r). And so on. Such instrumental principles of method can contain any value, thereby accommodating the position of pragmatists with their plurality of values and associated rules, and historically minded methodologists who allege that our values and rules have evolved over time.

There is an important requirement to be imposed upon such instrumental principles of method; this is that, whatever the value, the rule followed should reliably realise the value. To assess reliability we need to turn to the empirically testable assertoric claims associated with each hypothetical imperative. Thus for the rule 'follow r if you want v', there is an assertoric hypothetical of the form 'following r realises v', or a statistical claim of the form 'following r realises v n% of the time' (where n can range form some lower threshold of, say, 80% to 100%). Alternatively where the instrumental principle is comparative, there will be an associated comparative assertoric form: 'following r rather than some other r* will more likely realise v'.

As was discussed in section 1.2.2, such empirical claims play an important role in the legitimation of methodological rules within the programme of *normative naturalism*. The empirical claims can be tested against the historical record of successful and unsuccessful strategies employed in the history of science. If they pass their test then the truth of the empirical claims provides the ground on which the corresponding imperatives of method are to be adopted.[6] This is one important aspect of the legitimation project for the rules (categorical or hypothetical) of scientific method. We want to know which are the correct rules of method, or if they are not universally correct, then with what frequency are they reliable, if they are reliable at all.

Let this suffice to illustrate the idea of values and the two kinds of rules (either categorical or hypothetical) of methodology. In what follows all three items, values, categorical rules and hypothetical imperatives or instrumental principles of reasoning, will sometimes be called *principles of method*. And because they are all intended to have normative force, they will also be called *norms of method*, or sometimes *norms* where the context is clear.

Is there a link between epistemology and methodology? For some there is really no difference because they accept a claim like the following. Let us assume that knowledge involves at least true belief and some justification condition such as: belief that p is justified iff p is obtained by the use of some appropriate methodology M.[7] Thus epistemology's claims about justification are nothing but appeals to the norms of methodology in order to arrive at the truth. However others, such as pragmatists who adopt a plurality of values, might claim that methodology is not a compartment of epistemology but is much broader. One reason is that while epistemology is a truth-oriented discipline (since the very definition of knowledge must involve truth), there are some principles of method that have little to do with truth. For example consider the following situation. Suppose a theory T is correct in that all its test implications are correct; i.e., it is empirically adequate. But there is some anomaly A that T does not explain but is part of the domain of phenomena that T ought to deal with. Thus T is not refuted by A but its failure to explain A raises a serious question about the completeness of A. That is, there is a non-refuting anomaly for T. Let us say that T is an *incomplete* theory iff T has a non-refuting anomaly. But note that T is not false; in fact we presupposed that T was empirically adequate. Now if one were to reject T because it is incomplete, one would be adopting a rule that is not strictly epistemic in character in that it is not oriented towards the truth. Assuming that epistemology has to do with the truth in some broad sense, then a rule of method that says that we can reject incomplete theories (which are nevertheless true) is not strictly a rule of method that would be used within epistemology.[8] This suggests that the domain of methodology will be richer in the norms it employs than epistemology. As a consequence the task of naturalizing methodology will be a broader one than that of naturalizing epistemology.

There are a number of questions that can be asked about methodological norms. Which norms ought we adopt? Are there better formulations of these norms other than as the simple rules of thumb given above? (For example much machinery of belief revision theory is devoted to making Quine's idea of modesty more precise.) Which are the true, as opposed to false, or correct, as opposed to incorrect, norms? How are the rules and principles to be validated or tested? These questions will not be answered here; some possible answers are to be found in section 1.2 on the problem of the legitimation of methodological principles. Rather there is a different question upon which this and the remaining sections will focus – that of the location of such norms within the scientific worldview. What this means is the following. Given some principle of methodology such as 'one *ought* to follow r if v is to be realised', it is the normative force of the 'ought' that is up for consideration. We can take it that the values we strive to realise, such as simplicity, or fruitfulness, truth, and the like, can be characterised purely descriptively, as can the means expressed by r for arriving at v.

In section 3.1 the subject called *naturalised epistemology* was introduced. For some this subject is a non-starter; they claim that there is no way to conjoin the 'is' claims of naturalism with the 'ought' claims of epistemology. In particular some of the central notions in epistemology are normative and they cannot be got from descriptive notions. Belief is not a normative notion; but the notion of, say, *rational belief* employed within Bayesian scientific methodology is. Knowledge is also a normative notion because it is bound up with other normative notions such as a belief's being *justified*, having *good reasons*, being *warranted*, being *well* founded, being *rational*, and so on. These normative notions are to be contrasted with non-normative notions such as being an implication of, being probable, having some positive evidential support, and so on. This suggests that one can draw up two lists of notions within epistemology, one containing only normative notions while the other contains only non-normative notions (any contested cases being relegated to the second list). As far as these two lists are concerned, the values we pursue, such as simplicity, truth, and the like, will not strictly be normative. Even though we do speak of them as values, and sometimes as norms, they are not intrinsically normative. Thus realists will value theoretical truth in the sense of wanting their beliefs to be in accord with the unobservable world. Others of an empiricist persuasion deny this and claim that theoretical truth is not a value at all; it is not even worth pursuing because it is an unrealisable value. If truth were intrinsically normative this empiricist denial would not even make sense. Truth can still be a value that some wish to realise and others shun without being a normative concept (like *justified*) in the sense intended here.[9]

Concerning these two lists some say that the twain will never meet since, according to Hume, there is an is-ought divide, or a fact-value divide. Despite the wide adoption of Hume's dictum by ethicists and others, the common understanding of the claim that there is a logical divide between the *is* of facts and the *ought* of norms is not logically tenable as standardly expressed.[10] This is a matter that will not be pursued here, though the project of overcoming Hume's alleged divide is of a piece with the view of norms set out in later sections (understanding norms not merely as those of ethics but also of reason and of methodology).

Now the scene is set for the remaining sections of this chapter. The problem to be focused upon about methodological norms can be put as follows. Where are they to be located? Are they to be located outside the scientific world picture so that they are not part of science, yet they exert some kind of external influence upon science? Or are they to be located within science – whatever this might mean? The first view has been adopted by a number of philosophers: 'Now the idea that epistemic facts can be analysed without remainder – even 'in principle' – into non-epistemic facts ... is. I believe, a radical mistake – a mistake of a piece with the 'naturalistic fallacy' in ethics.' (Sellars (1963), p. 131); 'Justification, coherence, simplicity, reference, truth and so on, all exhibit the *same* problems that goodness and kindness do, from an epistemological point of view. None of them is reducible to physical notions' (Putnam (1990), p. 141). A different view is adopted here in which the normative in methodology does have a location within naturalism. Note that this is not the project of finding a better formulation of the principles of method, or of testing them – though this is something naturalizers will also want to do. The task of locating the normative within the natural will be carried out using an appropriately refurbished notion of definition (based on the Ramsey sentence as modified by Lewis) of normative notions employed in methodological principles, and an account of the supervenience of the normative on the non-normative. In establishing both of these claims, then the normative will have been located within naturalism.

3.4   NATURALISM AND NORMS OF REASONING AND METHOD: MAPPING THE TERRAIN

This section considers some theories about the status of norms. By this is meant whether norms are ultimately subjective and expressive of desires, whether they are *sui generis* transcendent objective facts in their own right, whether they are still objective but can be given a naturalistic understanding, and so on. As will be seen, many take the view that norms must in some sense be transcendent. But critics of such a conception of norms assume that, in rejecting the transcendence of norms, they are merely left with the view that norms are expressive of desires, or that one is entitled to take a thoroughly sceptical or nihilistic attitude to the force of any norms. They overlook the possibility that norms can still be objective and yet have a place in a naturalistic account of the world. It is this third position that will be advocated here. A useful fourfold distinction has been made by Jean Hampton ((1998), pp. 2-5) that provides an illuminating classification of the different ways in which normativity and naturalism can be related. The four positions are initially illustrated by examples from ethics where most work has been done in this area. But the main point of the classification is to illustrate the way it applies to the norms of reasoning, logic and method within the sciences. Hampton's 4-fold way does not fully capture all the positions that need to be demarcated; so what will be proposed is a modified "four-and-a-half-fold" classificatory scheme.

3.4.1 Position (1): *Norm Anti-Objectivism*

This is best described with respect to moral norms first. Norm anti-objectivists argue as follows. The commitments of scientific naturalism and the commitments of an objectivist moral theory are incompatible. Conclusion: so much the worse for objectivist moral theory; morality is not objective in the way that science is. There are several importantly different positions that fall under this heading, only a few of which will be mentioned. The lack of objectivity in the case of ethical sentences at least involves the following: there is no special domain of non-natural moral fact to which ethical sentences might pertain, in the same way as there is a domain of fact to which scientific sentences do pertain. From this lack of objectivity (so characterised) some draw (perhaps too quickly) the conclusion of moral nihilism, viz., the view that, contrary to linguistic appearances, there are no moral norms. Less strong is the associated view of moral scepticism, viz., that we cannot know what moral norms we ought to hold (and not that there are none).

Somewhat different are the claims of moral non-cognitivism that at least acknowledges the existence of moral norms. To characterise this, it is necessary to distinguish two theses. (i) *Semantic Thesis*: ethical sentences have truth values; (ii) *Psychological Thesis*: ethical sentences describe or report certain approving or disapproving attitudes or states of mind. Granted this, distinctions between different non-objectivist accounts of ethical sentences can be sketched.

(a) Expressivist, or prescriptivist, theories of ethical sentences deny both of (i) and (ii). Ethical sentences are truth valueless, since they do not report or describe, but rather expresses attitudes or prescribe actions (and thus do not make statements). This is to be contrasted with (b), some subjectivist theories within ethics. These accept both (i) and (ii); for some versions of subjectivism there are truth-valuable claims about approved or disapproved attitudes or states of mind. Understood in this way, some subjectivist theories need not be inconsistent with naturalism. (c) The position of Mackie (1977) on the status of ethical sentences can be characterised as accepting the semantic thesis but questioning the psychological thesis. Mackie "error" theory rejects the view that there is an ontology of queer moral fact that makes ethical sentence true or false, but they still have truth values. Alas, they are all false since there is a false, but pervasive, non-naturalist belief concerning what ethical sentences are about.

A parallel can be drawn with theoretical statements of scientific theories; these can be understood non-realistically in that they are not declarative and do not have truth values. Instrumentalism takes the theoretical sentences of science not to have truth values and to be akin to rules of inference. In contrast other kinds of anti-realism, such as constructive empiricism, allow that theoretical sentences do have truth values. But according to van Fraassen's version of constructive empiricism, we are to remain sceptical about what the truth values are (van Fraassen (1989), chapter 8.3). Finally scientific realism allows that theoretical sentences do have truth values; however scepticism about the truth values is not appropriate in that we can come to have good reasons for assigning truth values to theoretical claims.

(d) Finally there is the position of ethical relativism. Whatever else might be said of this position, it falls within the broad category of anti-objectivist theories of ethics. There is no special realm of fact that makes ethical sentences true or false in any absolute sense; nor is there any sense in which they could be true or false *simpliciter*. Rather their truth value is relative to a culture, epoch, moral system, or whatever. This brief list does not exhaust all the stances within moral theory concerning norm anti-objectivism, but it will provide a useful grounding for a parallel classification of the various stances concerning methodological norms, understood non-objectively.[11]

As applied to the norms of reason, *norm anti-objectivism* emerges in the following way. The commitments of scientific naturalism and the commitments of objectivist norms of reason are incompatible; conclusion – so much the worse for an objectivist account of the norms of reason. In particular, methodological norms are not objective in the way that science is. A nihilistic or sceptical view about methodological norms has been adopted by those who would dethrone the authority of science, in particular the authority of the norms as adjudicators of the claims within the various sciences. In talking of the authority of norms a number of things can be meant. First the norms are, in some sense, right, valid, correct, sound or warranted. Second, if we are aware of them, they impressive themselves upon us with a certain "force" as directives which we should follow in the situation in which we cognitively find ourselves; that is, they exert a "pull" upon us and are binding. Thirdly, in a sense yet to be filled out further, they are objective. While many philosophers can agree that norms can have authority in this sense, they disagree over how their authoritativeness and objectivity is to be cashed out. In some cases the account given of norms leaves only a very thin story about their authoritativeness, or none at all.

Though *position (1)* of norm anti-objectivism will be discussed more fully in the next section 3.5, some illustrations might be helpful here. Lyotard (see section 1.3.4) appears to adopt a form of norm scepticism. If we describe some story about how methodological norms are to be warranted in terms of Lyotard's word 'metanarrative', then postmodernism characterised as 'incredulity toward metanarratives' (Lyotard (1984), p. xxiv) is a sceptical or nihilist position about the justification of any methodological norms. Sociologists of scientific knowledge commonly adopt a similar stance. Their central claim is that all scientific beliefs are caused either by socio-political factors, or a scientist's interests in these. There is no room for the traditional norms of scientific method which, when applied, bring about our various beliefs in science. Given the socio-political story that such sociologists like to tell about the causes of scientific belief, then all the demands of naturalism have been met, albeit a naturalism that only requires that a scientists' social, cultural and political situation be causally efficacious in bringing about their scientific beliefs. Importantly there is no role for methodological norms, whatever their status, in the sociologists' story. So there is either a scepticism about such norms, or a nihilism in which, since we do not need them, we can do without them – and so wield Ockham's razor.

Not everyone who adopts norm anti-objectivism need be a postmodernist. Thus if one accepted the view of Quine that the norms of science are to be given a purely descriptive account as a chapter of psychology, then such eliminativism would be tantamount to norm nihilism. However, as was seen in section 3.1, this is not Quine's final position. Alan Gibbard is not a norm nihilist or sceptic; but he does adopt an anti-objectivist account that he calls 'norm-expressivism' (see the next sub-section and section 3.5.2).

3.4.2 Position (2): *Norm Anti-Naturalistic Platonism*

Again this position will be sketched first in the case of moral norms. Supporters of this position would claim that the commitments of scientific naturalism and of an objectivist moral theory are incompatible. From this they conclude: so much the worse for the narrowness of scientific naturalism. There are more things in Heaven and Earth, including moral facts, than are dreamt of in the philosophy of naturalism. This position can be appropriately called 'anti-naturalism'. But it is also strongly objectivist in appealing to a special realm of fact. So let us call *position (2)* 'norm Platonism' after the philosopher who was not shy in expanding ontology into realms which were overtly non-naturalistic. In ethics the most famous non-naturalists about 'good' and 'right' are Henry Sidgwick and G. E. Moore; they are commonly understood to have held that there is a non-natural realm of normative fact in ethics. Given such non-natural facts, there has been a tradition since Plato of developing an epistemological theory of how we can have knowledge of facts so different from the everyday and scientific facts of naturalism, knowledge of which we acquire through the familiar methods of perception and inference. Our powers of *Intuition* are said to enable us to apprehend moral facts. Though the manner of such apprehension may be obscure, it is commonly based on an analogy with our perceptual apprehension of natural facts. Platonists about moral facts resist any attempt to either eliminate such facts from ontology or to relocate them within naturalism; this they claim would change the very topic of moral discourse.

But the postulation of moral facts introduces two sorts of queerness into the world. The first concerns a special faculty that has to be attributed to us in order to discern such facts. There is little evidence that we have any such faculty for putting us in touch with such items. The second queerness concerns the very nature of the non-moral facts themselves. They appear to be an occult, or supernatural, kind of fact, different from all the natural facts with which we are familiar. Their oddness lies in the prescriptive character of this sort of non-natural fact, or the imperatival "force" they might be said to exert on us to do one thing rather than another. If we are merely to cash out how it is that moral judgements have the objective pull they do, then there is no need to postulate a realm of odd fact to account for this; there are alternative ways in which this can be done without a heavy cost in extra ontological baggage.

Shifting from ethical to methodological norms, norm Platonists argue in the following way. The commitments of scientific naturalism and of an objectivist theory of methodological norms are incompatible. Conclusion: so much the worse for the narrowness of scientific naturalism. Though intuitionism has played a large role in moral theory, counterparts to it in respect of methodological norms are hard to find. Perhaps Sidgwick might be construed as having held this position in his talk of 'Rational Intuition' which concerns 'the apprehension of universal truths, such as the axioms of Logic and Mathematics' (Sidgwick (1884), Book I Chapter III section 4). Perhaps it is some such view of an independent realm of logical facts for logical truths that Russell resists in his 1918 lectures on Logical Atomism (Russell (1918), Lecture III, pp. 209-14). Russell sets out to provide an inventory of the different kinds of fact that we need to postulate. There are atomic facts that can be of subject-predicate or relational form. As well as these, Russell argues that we also need general facts, since these are not just a conjunction of atomic facts. But he remains undecided as to whether we also need negative facts, such as 'there is no snake now crawling over this page'. Important for our purposes is that he shows that we do not need *sui generis* logical facts as truthmakers for logical truths. All such molecular propositions can be decomposed into atomic propositions which have truth values; and to establish the truth value of such molecular propositions all we need, besides the notion of truth, are truth tables for logical constants and the idea of a proposition (or more austerely a sentence). Whatever ontology rules of reasoning need,[12] they do not need a special realm of weird non-natural fact to validate them.

Usually the Platonist's approach is challenged in strong terms. There is Ramsey's famous quip: 'Theology and Absolute Ethics are two famous subjects which we have realised to have no real objects.' (Ramsey (1990), p. 247). And Alan Gibbard helps us grasp the strong Platonistic position by contrasting it with his own expressivist stance:

> It might be thought that ordinary conceptions of rationality are Platonistic or intuitionistic. On the Platonistic picture, among the facts of the world are facts of what is rational and what is not. A person of normal mental powers can discern these facts. Judgements of rationality are thus straightforward apprehensions of fact, not through sense perception but through a mental faculty analogous to sense perception.
>
> ...
>
> If this is what anyone seriously believes, then I simply want to debunk it. (Gibbard (1990), p. 154)

Gibbard endorses a position he calls 'norm expressivism':

> "To call something rational is to express one's acceptance of norms that permit it." ... Normative talk is part of nature, but it does not describe nature. In particular, a person who calls something rational or irrational is not to describe his own state of mind; he is expressing it. To call something rational is not to attribute some particular property to that thing... (*ibid*., pp. 7-8)

In setting his face against Platonism about methodological norms, Gibbard locates his norm expressivism squarely within the non-cognitivism of norm anti-objectivism. Those who reject the norm Platonism of *position (2)* often argue that the only option left is *position (1)* of norm anti-objectivism. And this appears to be the position of Gibbard.

But as we will see there are yet other positions that can be adopted. One way of overcoming the Platonist's appeal to non-natural facts in establishing our methodological norms that does not fall easily into the 4-fold classification adopted here is some kind of Kantian approach (i.e., the "half" position mooted earlier). This attempts to show that our methodological norms can be shown true, or false, in an *a priori* fashion without begging any embarrassing ontological questions. *A priori* proofs should at the same time provide an account of the grounds for the authority of our methodological norms without appeal to a special realm of queer non-natural facts. Here the objectivity of norms is not cashed out in terms of queer facts but in terms of a strong *a priori* justification for norms within reasoning itself. Call this quite modified version of *position (2)* a *Kantian a priori-ism* about norms. It recognises the objective pull of norms, not by locating it in a realm of non-natural fact, but in non-empirical (and thus non-natural) *a priori* considerations. Such a modified *position (2)* would not sit happily with a naturalism that rejects all attempts to establish on *a priori* grounds a first philosophy, or which would contain *a priori* warranted methodological norms.

The first two positions turn on the incompatibility of scientific naturalism and the objectivity of norms, the incompatibility being resolved by either discrediting norm objectivity or discrediting the claims of scientific naturalism to completeness. The next two positions turn on a compatibility between scientific naturalism and norm objectivity, either by showing that morality is just like science, or by reconstruing the nature of science so that it is more norm-laden than has been commonly supposed by positivists or traditionalists about science.

3.4.3 Position (3): *Norm Objectivist Naturalism*

Characterising this position first in terms of the norms of ethics, this is the view that there is no incompatibility between scientific naturalism and objective morality because the claims to objectivity in morality are just like the claims to objectivity in science. This position Hampton calls 'moral realism'; but because of the ambiguity in such a characterisation the term 'objectivist' will be used here to characterise the status of both moral and methodological norms. In ethics the best occupant of this position is Boyd's account of moral norms (Sayre-McCord (1988), chapter 9). He argues that both scientific and moral claims have truth values independently of our thoughts and attitudes, and that our canons of reasoning provide reliable methods for obtaining and improving both our moral and scientific knowledge. In the same way as our scientific claims answer to an objective reality, so do our moral claims. But is this an appeal to a non-natural kind of fact? Not so. Accompanying this is an identity claim to the effect that that moral facts or properties are nothing but naturalistic facts; and the identity between moral and natural facts is necessary but knowable only *a posteriori*. In this manner moral facts are located within naturalism and are not eliminated.

When expressed for the case of methodological norms, *position (3)* of norm objectivist naturalism is as follows: there is no incompatibility between scientific naturalism and objective norms of methodology because the claims to objectivity for such norms are just like the claims to objectivity in science. The task of the final three sections of this chapter is to show how methodological norms have a location with a broadly construed naturalism. There it will be shown how to "define" the normative in terms of the non-normative exploiting the device of the Ramsey-Lewis sentence (section 3.7). In section 3.8 it will be argued that the normative terms contained in methodological principles supervene on the descriptive properties and relations of our beliefs.

3.3.4 Position (4): *Norm Anti-Naturalistic Objectivism*

This is the view that there is no incompatibility between scientific naturalism and norms because the claims to objectivity in scientific naturalism, when properly construed, turn out to be just as deeply implicated in normativity as are claims found in ethics and elsewhere. Hampton calls this position 'sophisticated compatibilism'. Here the received view of the sciences within naturalism must undergo considerable transformation so that the sciences become much more value-laden than many philosophers of science, such as positivistic advocates of value-free science, have traditionally supposed. Hampton distinguishes two forms that this version of compatibilism might take. The first is that of Feyerabend who, in criticising the alleged objectivity of science, shifts the characterisation of science more in the direction of the normative (especially the ethical). Under such a reconstrual science becomes much less objective and more subjective; that is, if one adopts a subjectivist view of ethics, then science takes on more of the characteristics of such an ethics than would have been envisaged in more traditional conceptions of science. In addition the norms of method are not taken to be absolute but are claimed to be relative to a culture, a given stage of the development of science, or whatever. This Hampton calls 'sophisticated anti-objectivist compatibilism'. It is anti-objectivist because, on one understanding of Feyerabend's view, the norms of scientific method are alleged not to have any over-arching authority and cannot establish a body of knowledge that is in any way superior to myth, religion, or morality.

The position Hampton wishes to argue for in her book is 'sophisticated objectivist compatibilism' which she describes as follows: 'we can still see science as objective, but not in the traditional naive way, so that scientific objectivity is no different from the form of objectivity exemplified in the moral domain' (Hampton (1998), p. 4). She calls this account of methodological norms 'normative sophisticated compatibilism'. On this account methodological norms are objective in that Hampton views them as dependent on irreducible normative facts: 'science itself requires, and is undergirded by, a commitment to normative objectivism, meaning that any science-based argument against moral objectivism is ultimately self-defeating' (*ibid.*, p. 5). Hampton's own account remains underdeveloped and will not be discussed here.[13]

The above "four and a half" positions provide a useful setting in which to discuss the possible relations between the natural and the methodologically normative in science. It also helps distinguish friends and foes in the "science wars" whose views differ greatly on the status of such norms. But there are some friends whose help we do not want – those who invite us to acknowledge a non-natural realm of normative fact as their truthmakers. Though there is agreement that there are norms of rationality and science, it is *position (3)* rather than *(2)* which objectivists should adopt to underpin norms in a non-*a priori* fashion. So who are the enemies?

3.5   NATURALISM AND NORMATIVE ANTI-OBJECTIVISM

In this section we will set aside the ontologically dubious *position (2)* of normative Platonism with its appeal to independent normative facts and *position (4)* (second interpretation) with its radical reconstrual of what science might be like. However it should be recognised that *positions (2)* (including the "half" position) and *(3)*, along with the second interpretation of *(4)*, espouse an objectivism about norms while *position (1)* and the first version of *position (4)* are anti-objectivist. If we set aside *positions (2)* and *(4)*, then the real focus of contrast is between *positions (1)* and *(3)*. The real focus is not between *(1)* and *(2)* as some have supposed, e.g., Bloor (to be discussed later) and Gibbard as cited in section 3.4.2 who seems to assume that the only two viable positions on the status of norms are that of the norm Platonism of *(2)* and the norm expressivism of *(1)*. In rejecting the Platonism of *(2)* these theorists are left only with the expressivism of *(1)*. However *position (3)* of norm naturalistic objectivism is a thriving, viable position to adopt and not one to be overlooked or shunned. And *(3)* contrasts just as strongly with *(1)* as does the dubious Platonism of *(2)*.

The contested *positions (1)* and *(3)* encapsulate important rival philosophical stances. The advocates of the objectivism of *position (3)* are the contemporary heirs of the Enlightenment who, in their account of reason, urge that there is some objectivity to methodological norms, even within the context of naturalism. Many postmodernists and sociologists of science who advocate the anti-objectivism of *position (1)* wish to turn out the lights of the enlightenment. They come in a number of different hues of darkness, the darkest of all being the nihilists, sceptics and relativists who deny all objectivity to methodological norms. As already illustrated in section 3.4.1, Lyotard's postmodernism characterised as 'incredulity toward metanarratives' is highly sceptical of the force and correctness of methodological norms. This position is well characterised by Cahoone, an historian of postmodernism (see 'Introduction'), who says that one of postmodernism's characteristic views is that norms are immanent in, rather than transcend, the very social context and social interests which they are supposed to judge. Similarly, postmodernists allege that norms of reason and method arise out of a given social context and serve certain social interests. In denying that norms are transcendent postmodernists might be understood to have in mind the two notions of transcendence found in the two versions of *position (2)*, viz., either norm Platonism with its commitment to a realm of weird fact, or some *a priori* account of the justification of norms. And in denying either version of *position (2)*, they assume that only some version of *position (1)* is viable – and within this they espouse norm nihilism or norm scepticism.

3.5.1 Sociology of Scientific Knowledge and Norm Anti-Objectivism

Sociologists of scientific knowledge occupy highly ambiguous positions within the penumbra of anti-enlightenment. Their central claim is that all scientific beliefs are caused by socio-cultural factors, or by scientists' socio-political interests. In Part III one version of this claim emerges in Foucault's 'power/knowledge' doctrine; there is simply no room for methodological norms in the nexus of power relations in which all knowledge is allegedly enmeshed. Part IV addresses Nietzsche's norm nihilism in which much the same kind of role is played by the obscure "will to power" as is played by Foucauldian power relations. Much harder to classify are those sociologists of science who admit that there are methodological norms but who think that they are to be naturalised in terms of our social and psychological dispositions. Friends, one might think? No, foes, because they give a quite attenuated account of the objectivity of methodological norms.

Sociologists of scientific knowledge, as will be seen in Part II, hold that all scientific beliefs arise and are maintained by socio-political causes, or by scientists' interests in these. Case studies such as those set out in section 5.6 and section 5.7 purport to show that in Weimar Germany physicists adopted acausality because of their interest in the intellectual currents of their anti-scientific cultural milieu. And they purport to show that Robert Boyle adopted, within his mechanics, the view that matter was passive and lacked active powers largely because of his socio-political and religious stance during and after the English Civil War. Since these sociologists also insist that no methodological norms need be invoked in the sociologists' explanation of why scientists hold their beliefs, then there is either a scepticism, or a nihilism, about methodological norms. If such norms exist, they are otiose in any explanation of scientific belief. At worst such norms are a chimera foisted upon the world by misguided philosophers who wish to keep alive yet another illusion of the enlightenment. That methodological norms have no explanatory role to play is explicitly said to be a requirement of the important Symmetry Tenet imposed on explanations in the Strong Programme; this is a matter to which we will return in chapter 6.

Also important is that sociologists of science view their project as one that falls within the scope of naturalism. All scientific belief is to be explained naturalistically in terms of the co-operation of causally efficacious social, psychological and other non-social but naturalistic factors – with the social factor always being present and dominant. But their naturalism eschews physicalism given the prominence their Strong Programme assigns to an autonomous sociology that postulates social properties with strong causal powers which bring about belief. Given such an eschewal of norms in any explanation of scientific belief by sociologists, then the project of philosophers of science with their appeal to such explanatory norms is debunked.

Some sociologists of science might be taken to adopt a less nihilistic stance in that there is some role to be assigned to methodological norms of reason. Thus Bloor tells us of the Symmetry Tenet (more fully discussed in section 6.3) of his Strong Programme for the sociology of scientific knowledge: 'The Symmetry requirement is meant to stop the intrusion of a non-naturalistic notion of reason into the causal story. It is not designed to exclude an appropriately naturalistic construal of reason, whether this be psychological or sociological.' And he chides a critic for 'mistaking the sociologists' rejection of a non-naturalistic notion of reason as a rejection of reasoning as such' (Bloor (1991), p. 177). The last remark assures us that reasoning does go on. But when it goes on correctly, does it go on according to norms? We have good reason to think not since, in the case studies of episodes in the history of science such as the two just cited from the Strong Programme, nothing is said about the role of norms in the explanation of scientific belief. All emphasis in the studies is on the allegedly causally efficacious socio-political circumstances of believers, or the causal efficaciousness of their interests in socio-political factors. Setting this lacuna aside, let us suppose on behalf of the sociologists of scientific knowledge what philosophers and others have always supposed, viz., that scientists do often come to their beliefs by employing the norms of reason.

In the light of this supposition there are two important issues raised by the above quotation. The first is that we are to reject the idea of norms in which their operation is an 'intrusion of a non-naturalistic notion of reason into the causal story'. But what could this mean other than an account of norms understood non-naturalistically, as in the norm Platonism of *position (2)* (see section 3.4.2), in which norms are a weird kind of imperatival fact. Moreover, they are somehow able to intrude on the naturalistic causal order causing belief. Advocates of the Strong Programme in the sociology of scientific knowledge hold a contrasting view; their explanatory causal account of scientists' beliefs is one in which the causal factors are entirely within a naturalist framework, viz., causally efficacious socio-cultural factors. But have they not overstated the view they reject? One might well ask: how could non-natural norms ever intrude upon the natural order? Or intrude upon anything for that matter? There is an important category error here which can be made clearer by an example. Let us grant for the sake of the argument that the norm of inference *Modus Ponens* is to be understood as some kind of queer non-natural fact. Now as queer as this might be, it does not suppose the additional, also queer, view that *Modus Ponens*, of and by itself, could be causally active so that it intrude upon naturalism's causal order. Even if *Modus Ponens* is a queer non-natural fact, it is impotent to intervene in naturalism's causal order. If it could then perhaps we reasoners might have been better off.

So what does intrude in the naturalists' causal order? One obvious candidate is people and the cognitive attitudes they possess leading them to act in particular ways; they are part of the naturalists' causal nexus. Often people act within the causal order in accord with their beliefs and desires – which might include acting in accordance with the norm *Modus Ponens*. But *Modus Ponens* itself is not part of the causal order. Thus suppose that a person actively believes premises ' A' and 'if A then B'; and then, through the (conscious or unconscious) application of *Modus Ponens*, they come to have a further belief in conclusion 'B'. Here it is the person's inferring in accordance with *Modus Ponens* that is active in the causal order and not *Modus Ponens* itself. As will be seen, the Symmetry Tenet is important for the sociologists' of knowledge Strong Programme. But if all it is meant to do is stop the intrusion of a non-naturalistic *Modus Ponens* into the causal story, then it is already amongst the ranks of the unemployed. Understood this way, *Modus Ponens* is a peculiar *deus ex machina* which, unlike people who hold beliefs, cannot intrude on anything, including the naturalists' causal order.

This brings us to the second point about the above quotation, viz., that the Symmetry Tenet 'is not designed to exclude an appropriately naturalistic construal of reason, whether this be psychological or sociological'. What might this mean? It appears that we are to view people within naturalism's causal nexus as having cognitive powers of reasoning due either to the evolution of their cognitive powers, or due to their social circumstance. Now there are no grounds for denying that people are *natural reasoners*; that is, they have acquired an ability to reason either due to their evolutionary inheritance, or due to their acculturation in the reasoning practices of their community. However it should be noted, as much empirical research shows, evolution and our culture have not done a very good job in making us rational agents. We are prone to commit fallacies, such as the gambler's fallacy, and we have a well attested inability to correctly use the norms of reasoning (such as *Modus Tollens* in Wason's selection task) in simple test tasks. (For more on this, see Kahneman, Slovic and Tversky (eds.) (1982), or the recent review paper by Samuels, Stich and Tremoulet, (1999)). Setting aside these disturbing matters uncovered in cognitive psychology, the fact that we might be natural reasons (either good or bad) tells us nothing about what norms of reason we *ought* to follow, or which we *ought not* to follow. Evolution may select us as the users of one set of norms rather than another; and peer pressure or social conventions might constrain us to adopt one set of norms rather than another. But such a story tells us nothing about the soundness, authority or warrant of norms.

There is not just a location problem for the sociologist's norms of reason within naturalism. There is also a problem about the authority of norms which remains unresolved. Also given that norms are not to be understood as an outside intrusion of Platonistic items upon the natural order, the only fall-back position for the sociologists' account of methodological norms is a nihilism or scepticism within the anti-objectivist *position (1)*. The issues broached here concerning the account of norms adopted by sociologists of scientific knowledge will become important in chapters 5, 6 and 7.

3.5.2 *Quine, Kuhn, Feyerabend, Gibbard and Norm Anti-Objectivism*

Not everyone who adopts *position (1)* of norm anti-objectivism has postmodernist tendencies, or endorses the stance of sociologists of scientific knowledge just mentioned. Quine might be said by some to be a normative nihilist on the basis of his earlier naturalistic stance that principles of evidence and method are just a chapter of psychology. We can leave it to scholars of ethics to determine whether or not this is so when it comes to Quine on moral objectivity. But as we have seen, given sufficient Quinean hermeneutics and a downplaying of certain passages that are commonly cited in favour of norm nihilism, it is pretty clear than Quine is not a nihilist about methodological norms. As unclear as their status might be, there are norms governing the scientific enterprise that are to be understood as part of a technology of engineering for the anticipation of sensory stimulation (i.e., true predictions). Stripped of its metaphor, the Quinean account supports the idea of an independent normative epistemology. But he says little of the manner in which such norms are to be established or what comprises their naturalisation. Much more is said about this in Laudan's normative naturalism. Nor does Quine give an account of normativity within naturalism, something attempted in the last two sections of this chapter.

As was noted at the end of section 3.1.2, Quine is aware of the nihilistic tendencies in epistemology of Polányi, Kuhn and Hanson 'who belittle the role of evidence and accentuate cultural relativism' (Quine (1969), p. 87). But apart from the purple passages that can be cited from the 1962 *The Structure of Scientific Revolutions* (see section 1.4.1), Kuhn's more mature position is like Quine's: there are values that our scientific theories should exemplify. But whereas Quine seeks no further justification for either the values or the methodological norms he adopts, the later Kuhn sought, in an un-Quinean manner, a justification for these in terms of the analytical link they have to the concept of science, or as he puts it, the local holism of the term 'science' (see section 1.4.3). Where should the position of the later Kuhn be located with respect to *positions (1)* to *(4)* of the previous section? He does not advocate appeal to a realm of non-natural fact to underpin his values, expressed as methodological norms. Nor does he advocate *position (1)* of norm nihilism, scepticism or non-cognitivism; nor does he advocate naturalism. Rather he seeks a non-empirical justification for norms in terms of an *a priori* theory of holistic meaning. Such an *a priori* warrant for values suggests that Kuhn might be categorised within a quite modified version of *position (2)* – *a priori-ism* about norms that does not appeal to norms as a queer species of fact. That is, there is a special mode of *a priori* argumentation, not based on any appeal to non-natural facts, which underpins the objective pull that Kuhnian methodological norms have; in this case it is an appeal to holistic meaning networks.

The above four positions suggest a number of places for the location of Feyerabend's account of methodological norms. With respect to *position (1)* he could be characterised as a nihilist, if one places emphasis on his claim 'anything goes' and his advocacy of the relativity of norms to epochs in the history of science, or cultures, or the kind of science under consideration. But as has been argued at the end of section 1.5.1, Feyerabend does not endorse the claim 'anything goes'. And in the light of his advocacy of defeasible norms it is not clear that he is a relativist. Rather the norms of method have conditions of application which admit defeat without their being falsified. Nor at the meta-level does he always adopt relativism when it comes to the justification of defeasible norms; his position is more accurately characterised as that of a pluralist.

However Hampton ((1998), p. 4) does point to one important theme in Feyerabend when she locates him with in *position (4)* in which there is a compatibility between the naturalism of science and norms. The compatibility arises because the much vaunted objectivity of science is downplayed by rejecting both realism and a firm foundation for the norms of methodology it employs. Science is to be understood as much more value-impregnated enterprise than is commonly advocated in most received views of science, including (in Feyerabend's view) the rather bloodless account of it given by positivists, overzealous rationalists and the like. Given Feyerabend's revisionary stance towards science, a much more humane value-impregnated science is envisaged. So understood, science drops its highly problematic pretences to objectivity and becomes much more like morality, but a morality that is understood to be strongly non-objectivist along lines characterised by *position (1)*. By placing Feyerabend in *position (4)* rather than *(1)*, emphasis is placed upon the compatibility between a revisionary conception of science and a subjectivist (sometimes relativist) morality. Such a characterisation excludes an account of the objectivity of an unreconstructed science and its norms as depicted in *position (3)*.

Finally, a comment on Gibbard's norm expressivism. It might seen unfair to lump his views with nihilists, postmodernists and the like. He endorses naturalism and looks for an account of rationality and other epistemic norms within naturalism; in particular he wants an account of rationality that is consistent with the theory of evolution. However his avowed norm expressivism does locate him squarely within the anti-objectivism of *position (1)*. What is not up for question, in Gibbard's view, is that there are norms of rationality. But what is questioned is the status he assigns to them in his meta-theory about norms. Gibbard calls his position 'norm expressivism': 'My own analysis is not directly a hypothesis about what it is for something to be rational at all. It is a hypothesis about what it is to think or believe something to be rational, to regard it as rational, to consider it as rational' (Gibbard (1990), p. 46).

Gibbard's norm expressivism might be best expressed in the following way, which mentions a speech act expressing acceptance: S's utterance of the sentence 'A is rational' (in normal circumstances) is (by definition) S's expression of his/her acceptance of norms N that permit A. Then we may derivatively say of S calling or believing or judging, etc, that A is rational, that these are expressive of S's acceptance of norms which permit A.[14] For Gibbard, 'A' can stand for an act, emotion or belief; but in the context of the discussion here, 'A' concerns beliefs and their rationality as determined by norms N of method and logic.

Gibbard argues that norm expressivism is not the same as subjectivism. An example of a subjectivist thesis is: "'X is rational' means 'I accept a system of norms that permits X'" (*ibid.*, p. 153). The difference (in words at least) is that subjectivists merely *accept* a system of norms whereas expressivists *express their acceptance*. According to Gibbard the right hand side of the subjectivist thesis is merely reportive and states what S accepts (and so has a truth value). It does not express S's acceptance. As Gibbard commonly puts the difference: 'To express a state of mind is not the same as saying that one is in it [that state of mind]' (*ibid.*, p. 154; see also p. 84). The difference between subjectivism and expressivism is nonetheless subtle and important. One important difference is that epistemically evaluative sentences do, for the subjectivist, have truth values in that they purport to report or describe a statement of mind of a subject. In contrast, for the expressivist they do not have truth values because they do not purport to report or describe anything but, rather, *express* a state of mind. Gibbard recognises the controversial nature of his account of norms when he says the following. 'The controversial elements of the norm-expressivistic analysis are these: first, its account of what it is to make a normative judgement, and second, its claim that the meaning of normative terms is to be given by saying what judgements normative statements express – what states of mind they express' (*ibid.*, p. 84).

The above is a sketch of Gibbard's norm subjectivism, the only alternative envisaged to Platonism about norms: 'Norm expressivism is meant to capture whatever there is to ordinary notions of rationality if Platonism is excluded' (*ibid.*, p. 154). But Gibbard recognises that there is still an issue of objectivity to be addressed, as he continues:

> Even with Platonism excluded, our ordinary thought about rationality involves strong claims to a kind of objectivity, and it is these non- Platonistic claims that I want to elucidate. My hope, then, is to save what is clear in ordinary thought about rationality, and to find our reflective thinking about rationality reasonably clear and fully rectifiable, with one exception; our wavering penchant for Platonism. (*ibid.*, pp. 154-5)

Gibbard's task is, given his norm expressivism, to answer the following question: 'How, then, can we explain the objective pretensions in our normative talk?'(*ibid.*, p. 155). Objectivity turns out to be nothing but inter-subjective agreement about whatever we express acceptance. Thus if a person says: '*Modus Ponens* is rational', then they merely *express* their acceptance of norms that permit *Modus Ponens*. Any objectivity that is to be ascribed to *Modus Ponens* is just its inter-subjective acceptance. But this leaves untouched exactly what features of *Modus Ponens*, and other principles, underpin its rationality, other than its mere inter-subjective acceptance. There are difficulties in both the expressivism and the attempts to reconstruct objectivity that cannot be gone into here. Instead Gibbard's espousal of the subjectivism of *position (1)* can be viewed in contrast with the much more objectivist position set out in the following sections. The rationality of principles like *Modus Ponens* will be located in its descriptive characteristics and not in any inter-subjective agreement over what we mutually express our acceptance.

3.6   FOLK SCIENTIFIC RATIONALITY

Just as many philosophers of mind make a case for there being such a thing as folk psychology, so it will be argued here that there is such a thing as folk scientific rationality (FSR). But there are a few small differences. First, folk psychology is the indefinitely large number of commonplace judgements that most of us folk make about the everyday behaviour of one another in terms of mental states such as beliefs, desires, intentions and the like. In FSR the folk are to be restricted to scientists (and perhaps some philosophers) who have at least some training in the science in which they engage. By narrowing down the folk to the scientists we can hope to avoid, or at least minimise the effect of, some of the rather depressing results of empirical investigations into human rationality which show that when it comes to the application of some principles of reasoning or method we in fact reason quite badly. In any case what we are interested in is not what judgement anybody might make about some hypothesis in science, but rather what trained scientists in the field will judge. The second difference between folk scientific rationality and folk psychology is in the content of the judgements made. Let us turn to these.

3.6.1 *Value-Laden Judgements of the Scientific Élite as a Source of FSR*

Popper is well known for arguing that one of the features we value in our theories is their openness to criticism and their susceptibility to revision, and even abandonment, in the light of contrary evidence. This virtue he characterises in terms of the falsifiability of theories. In addition he proposed some rules for how we are to treat hypotheses that would ensure that this value would be realised in our theories (for some of his rules see Popper (1959) section 11). These rules and values constitute Popper's "definition" of science, or account of scientific rationality, or demarcation criterion, as he puts it. The question then arises: how might we test such a definition for its adequacy? Popper in fact gives several different answers. The answer to be focused on here is the one he gave in 'Reply to My Critics' (in Schilpp (ed.) (1974), pp. 976-87). There he draws up a list of what he calls 'great' or 'heroic' science, or as will be said here, a list of the *good* sciences, a few of which are exemplified by the scientific work of Galileo, Kepler, Newton, Einstein and Bohr (the list can be extend further). Then he draws up a list of non-sciences or pseudo-sciences or unacceptable sciences; these will be called the *bad* sciences. They are exemplified by the work of Marx, Freud and Adler (and the list can be extended).

These two lists comprise the empirical base against which Popper's demarcation criterion for science is to be tested, using the hypothetico-deductive schema as a meta-criterion for testing the demarcation proposal. Suppose in such a schema that Popper's definition of science, including his value of falsifiability and the set of rules alleged to achieve it, comprise the hypotheses under test. Then the definition will pass its test if it captures all the sciences on the good list and none on the bad list. It will be inadequate (and so falsified) if it leaves out some of the good sciences even if it captures none of the bad. And it will definitely be falsified if it captures some of the sciences on the bad list. I do not wish to discuss this test procedure; rather I wish to focus on the test basis, the two lists of good and bad science.

Where do the two lists come from? They are Popper's own pre-analytical value judgements about the good and the bad in science. The idea of such a list can be made more precise. First, instead of merely listing the names of scientists one could select some of their more characteristic hypotheses or theories and place them in the good or the bad list. Second, one could conduct a sociological survey of other scientists working in the field to determine their collective judgement about which of the two lists some theory or hypothesis might be placed. Lakatos proposed such a modification when he appealed to the value judgements the "scientific élite" make about the scientificness or otherwise of particular hypotheses or theories. In this way a less fallible and more empirically well-founded list of good and bad theories and/or hypotheses can be drawn up. Lakatos extends this idea even further, saying:

> While there has been little agreement concerning a *universal* criterion of the scientific character of theories, there has been considerable agreement over the last two centuries concerning *single* achievements. While there has been no *general* agreement concerning a theory of scientific rationality, there has been considerable agreement concerning whether a particular single step in the game was scientific or crankish, or whether a particular gambit was played correctly or not. A general definition of science thus must reconstruct the acknowledgedly best gambits as 'scientific': if it fails to do so, it has to be rejected. (Lakatos (1978), p. 124).

Here we are to set aside disputes over what is our more general theory of scientific rationality, or our demarcation criterion, or whatever one would like to call our best theory about the nature and methodology of science. In contrast, says Lakatos, there has been considerable agreement, amongst members of the scientific élite, over single achievements, or particular gambits, in science. Taking these gambits into account we can considerably expand the items in the good and the bad lists. Thus not only may the scientific élite say that the acceptance of some hypothesis or theory was good or justified or warranted; but also enough of the élite might agree over what are the good or bad moves in the game of science, or the warranted or unwarranted moves, or the justified or unjustified moves.

In order to illustrate, consider the following value judgements of particular theories or moves in the game of science. (1) "Neils Bohr was *justified* in ignoring the inconsistency between his planetary model of the atom, in which electrons orbited a nucleus, and the Maxwell-Lorentz theory of electromagnetism which said that the orbit is unstable and should collapse." In this example most of the terms are descriptive, including talk of inconsistency. The evaluative term is 'justified' because, despite the inconsistency, some novel empirical confirmation of the model was found. (This, we may suppose, is the value to be realised; and it can be descriptively specified as well.) Note that this particular value judgement violates any methodological principle (or value) against inconsistency, such as Kuhn's. There are various ways of resolving the inconsistency, such as rejecting the particular value judgement, rejecting the methodological principle, or refining it in some way so that there is a time limit on how long an inconsistency can remain and under what conditions, and so on. These manoeuvres need not concern us here; the focus is on the character of the particular value judgements.

(2) "Copernicus was *correct* in rejecting the existence of the equant point as a centre about which rotation could occur." This was a device used by followers of Ptolemy for which Copernicus could find no *justification* within Plato's and Aristotle's theory of centres about which eternal motion could occur. (3) "Copernicus was *justified* in claiming that his early model of the planets (set out in the 1512 *Commentariolis*) was simpler than extant Ptolemaic models in respect of the number of epicycles added to deferents. He was not so *justified* for the final version of his theory published in the 1543 *De Revolutionibus*."[15] (4) "We have *good reason* to accept Einstein's general theory of relativity because it was more highly confirmed than Newton's; the former could explain what the latter theory could not (even though it was used to uncover this fact), viz., the additional precession of the perihelion of Mercury by 43 seconds of an arc per century." And so on, for a host of other methodologically evaluative terms that may occur in the particular judgements of the scientific élite, such as 'justified', 'warranted', 'has good grounds', 'has good reason', and so on. On a scale of thick to thin, many of these normative epistemic concepts are at the thin end. But whether thick or thin they give methodological principles their normative bite.

We now have a better account of FSR than that provided by Popper. We can adopt Lakatos' proposal and say that FSR is (at least) the large conjunction of all the value judgements of the scientific élite within each science, as illustrated in the particular judgements above. The value judgements will include even the gambits played within science, such as Ptolemy's use of, or Copernicus' decision not to use, some *ad hoc* hypothesis such as the existence of equant points. Moreover, it is relatively easy to distinguish between what we might call the norm-laden terms in the sentences of FSR, and those which are not norm-laden or descriptive, even though on the descriptive side there may be notions which need much analysis such as the values we adopt, e.g., explanatory power, simplicity, fewer *ad hoc* hypotheses, and so on.

One question we can ask is this: are such value judgements of the scientific élite unproblematic and readily available? Lakatos cannot relegate all sociological considerations in science to external factors; some sociology is needed to survey the scientific élite to discover what are their judgements about particular moves in the game of science. Nor should it be assumed that there would be unanimity amongst the élite; a sociological survey might show that the views of scientists ranged from strong consensus, or dissensus, to roughly equal division for and against. Nor is it clear what the lowest threshold for agreement might be. If there is less than, say, 80% agreement then some of the value-judgements might not be useable to decide important methodological matters. We should settle for whatever might be the largest area of agreement about what are the uncontroversially good, or bad, moves in the game of science. (Perhaps the four example used as illustrations of judgements are controversial and do not appear in any final list upon which the scientific elite can agree.) In this way inconsistent judgements can also be eliminated. And some allowance might have to be made for the contents of the list to change if the scientific community were to change its views about some episode over time.

A further matter concerns how the scientific élite is to be determined. We should not admit that all scientists can make value-judgements about all moves in all the sciences, including the many sciences with which they are unacquainted. Nor should we allow the élite to be chosen by the fact that they are *good* scientists, for what counts as a "good" could well turn on whether they are appropriate users of the norms of some scientific method – the very matter over which the judgements of a scientific élite are to be invoked in order to make adjudications. But the scientific community should not be so finely individuated that it picks out, say, only the adherents to Freudian psychology in the various institutes that promote his kind of psychoanalysis. It should be broad enough to include those working in related fields but using different theories, such as other researchers in psychology and psychiatry.

What of the objection that some of the scientific élite might have invoked some methodology as the grounds on which their value judgements are made about particular moves in the game of science? If such judgements were to be used to test methodological claims, as Popper and Lakatos wish to do, then there would be a threat of circularity. However the list of good and bad particular value judgements in science which comprise FSR will not be used here to test some theory, definition or demarcation criterion of science. So this circularity threat will not be a problem. In fact that the scientists make their judgements in accordance with some broad methodological principles is merely more in the way of value judgements that can enter into folk scientific rationality. However what we wish to focus on here is not the way in which scientists apply the norms of methodology in some particular case. What we wish to focus upon are their pre-analytic judgements, in the absence of any well-articulated set of methodological norms, about what are particular good and bad moves in the game of science.

3.6.2 The Maxims of Epistemic and Methodological Action in the Game of Science as a Source of FSR.

There is another approach that we can take concerning the construction of FSR. The earlier Laudan of *Progress and Its Problems* had assumed, like Lakatos, that we have *'our preferred pre-analytic intuitions about scientific rationality'* (Laudan (1977), p. 160). These are based on episodes in the history of science and are much more firm than any intuitions we have about the theories of scientific method that embody such rationality. Later Laudan abandoned any such role for pre-analytic intuitions for the following, and other, reasons. Scientific methodologies are deprived of any substantive critical role; the critical role is to be played only by the intuitions that no methodology should overturn. Nor is it clear that the intuitions will single out some preferred methodology above all others; it might well be the case that given all the widely acceptable intuitions, methodologies as disparate as those advocated by Bayesians, Popperians, Lakatosians and Kuhnians might fit them equally as well. That is, in Quinean fashion our intuitions, which play a role similar to that of observations in science, might underdetermine methodologies in that two or more scientific methodologies are tied for best fit with the intuitions (this is envisaged in Laudan (1984), chapter 2).

None of these points count against the role that FSR is to play here. Laudan's earlier approach is to test some principles of methodology against a test basis comprising Lakatosian pre-analytic intuitions. If that is their purpose then Laudan's criticisms of what he calls 'pre-analytic intuitions about scientific rationality', would have some force. But since the purpose here is not one of test, as will be seen, then these objections can be set aside. However the later Laudan has developed a different approach to methodology, *normative naturalism* as he calls it, which suggests another source of claims to include in FSR.

Laudan's approach within normative naturalism is to start with some alleged principle of method that scientists say they have adopted, or which philosophically minded methodologists have proposed, and then test it against particular episodes in the historical record of science. As already noted in sections 1.3 and 3.3, methodological principles are either hypothetical imperatives, or categorical imperatives with a suppressed value. And as has also been noted, these have associated assertoric claims that are open to empirical test. Their test basis is not the pre-analytic judgements of a scientific élite. Rather the test basis is provided by the entire history of science understood as a repository of particular cases of the application of some means to some epistemic end.

Let us take a particular example. Copernicus was well aware of Aristotle's argument against the motion of the Earth about the Sun that went as follows. If the Earth orbits the Sun, then we ought to observe stellar parallax in the course of a year's orbit; but we do not observe stellar parallax; so, the Earth does not orbit the Sun in the course of a year. Of course there are a number of assumptions here, such as that the stars are near enough for us to observe angular displacement. (Note that Aristotle already had independent grounds for claiming that the stars were near enough so that, if it occurred, we ought to be able to observe stellar parallax using naked eye observations). In order to save the claim that the Earth orbits then Sun, Copernicus' response to Aristotle's argument was to adopt the un-Popperian untestable (at the time) saving hypothesis, viz., that the stars were so far away that we could not observe stellar parallax in the course of the Earth's annual orbit of the Sun. Now Copernicus' heliostatic hypothesis was able to explain a number of phenomena that the Ptolemaic models did not explain. But it raised a problem about stellar parallax that these rival models did not even face. Given the evident progress in some areas made by Copernicus' heliostaticism, but also the new problems it faced in other areas that its rivals did not face, what ought any scientist in this epistemic situation do? Reject the theory because of this problem? Keep the theory because of its progress elsewhere and ignore the problem? Or introduce saving hypotheses?

In this example from Copernicus we have a particular case of an ends-means strategy, the end being the preservation of hypotheses and models to do with heliostaticism which have made some progress in the explanation of planetary phenomena. And what are the means? This is not entirely clear, but it is at least to bid Copernicus, in the light of the success his model had in explaining planetary motion, to adopt saving hypotheses elsewhere even if they are at the time beyond test. Also in this example there is some judgement about the trade-off to be made between increased explanatory· success in one area against *ad hocness* elsewhere.

In order to describe this situation more fully we can borrow some terminology from Kant's *Groundwork of the Metaphysic of Morals*, and employ it, not in the context of moral action, but in the context of epistemic action or decision. Kant draws a distinction between objective moral principles upon which moral agents would act if they were rational, and maxims of action which are subjective principles which individuals apply in particular cases of action. Moreover these maxims can be good or bad. The first distinction, viz., that of objective moral principles, is akin to the more objective and binding principles or norms that we seek in some theory of scientific method. But what in scientific method corresponds to the Kantian maxims of action? These, I suggest, are the particular maxims that epistemic agents, such as Copernicus, act upon in their particular circumstance (as they understand it) in which they use some means to some end, or to realise some aim or value. And these maxims may be good, or bad, in that the means employed does, or does not, realise the aim.

We can draw on the Kantian analogy of a maxim of action and apply it to the case of Copernicus in his epistemic situation. In his situation he acts as an epistemic agent in making some move in the game of science. And in so doing there is some maxim under which his particular action falls. If he were sufficiently reflective he might be able to formulate a maxim for himself; but if not, then there can still be a maxim on which he does act that historical research into the particular Copernican could reveal. Such a maxim might be the quite particular: in his epistemic situation S (S is a description of

Copernicus' scientific problem situation as indicated above), adopting some particular saving hypothesis (about great stellar distances) will achieve certain ends (here there needs to be supplied the ends Copernicus strove to realise, including the preservation of his heliostaticism). Given their particularity, such maxims are not principles of methodology. The focus here is upon the particular means-ends strategy employed by Copernicus in his particular situation. And this can be expressed in some maxim of action, a quite circumscribed hypothetical imperative applicable to his situation, which has normative force. Of course, some of these maxims might be bad or unsuccessful; or they may be unreliable with respect to their proposed end; or they may produce bad or unsuccessful moves in the game of science leading to stagnation or to pseudo, or crank science, or to non-science. For example, Copernicus might, in his problem situation, toss a coin, and this is his maxim of action in that circumstance. In contrast other maxims will be good in that they are involved in the generation of a successful move in the history of science.

Here is a second source of particular claims for FSR. They are the maxims on which scientists have acted when they use ends-means strategies at particular junctures in the historical sequence of our developing sciences. Thus there are two sources of FSR. One is the Lakatosian value-judgements of a scientific élite about the moves that have been made in the game of science. Let us refer to these as j_1, j_2, ..., j_m. The other arises from the successful means-ends strategies, or maxims, that scientists have adopted in particular situations. Let us call these m_1, m_2, ..., m_n. Could they be combined? Only if no inconsistency were to arise in doing so. Here it is supposed that there is a consonant relationship between (a) the value judgements of the scientific élite over moves in the game of science, and (b) the maxims upon which scientists have acted. Granted this, FSR can be specified in much the same way as ordinary folk psychology. FSR is given by a large conjunction of value-judgements of the scientific élite about episodes in the history of science, and the maxims of action of scientific actors in their particular means-ends situation. That is:

$$FSR = j_1 \& j_2 \& ... \& j_m \& m_1 \& m_2 \& ... \& m_n.$$

(Note: though the conjunction, as expressed, is finite it might be quite large.)

3.6.3 The Need for Folk Scientific Rationality

Are we guaranteed that there is such a thing as FSR, or in other words, is there such a conjunction? We have seen Lakatos' remark that while there have been, and still are, arguments about what are the high level methodological principles which comprise our theories of scientific method, there has been large agreement about what are the successful moves in the game of science. And this agreement is to be found in the judgements of the scientific élite. Where there is not wide agreement about some episode then no claim on behalf of the scientific élite can be listed in the j_is. What about the maxims? These can be discovered by historical research into what were, or are, the actual avowed maxims of scientists. Where these are not available they can be conjectured on the basis of further historical research into the scientific episode to uncover the ends-means strategy adopted in a particular epistemic circumstance, and what maxim might suit it best.

Both Popper and Lakatos require that there be something like FSR to test their claims about a universally applicable scientific methodology, even though they differ over what this universal theory of method might be. Many Bayesians also see themselves as proposing a broad theory of scientific method which derives from a particular understanding of the probability calculus, rules for up-dating, and the rest of their rather impressive technical machinery. They, too, try to compare their Bayesian methodology with what FSR yields. Despite the many optimistic Bayesians who claim that Bayesianism can capture all of FSR, there are those such as John Earman who thinks that subjective Bayesian confirmation theory will perform its task on only half the days of the week (Earman (1992), 'Introduction'). And there are those such as Ronald Giere (Giere (1988), chapter 6) who have abandoned Bayesian confirmation theory though they keep aspects of decision theory. But they, too, need something like FSR. It is not the task here to assess the performance of various theories of scientific method and their principles and norms against FSR. Rather the point is to show that methodologists have thought that FSR plays a necessary role in assessing the principles of any scientific methodology.

Even Feyerabend thinks that there is something like FSR. As has been argued (section 1.5.1), he never really endorsed the claim that 'anything goes' in methodology; rather he attributed this jokingly to his opponents. What he does say in many places is remarks like the following: 'I never "denigrated reason", whatever that is, only some petrified and tyrannical version of it' (Feyerabend (1995), p. 134). Or, 'science is not "irrational"; every single step can be accounted for (and is now being accounted for by historians ...). These steps, however, taken together, rarely form an overarching pattern that agrees with universal principles, and the cases that do support such principles are no more fundamental than the rest' (*ibid.*, p. 91). What Feyerabend rejects is the idea of a universal theory of method applicable to every move in the game of science of the sort proposed by Popper, Lakatos, the Bayesians, and others. What he instead endorses is the idea that each episode in science, each move or gambit in the game of science, does have a rationale; and this can be captured by maxims of action, or can be discerned by a scientific élite. As indicated in section 1.5.1, Feyerabend says: 'I argue that ... there is no comprehensive 'rationality', I do not argue that we should proceed without rules and standards. (Feyerabend (1978), p. 32) Thus, far from rejecting the idea of FSR, Feyerabend actually endorses, and argues for, its existence through his own and others' investigations into the history of science. Perhaps the best characterisation of the position that Feyerabend adopts is that he is a minimalist about the claims of any theory of methodology. All we might ever be able to establish is the existence of FSR and not any grander claims about a more general, or a universal, theory of method which would capture all of FSR.

Whatever verdict FSR gives concerning the existence of some ultimate theory of method, the task in which Popper, Lakatos, Laudan, the Bayesians and others have been engaged, the use of FSR here will be quite different. It is to show how the normative notions employed in FSR can be reconciled with naturalism. It is not to tell us what are the principles of a more general normative scientific methodology. In the next section 3.7 it will be shown how to use FSR to "define" its normative notions; in section 3.8 it will be shown that the normative in methodology and epistemology is supervenient on the non-normative.

3.7 RECONCILING THE NORMATIVE WITH THE NATURAL: RAMSEY-LEWIS DEFINITION.

In what follows I will draw on the work of Ramsey ((1990), chapter 6), Lewis ((1983), chapter 6 and (1999), chapter 16) and Jackson ((1998), chapters 5 and 6), using a Ramsey-Lewis approach to define the normative epistemic terms contained in our methodological norms by means of descriptive language only. This style of "definition" is a considerable departure from the common style of finding necessary and sufficient conditions for some *definiendum*. Rather it begins with the overall role that a term plays in a large number of contexts in which that term occurs and uses that context to fix the meaning of the term. Such a style of definition shows how the normative can in principle fit quite happily into a broadly construed descriptive-naturalistic framework.

Consider our folk scientific rationality FSR (abbreviated to 'S'):

(1) $S = j_1 \& j_2 \& \ldots \& j_m \& m_1 \& m_2 \& \ldots \& m_n$.

Each of the conjuncts on the right hand side will contain logical terms and non-logical terms that are either epistemically normative (such as 'justifies', 'warranted', etc) or non-normative, or descriptive terms (this including terms expressing epistemic values such as 'simplicity', 'accuracy', etc, which are in this context descriptive). We can regard the conjunction as an "implicit" definition of the normative terms it contains. Thus we can say that the meaning of each normative epistemic term is given by the role it plays in our folk scientific rationality, that is, the role each plays in the conjunction of all the judgements of the scientific élite and all the maxims of epistemic decision. Whatever other contexts these terms might occur, it is in the context of S that their meaning is fixed. We can explore this a little further.

The first task is to take all the epistemically normative terms which are syntactically not names and treat them as names of properties, relations and so on. For example the term 'justifiable' is not a name but a predicate in the following: adopting *ad hoc* hypothesis H in situation S is justifiable if the goal is to achieve v'. But it is possible to transform this sentence without loss of meaning so that a name of a normative property is introduced in the following way: 'making *ad hoc* move H in situation S to achieve v has justifiability (or has the property of being justifiable)'. The task here is to make all the normative terms referential so that logical inferences can be made, such as existential generalisation, substitution by co-referring names, and so on, entirely within the framework of first-order logic. Importantly with all the normative terms in a name position, it will be possible to use only first order logic to "Ramsify" the large conjunction of sentences of FSR in a manner recommend by Lewis. So in what follows suppose that all the conjuncts in FSR are suitably reformulated without loss in this manner.

Let $N_1, N_2, \ldots, N_k$ be the names of normative properties which occur in S; and let $D_1, D_2, \ldots, D_r$ be the non-normative, non-logical, descriptive terms in S. Then the left hand side of (1) can be written as:

(2) $S(N_1, N_2, \ldots, N_k, D_1, D_2, \ldots, D_r)$.

The next step is to replace all the names of normative epistemic properties by variables so that one obtains the following open sentence containing only variables, logical expressions and non-normative or descriptive expressions $D_1, D_2, \ldots, D_r$:

(3) $S(x_1, x_2, \ldots, x_k, D_1, D_2, \ldots, D_r)$

Each of the k variables in the above open sentence make clear the context, within S, of the normative expression each replaces, and thus the meaning role that each epistemic notion plays in the conjunction **S**.

Denote the Ramsey sentence of (3) by 'S^R'. This is got by placing an existential quantifier in front of the open sentence for each of the variables and forming a closed sentence:

(4) $S^R = (\exists x_1)(\exists x_2) \ldots (\exists x_k)[S(x_1, x_2, \ldots, x_k, D_1, D_2, \ldots, D_r)]$

For the sake of convenience, let us prescind from the generality in the above, and consider S as having just one name of a normative property, and thus one variable in its open sentence. For example the name 'N' could be 'justifiability'. Then we have respectively:

(2*) $S(N, D_1, D_2, \ldots, D_r)$.

(3*) $S(x, D_1, D_2, \ldots, D_r)$.

(4*) $S^R = (\exists x)S(x, D_1, D_2, \ldots, D_r)$.

Lewis' modification to Ramsey's approach is not to place an existential operator in front of (3*) to get a closed sentence (4*), but rather to introduce a definite description operator so that a generalised description is introduced. Thus if the definite description operator is '(¶x)' and 'N' is a name, then a name 'N' can be introduced *via* a description as follows:

(5*) N =$_{Defn.}$ (¶x)S(x, D_1, D_2, ..., D_r).

We need not be too concerned with the issue as to whether (5*) picks out a reference for N or gives the meaning of N. In one sense it does both. Thus it could be said to specify a meaning for the folk scientific norm 'N' as 'whatever typically fills the N-role in the conjunction S'. And the right-hand description picks out a reference for N. But a preliminary matter first.

The open sentence (3*) may be satisfied by no entity, or one and only one entity, or by two or more entities. In the first case, if nothing satisfies (3*), and so (¶x)S(x, D_1, D_2, ..., D_r) picks out no entity, then nothing fills the N-role in S, and the name N has no reference. In the second case, if (3*) is satisfied by one and only one entity, then (¶x)S(x, D_1, D_2, ..., D_r) picks out a unique entity, or there is a unique thing that fills the N-role, then N refers to that unique thing. Moreover 'N' simply means 'whatever fills the N-role in S'. In the third case, in which there are two or more satisfiers, we can proceed in one of two ways. Following the earlier view of Lewis we can say that (¶x)S(x, D_1, D_2, ..., D_r) picks out nothing and so N does not refer. Or following the later Lewis ((1999), p. 301), if two or more entities satisfy (¶x)S(x, D_1, D_2, ..., D_r), then we can say that N has multiple, or indeterminate, or ambiguous, reference to the two or more entities. In what follows, nothing will turn on which of these two ways we proceed.

A further modification is also in order. The satisfaction of an open sentence by an entity is usually understood to be an all or nothing affair; either there is perfect satisfaction or no satisfaction. However it need not be "all or nothing"; we can satisfice instead. We can look for not the perfect satisfier, but that entity which amongst the imperfect satisfiers comes the closest to satisfying the open sentence. Such maximal, but less than perfect, satisfaction is important for the following reason. Given that our FSR is a conjunction of claims, then the more conjuncts there are the greater the chance will be that there is no perfect satisfier of the conjunction. We could of course delete conjuncts until there is a perfect satisfier. But there may be no unique way of performing the deletion. Moreover we would strip away in an arbitrary manner some of the claims of our FSR. Rather the better strategy is to keep the claims of our FSR and look for the nearest best satisfier of the open sentence, if there is no perfect satisfier.

In the above we have, for the sake of simplicity, considered only one normative term. But there will be more than one normative term in any sufficiently rich S. These several terms also enter into relations with one another in the context of S so that each contributes to the implicit definition of the other. That is, not only is there a meaning role that each normative term plays in S in the context of descriptive terms only; there is also a further role that each normative term plays with other normative terms to take into account in specifying the meaning. This can be illustrated in the general case where there are k names for normative properties, N_1, N_2, ..., N_k, and there is a sequence of definitions for each name from 1 to k of the following sort:

$(5_{(1)})$ $N_1 =_{Defn.}$ $(\P x_1)(\exists x_2, ..., x_k)(\exists y_1, ..., y_k)[S(y_1, ..., y_k, D_1, D_2, ..., D_r)$
$\equiv (x_1 = y_1) \& ... \& (x_k = y_k)]$

. . . .

$(5_{(k)})$ $N_k =_{Defn}$ $(\P x_k)(\exists x_1, ..., x_{k-1})(\exists y_1, ..., y_k)[S(y_1, ..., y_k, D_1, D_2, ..., D_r)$
$\equiv (x_1 = y_1) \& ... \& (x_k = y_k)]$

Setting aside this more formal exposition, what has happened in proposing such a kind of definition? Consider first the original use to which the Ramsey-Lewis style of definition was put, viz., to give an account of the meaning of theoretical terms in scientific theories. Initially there was a separation of the vocabulary in which scientific theories are expressed into three sorts: the logical terms, and amongst the non-logical terms there was a further division into O-terms, the old, or original, or if you like the observational, terms which had their meaning antecedently fixed, and the T-terms, the theoretical or new terms, which are to have their meaning fixed by means of the O-terms. The specification of meaning is to be done by means of equations akin to those labelled as '(5)' which contain, as well as logical expressions, only O-terms (compare these with what are called above 'descriptive terms'), and no T-terms (compare these with what are called above 'normative terms'). The style of definition is not that of giving necessary and sufficient conditions for the application of some term, as in the case of 'father' means 'male parent'. Rather, the Ramsey-Lewis style of definition does something quite different. It takes the overall context (or most of it, or some chosen sub-context) in which a term occurs, and then specifies its meaning, and its reference, in terms of the role it plays in that context. In addition, a reference is fixed by expressions like (5) in much the same way as reference would be fixed by an ordinary definite description.

Much the same has been done in the case of the claims of our FSR. There are several points to note here. First, the vocabulary in which they are expressed is separated into the logical and the non-logical and the latter is further divided into the epistemically normative (compare with T-terms in the case of science) and the epistemically non-normative, or descriptive (compare with O-terms of science). In order to make this separation it is assumed that there is a ready way of identifying what are the epistemically normative terms of appraisal. And the best way of doing this is to simply draw up a list of the epistemically normative and relegate the rest to the non-normative. It is then assumed that the meaning of the descriptive terms is already antecedently given. (Note that in this context the so-called "values", such as 'simplicity', 'truth', etc, will be descriptive terms.) The task is now to specify the meaning (and reference) of "new" terms, in this case the normative terms employed in FSR, through the role they play in the context of FSR.

Second, the style of definition in equations labelled '(5)' is such that some normative term is chosen as the term to be defined on the left-hand side of a definitional equation. What does the defining is the (generalised) definite description on the right-hand side of the definitional equation. Importantly the right-hand side contains only logical vocabulary and non-normative, or descriptive, vocabulary. And it is this right-hand side which gives a definition of a normative expression in terms of purely descriptive terminology. Within the conjunction that is FSR, an implicit definition of normative terms is given by means of the context in which they occur in the conjunction, that is, by the role they play in FSR. The definitions given in the various equations labelled '(5)' merely spells this role out, but in terms of an explicit style of definition in which the notions in the *definiens* are purely descriptive.

Third, the right-hand side of the equation will be satisfied by a descriptive property (if it is to be satisfied at all). This is so because all the terms on the right-hand side are descriptive. But since it is a definition of a normative expression on the left-hand side, what we have is an account of the descriptive conditions for the application of the normative expression. Importantly, the right-hand side gives the truth conditions for the application of the term. In this manner there is a reconciliation between normativity in epistemology and methodology and a broadly conceived naturalism that admits descriptive properties of the sort indicated above. And the reconciliation is bought about while being realist about both. That is, we can adopt *position (3)* of normative objective naturalism (see section 3.4.3).

It should be clear now that there is one task that the above does not accomplish. It does not generate any substantive principles of a theory of scientific method of the kind sought after by Popper, Lakatos or the Bayesians, or of the sort denied by Feyerabend. This is not to be expected from a procedure that sets out to define only normative terms. Finding the substantive principles of method still remains a task to perform. What has been done is to specify the meaning of normative expressions, as they occur in our folk scientific rationality FSR. And if some set of methodological principles were to capture all (or most) of FSR, then the normative expressions in these principles would also have their meaning and reference specified through FSR.

The above can also apply to the principles of deductive logic (but they are not strictly part of FSR). Thus the deductive rule *Modus Ponens*, says: from premises A, and if A then B, one is *justified* in inferring B. The terms in this rule are either logical or descriptive, the only evaluate term being 'justifies'. So even the rules of deductive logic can be "Ramsified". The term 'justifies' occurs in the context of many rules of inference and can be defined using an appropriately formed definition along the lines of equations labelled as '(5)'.

There remain two further matters to be resolved. The first is to say what property the normative term N does pick out. It will not be some weird Platonic item such as a *sui generis* normative item of the sort sometimes envisaged by G. E. Moore or H. Sidgwick. Rather, as the account of definition just given shows, as well as the account of supervenience of the normative on the non-normative in methodology set out in the next section, it will be non-normative descriptive property which congenially fits with naturalism. But all this shows is *that* there is some descriptive property to be picked out; it does not say *what* property.

The second unresolved matter is to ask: what descriptive property, or properties, are picked out by whichever of the equations (5) are used? Let us suppose that whichever equations (5) we use, some descriptive property, or properties, are picked out. All we have shown is that, under specified conditions, something is picked out, and not what it is. However the equation does specify descriptive conditions which must apply for 'N'. This is a considerable advance in itself. We have an account, for all the normative expressions of FSR, of what are their descriptive conditions of application. But we need to go further and attempt to say which property, or properties, best fulfil the available roles within FSR.

Here we can make a conjecture about what is either the perfect satisfier of the open sentence given in the equations labelled as '(3)', or the nearest best satisfier of (3). This conjecture may not be the correct one, but it is at least one that has been canvassed in the literature. Suppose that amongst the various normative terms we focus upon 'justifiability'. And further, suppose that the property that is picked out according to one of the definitions given as (5) is: the property some hypothesis H possesses of having a much higher degree of probabilistic support on the basis of evidence than in the absence of E (that is, for some hypothesis H and evidence E we have prob(H, E) >> prob (H, not-E)). (Perhaps we could claim, independently of the conjecture, that this property is a pretty good satisfier of (3) though perhaps not the best or the most perfect satisfier.) Now note that the suggested property is one that is given only in terms of purely descriptive and logical vocabulary. And by supposition it is the descriptive property that best satisfies, or is the perfect satisfier, of an open sentence such as that indicated by (3). Moreover it will be what the term 'justifiability' names according to one of the definitions (5). That is, *being justifiable* just is *having a much higher degree of support on evidence than in the absence of that evidence.*

In so claiming we have done three things. We have specified the meaning of the term 'justifiability' through the role the term plays in our folk scientific rationality FSR (some equation (3) will do this). We have also given the truth conditions for the application of the term through equations like (5). And we have shown, granted the above illustrative case, what descriptive property the description on the right-hand side of (5) picks out. In this way the project, indicated at the beginning, of naturalising the normative expressions of methodology has been carried out. The normative concepts that we employ in judging, criticising and evaluating the claims of science do not stand outside science but have a place within it, given in terms of the descriptive vocabulary also employed in our methodological principles, and particular judgements and maxims of strategic epistemic action. Against this back-drop there is a guarantee of the reconciliation of the normative with the descriptive, and so with a broad naturalism.

3.8   THE SUPERVENIENCE OF THE METHODOLOGICALLY NORMATIVE ON THE NON-NORMATIVE

The previous section showed how the naturalisation of the normative by the non-normative arises through the "definition", in the sense specified, of normative expressions employed in methodology. In this section another aspect of naturalisation will be discussed, that of the supervenience of the normative on the non-normative.

The world contains a number of items such as objects, properties of objects, events, facts, and so on; it also contains cognitive items such as beliefs, etc. Many of these can change quite independently of one another. But the world also exhibits some structure and interconnection between such items. Causation is one such relation in which one event, the effect, bears a relation of dependence on another event, the cause. But more generally one can speak of many other kinds of relation of dependence, or independence, between items. One such non-causal relation is that of the dependence of place upon some co-ordinate system. Thus consider the place of some body at a given time within a space-time system of co-ordinates. Three co-ordinates, say one of time and two of position, will not suffice to determine what place some body occupies. The third co-ordinate of space can vary quite freely; and so the place of the body can vary quite independently of the three given co-ordinates. However if four co-ordinates are given then the place of the body is fixed and there is no independent variation of the place of the body. If there were still independent variation in place, then we would have to allow that the body could be in two (or more) places given the single set of four space-time co-ordinates; but this is impossible. The place of a body depends on the four co-ordinates; or alternatively, as might be said, place *supervenes* on the four co-ordinates.

Again there is no independent variation of the (average) density of a body with respect to its mass and its volume; once the mass and the volume are fixed then so is the (average) density. So we can say that (average) density of a body supervenes on the more fundamental physical properties of mass and volume of the body.[16] The temperature of a gas in a container does not vary independently of, but rather supervenes upon, the more fundamental complex property of the mean kinetic energy of the particles of the gas. The solidity of a body does not vary independently of the strength of the lattice of attraction relations between the atomic parts of the body; rather it supervenes on these attraction relations. Eddington ((1958), pp, xi-xv) is famous for supposing that there is not just one table but two. There is the table of our commonsense world that we say is solid. But there is the table of physics that, on the atomic theory of matter in which an atom has largely empty space between its nucleus and electrons, is largely empty; so the table is insubstantial and not solid. He thereby created the impression that there were two perspectives, that of our ordinary belief and that of our scientific belief, with the scientific story being the only one that tells us about reality. In this story solidity seems to have melted away. Many have balked at this conclusion. Claiming that the solidity of an object is supervenient on the microstructure of forces within the object is one good way of resisting Eddington's "two table's" conclusion.

3.8.1 Some Kinds of Supervenience

The Latin origin of the term 'supervene' is from 'super', 'on top of' and the verb 'venire', 'to come'. The idea of "coming on top of" is suggestive of lack of independent variation of one item with respect to another, but it needs a precise analysis. There are a number of analyses of the relation of dependence, and a number of different relata than stand in the relation of "coming on top of", or of supervenience. But as the examples illustrate, the place of an object, or its density or solidity, can be readily understood to "come on top of", and not be independent of, respectively, its space time co-ordinates, its mass and volume, or attraction forces between its parts. In what follows we will focus on particular cases of the schema 'S supervenes on B' where (bold) 'S' stands for one set of properties, the *supervenient* family, and (bold) 'B' stands for another set of properties, the *supervenience base*. The properties in each family can be taken singly or in combinations involving the conjunction, disjunction or negation of properties. In what follows the (not bold) letters 'S' and 'B' will stand for some member of the two sets of families of properties S and B respectively. Note also that supervenience is an asymmetrical relation; if the S supervene on the B then the converse does not follow.

The notion of supervenience entered recent philosophy through a discussion in moral philosophy of the relationship between moral properties, denoted by evaluative terms such as 'right', 'good', 'ought' and so on, and non-moral properties denoted by non-evaluative or descriptive terms such as those which describe the behaviour of a person, their mental states, traits of character, the consequences of their actions, and so on. It has been extended to non-moral evaluative contexts such as aesthetics and epistemology, and to non-evaluative contexts such as the philosophy of mind in which the mental is said to supervene on the physical. We will be concerned here with evaluative contexts only, and in particular the evaluations that arise in epistemology and methodology.

To illustrate some of the different kinds of supervenience consider an uncontroversial case of the grading of apples. Apples can have non-evaluative properties such as an appropriate oval shape, not being bruised, and being red, crisp, ripe, shiny, and so on (these properties and their combinations give us the family of properties which is the supervenience base **B**). We also say of apples that they are good (goodness being one property in the supervenient family **S**). But do we admit the independent variation of goodness from the supervenience base? Suppose we do. Then it would be possible that one apple x has the properties of being oval, unbruised, red, crisp, juicy and ripe, and of x we say that it is good; however another apple y has exactly the same properties and yet of y we say that it is bad (= not good). There is no formal inconsistency in admitting of two objects, the same in all relevant respects, that one is good and the other is not good. But there is a seemingly arbitrary variation in goodness with no variation in the supervenience base **B** of natural properties. If we were to ask why one apple was good but the other was not when the two apples are exactly the same in all other relevant respects, then we would be at a loss to point to any difference. Since there is a violation of a principle of sufficient reason, many have claimed that the independent variation of goodness from the class of properties in the base **B** is not to be admitted. Instead they claim that if one apple x is good then the other apple y is also good.

This gives rise to the first kind of supervenience:
S *weakly supervenes* on **B** = necessarily [i.e., in all logically possible worlds w] for any x, and for any property S in the supervenient family **S**, if x has S then there is some property B in the supervenient base **B** such that x has B, then for any other y, if y has B then y has S.
Putting this another way, for two objects [in the same world w] if they share all their **B** properties then they share all their **S** properties.

However this notion of weak supervenience does not capture many of the claims that some have made about the dependence holding between **S** and **B**. Applying the above example of apples to weak supervenience, what we have is a claim which considers a pair of apples, x and y, *within* a possible world. So, *within* any world w, if two apples x and y agree in their base properties **B** then they do not differ in their evaluative properties such as goodness. But what happens when we consider the very same apples, x and y, across other possible worlds? To make matters simpler, consider the actual world w_a which has two apples, x and y, in it. Supposing weak supervenience holds; then if x and y share their base properties **B** they are both good (i.e., they share some supervenient property from **S**). But this holds only in the actual world w_a. It does not say what happens when we envisage apple x (or y) being in some other possible world w with the *same* base properties **B** as it has in the actual world. Does apple x with given properties from **B** also have, in some other world w, the same supervenient property of goodness as it has in the actual world, or not? If not, then the association of **B** properties with goodness is merely a coincidence that holds only in our actual world.

We need to envisage the dependence of **S** properties on the base properties **B** to hold in counterfactual situations as well, and not to vary across possible worlds. This yields a stronger notion of supervenience:

S *strongly supervenes* on **B** = for all possible worlds (i.e., necessarily), and for any two objects x and y, either in the same world or in different worlds, and for any property S in the supervenient family **S**, if x has S then there is some property B in the supervenient base **B** such that x has B, then for any other y, necessarily if y has B then y has S.

Putting this another way we can say, of two objects, if they share all their **B** properties then they share all their **S** properties, and necessarily so. The difference in the two kinds of supervenience is bought out as follows; for weak supervenience the conditional 'for all y if y has B then y has S' is contingent while for strong supervenience it is necessary. Thus for example, it is necessarily the case that if an apple is unbruised, red, crisp, etc then it is good. For strong supervenience, there is a one-way necessary connection between base properties and a supervenient property; in the case of weak supervenience this conditional is not necessary true but only contingently true.

The above concerns, in the case of individual objects (such as apples), their base and supervenient properties. But some have wanted to talk not just of individuals in the world but globally of the world as a whole. Thus in the moral case they wish to say that there could not be two worlds which are the same in every non-moral feature yet differ in their moral features. Or in the case of global materialism with respect to the mental they wish to say of any two possible worlds, if two worlds are indiscernible physically then they are indiscernible psychologically.[17] Putting this more generally we have a further kind of supervenience:

S *globally supervenes* on **B** = worlds that are identical (indiscernible) with respect to the base properties **B** are also identical (indiscernible) with respect to their **S** properties.

We will not explore further the wide range of different notions of supervenience that can be distinguished, but rather focus on just one of them in what follows, viz., global supervenience. What global supervenience rules out is the following: there are two possible worlds which are such that they are indiscernible with respect to their base **B** properties but are discernible with respect to their **S** properties. That is, global supervenience rules out the possibility that there be differences in **S** properties with no difference in any **B** properties so that the **S** properties vary independently with no change in the **B** properties. There is an important consequence to be mentioned. In the case of global supervenience the following can be established. If B is from the base supervenience family **B** and S from the supervenient family **S** then, if S globally supervenes on B then S and B are necessarily co-extensive properties; that is, it is (metaphysically) necessary that, for all x, x is B if and only if x is S. This will not be proved here.[18] But it is significant for what follows.

In the case of the global supervenience of (ethically, or epistemologically or normatively) evaluative properties on non-evaluative or descriptive properties then, given necessary co-extensiveness, we have the following. The asymmetry of the dependence of the evaluative on the descriptive is retained and is not undercut; this is so because supervenience is an asymmetric relation. But it also turns out that, due to the above necessary co-extensivity, the following holds. All the possible ways we have of evaluating our beliefs are just all the possible ways in which descriptive and relational properties (in some base family **B**) can pertain to the beliefs. That is, there is no sameness or difference in evaluation of beliefs that is not also a sameness, or difference, in description of the beliefs. Or putting the point linguistically, all the possibilities of the use of evaluative language to talk about the world are just all the possibilities of the use of non-evaluative (descriptive) language to talk about the world. Neither can depict some possibility that the other cannot. But this does not mean that evaluative talk is somehow eliminable; far from it. For example, being bald is supervenient on hair distribution. But there are an infinite number of possible hair distributions that count as being bald. We would be hard pressed to list each possible description of hair distribution in an infinite disjunction and say that this is what 'bald' means. Similarly, we may be hard pressed to specify in a disjunction all the descriptive ways that count as fulfilling some evaluative notion. But given the supervenience, there are no possibilities left out so that the evaluative and descriptive possibilities do not mirror one another. Later we will want to say much the same about epistemic and methodological evaluations about our beliefs.

Pursuing the analogy with hair distribution will help further. There are a very large number of patterns of hair distribution D_i (i = 1, 2, ...) such that if one is bald then one exemplifies one of these patterns of hair distribution. Also being bald, it is said, is supervenient on hair distribution. What this means of that if there are any two instances of the same pattern of hair distribution, D_i, then one of these cannot be said to be bald while the other is said not to be bald. Putting these two claims together we get the following. If baldness is present somewhere then there will be some pattern of hair distribution, from the large number of possible patterns of hair distribution, say D_k, upon which the baldness supervenes. So in this respect talk of baldness can go along with talk of some hair distribution. In addition, by virtue of the supervenience we are guaranteed that talk of each of the many patterns of hair distribution can also count as talk of being bald. But we would not want to abandon our talk of baldness and replace it by talk of the vast number of patterns of hair distribution that count as baldness. However there is nothing that cannot be said in one mode of talk, bald talk or hair distribution talk, that cannot be said in the other.

Similarly for moral judgements. To use an example from the literature, if St Francis has a number of descriptive features which hold of his character, actions and thoughts (e.g., his character is honest and just) and we say of him that he is good, then any person with the same descriptive properties *must* also be good. Similarly we can say of Socrates that, even though he never fed the birds and seems on occasion to have harboured thoughts of sexual intent that St Francis did not, he had a courageous character, he never knowingly did anyone harm, and so on; so Socrates, too, is good (though the properties from the supervenience base are somewhat different from those of St Francis). Then on the supervenience thesis, any other person who exhibits the same descriptive properties (from base **B**) as St Francis, or Socrates, then they too will be good. If they are not good, then there must be some difference in respect of the supervenient base **B**. Thus that Hitler is evil indicates that there are descriptive properties of him that are not in the supervenient base for either St. Francis or Socrates (e.g., has a desire to exterminate all Jews and non-Aryans). In sum, talk of goodness and talk of any of large class of descriptive properties from base **B** go along with one another. But this does not mean that talk of goodness is eliminable any more than talk of baldness is eliminable in favour of talk of hair distribution.

3.8.2   The Supervenience of Normative Methodological and Epistemological Properties.

The significance of the above is that it can be usefully applied in epistemological and methodological contexts. Here the evaluate terms include 'knowledge', 'justification', 'good' (as applied to, say, argument), 'right' (as applied to, say, an inference or a conclusion), 'ought' (as in the case of the inferences one ought to make), and so on. Thus consider some belief p. We can make any of the following purely descriptive and non-evaluative claims about p: p is a logical consequence of another proposition q; p is indubitable (i.e., beyond all logically possible doubt); p is irrefutable (has no possible evidence against it); a person remembers that p; p is supported (above a minimum threshold) by evidence e; p is more strongly supported by e than is q; p is *ad hoc* with respect to hypothesis H; p has more empirical content than q; p has greater truthlikeness that q; p explains e; p explains more facts than q; p has been produced by a reliable belief forming process; p is a belief based on perception; p coheres with set of beliefs S; p is true; and so on. Suppose that these and other descriptive properties, either taken individually or in combinations, comprise the supervenience base **B**. The we can ask of belief p: if p is justified, in virtue of what properties B from the base **B** does this hold? We could then appeal to its indubitability, or its high degree of evidential support, or that it is based on memory or reliable experience, etc (or a combination of these).

These descriptive properties might cluster together and comprise the core of some theory of knowledge such as the coherence, reliabilist, justificationist or classical foundationalist theories. One epistemologist might appeal to some cluster of descriptive properties that another does not. Thus a classical foundationalist will appeal to the property of indubitability whereas a reliabilist or a coherentist would not; they make appeal to other descriptive properties. So on some theories within epistemology some of the above descriptive properties (and their combinations) will turn out to be inadmissible. This can be taken into account in specifying what supervenience base **B** a particular epistemological theory requires.

These points are also usefully made in Sosa ((1991), chapter 9, section II). He sets out some of the requirements of supervenience of a normative property as follows. A belief may be said to be justified if the belief has the following descriptive property S (which may be complex relational): (i) the belief has property S; (ii) S is not a normative property [this is the descriptive requirement], and (iii) necessarily whatever belief has S is also a justified belief [this is the supervenience requirement]. To illustrate further, we may take as an example the descriptive properties of some classical foundationalist theory of knowledge to involve, say, indubitability and inference, so that the following holds: a proposition is either indubitable or can be logically inferred from indubitable propositions. Then within classical epistemology the norms of knowledge or justification supervene on the descriptive features of indubitability and inferential relations. For other epistemological theories other supervenience bases can be specified. (Which is the correct supervenience base is a matter to resolve concerning the various rival epistemological theories.)

These considerations based in epistemology carry over directly into methodology. The descriptive properties will be part of some theory of scientific method. For example a theory, such as Popper's, puts emphasis on non-*ad hocness*, absence of irrefutability, high comparative truthlikeness, etc; or a theory such as Lakatos' puts emphasis on whether a programme is progressive, stagnating, or has degenerating problems shifts; and so on. And the evaluative or normative notions in each methodology will be supervenient in just the same way as they are in epistemology. If a belief p is *justified*, or *warranted* or *ought* to be accepted, etc, then there will be some (combination of) descriptive properties in virtue of which it is so evaluated. For example, p might be observed to be so, has weight of evidence in its favour, is not *ad hoc*, etc. Suppose some other belief q also has exactly the same descriptive properties. Then to say that it would *not* be similarly evaluated would be to admit that there can be independent variation of methodological evaluation of belief with no variation in any of its descriptive properties. And as in aesthetics or ethics, such evaluations would appear to be arbitrary in that they are founded in no difference in the descriptive base **B** whatever. To rule out such arbitrariness, the evaluation of a belief must be supervenient on some descriptive base. That is, for an evaluative (normative) property N of the supervenient class **S** there are descriptive properties B in a supervenience base **B** which are such that if some belief has B then it must necessarily have evaluation N. This is an expression of strong supervenience.

If we consider epistemic and/or methodological evaluation as a case of global supervenience then the following will hold: worlds in which (sets of) beliefs are identical (indiscernible) with respect to their descriptive base properties **B** are also identical (indiscernible) with respect to their epistemically and/or methodologically evaluative properties. And there is an additional boon of the necessary co-extensiveness of properties. Thus we can say of our beliefs that any evaluative way in which they can be characterised (such as being justified, etc.) is also a way in which they can be characterised by non-evaluative or descriptive properties; and conversely. Any claims we can make about the evaluation of our beliefs (or system of beliefs) cannot make distinctions that cannot also be captured by the non-evaluative or descriptive characterisation of our beliefs. Put linguistically, the possibilities of evaluative talk about our beliefs just are the possibilities of descriptive talk about our beliefs. Neither can make distinctions that the other cannot.

Thus far we have argued for the bare claim (concerning our beliefs, theories, hypotheses etc), that the evaluative properties of our beliefs supervene on the descriptive properties of our beliefs. We have not said in what way they do supervene. But this is not a matter that we need address here. We can leave it up to each epistemologist, or to each methodologist, to spell out the way in which they think that the supervention holds. All we have given is a reason for thinking that some supervention holds, not what it is. And this is all we need to do to make possible the reconciliation of the normative with the natural. What we have shown is that there can be epistemic norms that do not appeal for their correctness to any kind of weird fact; but they do retain their objectivity. Nor need we reconstrue them so radically that they do not have truth values, as in the case of Gibbard's norm expressivism. Finally such an account of norms is not inconsistent with naturalism. What we require is that the norms have a basis in non-normative descriptive properties of our beliefs, theories and hypotheses. And such descriptive properties of our beliefs fit happily with a broad naturalism.

Finally, what is the connection with the definition of normative properties given in the previous subsection? The two sections come together nicely. In the previous section it was shown how the meaning of normative terms can be determined by the functional role they play in the context of folk scientific rationality. And the Ramsey-Lewis style of definition shows how to give the truth conditions of normative terms using only non-normative or descriptive language. It also shows how to pick out a property, understanding the definition as illustrated in equations (5), as one that can fix the reference of a term. The reference fixer does this using only descriptive language. And what is referred to is itself a descriptive property. This position is equally supported by the above discussion of the global supervenience of normative notions in methodology upon the non-normative. And as pointed out, the possibilities of normative and non-normative talk in methodology are the same; so, nothing cannot be said in one kind of talk that cannot be said in the other. And this much does effect a reconciliation between the natural and the normative. The reconciliation retains objectivity without appeal to an independent realm of weird normative fact. And it shows us how to avoid norm nihilism, norm scepticism, or non-cognitive accounts of normativity such as norm expressivism. The norms remain with their full force of authority and objectivity, and in a way that sits happily within naturalism.

NOTES

[1]   See the remarks of Cahoone cited in the 'Introduction' to this book, p. 2, where he says that one of the central features of postmodernism is the view that norms of reason are immanent in social processes and not transcendent.

[2]   For a much clearer account of the connection between the hypothetical imperatives of scientific methodology (linking means with epistemic ends) and empirical claims that are open to test, see the following section 3.4.3. There will be found a brief discussion of Laudan's account of method that is more sophisticated than Quine's. See Laudan (1996) Part IV, and Nola and Sankey (eds.) (2000) section 11, for more on the connection between Quine and Laudan on methodology. There are important differences between Laudan and Quine that go

unmentioned in the above but are set out in Laudan (1996) Part Two, even though both might, in a broad sense, be said to be pragmatists.

3   Talk of location and elimination can be found in Jackson (1998) chapter 1. Jackson's book has strongly influenced some of what is said in this section, as also has Papineau (1993) who addresses similar matters in relation to naturalism. David Armstrong's writings, such as Armstrong (1981), contain several characterisations of naturalism, aspects of which have been taken on board here.

4   See Cartwright (1999) chapter 1 section 5, for her denial that the rejection of the supervenience of the physical on the non-physical need commit here to emergentism. Her pluralism in ontology is also echoed in Dupré (1993) with its denial that the sciences could ever constitute a unified project, and the positive claim that there is a plurality of things within ontology that do not stand in any relation of reduction or supervenience to one another. In the text above, a range of positions about naturalism, from an austere physicalism to a disunity of all sciences within naturalism, has been set out in order to depict the terrain of possibilities; however neither of these extremes will be argued for despite the author's own predilection for physicalism.

5   There are a wide range of contexts in which norms are relevant beside epistemology, such as ethics, language, law, games, our ordinary human behaviour and action, etc. These other norms are not of concern in this book. For a more general account of these and other areas in which norms are applicable see Raz (1975). For our purposes methodological norms may contain one or more of the following elements: a deontic operator of obligation, permission, etc; a specification of those persons (scientists, reasoners, inquirers, etc) who are to follow the directive of the norm; the action required of any person; the conditions under which the directive will come to apply. These elements are suggested in Hampton (1998), pp. 49-53; much else of what is said by Hampton in these pages can apply here.

6   Some of the main ideas behind Laudan's normative naturalism are employed in the above. However not all aspects of normative naturalism will be adopted here. For some criticisms and replies to criticisms, especially concerning Laudan's test procedure for normative naturalism, see Laudan (1996) chapter 9; see also some problems raised in Nola (1999).

7   Such a straightforward link between the justification required of epistemology and the norms of scientific method is proposed in Sosa (1991) chapter 14 'Methodology and Apt Belief'.

8   The example of non-refuting anomalies and incompleteness, along with the claim that methodology needs superior tools than epistemology, is discussed by Laudan in 'Is Epistemology Adequate to the Task of Rational Theory Evaluation?', in Nola and Sankey (eds.) (2000), pp. 165-175.

9   See van Fraassen (1980), chapter 2, who rejects the realist value of theoretical truth, and Laudan (1984) chapter 5 who also argues against it.

10   That one can deduce from a purely factual claim an ought claim was first shown in Prior (1960). This is also the topic of a whole book, Schurz (1997), in which the problem of the very formulation of Hume's dictum is pursued through a range of deontic logics. Schurz shows that there are some special versions of Hume's thesis but that it cannot hold generally.

11   In drawing the parallel suggested, it is not claimed that methodological and ethical norms are similar in all respects. Clearly their subject matter is different. And norms of logic and method can have the property of (in)validity or (un)soundness attributed to them while ethical norms cannot. But for both sorts of norm we can speak of their correctness or incorrectness, or their rightness or wrongness – though on certain strong anti-objectivist views of the status of ethical norms even this is mistaken. Both kinds of norm also share some of the same elements mentioned in footnote 5 above, especially the deontic operator.

[12] Predicate logic will need in its ontology at least sentences, truth, truth tables for logical constants and a domain for quantifiers. How this sits with the naturalist programme is an issue with which the philosophy of logic must deal. Important here would be the paper by Field (1972) which investigates Tarski's theory of truth within the context of naturalism.

[13] Hampton (1998) was compiled from the computer files that remained at her untimely death. Though she provides a useful framework in which to place the issues under consideration here, and develops criticisms of others, her positive view of her position is largely in the incomplete chapter 6. That view would have attempted to show any science-based argument against the objectivity of methodological norms is self-defeating.

[14] The definitions given in Gibbard (1990) of his norm expressivism on pages 7, 47, 83, and elsewhere, which use the formulation with intentional verbs such as 'thinks', 'judges', 'believes', etc, can also be construed as derivative upon the speech act version given in the text.

[15] For the basis of this claim see Gingerich (1975).

[16] These suggestive examples are from Jackson (1998), p. 9. The text is also indebted to chapters 1, 5 and 6 of Jackson's book where arguments for the supervenience of the evaluative (in his case the morally evaluative) on the non-evaluative are given. The above is also indebted to Kim (1993), chapters 4, 5, 8 and 9, where various kinds of supervenience are also distinguished.

[17] Note that such a global materialism rules out the possibility of a physicalism that leaves conceptual room for dualism, though such dualism is not true of our world. For a more adequate formulation of physicalism with respect to the mental see Jackson (1998), pp. 9-14. The above examples are intended as only illustrations of global supervenience and not of how an adequate physicalism with respect to the mental may be defined that leaves room for dualism.

[18] A proof is given in Jackson (1998), pp. 121-3. A similar set of considerations can be found in Kim (1993), chapter 4 section IV, and chapters 5 and 9. Bacon (1986) also provides an argument; but he also shows that many other notions of supervenience are also committed to co-extensivity in a way that might lessen the use of a notion of supervenience as distinct from that of reduction. However the necessary co-extensivity that accompanies global supervenience will be exploited to a particular end in what subsequently follows.

PART II

THE POVERTY OF THE SOCIOLOGY OF

SCIENTIFIC KNOWLEDGE

SYNOPSIS OF PART II

Part II argues that the sociology of scientific "knowledge" provides a poverty stricken approach to an understanding of both the sciences and knowledge, especially from a methodological point of view. It begins with chapter 4 on early intimations of a sociology of "knowledge", and then turns to one of its founders, Karl Marx, who alleged that much of what we believe in science depends, in some unspecified way, on forces and relations of production (section 4.1). It will be argued that Marx, along with his followers, had no clear idea of how the content of science could so depend; but he did have some useful things to say about how the sciences can, quite obviously, be externally influenced by the productive relations in which they occur. Section 4.2 deals with the quite limited scope that Mannheim claimed for any sociology of "knowledge". Section 4.3 briefly discusses the quite different sociological approach adopted by Merton, now out of fashion, in which there are sociological norms that prescribe an ethos for science quite different from the claims made by the sociology of scientific "knowledge", which usually have no prescriptive force at all.

Chapters 5 and 6 concern one of the leading contemporary theories within the sociology of scientific knowledge, two of its advocates connected with the University of Edinburgh being David Bloor and Barry Barnes. Hence the "Edinburgh Connection". They have done more than most to develop, articulate and defend what they call 'The Strong Programme' (SP). They have high hopes for SP as 'one of the heirs to the subject that used to be called philosophy' (Bloor (1993), chapter 9). But it will be argued that the heirs have squandered their legacy and left us bereft of a clear understanding of scientific knowledge. Section 5.1 sets out the central tenets of SP, the main focus being on its central Causality Tenet; this claims that, for all scientific belief, socio-political factors, along with some non-social factors, are jointly the causes of such belief. The strength of SP is in both its wide scope of application to all beliefs (unlike Mannheim's "weak" programme), and its requirement that social factors be omnipresent in the causal process. SP is an entirely naturalistic programme, extending the sciences into the very realm of scientific belief itself. But here serious pitfalls lie, as will be seen, especially in SP's eschewal of any role for the norms of scientific rationality.

Section 5.2 explores some of the types of social and non-social factors that SP alleges cause scientific belief. This leads to SP's causal explanatory model for why we hold the scientific beliefs we do (section 5.3). This model is contrasted, in section 5.4, with a different rational explanatory model that makes explicit appeal to norms of reason, a feature not present in SP's causal model. It is here that there is an important clash between the kinds of explanation of belief sought by advocates of SP, and philosophers and scientists who have maintained that there must be some role for the rational in explanations of scientific belief. Section 5.6 introduces a variation on the causal explanatory model, viz., that in which social and political *interests* are said to be the causes of scientific belief. This leads to a different "interests" explanatory model within SP. Throughout these sections difficulties with the scope of SP's main thesis are encountered. More narrow versions of its main causality tenet, advocated by Mannheim, have some plausibility; but the quite general form of the Causality Tenet makes SP both strong and wrong.

Evidence for the claims of SP often turn on the large number of case studies based in episodes in the history of science which allegedly instantiate it. But this evidence, it is argued, is weak. Section 5.6 critically appraises Forman's study of how the cultural milieu of Weimar Germany, with its intense opposition to science and notions of causality, led some German physicists of the time to adopt acausal theories of physics. The main defect of the study lies in the improper application of some of the very principles that sociologists of scientific knowledge attempt to bypass or debunk, in this case principles concerning causal methodology. SP claims to extend scientific investigation to scientific belief itself; and it does so in a way which is strongly naturalistic. But at the same time its eschewal of the traditional norms of science leaves it bereft of means for assessing the very causal claims on which its own theses depend. Related difficulties also occur for Bloor's own account of Robert Boyle's belief in the inertness of matter, which is allegedly caused by his socio-political circumstance; this is discussed in section 5.7. If these case studies fail, as they do, then some of the important grounds for accepting SP fall away.

There are other grounds for not accepting SP. Section 5.8 endorses a common complaint, even raised by other sociologists of science, that SP has established no laws linking belief with socio-political circumstance. This objection has some force; but it can be countered if SP adopts a different counterfactual account of causation, and thereby a different model for the causal explanation of scientific belief. This is outlined in section 5.9. But there is a high cost. The central distinctive requirement of SP that socio-political factors must be omnipresent must be abandoned. The final section points out that, even though SP attempts to give a naturalistic account of why scientists believe what they do, it lacks all normative force and so cannot lead to knowledge. Knowledge, being a normative, critical and evaluative, notion can have no place in the account of SP developed so far.

Chapter 6 begins by revisiting the differences between the causal and rational models of explanation of belief, adding a further refinement to the rationality model defended here. Chapter 5 was devoted to the central Causality Tenet of the SP. Chapter 6 turns to the other tenets. Section 6.2 investigates the Impartiality Tenet, which makes clear that the Causality Tenet applies to all beliefs whatever, regardless of their truth value, or whatever other epistemic status they might have. Section 6.3 examines the Symmetry Tenet imposed on all explanations; it argues that there is no ground for imposing the condition that all kinds of explanation of scientific belief must be of the same sort, viz., causal, regardless of the epistemic status of the belief. Since the Causality Tenet is intended to be wide in scope, then it must also apply to the very doctrine SP itself. This is expressed in the Reflexivity Tenet of SP. Section 6.4 argues that reflexivity leads to serious difficulties for SP. SP is often said to be committed to relativism; section 6.5 examines what this might mean, how it applies to SP and in what way it is damaging for SP.

Sociologists of science have adapted some of the philosophical doctrines of Wittgenstein to give their viewpoint greater force, and SP is no exception. Advocates of SP claim that inference is a social relation and not a logical relation at all. Section 7.1 explores ordinary inference as a counterexample to SP's claim that social factors must always be present, even when we form beliefs based on inference from other beliefs. Section 7.2 investigates whether or not the emergence of alternatives to classical logic, such as theories of strict implication developed by C. I. Lewis, was due to social causes or social interests. Evidence is provided that Lewis' had interests in proposing rival logics, but these were not social. Rather he was concerned to satisfy certain *intellectual* interests and to solve certain *intellectual* problems. Section 7.3 focuses directly on the claim that logical relations are really social relations of constraint, and the ambiguous Wittgensteinian texts which, in the end, do not offer this much support. Section 7.4 extends this discussion into the realm of our inferential practices and their associated Wittgensteinian "forms of life"; however little is to be found there in support of the idea that logical relations are really social relations.

It is well known that the later Wittgenstein developed a somewhat hostile attitude to science. So it is surprising that he is co-opted in support of the causal and scientistic account of our beliefs of the sort found in SP. Section 7.5 explores this conflict and Wittgenstein's claim that in the end explanations run out and only description can, and should, prevail. This might suggest that a descriptive ethnomethodology might be a better sociological theory on which to graft Wittgenstein's philosophy; however not even this proves viable. The central aspect of Wittgenstein's theory of meaning that SP takes over is his account of rule following and the communitarian solution to the problems it raises. Section 7.6 argues that the meaning finitism associated with SP, and the communitarian solution to rule following, are not viable. Section 7.7 raises problems with the nominalism that is implicit in SP and its account of kinds that turns on the doctrine of meaning finitism. The final section returns to the anti-scientism of Wittgenstein that would lead him to reject any scientific sociology along the lines of SP.

Does SP really eschew all kinds of rational explanation of scientific belief that appeal to norms of methodology and rationality? In the light of chapters 5 and 6 this would appear to be the case. Methodological principles have no role in determining what theories or hypotheses scientists adopt, or ought to adopt. The Wittgensteinian turn of chapter 7 puts this question in a different light. SP now endorses a rival theory of scientific rationality that is social in character. This arises from the communitarian solution to problems to do with rule following. In so far as methodological principles are to be understood as rules, then they, too, are to be understood along the lines of Wittgensteinian rule following. However the communitarian response to rule following that advocates of SP adopt, is a highly controversial interpretation of the admittedly obscure remarks of Wittgenstein. Moreover, critics of this position claim that this is not so much a rival novel theory of the rationality of science but, rather, such a massive change in the very subject matter under discussion that no recognisable theory of rationality remains to be discerned.

CHAPTER 4

SOME GERMAN CONNECTIONS:
MARX AND MANNHEIM

Broadly understood, the sociology of knowledge (SK) is that field of sociology which investigates the alleged relationships between the thoughts, ideas, beliefs, knowledge and/or other cognitive activities of the members of a given society and their social, historical and cultural situation. It is a highly contested field – including its very name. That there is even a sociology of *knowledge*, as distinct from, say, a sociology of *belief*, is resisted by those who adopt philosophical accounts of the nature of knowledge. The focus of Part II is on the very viability of not just a sociology of *knowledge* but of the sociology of *scientific* knowledge (SSK). The second section of this chapter contains a discussion of Karl Mannheim,[1] a leading founder of twentieth century SK. In his 'Brief Survey of the History of the Sociology of Knowledge', Mannheim mentions Marx and Nietzsche as the two important nineteenth century precursors of SK (Mannheim (1936), pp. 278-9). Marx will be discussed in the first section of this chapter while the discussion of Nietzsche will be left to Part IV. The final section of this chapter draws a contrast with Merton's quite different conception of the ethos of science that does not depend on any account of SK.

The idea that there are connections between what we know, or believe, and our social context is quite old. In the first book of his *Metaphysics*, Aristotle notes that the development of science for its own sake rather than for its utility occurred 'first in the places where men first began to have leisure' (*Metaphysics* 981b 22), his example being that of the priests of Egypt who had the leisure to develop mathematics for its own sake. But the first epistemologist and sceptic, the Presocratic Xenophanes, was probably also the first sociologist of "knowledge". He claimed that even if we manage to believe something that is also true, we still do not know. In illustration of our beliefs about the gods he says:

> If cows and horses or lions had hands, or could draw with their hands and make the things which men can, then horses would draw pictures of gods like horses, and cows like cows, and they would make bodies in just the form which each of them has itself. (Barnes (1979), pp. 92-3)
> Each group of men paint the shape of the gods in a fashion similar to themselves; the Ethiopians draw them dark and snub-nosed, the Thracians red-haired and blue eyed. (*ibid.*, p. 142)

In his commentary on the second passage, Barnes (*loc. cit.*) points out that it is not the gods who are the cause of the beliefs. Rather it is something quite unconnected with the gods. It is the circumstance in which each believer finds themselves, in particular their own social group and the group's features. Normally for knowledge it is assumed that what is known plays some role in forming the belief; this is a feature of the reliabilist theory of knowledge (section 2.3). Where there is no connection with the fact that p and the belief that p, then there cannot be knowledge. And this is exactly the case with the gods; the gods (whether they exist or not) have no role in causing the belief contents about themselves. However the beliefs are not uncaused; they are caused by something. And this something is the circumstance in which the believers find themselves, viz., their social group. The same is even alleged by Xenophanes about the beliefs that animals might have about their gods. This is an important point about the sociology of "knowledge". What sociologists allege cause beliefs, as will be seen, has nothing to do with what the belief content is about; rather it has to do with quite other, extraneous, social circumstances. If Barnes is right, then Xenophanes was the first sociologist of knowledge.

In 1605 Francis Bacon anticipated aspects of the sociology of "knowledge" when he said:

> Of much like kind are those impressions of nature, which are imposed upon the mind by the sex, by the age, by the region, by health and sickness, by beauty and deformity, and the like, which are inherent and not extern; and again, those which are caused by extern fortune; as sovereignty, nobility, obscure birth, riches, want, magistracy, privateness, prosperity, adversity, constant fortune, variable fortune, rising *per saltum, per gradus*, and the like.' (Bacon (1973), Second Book XXII, 5, p. 170).

What Bacon envisages is an investigation into 'impositions on the mind' that arise from external social factors. A later influential theory within SK was proposed by Durkheim and Mauss in 1903. They alleged that even the basic categories of thought about space, time and causality, for which Kant had hoped to provide an *a priori* justification, are, instead, to be given a sociological explanation. They also alleged that our very systems of classification of natural objects reflect social relationships. Notoriously they say: '*the classification of things reproduces [the] classification of men*' (Durkheim and Mauss (1963), p. 11).[2] Though much vaunted in the twentieth century, there are grounds for doubting whether SK has got much beyond the obvious suggestions of Aristotle and Bacon when it comes to the details of scientific knowledge.

It is important to distinguish two branches of sociology related to SK: the sociology of *science* (SS) and a sub-branch of SK, the sociology of *scientific knowledge* (SSK). The difference lies in the different kinds of "objects" that are to be explained by socio-historico-cultural context. Traditional SS seeks to explain "objects" such as particular hierarchies in the scientific community, its reward structure, priority disputes, multiple discoveries of the same item, the reception of discoveries, science's funding arrangements, the creation of new fields of study, and so on. SSK is more radical, and at the same time more narrow, in that it focuses on the propositional content of the very sciences themselves and attempts to provide a socio-historico-cultural explanation of why these contents are believed (or not believed as the case may be) within a given community of scientists, and how and why these beliefs change over time and place. It is SSK, and not SS, which comes into conflict with traditional epistemological theories of scientific knowledge, and attempts to replace philosophy with sociology.

Not all sociologists have been sanguine about the prospects of SSK. The eminent sociologist of science, Joseph Ben-David in his essay 'Sociology of Scientific Knowledge', identifies the theories inaugurated by Marx and Mannheim as those that have most strongly maintained SSK as a programme of investigation in the twentieth century. But to no avail, says Ben-David: 'No success can be claimed for the new Marxian-Mannheimian attempts to find a systematic (that is, permanent and regular, not just occasional) relationship among macrosocial location, ideology, and scientific theory. Indeed there is little reason to expect that there should be such relationships.' (Ben-David (1991), p. 462). This negative view is explored and endorsed throughout Part II. It is directed against the claims of SSK; it is not directed against the claims of SS which remains a quite viable area of study.

In turning to the sources of SSK in Marx and Mannheim, it will be seen there is much perplexing unclarity about three major aspects of their views. There is confusion about, first, the *explanandum*, the kinds of "object" which are to be explained, and, second, the *explanans,* what does the explaining, *viz.,* socio-historico-cultural context. Also the *explanans* is often inexplicit about what kind of explanation is being proposed and whether the explanation is itself successful and has empirical support. Third, there is also much obscurity concerning the explanatory link that is said to hold between *explanans* and *explanandum*. As will be argued, these difficulties are endemic to SSK and vitiate not only the doctrines of its founding fathers but also all descendant doctrines.

4.1 MARX AND THE SOCIOLOGY OF SCIENCE AND SCIENTIFIC KNOWLEDGE.

The writings of Marx and Engels relevant to SS, SK and SSK contain many of the difficulties that infect these subjects even now. In their early *The German Ideology* of 1845-6, Marx and Engels distinguished between: (a) what is to be explained, viz., aspects of cognition which they call 'forms of consciousness', examples of which are ideas, conceptions, religious and metaphysical beliefs, the norms of law, etc; and (b) what does the explaining, viz., social forms of life, or as they also say, 'material life' or 'material activity', or 'real' life – later Marx would also refer to these as 'relations of production', 'economic basis' or 'real foundation'. In addition they importantly claim that there is a one-way relationship between the two relata (a) and (b); the former are 'dependent' upon, or 'interwoven' with, the latter, or the former are 'conditioned' or 'determined' by the latter. That the relationship is one-way, and in the direction specified, is foundationally important for Marx's doctrines and all SK; if the relation were to also hold in the other way then they would be subjects without and special significance.[3]

The precise specification of the two relata and the relation that allegedly holds between them, has been at the heart of confusions surrounding what will be called 'the Marx-Engels Thesis' which gets its first expression by Marx and Engels as follows:

> The production of ideas, of conceptions, of consciousness, is at first directly woven with the material activity and the material intercourse of men .... The same applies to mental production as expressed in the language of the politics, laws, morality, religion, metaphysics, etc., of a people. Men are the producers of their conceptions, ideas, etc, that is, real, active, men as they are conditioned by a definite development of their productive forces and of their intercourse corresponding to these, up to its furthest forms. ...
>
> Morality, religion, metaphysics, and all the rest of ideology as well as the forms of consciousness corresponding to these, thus no longer retain the semblance of independence. ...
>
> It is not consciousness that determines life, but life that determines consciousness (Marx and Engels (1976), p. 36-7).

The words 'at first' in the first sentence might be taken to suggest that the alleged relationship holds only for early forms of society; but this impression is dispelled when in the third sentence the authors speak of the relationship holding 'up to the furthest forms' of development of the forces of production and their 'intercourse' in relations of production. By the end the quotation the Marx-Engels Thesis is expressed quite generally and without qualification. This can be set out as the following schema (ME):

(ME): *For all forms of consciousness, C, there is some form of social and/or productive life, S, such that C is dependent (in some way) on S.*

In (ME) the precise nature of the dependence link is left open but it can include such relations as determines, conditions, causes, etc, as well as functional relationships[4]. However once some relationship of dependence is specified, it is a further empirical matter to discover whether the instance of (ME) is true or false, or has some evidence in its support. Some version of this general claim, and its associated problems, can be found not only in Marx but also in Durkheim and Mauss who speak of a relationship of *reproduction* when they claim that 'classifications *reproduce* more basic social relations'. As will be seen, Mannheim and advocates of the Strong Programme, endorse a version of (ME).

4.1.1 Science and the Marx-Engels Thesis

Is science, or some aspect of science, to be included under the admittedly vague umbrella term '(forms of) consciousness'? Scientific theories and laws are at least general beliefs, or 'ideas' or 'conceptions', about the world. And we have subject them to examination by scientific methods – the methods themselves being a further class of belief or 'ideas' which, depending on how general (ME) is taken to be, might also fall within its scope. Considered as beliefs, ideas, or conceptions, it would not be inappropriate to think of scientific theories, laws, and even methods, as one of the 'forms of consciousness'. Granting this, the very content of the sciences then falls within the scope of a quite general version of (ME) alongside other examples of forms of consciousness such as religion, law, metaphysics, etc. Understood this way, Marx and Engels provided the first formulation of a central tenet of a strong version of SSK.

But there is evidence that Marx and Engels do not intend (ME) to be about *all* forms of consciousness, but only *some* (though why it applies to one but not another form of consciousness lacks clear demarcation). Such evidence comes about in two ways. First, despite the open-ended list of forms of consciousness, nowhere in the above remarks is science explicitly mentioned, whether it be theories, laws, observational reports, practices, technological applications, or whatever else. Nor are the very methods of science mentioned. Despite the lack of an explicit endorsement by Marx and Engels, some practitioners of SSK have extended (ME) (along lines suggested in the previous paragraph) to include the very content of the sciences as forms of consciousness that are dependent on social conditions.

Second, we need to consider Marx's response to what might be called the 'reflexivity problem' that arises in relation to the quite general (ME). What of belief in Marx's and Engels' new science of Historical Materialism (HM) which holds that there are specific one-way links between *some* forms of consciousness and their social context? Surely belief in (HM) is itself an instance of (ME):

For a particular form of consciousness, such as a person's belief in (HM), there is some form of social and/or productive life, S*, such that a person's belief in (HM) is dependent in some way on S*.

That is, there are particular social conditions S* which bring about belief in (HM); further, one may suppose that in their absence there is no belief in (HM). Some Marxists willing embraced the mutability of belief in (HM) according to social circumstance. There is alleged to be a certain propitiousness in the social circumstances which lead Marx to advance (HM), but in the fullness of time, well after the ripe development of communism (let us suppose), these will pass away thereby removing the conditions which enabled the very possibility of envisaging (HM) in the first place. However others reject this view, preferring to view (HM) itself as a science open to empirical investigation; they thereby reject (ME) in its full generality, and in particular the reflexivity that follows from it. Marx, it will be argued, adopts the second position – but this entails a reconsideration of (ME) and the place of science and its methods within it. (The problem of reflexivity will become important later in section 6.4 as it is embraced as a central tenet of the Strong Programme for SSK.)

Consider what Marx and Engels say of ideologically held beliefs: 'The phantoms formed in the brains of men are also, necessarily, sublimates of their material life-process, which is empirically verifiable and bound to material premises' (*loc. cit*). The chemical metaphor of sublimation (the conversion of a solid into a vapour) should not mask the fact that this is a further instance of (ME) which can be expressed as: 'ideological beliefs are causally dependent in some way on (they are "sublimates" of) the material life processes of people'. Marx and Engels also add that this claim of dependence can be tested by empirical means; i.e., the methods of science can be applied to confirm, or falsify, this alleged instance of (ME). But for this to be possible there must be scientific 'forms of consciousness' not dependent on some material life-process which are used to test any alleged connection between material life-processes and forms of consciousness.

Looked at this way the methods of the sciences are outside the scope (ME), which then must be construed less generally. But if the methods whereby we produce knowledge in the special sciences lie outside the scope of (ME), then what these methods produce, the special sciences themselves, are also outside the scope of that thesis. For example, at least two factors produced Newton's belief in his Law of Universal Gravitation. Contrary to what advocates of SSK might say, the belief did not arise in Newton's mind as a result of some aspect of his social circumstance. The first factor involved methods of science, such as the 'Rules of Reasoning in Philosophy' that Newton explicitly set out before inferring his Law. The second factor concerns the 'Phenomena' he employed about planetary motion that were crucial starting points for the inference to his law.[5] Here both belief in the methods of science, and belief in what they produce (e.g., the inverse square law of gravitational attraction) are outside the scope of the Marx-Engels Thesis.

A little further on Marx and Engels reiterate the same point: 'Where speculation ends, where real life starts, there consequently begins real, positive science, the expounding of the practical activity, of the practical process of the development of men' (*loc. cit*). In the paragraph before this remark, the authors castigate both empiricists who treat the active life process as 'a collection of dead facts', and idealists who treat this process as 'an imagined activity by imagined subjects'. Such empiricist or idealist approaches to the connection between consciousness and active life processes are to be eschewed, as they say, by the application of 'real, positive science'. It was Comte who first introduced the epithet 'positive' to describe his conception of science; and this is something that Marx and Engels take over without necessarily endorsing Comte's characterisation of 'positive science'. What is clear from these remarks is that they did believe that instances of (ME) could be examined in a scientific manner independently of those who take a limited view of the character of science and dismiss it out of hand (idealists) or fail to notice it (empiricists).

The same position is maintained by the later 1859 Marx in his often-cited account of what he referred to as 'the guiding principle for my studies':

> The mode of production of material life conditions the general process of social, political and intellectual life. It is not the consciousness of men that determines their existence, but their social existence that determines their consciousness. ...
>
> The changes in the economic foundation [the totality of the forces and relations of production] lead sooner or later to the transformation of the whole immense superstructure. In studying such transformations it is always necessary to distinguish between the material transformation of the economic conditions of production, which can be determined with the precision of a natural science, and the legal, political, religious, aesthetic or philosophic – in short, ideological forms in which men become conscious of this conflict and fight it out. (Marx (1970), pp. 20-1)

In the above Marx again sets out his two-tiered superstructure/foundation view of the nature of any society in which superstructural items stand in some dependent relationship to the economic foundation of relations and forces of production. Much ink has been spilt in the attempt to spell out the two relata and the relation in which they stand – and no further spillage is required here. Suffice to note that though the first two sentences are general enough to include science, or some aspect of science, within its scope, the final sentence tells us that the very material transformations that produce the superstructure, and changes in it, can be determined 'with the precision of a natural science'. However note once more the open-ended list of superstructural forms of consciousness fails to explicitly mention any science, while it is explicitly mentioned that science is to be used to determine the foundation/superstructure dependencies. Again fervent advocates of SSK might rush in where Marx himself failed to explicitly tread – and the reason can only be that he treated scientific methods and its products as outside the scope of his thesis.

4.1.2 Freeing Science from the Marx-Engels Thesis

Setting a textual investigation aside, it will now be argued that it is appropriate for Marx to hold a restricted version of (ME) that excludes science. To show this we will investigate whether both of the following claims can be held. (1) Theories and methods of science are examples of forms of consciousness; and scientific theories and methods are superstructural items open to change with transformation of society's economic foundation. (2) Science, especially its methods, can be pressed into service to determine both of the following: (a) the characteristics of each economic foundation and their transformation from one set of characteristics F to another F* (denote this transformation relation by 'F≈F*' for short); and (b) the nature of the dependent relation of superstructure S (including science), upon economic foundation F (denote this dependent relation by 'F→S' for short). There are two levels at which the matter can be investigated. The first level concerns whether the transformation F≈F*, and the relational dependence F→S, are objective facts to be uncovered within some science (in this case Marx's Historical Materialism). The second level concerns the objectivity of the methods employed to establish these facts.

Concerning the first level, Marx says that relations such as F≈F* and F→S, 'can be determined with the precision of a natural science'. We can understand this to entail that their determination uncovers some objective relational facts about our social structure (which can vary over time). And they are objective at least in the sense that such factual relations will remain invariant whatever superstructural forms of consciousness we happen to adopt in thinking about and describing these relations. (This presupposes that Marx's account of the transformation relation F≈F* and the dependent relation F→S are correct; such claims of Marx's Historical Materialism are not assessed here.)

Suppose for the sake of the argument the opposite of such objectivity. We allow that the very determination of the relations themselves, using the methods of science, differs according to what superstructural forms of consciousness are adopted in thinking about and investigating the relations. Then there is no fact of the matter as to what the relations are. We fall into a vicious form of relativism which says that what relations hold is relative to the superstructural forms of consciousness which prevail and which are used to think about and describe them, and that these very relations change with change in the superstructural forms of consciousness we use to think about, investigate and describe them. To adopt a term of Thomas Kuhn, we have incommensurable accounts of what are the alleged relations that make up our social structure that depend on the 'forms of consciousness' in the light of which we spell out these "relations". The very relational claims are incommensurable because they are described very differently from within quite disparate forms of superstructural consciousness (including the very languages of the sciences in each superstructure). Such a relativism is vicious because it inverts one of the essential features of Marx's Historical

Materialism, viz., that the economic foundations determine the superstructure, and not the other way round as just envisaged.

Concerning the second level, let M be some method of science which is used to determine the truth value of the F≈F* transformation and the F→S dependent relation. Is M independent of both the foundations and the superstructure, even though it is used to determine the characteristics of both? Or is M is to be located among the superstructural forms of consciousness which are open to change with change in the economic foundations of society? We may say 'yes' to the first question while denying that the second alternative is coherent. Suppose that the transformation F≈F* is so radical that two things occur with new F*: first a new S* stands in a dependent relation upon F*; and second a new M* emerges within S* which rivals and replaces the old M within the old S. What difficulty is there in this supposition?

If there are genuinely rival scientific methods, M and M* then, as above, we can no longer believe that our science has uncovered objective facts such as the transformation F≈F* or the dependent relation F→S. If M and M* were methods which arrived at the same facts or theories time and time again, then they could hardly be said to be rival theories of method in the sense intended. We would have no grounds for claiming that there is a significant difference between M and M* as methodological principles of science, and they would not be the rivals they are supposed to be. A necessary condition for genuine rivalry at the level of scientific method is that M determines one set of facts or theories and M* determines a quite different set over a wide range of cases. In particular both M and M* would have to produce different sets of relational facts about the transformations that have occurred and the dependencies which obtain. Thus there would be no fact of the matter as to which set of relations hold; what relations hold are relative to the scientific methods, M or M*, embodied in each superstructural form of consciousness, S or S*. Once again we have a vicious kind of relativism (this time with respect to scientific methodology) that inverts one of the essential features of Marx's historical materialism, viz., the priority of the economic foundation over superstructural forms of consciousness.

If science is alleged to be a superstructural item then there is a further reason for regarding it as distinct from ideological forms of consciousness. Ideological beliefs are commonly said to stand in contrast to scientific beliefs, for two reasons. First, ideological beliefs are often simply defined as those which are unscientific. Second, ideological beliefs concern overt or latent interests of various groups in society. However science can be employed to uncover what are the ideological factors at work and what interests and social relations they support. In Marx's view, the operation of ideologies as special class interests can be 'determined with the precision of a natural science'. Thus it is misleading to lump science together with distorting ideologies.

What other reason might there be for placing science within the superstructure rather than the foundation of Marx's two-tiered model? The following inference suggests a clue:

(1) the laws, theories and methods of science are a special class of beliefs (ideas, conceptions) which we hold about the world (independently of whether they are true of false of the world);

(2) all beliefs (ideas, conceptions) and their contents are something mental;

(3) whatever is mental must pertain to the superstructure since it has no place in the economic foundation, the very 'material' basis of society (as Marx puts it);

(4) therefore, science must be superstructural.

If this is a valid argument and we reject the conclusion, as Marx does, then we must reject one of the premises – the most likely candidate being premise (3). As Marx outlines in his 1859 statement of the 'guiding principle of his studies', the economic foundation of society comprises two further elements: the *forces* of production and the *relations* of production (of which ownership relations are a major subclass). Following the analysis of Marx's materialist theory of history as set out by Cohen (Cohen (1978), chapter 2), the forces of production comprise two further elements: *labour power* and *means of production*. The means of production also comprise two further elements: the *raw materials* on which labour power is exercised to yield a product, and the *instruments of production* labour power uses in the production process. These latter may vary from the adze used by early humankind to the use of any other tools and machinery such as lasers and computers. Thus any of the technology employed in scientific investigation and their technological spin-offs, such as non-stick fry pans and plaque-resistant toothpaste, can be either an instrument of production (i.e., the pan as used in a restaurant to prepare meals) or a product arising in another process using other instruments of production (i.e., the toothpaste). As such the products and processes arising out of science turn out to be fundamental to the forces of production and are not superstructural items.

What of labour power, the abilities exercised by people in the process of production? Clearly the degree of development of labour power depends upon various kinds of mental capacities labourers possess, which in turn depend on the extent to which people are required to master some science or technological *know-how* in order to use, design, build or repair instruments of production. In some cases this can be quite low, as in the case of routine circuit-board assembly for computers, but high as in the creation of software programmes for a firm. And the latter will require much cognitive scientific *knowledge that* ... (such-and-such is the case). Thus the very content of science can appear as items of belief held by those who exercise their ability to labour in various ways thereby expressing their know-how.

What is obvious from the above is that our scientific theories and laws, even though they may be (in a broad sense) mental items, they are also an important part of the 'material' forces of production. And it is the productive forces which, as Marx says concerning the 'guiding thread for his studies', can exist in relations of production which are either conducive to their growth, or, if fettered, lead to the transformation of the relations in which they exist. In regarding science as a force of production, it is necessary to avoid confusion due to Marx's inclusion of a 'mental' item (scientific beliefs and know how) in the 'material' foundation of society. The 'material' foundation is not necessarily *materialist* in the philosophical sense in which it is alleged that there are no mental phenomena or minds but only matter. What view one holds about the body-mind relationship is independent of Marx's talk of the 'material' economic foundation of a society. So the 'material' foundation can include the scientific beliefs and scientific know-how of the people who are part of the forces of production. Thus it is possible to remain an Historical Materialist in Marx's sense and to reject conclusion (4); and the reason for this is that there is no need to accept premise (3). (It is important to note that it is possible to be both materialist about the mind-body relation (and thus a materialist about beliefs) and adopt Marx's materialist theory of history – a position which Marx himself presumably endorsed. Being a materialist in both senses does not affect the above point.)

Though some science might be part of the forces of production, not all science need be. Mendelian genetics and the biochemistry of DNA molecules have only recently been pressed into service for the various gene technologies that are now being integrated into the forces of production. In contrast Einstein's general theory of relativity still remains largely outside the sphere of the forces of production (though much testing of the theory does rely on technological developments arising out of developing forces of production, e.g., radar to test the theory's predictions about the gravitational bending of light passing the Sun). In sum, scientific theories are not dependent superstructural items but can be either important items driving the very transformation of the material foundation of society, or are quite outside, and independent of, Marx's two-tiered framework

4.1.3 The Marx-Engels Thesis and the Claims of SS versus the Claims of SSK

Once science is freed from over-strong claims about SSK based on the Marx-Engels Thesis, there are some much weaker obviously true claims that can be made about the connection between science and society that pertain to SS rather than SSK. The first, is that experimentation in science often requires a sufficiently developed technology to produce the requisite apparatus; without this there would be no experimental science and no experimental testing of theories. This is an obviously true claim that no one should contest. To illustrate, Newton said in a letter to Oldenburg (February 6, 1672) 'I procured me a Triangular glass-Prisme to try therewith the celebrated *Phaenomena of Colours*'; and later he says that he purchased prisms at a fair to investigate the colour spectrum and refraction. This suggests that there was, already locally available, a sufficiently well developed technology of glass-making that would suit Newton's experimental purposes; without such a level of development of prism-making he would have not been able to conduct any investigation at all. Clearly, the development of glass technology had proceeded without any input from theoretical science, whether of optics or chemistry. But equally clearly, such technology can tell us nothing about the way Newton was to conduct his experiments, or what theory of reflection and refraction should be adopted. Micro historico-social studies can inform us about the detailed ways in which some technological development enables growth of some aspect of science. As interesting as the story at the micro-level may be, the macro historico-social claim 'technology shapes science' remains a dull platitude.

The same can be said about the extra-scientific social interests that shape the way in which science develops. Scientists might be interested in acquiring theories with some or all of the following features: observational correctness, explanatory power, high probability on evidence, truth, precision, simplicity, a large number of predictions, and so on. These are the values of scientific method we have already encountered, some of which are arguably intrinsic to science. But scientists can also have extra-scientific interests such as: personal interests in pursuing some science (because it might lead to a lucrative patent); professional interests (in that it might lead to peer recognition through the award of a prize); or social and humanitarian interests (in that the science could increase the productive capacity of a nation, or enhance social control). And non-scientists, from politicians to businessmen, can have an extra-scientific interest in ensuring that scientists pursue particular kinds of research rather than others because they are believed, if successful, to increase employment or profit.

Such interests are amply illustrated in *Capital* (Volume I, Chapter XV, Part I, Section 1). There Marx discusses, along with other examples, the way in which the growth in production of cotton fabric due to the introduction of mechanical looms far exceeds the ability of traditional methods of dying using vegetable extracts to meet production needs. Hence an interest on the part of capitalists in developments in the science of chemistry to produce new processes of mass bleaching, or dying. But here the "object" to be explained within SS is the following; *why there was growth in knowledge*, at a given time and place, in the application of chemistry (which in turn provided new bleaching and dying processes). This is quite distinct from a second "object" which also calls for explanation, viz., *what that growth in knowledge was*. The new piece of knowledge that has been acquired is: that such-and-such processes provide new ways of bleaching or dying. An explanation of this is to be found in chemistry using the methods of science. That this second "object" can only be explained or established by sociology is part of the hubris of SSK. As interesting as the story at the micro-level of historico-social research may be, the macro historico-social claim 'social interests shape science' remains a platitude from which no one would dissent.

Marx gives us many micro historico-social stories about the way in which technology or extra-epistemic social interests shape particular aspects of science, or the way in which science can sometimes act as a force of production. This yields a sociology of science which may be controversial in its details. But the stories raise no special issues within the sociology of scientific knowledge. In contrast the Marx-Engels Thesis (ME) make important claims about the way in which the very content of the various sciences (their language, theories, laws, methods, etc) are allegedly shaped by the economic foundations described by Marx rather than the epistemic factors described by philosophers. But (ME) faces difficulties of its own when understood in the context of Marx's materialist conception of history. Some Marxists endorse (ME) with respect to science; but it is not clear that Marx did. If he did not, then Marx has no contribution to make to SSK (apart from the negative contribution of muddle due to inexplicitness about (ME)). But he does have a contribution to make to SS through the micro historico-social investigation of aspects of particular episodes of the development of science within its social nexus.

Finally, what of some of the subsequent applications of (ME) to science? Perhaps the most famous is that of Boris Hessen's 1931 paper 'The Social and Economic Roots of Newton's *Principia*'. Hessen argues that Newton was a 'child of his class' (Hessen (1931), p. 183), and of his time, in that his work was used to solve technological problems that arose with the rise of capitalism. One should not deny whatever truth their may be in this claim; but one should also note that at best it is a claim within SS about the uses to which some science can be put, or the urgency with which some problems were required to be solved by a newly developing science. But none of this bears on the actual content of Newton's science (e.g., what laws hold), and so provides no support for SSK. To illustrate, in the early parts of Book I of *Principia* Newton explored mathematically models of the solar system in which forces act between bodies either inversely with the distance between them, or as inverse square, or inverse cube, etc. But in Book III he establishes that in our world the force must be inverse square, thereby eliminating the other possibilities he had envisaged in Book I. Whatever external socio-economic influence their may have been on Newton, or any other scientist preoccupied with these issues, none of this shows that belief in the inverse square law rather than belief in any of its rivals, is itself dependent on socio-economic factors. What is crucial here is the evidence of planetary motions that led to the inverse square law rather than one of its rivals.

Hessen provides various summaries of Newton's *Principia* (e.g., p. 174), and he lists various problems in Part 2 of his paper (under various headings such as 'Communication', 'War', 'Industry' etc) for which non-scientific people required particular solutions, e.g., navigation and the problem of longitude, problems in ballistics and of projectile paths that arose in canon making, etc. One can readily agree that extra-scientific needs, as well as problems that arise quite naturally in the course of the development of a theory (what Kuhn would call 'normal science'), can provide a context in which there is a need for theoretical development. But none of this determines *what* theoretical development *ought* to take place. All that Hessen says in relation to this are bland remarks like the following: 'Newton's laws provide a general method for the resolution of the great majority of mechanical tasks' (*ibid.*, p. 174). One cannot disagree with this. But then it only tells us that Newton provided solutions to some pressing problems. Whether these problems were the, or a, reason for Newton's engaging in the investigation of mechanics is something that historians could tell us. But whatever the answer, it has no bearing on *what are* the laws Newton *should* use to solve the problems given the various contending hypotheses, and *what evidence* there might be for them. This is particularly the case for Newton's considerations in Book III of the *Principia* concerning the laws governing planetary motion. This is a matter of scientific method upon which Hessen has little to say. But not everything Hessen says about Newton's theoretical views is without interest, for example Newton's resort to theological explanations when his physics runs out. But again, this need not be a matter on which something like the (ME) thesis need necessarily have a bearing. Once more the legitimate scope of SS is not to be confused with the hubris of SSK – a matter on which Hessen and his ilk are not very clear, given the confusions they inherited from (ME).

4.2   MANNHEIM AND THE SOCIOLOGY OF SCIENCE AND SCIENTIFIC KNOWLEDGE

One of the more prominent twentieth century progenitors of the misnamed sociology of "knowledge" is Karl Mannheim, who envisaged the eclipse of epistemology by sociology: 'Theoretical and intellectual currents of our time seem to point towards a temporary fading out of epistemological problems, and towards the emergence of the sociology of knowledge as the focal discipline' (Wolff (ed.) (1971), p. 61) – though the word 'temporary' might indicate misgivings about such hyperbole. The core of his theory of SK is expressed in his 1936 *Ideology and Utopia*[6] as follows: 'The principal thesis of the sociology of knowledge is that there are modes of thought which cannot be adequately understood as long as their social origins are obscured' (Mannheim (1936), p. 2). The claim is modest; the words 'there are', instead of 'all', indicate that not all modes of thought lack the requisite understanding, but only some. For Mannheim the distinction is between modes of thought which 'can be explained solely on the basis of [an individual's] own life-experience' and those which need to be explained on the basis of the 'historical-social-situation out of which individually differentiated thought only very gradually emerges' (*ibid.*, p. 2 and p. 3). However as one pursues Mannheim's remarks in these pages, it is not only *modes of thought* (perhaps to be understood as propositional contents) that are to be explained, but other quite different entities such as the *activity of thinking* and *languages*, including words and sentences and the specific *meanings* we give to them. We can grant the claim that languages, the words that comprise them and the meanings assigned to words and sentences, are based in group, not individual, activities. But it remains to be seen in what way an individual's thought activity and the very propositional contents each entertain (which are expressed by meaningful sentences of a socially made language), 'cannot be adequately understood' (by whom?) without appeal to the historical-social-situation of the individual entertaining the thought.

Thus far the principle thesis of SK involves two relata. The first relata is an individual's 'historical-social-situation', or some aspect of it. The second relata includes a broad class of items, four of which have been mentioned, viz., the act of thinking, thinking's content, languages and meanings. Being relata, there must be some relation which holds between them. Though hardly spelled out, there are enough suggestions that it must be some relation of dependence of the second relata on the first, and not the converse (a view which would be anathema to all SK and SSK). One prime candidate is the relation of counterfactual dependence. Thus Mannheim's 'principle thesis of the sociology of knowledge' could be read as the counterfactual dependence of modes of thought upon (aspects of) society. Expressed this way it says: if some modes of thought *were* to be adequately understood then their social origins *would* not remain obscured.[7] It is obvious that the very same problems that surfaced for the Marx-Engels Thesis (ME) resurface at this point for Mannheim. But not only do the two relata and the relation between them need to be made much more specific; a minimally adequate formulation needs to be provided for Mannheim's SK if it is to be testable.

Turning to the final section V of Mannheim's book entitled 'The Sociology of Knowledge' does not clarify matters concerning the relata and the relation.

Of the various projects in which SK could be involved Mannheim tells us that he 'will present the sociology of knowledge as a theory of the social or existential determination of actual thinking' (Mannheim (1936), p. 239). Here the range of items covered by the second relata is expanded to include the activity of 'actual thinking', the process of 'knowing' and its product, 'knowledge'. Mannheim wisely continues: 'It would be well to begin by explaining what is meant by the wider term "existential determination of knowledge" (*Seinsverbundenheit des Wissens*)'.

Though not fully precise, the following passage introduces Mannheim's distinction between 'theoretical' and 'extra-theoretical' or 'existential' (*Seins*) factors:

> The existential determination of thought may be regarded as a demonstrated fact in those realms of thought in which we can show that the process of knowing does not actually develop historically in accordance with immanent laws, that it does not follow only from the 'nature of things' or from 'pure logical possibilities', and that it is not driven by an 'inner dialectic'. On the contrary, the emergence and crystallisation of actual thought is influenced in many decisive points by extra-theoretical factors of the most diverse sought. These may be called, in contradistinction to purely theoretical factors, existential factors. (Mannheim (1936), pp. 239-40)

There are a number of points to note. (1) The 'existential determination' of thought is contrasted with 'theoretical factors' by which 'knowledge' has *sometimes* historically grown in drawing upon its own *inner dialectic* or *immanent laws*. The last two italicised phrases are unclear, but they can be readily understood to refer to the principles of reasoning and method we often employ in science and everyday life when we advance from one set of beliefs to another (see note (3) following). But what contrast is being drawn here? The contrast between the theoretical and the extra-theoretical is obviously logically exhaustive. However the contrast between the theoretical and the existential is not exhaustive if 'existential' is too readily identified with 'historical-social-situation'. There remain extra-theoretical items such as cognitive and/or biological factors which are not obviously part of an individual's 'historical-social-situation'. The conflation of the extra-theoretical with historical-social-situation can only generate confusion within SK. (2) Mannheim also expands the number items that fall under the second relata when he says that the first 'existential' relata concern 'not only the *genesis of ideas*, but penetrate into their *forms* and *content*' and also 'determine the scope and intensity of our *experience* and *observation*' (*ibid.*, p. 240, my italics).

(3) It is important to note that the quotation allows that there are some modes of thought that do have an *inner dialectic* of their own, independent of any existential/extra-theoretical circumstance, while other modes of thought are not so independent. For Mannheim's SK, and his SSK, only *some* items falling under the second relata (e.g., knowledge, belief, thought, etc) have an

existential determination (*Seinsverbundenheit*); others are independent of any such determination. This is an important restriction upon Mannheim's various theses about SK. In this context Mannheim relies on a distinction, now commonplace in the philosophy of science, between 'internal' and 'external' accounts of change and growth in scientific knowledge.[8] In the case of science, explanations of the change in, or growth of, scientific knowledge are internalist if what does the explaining is restricted only to some theory of scientific method (Mannheim's 'inner dialectic' or 'immanent laws'). Externalist explanations appeal to extra-methodological factors such as the social circumstance of the scientists who accept, or believe, the theory. Both styles of explanation can be consistently adopted even where the explanandum is the same kind of item, such as the acceptance, or non-acceptance of the same laws of science. Thus an internalist explanation is commonly offered as to why, say, the laws of Mendelian genetics have wide acceptance in the western scientific community in the 1940s (the grounds for the acceptance being, let us suppose, internalist because they are based on an appeal to confirming evidence). In contrast an externalist explanation is commonly given as to why the very same laws of Mendelian genetics did not have wide acceptance in the Soviet Union scientific community of the 1940s (Stalin's advocacy of the non-Mendelian views of Lysenko). The difference lies in the socio-historical circumstance of the scientists.

(4) Finally to the term 'Seinsverbundenheit' which Mannheim introduces. He tells us in a footnote intended to clarify matters: 'Here we do not mean by "determination" a mechanical cause-effect sequence; we leave the meaning of "determination" open, and only empirical investigation will show us how strict is the correlation between life-situation and thought-processes, or what scope exists for variations in the correlation'. To this the translator adds: 'The German expression *"Seinsverbundenes Wissens"* conveys a meaning which leaves the exact nature of the determination open' (*ibid.*, p. 239). The term 'verbundenheit' means little more than the bare idea of a relation, connection or link without more specificity. Since Mannheim's thesis is not general, there will be cases where there is no relation between the first existential relata and the various second relata such as thought, knowledge, ideas, etc. But where Mannheim admits that there is a relation, it might be as weak as a low statistical correlation. However he has been too hasty in identifying all causation, or causal dependence, with mechanical 'push-pull' models of causation; in fact he uses the term 'Seinsverbundenheit' to exclude any connotation of mechanical causation by existential factors. But such total exclusion of all notions of causality vitiates his theory. As will be seen in the next chapter (section 5.9), relations of causal dependence analysed as counterfactual dependence are an appropriate way to understand the relation involved here; this carries no connotation of mechanical causation.

Given the above, Mannheim's thesis turns out to be none other than a more restricted version of the Marx-Engels Thesis (ME) in which the explanation of some aspects of science are to be excluded from its scope. But Mannheim's account of what he intends by the two relata and the relation has lead to some devastating criticisms of his theory of SK, one of the most devastating being Robert Merton's important 1943 paper 'Paradigm for the Sociology of Knowledge' (Merton (1973), chapter 1). Merton provides a long list of interpretations of what the two relata might stand for, and what relation might hold between them, in order to find some interpretation of the principal thesis of Mannheim's SK which might withstand empirical investigation.

In contrast other interpreters invite us to reconstrue Mannheim's project and what he means by *Seinsverbundenheit des Wissens*. For example Simonds ((1978), chapter 5) invites us to place Mannheim's project of SK within the hermeneutic camp in which SK is understood as an interpretative method. Simonds puts his point this way:

> At no point in Mannheim's argument does he posit a link between un-meaningful phenomena and the meaningful content of expression. It is absolutely essential to the sociology of knowledge, as he develops it, that the 'existence' of *Siensverbundenheit* be understood to have a conceptual structure: his method ... relates thought to "social existence as a context of meaning", not to some conception of "social existence as brute data". (Simonds (1978), p. 119)

Though much more needs to be said of this, the phrase '*relates* thought to' raises once more the problem of the two relata and their relation. For Simonds the two relata are thought on the one hand, and social contexts of meaning on the other – but the relation between these two relata is left quite unspecific. Once again a too mechanistic understanding of causation dominates Mannheim's, and Simonds', rejection of the possibility that the relation might be causal. As suggested, there are non-mechanical accounts of causation in which Simonds' talk of a relationship can be understood as the counterfactual dependence of an individual's thought upon their meaningful social context (rather than the brute social context itself). Putting matters this way we would say: *if* one *were* not situated within a particular social context of meaning *then* one would *not* have those particular modes of thought.

Simonds' attempted rescue of Mannheim's version of SK raises further issues that have to do, once more, with how the two relata are to be understood, viz., the alleged "objects" that are to be explained (the *explanandum*), and what is to be appealed to in the explanation (the *explanans*). Taken at its face value SK, the sociology of *knowledge*, has to do with the explanation of why someone *knows* that p, or better, *believes* that p (where 'p' is some propositional content). However for Simonds the "object" to be explained by SK appears not to be *knowledge* but the *meanings* we share (presumably of the words and sentences of our common language). And the "object" that does the explaining is not brute social existence, but rather what is called 'social existence as a context of meaning', which, as Simonds says to add to the confusion, Mannheim simply refers to as 'social existence' (*ibid.*, p. 119). It adds to the confusion because there is a big difference between our quite objective social existence and the meanings we attribute to our social existence – a distinction that Marx would be at pains to make.

Given this understanding of *Seinsverbundenheit,* SK is now said to deal with the connection between two kinds of meaning: the meaning we give to our linguistic expressions and the "meaning" attributed to our social existence. In a passage not entirely consistent about the "object" to be explained, Mannheim puts this interpretation as follows: 'A position in the social structure carries with it ... the probability that he who occupies it will think in a certain way. It signifies existence oriented with reference to certain meanings.' (Mannheim (1936), p. 264). But note how Mannheim now characterises the link between the two meaning relata; there is alleged to be a *probabilistic* connection between a position a person occupies in a (presumably meaningful) social structure and the way that person thinks.

If the focus of Mannheimian SK is on the meaning of words and sentences, and not knowledge, then his views do not have much to do with recent discussions within SSK. Rather, some have seen a connection between Mannheim's views, so understood, and issues in contemporary 'analytic' philosophy about meaning – though Mannheim's obscure pronouncements have made no contribution to their discussions. One obvious connection that could be explored would be to view Mannheim in the light of Wittgenstein's account of the meaning we attach to words through following the rules of language games, which in turn are related to a form of life. Though 'form of life' is an obscure notion in Wittgenstein, it seems no more or less obscure than Mannheim's appeal to 'social existence as a context of meaning'.

Another connection made by D'Amico ((1999), pp. 111-2) between Mannheim and 'analytic' philosophy takes us to the debate about whether wide or narrow content is to be attributed to our beliefs, as illustrated in Putnam's classic 'Twin-Earth' story (Putnam (1975), pp. 223-4). Suppose that there are two identical Earths with identical inhabitants $Oscar_E$ in our Earth and $Oscar_T$ who inhabits Twin Earth. Suppose also that their experience of water is the same, that they speak the same language English, and that their mental states are exactly the same when they entertain the belief expressed by utterances of the same sentence type 'water quenches thirst'. The only difference is that on Earth the chemical composition of water is H_2O while on Twin Earth it is XYZ (something unknown by both Oscars). Now, if meaning is a matter of narrow content only, and is just "in the head", there is no difference in meaning of the expression 'water quenches thirst' for both Oscars. But the truth-condition of the sentence uttered by $Oscar_E$ is H_2O

quenches thirst, while the truth-condition of the sentence uttered by Oscar$_T$ is *XYZ quenches thirst*. If belief content is wide rather than narrow, so that 'meanings' are not just in the head, then external circumstances must be taken into account in meaning specification.

The connection to Mannheim is allegedly this. Meanings are not merely in the heads of people. To establish the meaning of terms we must also invoke the external circumstances which give rise to thoughts such as 'it is water'. And it is these external circumstances which play an essential role in Mannheim's sociological account of the meaning we give to language and thought. But the connection is tenuous at best. The external circumstance in the case of Putnam's Twin Earth example is either the kind of stuff H_2O or the kind of stuff *XYZ*. But neither kinds of stuff are anything like Mannheim's 'social existence as a context of meaning' which in turn underpin the meaning we attribute to language. As suggestive as the connection might be, if Mannheim's SK is understood to be about meanings, then it is not obvious that he addressing issues about meaning such as those raised by Wittgenstein or Putnam. This would be to attribute an excessive amount of philosophical prescience to Mannheim's obscure remarks about SK, understood to be about meaning.

Let us set aside issues to do with the sociology of meaning and instead focus on the sociology of knowledge (SK), or more properly belief. What does Mannheim say about the overlap between SK and the more general sociology of science (SS)? This is the field of the sociology of scientific knowledge (SSK). David Bloor, the advocate of the Strong Programme for SSK, (SP), views Mannheim as adopting a 'weak programme' within SSK. We have already seen that Mannheim is willing to admit that there may be systems of belief that allegedly grow through their own "immanent laws" or have their own "inner dialectic" and thus have no "existential determination". That is, Mannheim endorses some version of the internal/external distinction which entails that there are two very different kinds of explanation to be given of a scientific community's acceptance of change or growth in scientific knowledge; one is in terms of the canons of some theory of scientific method and the other in terms of the sociology of belief. Mannheim's liberalism concerning kinds of explanation is not endorsed by advocates of SP. Bloor cites the very passage discussed above in which Mannheim allows that some "modes of thought" might have their own "immanent laws" or "inner dialectic" and have no "existential determination". But he pointedly adds that Mannheim's 'nerve failed him when it came to such apparently autonomous subjects such as mathematics and science' (Bloor (1991), p.11). This indicates one way in which SP is strong and Mannheim's programme is weak. All of mathematics and science is to be included within the scope of SP while they are excluded from the scope of Mannheim's SK; in fact SSK is a quite narrow discipline for Mannheim.

Advocates of SP have suffered from no such lack of nerve. For them all 'modes of thought' have existential determination and there are no autonomous internalist explanations (or, if they are given, they are merely ideological). Advocates of SP do not attack Mannheim's version of SSK in order to show where it went wrong. Rather they think he was right as far as he went; what they attack is the limited scope Mannheim claimed for it by expanding it into areas where he thought it not applicable. But the truth of the generality of SP is highly contested, as will be shown in the next three chapters.

4.3   MERTON AND NORMS FOR THE ETHOS OF SCIENCE

What other approaches were there to SS and SK at the time Mannheim's work became available in English? Robert Merton developed in the late 1930s and 1940s a version of the sociology of science that allowed that there might be empirically discovered instances of Mannheim's SK applied to SSK. But since Merton, and Mannheim, allowed that there can be adequate internalist explanations of scientific change and growth, this only gives support to a quite weak programme within SSK and not any stronger programme. However there is much more to SS than SSK. It is here that Merton made a significant contribution in matters concerning priority disputes in science, patterns of peer evaluation in science, the social character of the scientific profession, and so on. Of importance in the present context is Merton's work on the normative structure of science, and in particular his characterisation of the 'ethos of science'.

Though this work was done before and during the Second World War period and had its immediate concern with the role of science and scientists in non-democratic societies, Merton's ethos is of more general interest with its focus upon those norms which allegedly bind any scientist in any social context and at any time. Merton's approach is functionalist in outline. He assumes: 'The institutionalised goal of science is the extension of certified knowledge' (Merton (1973), p. 270). Granted this, one task of SS will be to discover what norms ought to bind scientists if this institutionalised goal is to be best realised. Merton's norms are sociological in character and are quite distinct from any epistemic norms that philosophers of science might propose as part of the methodology of science. But as will be seen, in order to express his norms, Merton must appeal to some norms of a theory of scientific method.

One norm arises in his discussion of the Nazis and their advocacy of 'Aryan science'. This is the norm of the *Autonomy* of scientists from external political authority: 'The ethos of science involves the functionally necessary demand that theories ... be evaluated in terms of their logical consistency and consonance with the facts. The political ethic would introduce the hitherto irrelevant criteria of the race or political creed of the scientist.' (*ibid.*, p. 258). Mixed with this sociological norm is a contested epistemic norm of scientific method requiring that any theory be internal consistent and be externally consistent with known facts. To avoid appeal to this particular epistemic norm, Merton's *Autonomy* norm can be re-expressed: 'whatever are the appropriate epistemic norms that scientists ought adopt concerning theory evaluation, they ought not also adopt criteria for theory evaluation based on the race, creed, social standing, gender, etc, of the scientists who propose or advocate the theory'. Merton is aware that such a norm may not always be obeyed in practice. What is of concern to him is whether this norm, rather than its opposite, would realise better the institutional goal of science of 'extending certified knowledge'. Not much empirical research is needed to show what the answer would be.

Other norms of Merton's ethos are of a similar character. Thus the norm of *Communalism* (Merton sometimes says *'Communism'*) requires that the results of all scientific investigation should be publicly accessible. In contrast, if the results of science could be kept in private ownership or kept the secret of the organisation that paid for the research, then such an ethos would tend to detract from the institutional goal of the 'extension of certified knowledge'. The very scientific discoveries kept in secret might be those that are necessary to make the next advance in science. Newton's remark 'if I have seen father, it is by standing on the shoulders of giants' (Merton (1993), p. 1) requires the ethos of communalism.

Another norm of *Universalism* says that theories 'are to be subjected to *preestablished impersonal criteria*: ... the acceptance or rejection of claims entering the lists of science is not to depend on the personal attributes of their protagonist' (*ibid.*, p. 270). And the norm of *Disinterestedness*, while allowing that scientists may be quite passionate advocates of their theories, says that they ought to be quite impartial when it comes to the testing of their theories and not judge theories in ways proscribed by the above norms, or indulge in fraud. A further norm of *Organised Scepticism* requires that all beliefs and theories be subject to critical scrutiny regardless of the support these beliefs might have from religious, political, business or other organisations in society. For the scientist, if a belief is testable then, when it is tested, the result of the test ought to be accepted regardless of what are the vested interests of others in maintaining the belief. In Merton's view, scientists ought to offer a strong challenge to authority and tradition within society.

A number of questions have been raised about Merton's norms by his critics. Did he pick out *all* the norms of the ethos of science? Are scientists always, either now or in the past, bound by these norms? Would norms counter to these be equally efficacious in realising the goal of science? To what extent are the norms violated? If they are, to what extent could they be binding if science continues to thrive when they are violated? But the critics often forget that the norms are to be viewed functionally and proposed with respect to the 'institutionalised goal of science', viz., 'the extension of certified knowledge'.[9] Many of the sociological norms listed above (often referred to by the acronym CUDOS, though this leaves out the goal of *Autonomy*) obviously do, if followed, achieve the goal of extending certified knowledge while following the counter-norms would not. However in order to give a fuller account of the norms governing science Merton himself toyed with the idea of counter-norms, that is pairs of norms which appear to offer contrary recommendations because they embody allegedly incompatible values. One pair Merton suggests is: 'the value set upon originality, which leads science to want their priority to be recognised, and the value set upon due humility, which leads them to insist on how little they have been able to accomplish' (*ibid.*, p. 383) Even if these are acceptable general norms (there may be doubt about this), they are hardly counter-norms in that both could be realised at the same time. Newton's remark about standing on the shoulders of giants before him is an indication of humility; however Newton entered into quite savage priority disputes in which he clearly set much store on his own originality. In Newton's case humility extended only so far.

Broadly construed Merton's norms do give us a partial picture of the ethos of science. Important to this picture is a commitment to a number of theses of a quite traditional conception of science: that there is such a thing as scientific method of the sort philosophers have described (though, as noticed, Merton often works with a naive view of what this is); that there is an internalist explanation of the change and growth in science; that the criteria of theory acceptance and rejection in science ought involve only those sanctioned by some theory of scientific method and that one ought to eschew any extra-methodological criteria. The last of these brings Merton into conflict with what might be dubbed "new wave" SSK that arose in the 1970s. One of the prominent new wave theories of SSK is the Strong Programme (SP) with its denial that there is a purely internalist account of the change and growth of scientific knowledge and its strong counter claim that always there is, or must be, present social factors or social interests (i.e., Mannheim's 'existential determination of knowledge') influencing theory choice, and thus the very content of science itself. On the face of it, if such elements of a scientists' historical-social-situation enter into the existential determination of scientific knowledge then the Mertonian ethos will have been violated.

The post-Mertonian "new wave" SSK of the 1970's is still with us despite claims, even amongst some of those still surfing the new wave (ethnomethodologists, actor-network theorists, discourse analysts, etc), that some of the original surfers have now disappeared from view. This is sometimes said by those who now think that Bloor's SP has gone out with the tide. New wave SSK has a broad number of adherents with differing views.[10] Since it is of major epistemological interest, the next three chapters explore SP while containing only passing reference to other features of new wave SSK.

NOTES

[1]   Though Mannheim (1893-1947) was born in Budapest he counts as a 'German Connection' since the philosophical context of his work is within the many streams of German philosophy of the time, especially Marxism.

[2]   See Rodney Needham's introduction to his edition of Durkheim's and Mauss' book with its severely critical account not only of the leading thesis of the book but also much of the anthropology on which it is based.

[3]   Some praise Marx's deliberate vagueness here saying that he suggested 'a rich field of interactions and interrelations' (Lukács (1978), p. 32) for us to consider. But for many commentators there must always be some minimal one-way dependence; thus Lukács' goes on to talk of 'the predominant moment' and Althusser's talks of "the determinate principle 'in the last instance' of the economy" (Althusser (1969), p. 117). However if one allows such a rich field of interactions, including action in the opposite direction from superstructure to base, then Marx's theory is in danger of losing its bite and becoming merely a many-factor theory in which there is no preferred one-way dependence. Even allowing for many particular instances of different kinds of interaction, there remains obscurity about the alleged relation in schematic formulations of Marx's view. This obscurity is endemic to the whole subject of SK, and contributes to its lack of falsification and/or empirical conditions of test.

[4]   That some forms of consciousness provide a functional explanation of why particular forms of social and/or productive life obtain is a central thesis of Cohen (1978) chapters IX and X; this is also outlined in Elster (1985), pp. 27-37.

[5]   Newton sets out in his *Principia* Book III 'System of the World' not only some of the *Rules of Reasoning in Philosophy* that he employed, but also the informational base (called by him *Phenomena* which included two of Kepler's laws of planetary motion) that he employed to arrive at the various propositions which culminate in his Law of Universal Gravitation. The exact manner of how principles of inductive and deductive inference were applied to the data to arrive at a version of his law is a matter of scholarly investigation. For one account of the complex character of Newton's inductive procedure using his *Rules* and *Phenomena*, see Glymour (1980) pp. 203-26. What is clear however is that social circumstances do not enter into the picture as a cause of belief in the law, as required by the Marx-Engels Thesis (ME). This last claim is one that will be contested by practitioners of SSK. However they do not discuss this example; and any case they might attempt to make would be extraordinarily weak given the copious mathematical reasoning Newton advanced for his inverse Law and the complexity of the inferences as set out in Glymour.

[6]   This is a 1936 English translation of writings published in German in 1929 as *Ideologie and Utopie* with a translation of a 1931 paper called 'Wissenssoziologie' added as Part V and a new Part I for the 'Anglo-Saxon' reader as an introduction to the book. In what follows the book will be treated as a whole and no distinction between its parts will be drawn, as some authors advise.

[7] In the continuing SSK saga of what relationship the relata stand in, the above counterfactual dependence relation will be taken as the fundamental one-way relation that is needed in all SK and SSK. In the next chapter the relation of counterfactual dependence will be linked to one prominent theory of causality, first suggested by Hume and developed in Lewis (1986), chapter 21, in which causation is defined in terms of counterfactual dependence.

[8] The important internal/external distinction which Mannheim adopts has been drawn in various ways. For two ways see Kuhn (1977) chapter 5, and Lakatos (1978) chapter 2. Of course radical sociologists of 'knowledge', as we will see, deny that any such significant distinction can be drawn.

[9] One critical response is that of Barnes and Dolby (1970). However they do not always note the import of the functional role that Merton's norm's play. And the fact that they are norms expressing imperatives means that they are not readily susceptible to counterexample if the norm has been shown to be violated.

[10] For a sympathetic survey of recent theories of SS and SSK see Lynch (1993). For a critical survey see Bunge (1991) and (1992).

CHAPTER 5

THE EDINBURGH CONNECTION I:
THE STRONG PROGRAMME AND THE SOCIAL
CAUSES OF SCIENTIFIC BELIEF

5.1   INTERPRETING THE STRONG PROGRAMME

The formulations of the main theses of the sociology of knowledge (SK) are somewhat crude in Marx and Mannheim, even though their applications are less so. Since the 1970s much more has been done to state the main theses of SK through its application to the sociology of *scientific* knowledge (SSK). A number of people associated with the University of Edinburgh Science Studies Unit, such as David Bloor, Barry Barnes, Michael Mulkay, Harry Collins and David Edge amongst others, have done much to shape what a thoroughgoing SK and SSK would be like, and to explore, more fully than others before, what are the implications of a sociological approach to scientific belief. In particular David Bloor's formulation of the *Strong Programme* for the sociology of scientific knowledge (SP) has provided the main themes that research projects into scientific belief have come to address. A number of variations on these themes have been played by investigators who acknowledge varying degrees of kinship to, or distance from, SP.

SSK, and SP in particular, have excited much negative critical comment by both philosophers and those sociologists not enamoured with the path along which SSK was being taken by SP. The ground for the contest between philosophers and sociologists was laid out by Bloor when he provocatively declared, adapting a remark of Wittgenstein's, that sociological studies of science are 'the heirs to the subject that used to be called philosophy' (Bloor 1983, chapter 9). But reports of the death of philosophy in connection with the study of science have been greatly exaggerated. The next three chapters continue this critical debate through an investigation of SP, given the central role it has played in the formulation of one of the most influential versions of SK and SSK to date. This chapter deals largely with how we are to understand the first crucial tenet of SP, its ramifications for the explanation and understanding of belief in science, and the stark contrast it makes with explanations which appeal to methodological and epistemological norms of science outlined in Part I. Chapter 6 deals with the other three tenets of SP. Chapter 7 investigates the influence of Wittgenstein on SP and its social characterisation of rationality, especially its social account of logical relations.

5.1.1 The Knowledge/Belief Confusion Again, and its Significance

As a first salvo from one side of the critical debate, both SK and SSK are subjects conceived in sin; their progenitors failed to guard against the knowledge/belief confusion. Most practitioners of SSK ignore the distinction. The "object" to be explained within SP is often said to be knowledge, or the variability of knowledge across different groups. 'The sociologist is concerned with knowledge, including scientific knowledge, purely as a natural phenomenon.' (Bloor (1991), p. 5). But even if both knowledge and belief are to be treated as "natural phenomena", it does not follow that the distinction between them is to be ignored, as chapters 2 and 3 show. Bloor expresses well the sociologists' idiosyncratic conflation of knowledge and belief:

> The appropriate definition of knowledge will therefore be rather different from that of either the layman or the philosopher. Instead of defining it as true belief – or perhaps justified true belief [added in second edition]- knowledge for the sociologist is whatever people take to be knowledge. It consists of those beliefs which people confidently hold to and live by. In particular the sociologist will be concerned with beliefs that are taken for granted or institutionalised, or invested with authority by groups of people. Of course knowledge must be distinguished from mere belief. This can be done by reserving the word 'knowledge' for what is collectively endorsed, leaving the individual and idiosyncratic to count as mere belief. (*ibid.*, p. 5)

And in the next paragraph Bloor flits indifferently between talk of ideas, beliefs and knowledge. In the quotation, the parenthetic second edition revision of knowledge to "justified true belief" at least brings us up to Plato's account of knowledge in the *Meno* discussed in section 2.1. But it does not take us as far as Plato's later criticisms of this in the *Theaetetus,* or later Hellenistic broader articulations of the theory and its problems, such as those of Agrippa, set out in section 2.2. Nor is anything said of other twentieth century theories of knowledge, for example, the reliabilist theory set out in section 2.3. Theories within epistemology from Plato to the present are a rich field with which advocates of SSK have not come to terms. Philosophers can rightly complain that, in his account, Bloor has simply redefined the subject under consideration, viz., knowledge, and is not talking about anything that could recognisably taken to be knowledge.

What the quotation endorses is a version of group belief but without any qualifications, such as those set out in section 2.4.1, which mark the difference between group belief and group knowledge. In section 2.4.5 some grounds were set out for distinguishing social theories of knowledge from the sociology of (scientific) "knowledge". (Heavy use will be made of scare quotes to distinguish the sociologists' misuse of the term 'knowledge' from the philosopher's use.) The quotation also endorses the "inversion thesis" in which group "knowledge" is taken to be more fundamental than individual "knowledge", and under which individual "knowledge" is subsumed. If an individual deviates from what the group "knows" then the individual cannot "know" but, at best, merely believes. But there is fast and loose play here with epistemological notions. First, "knowledge" is taken to be just which a *group* takes for granted, or invests with authority, or collectively endorses. But this is merely group belief. No mention is made of the truth or justification conditions gestured at earlier in the quotation. Second, such a view has the unfortunate consequence that dissenters cannot be "knowers" (in the sociologists' above use of the term). But they can be knowers (in the epistemologists use of the term), as is obvious from the discussion of knowledge in chapter 2.

In the quotation, lip service is paid to Plato's problem of distinguishing knowledge from belief; but its significance is ignored. The point being laboured here might seem to be a verbal dispute about the use of terms like 'believe' and 'know'. But to run them together is to ignore the vital distinction, emphasised in chapter 3, between a purely descriptive notion such as belief, and a normative notion such as knowledge. To strip knowledge of its normative evaluative aspects is to reduce it to a descriptive notion such as belief. Apart from being ruinous itself, this has two important consequences. First, there is an important difference between the "objects" to be explained, as when we want to explain (causally or in some other fashion) why a person believes something (belief being a naturalistic state of affairs), as opposed to why a person knows something (a normative state of affairs). Both kinds of explanation can be causal (broadly understood), but what does the explaining must be different. What explains why a person knows, as opposed to why they believe, must make some reference to the norms of reason. No satisfactory explanation is possible if the normative aspects of knowledge are ignored, as is too often the case in sociological explanations of why a person knows. And all of this holds even if, as was argued in the final two sections of chapter 3, there is no incompatibility between naturalism and normativity and that the norms of method and epistemology can be naturalised. But advocates of SSK do not take the view of normativity set out in chapter 3; they often think of it as a form of norm Platonism, as was seen in Bloor's account of the norms of reason in section 3.5.1. And of course, for any naturalist, norm Platonism is to be rejected. The second dubious consequence is that naturalism concerning our knowledge is too cheaply realised. Just downgrade any claims about knowledge to those about belief, as in the above quotation, by striping away all normativity; then one need appeal in explanations to nothing but non-normative descriptive items.

Such blindness about knowledge is endemic to sociologists of knowledge (though not only them), three good cases being Bloor, Mannheim (see section

4.2) and the economist of knowledge, Fritz Machlup[1] (who is not always without a feel for epistemological notions). The conflict between sociologists and epistemologists would be vastly reduced if the former were to study the emergence, maintenance and epidemiology of *belief* in a population over time (as Machlup and others sometimes do), while epistemologists are left to deal with the problem of *knowledge* bequeathed by Plato concerning the reasons or justifications for our beliefs, an enterprise which is normative. Bloor's SP is intended to be a contribution to a naturalistic theory of knowledge. But it is best viewed as a contribution to a naturalist theory of the explanation of *belief*, since it ignores the normative conditions required of knowledge by epistemologists.

However sociologists might not be content with the suggested division of labour and see their job extending even into the epistemologists' domain of the norms that yield reasons for, or justifications of, our beliefs. For radical advocates of SP who wish to usurp the task of philosophers, there is alleged to be no independent theory of reasons or justification to be given that would underpin the autonomy of knowledge. Instead there is merely a further sociological investigation to be carried out even in the domain of reason-giving, or justification-making, which thereby extends the dominion of the sociology of belief. This issue will get more airing later since it involves the question of whether social factors could be constitutive of epistemological and methodological norms and whether they could even determine what norms of rationality we ought to accept. From the account given in Part I of how the norms of rationality can be tested, the involvement of the social is either non-existent or innocently minimal; but the case needs to be considered.

5.1.2   *The Four Tenets of a Sociology of Scientific "Knowledge"*

Consistent with his conception of naturalism, Bloor sets out his version of SP as a causal-explanatory theory which is to account for all our beliefs, including scientific beliefs (but note that Bloor's formulations are sometimes in terms of "knowledge", and sometimes in terms of belief). Though SP is stated succinctly in the form of four theses, as will be seen there is much ambiguity in how they are to be understood – so much so that quite disparate claims are said by other sociologists to be in conformity with SP. The theses of SP are not part of any piece of empirical research into particular areas of the sociology of science, though evidence culled from sociological studies of science is cited which inductively support the theses.[2] Rather, they are a 'meta-sociological manifesto'[3], a set of theses outlining a programme to which any version of SSK allegedly worth its salt should conform. SP is not strictly part of empirical sociology; rather it is a meta-theoretical schema intended to guide empirical studies in uncovering the social causes of belief. As such SP can be of interest to epistemologists as much as to theoretically-minded sociologists. But epistemologists would reject the more radical pretensions of advocates of SP who wish to overthrow normative epistemology.

SP comprises the following often-cited four tenets:

(1) *Causality Tenet (CT):* It [i.e., SP] would be causal, that is, concerned with the conditions that bring about belief or states of knowledge. Naturally there will be other types of causes apart from social ones which will co-operate in bringing about belief.

(2) *Impartiality Tenet (IT):* It would be impartial with respect to truth and falsity, rationality or irrationality, success or failure. Both sides of these dichotomies will require explanation.

(3) *Symmetry Tenet (ST):* It would be symmetrical in its style of explanation. The same types of cause would explain, say, true and false beliefs.

(4) *Reflexivity Tenet (RT):* It would be reflexive. In principle its patterns of explanation would have to be applicable to sociology itself. Like the requirement of symmetry this is a response to the need to seek for general explanations. It is an obvious requirement of principle because otherwise sociology would be a standing refutation of its own theories. (Bloor (1991), p. 7)

On the surface the four tenets seem to be clear enough. But this obscures difficulties that will be explored and critically evaluated in this and the next chapter.

Before giving the first tenet a more explicit formulation, four important points need to be made. First, the four tenets will be expressed in terms of belief rather than knowledge. However the question as to whether there is, as well as a sociology of belief, also a sociology of the reasons and justifications given in knowledge, will have to be addressed later. .

Second, though a case can be made for talk of disembodied, or objective, knowledge as when it is said 'it is known that', it is primarily persons that are the holders of belief, and knowledge. So, following the general practice of epistemic logicians, it is primarily persons (denoted by the variable 'x', or the names 'a', 'b', 'c', etc.) who are believers, and propositional contents that are believed (denoted by the letters 'p', 'q', 'r', etc.). Thus the claim that Plato believes *that knowledge is more than mere belief* can be rendered as 'aBp' where 'a' stands for Plato, 'B' is the belief relation and 'p' stands for the propositional content, *that knowledge is more than mere belief.* Though persons are primarily the bearers of belief, groups, or collectives, can also be said to hold beliefs (see section 2.4.1). So groups of scientists as well as individual scientists can be included within the range of 'x' or named by 'a', etc (e.g., astrophysicists, algebraists, acupuncturists, etc). But note that there is now a distinct sort of object to be explained, viz., why some group believes that p as opposed to why some individual believes that p. These can have quite distinct explanations because (as was indicated in section 2.4.1) group belief is not necessarily the same as the sum of the individual beliefs held by members of the group; nor is group belief "reducible to" individual belief.

Third, it is important to distinguish between, on the one hand, the *content* of a person's belief, viz., *that p*, and, on the other, the *act*, or activity, of a person believing that p. People are active both physically and mentally, typical mental acts being that of a person believing that p, or knowing, hoping, desiring, entertaining, dreaming, accepting or grasping that p. Acts of believing start, stop, can be caused or manipulated, last a long time, and so on. Since a dispositional account of belief will be presupposed here,[4] we may also be interested in how belief dispositions are set up, maintained or cease. Understood this way, *acts* of believing are part of the causal nexus of the natural world. In contrast *what* people believe, the *propositional contents* of their beliefs, are quite distinct since they do not stand in causal but in logical relations, such as that of contradiction, consistency, implication, and so on.[5] The distinction is important because, in a causal-explanatory theory such as SP, it is people's acts of believing (or their dispositions to believe) that are caused or need explanation and not the propositional belief contents.[6] SP will be understood to give a causal-explanatory account of *acts* of believing and not the propositional *content* of the belief. This point is obscured in much sociology of science yet it is fundamental to an understanding of the proper scope of the tenets of SP.[7]

To illustrate, one may ask for an explanation of *why* (say) the Earth is an oblate sphere. But this is quite different from asking *why* (say) *Albert believes that* the Earth is an oblate sphere. In each case a different "object" is to be explained: (i) an alleged state of affairs of the world, *that* the Earth is an oblate sphere; and (ii) another different state of affairs which is a person's (say Albert's) act of belief, viz., *that Albert believes that* the Earth is an oblate sphere. Importantly, the latter can be caused in ways in which the former cannot. For example, Albert might be caused to believe that the Earth is an oblate sphere in a number of distinct ways: he might have investigated the evidence for this to a reasonable degree of satisfactoriness and believes it on this basis; or he may have written it out 500 times at school, the proposition thereby remaining lodged in his mind; or he may believe it because his local guru, in whom he fervently believes, told him so; or he may have had the proposition put in his mind during an hypnotic trance; and so on.

Fourth, SP is expressed in terms of causal relations. Some philosophers think of causal relations as holding between events, others think of them as holding between facts. Nothing in what follows depends on this dispute about the analysis of causality. We may think of 'aBp' as expressing a fact, for example, the fact that Albert believes that the Earth is an oblate sphere (over some period of time t). Or we may think of 'aBp' denoting an event or activity, for example, the event or act of Albert's believing that the Earth is an oblate sphere (which occurs over a stretch of time t).

The first tenet of SP is expressed in terms of causation, while the other three are expressed in terms of explanation. So the task of elucidating SP will be in terms of both causal relations and causal-explanatory relations. Since the Causality Tenet (CT) is a causal claim, the core idea it expresses has the following form: S_a causes aBp. In this schema 'a' stands for some person (say Albert) or group (say astrophysicists), 'S_a' stands for those aspects of a's historical-social-cultural conditions which are causally efficacious; 'aBp' stands for the effect brought about (e.g., Albert's act of belief that the Earth is an oblate sphere). Finally 'causes' stands for some causal relation between S_a and aBp. Advocates of SP say little about the theory of causation they presuppose. In section 5.9 this defect is remedied by arguing that the counterfactual theory of causation best suits SP. In the intervening sections a number of explanatory models will be explored.

5.2 SOCIAL AND NON-SOCIAL FACTORS IN BELIEF CAUSATION

Within SP, what is to be causally explained? As discussed, it is states or acts of believing p by person a, or for short 'aBp'. What does the causal explaining? The Causality Tenet (CT) is clear that there are always two broad kinds of causes operating in conjunction. The first is a social cause. Let us denote the social-historical-cultural condition of some person 'x' by 'S_x'. The second broad kind of cause is 'other types of causes apart from social ones'. To make the two sets of causes exhaustive and exclusive, let us refer to these "others" as *non-social* factors or conditions which co-operate with the social conditions to causally bring about the effect of believing. Let us denote the non-social conditions of person x by 'N_x'.

5.2.1 Some Examples of Non-Social and Social Factors

What sorts of thing are to be included in the non-social factors N_x? Since acts of believing are mental, then mental features of our minds, and also our brains, must be causally at work. One kind of cause would be (a) mental states such as other beliefs, or our desires and wants (as will become evident in the 'interests' version of SP). But we can also appeal to the sciences that have been developed in the study of our minds/brains. They yield the following kinds of non-social causes: (b) our brain physiology with its chemical and electrical processes; (c) whatever 'natural' cognitive capacities we have; (d) the perceptual apparatus that we possesses that has been bequeathed, as have our cognitive capacities, to us by (e) our historical evolution as a perceiving and cognising animal; (f) the genetic characteristics that we possess as a member of the human species. All of these psychological, physiological, genetic and evolutionary characteristics are non-social factors which make some causal contribution to any act of believing. They are common to humanity and causally contribute to acts of belief in each of us in similar ways.

Since they are common, they cannot explain why acts of belief on similar subjects differ across humanity. One thing that might do this is: (g) the differing histories of sensory input each of us has which arises from our particular perceptual apparatus and its spatial and temporal orientation in the world, the input arising from the stimulation of our nerve endings mainly by more distant causes such as external objects, properties and states of affairs. Even though each of us has different histories of sensory inputs, these histories are simply more of the non-social features that bring about some of our acts of believing; they differ from other non-social factors only in that they are not common to, but are variable across, humanity. Those of us that live close to one another will have largely qualitatively similar individual histories of sensory input while others who live in, say, jungles, deserts or amongst arctic ice, will have histories that do differ qualitatively in many (but not all) ways. Such histories are not to be confused with the running reports we might give of our sensory input, whether it be of the experience itself (of, say, the way the Sun at dawn *looks* to each of us) or of the things that give rise to the experience (the fact that the Sun is rising). Advocates of SP hold to the doctrine of the theory-ladenness of observation, viz., the view that our common and scientific languages are laden with some theory or other, either an explicitly adopted scientific theory or one which is deeply embedded in our ordinary ways of talking (as when we say at dawn 'the Sun is rising'). In the shift from sensory input to *reports* on that input we have moved from a cause of acts of believing that is non-social to something that has social aspects, viz., reports expressed in our socially made languages and theories.

SP invokes a very broad range of social factors S_x which can feature in any causal explanation of our acts of believing. Unlike the non-social factors which (apart from our histories of sensory input) are common to humanity, the social factors can differ from one social group to another. It is this social variability which allegedly does most of the work in explaining most of the differences in our acts of believing (but always in conjunction with non-variable non-social factors). Advocates of SP have not drawn up a comprehensive typology of social factors S_x which could enter into the causal explanations; and this will not be done here. But the following would be included in any list of types of social factor. (a) One already cited factor is the natural language we learn (as distinct from, say, the Chomskyan deep linguistic structures that we all possess as a result of our evolution), in particular, the language we use to talk about the world (including our sensory stimulation and experiences) and in which we express our beliefs and theories. (b) Any theory or general point of view or set of beliefs, from mythical and religious to scientific, that a particular community of people might adopt. Such general beliefs, according to the doctrine of the theory ladenness of observation, load themselves, in some way, onto our observational reports.[8] (c) The history of each person's learning of, and education within, the cultural traditions, beliefs and practices of their community, i.e., their processes of acculturation. (d) Each person's social and political context, or their class or status, or the interests they may have which either result from, or support, these.

(e) Advocates of SP often talk of our scientific beliefs arising from the negotiations scientists enter into with one another that result from their differing stances over some scientific issue and the social processes in which they engage in order to reach an overall consensus about what to believe. The view of SP is that evidence always underdetermines any theory, or set of theories, and that social factors are typically invoked to fill in the gap between evidence and the theory that is accepted (believed). The sociality of negotiation leads to an overall consensus about which theory to believe in the face of the inability of evidence to point directly to that theory. Evidential underdetermination of all scientific belief is one of the cornerstone doctrines of SP from which sociality gets much of its purchase as a leading cause of scientific belief. In the light of the scientism and determinism of SP, if evidence does not make our scientific beliefs fully determinate then social factors can, and do, fill the gap between evidence and the theory it underdetermines. This is a matter not discussed in this book, as it has been amply criticised elsewhere.[9]

These, and other types of social factor, can vary from society to society. That is, we can differ in our languages, our theories, points or view and mores, our modes of acculturation, our socio-political context, our socio-political interests, and so on. It is these variable social factors that advocates of SP argue cause us (but always in conjunction with non-social factors) to have one belief rather than another, or sustain or change our beliefs. This conjunction thesis of SP will be set out in the next section. But before this, one further distinctive social factor shaping meaning as well as belief.

5.2.2   Social Aspects of Meaning: Meaning Finitism

Advocates of SP introduce a further important social factor into considerations about language ((a) above) that deserves consideration by itself. They adopt a quite particular theory about the meaning of the classificatory terms of our language such as 'swan', 'rose', 'water' (and in fact all terms on our language), called the doctrine of *meaning finitism*. This theory of meaning has its foundations in Barnes' adaptation of some views of Kuhn on classification, but more importantly in Bloor's account of Wittgenstein on rule following (discussed more fully in section 7.6).[10] Consider the use of "observational" terms like 'swan' in observational statements such as 'this is a swan'. What determines whether, or not, we say 'it is a swan' when we are confronted with some new item? Some might say that the meaning of the word 'swan' constrains us here; or they might say that there is a (objective) rule that we follow when we say ' it is a swan'. But according to the doctrine of meaning finitism, nothing constrains us to say either 'it is a swan' or 'it is not a swan'. The future is open and a term's 'meaning is constructed as we go along' (Bloor (1991), p. 164). Elsewhere we are told: 'According to meaning finitism, we create meaning as we move from case to case. We *could* take our concepts or rules anywhere, in any direction, and count anything as a new member of an old class, or of the same kind as some existing finite set of past cases.' (Bloor (1997), p 19). On an analogy with the number sequence '2, 4, 6, 8, …', if asked 'What is the next number?', then on this view any number could be next. There is no unique rule, or meaning, to constrain us. Rather, there is an infinite number of rules that generate the first four numbers of the sequence; so any rule could be used to generate any further number as the next number in the sequence. But we all agree that not anything could, or does, follow in the sequence of swans, or numbers. So, what are the constraints which determine what comes next in the number sequence, or whether we say of the new item 'it is a swan'?

Non-social factors, due to our sensory experience, will enter into saying 'it is a swan'. But these are not enough since, according to advocates of SP who also adopt some version of the underdetermination of all theory by experience, many different classifications can fit our actual history of experience (and even all possible experience past, present and future). Given the range of possible classifications that any group might adopt, what determines the classificatory system any one group does actually adopt? According to meaning finitism the underdetermination is resolved by social factors due to, for example, our past training in the use of the word in the presence of similar items, or the pressure of what our peers say and do in similar circumstances, or the collective authority of tradition, and so on. The doctrine the sociologists adopt here is one they attribute to Wittgenstein's account of rule following and meaning. The real constraints that are alleged to lie at the heart of 'Wittgensteinian meaning finitism' that stop us from going anywhere in the move to the next case, are a conjunction of quite local circumstances which impinge on us, as Bloor says in his recent book on Wittgenstein:

> The real sources of constraint preventing our going anywhere or everywhere, as we move from case to case, are the local circumstances impinging on us: our instincts, our biological nature, our sense experience, our interactions with other people, our immediate purposes, our training, our anticipation of and response to sanctions, and so on through the gamut of causes, starting with the psychological and ending up with the sociological. That is the message of Wittgenstein's meaning finitism. (Bloor (1997), p. 20)

Bloor's story is causal; it lists all the causal conditions that lead one to say of the next item 'it is a swan' (or deny it as the case might be). The story is also a naturalistic one that does not appeal to abstracta such as meanings or rules as constraints. What it does appeal to is a range of non-social factors including our biology and our sensory experience, and, importantly, a range of social factors including our past training and our interaction with others, what they would say, and our anticipation of what they would say and the ensuing consequences. All these social and non-social factors come together to causally bring about our response, and the response of others, when we go on to say of the next case 'it is a swan'. Such a naturalistic causal account of the constraints proposed by meaning finitism for our everyday classificatory concepts is extended to all our concepts, including those of science, mathematics and logic. Here social causes not only bring about belief in accordance with the Causality Tenet CT; they are also the determinants of the meaning of 'swan' and our classification of some item in front of us as a swan. Here we are to understand CT not as a claim about the causes of a's believing that p, but something quite different, viz., the meaning attributed to a term. Not only is belief socially caused; the very meaning of a term is socially caused, including whether or not some item falls under its extension. We will return to the issues raised by the above theory of meaning in chapter 7 which largely deals with the influence of Wittgenstein on SP and the way in which advocates of SP understands his notion of rule following.

To sum up, there is a deep well of alleged social factors that advocates of SP can appeal to in giving a causal explanation of any act of believing. We could appeal to social factors of type (a) to (e) of the previous sub-section. If none of these yield any causal factors for belief, then they can appeal to social factors due to the specific doctrine of meaning finitism applied to the use of the very language we employ in formulating our beliefs. Since this type of factor is a pervasive feature of our language, there would seem to be an inescapable social element in all our acts of belief. In fact the first five types of social factor presuppose the sixth of meaning finitism. In order to claim that, say, acculturation has been a social cause of some belief, that p, we need to be able to specify the meanings of the terms in the sentence that expresses the belief. Here the determinants of meaning are a quite broad range of social causal factors shaping our very use of language. And the object caused, viz., our use of language, is a quite different kind of object from the beliefs mentioned explicitly in SP. This understanding of CT will need different treatment (see chapter 7). A final point. Given this disjunctive list of factors, it is hard to see how any act of believing could ever fail to have a social cause, particularly because of pervasive social factors shaping the very meaning of what we say. That this can lead to CT lacking any test and/or falsification conditions is a point returned to at the end of section 5.4

The above lists only six broad types of social factors that could be invoked in formulations of SP. In the light of this CT tells that at least one (or more) type(s) of social factor in conjunction with at least one (or more) type(s) of non-social factor combine to cause acts of belief. Could social and non-social factors be given different causal weightings? Non-social factors N_x (except for sensory input) remain largely fixed across human geography and history; in contrast the non-social factor of sensory input and the social factors S_x are not fixed and can be highly variable across human geography and history. It is these variable factors that are to play the role of explaining any difference between acts of belief. But the relative invariance of non-social factors does not mean that they must always be relegated to background causal factors with little weighting. It remains to be seen what role can be given to purely cognitive capacities when we, say, think our way to new beliefs, as for example drawing a new conclusion from old premises. In such a case, the social factors might be relegated to the background with relatively little or no weighting. Could no social factor ever be present as we think our way to new beliefs? If this were the case then SP would be refuted by counterexamples (this is discussed in section 7.1). To rule this out, SP insists that some social factor must always be present. This suggests one way in which SP is *strong*. In all cases of belief, at least one type of social factor S_x must (allegedly) be present (though its weighting can vary); and if no other social factor can be found then there is always the doctrine of meaning finitism to come to the rescue.

5.3 THE CAUSALITY TENET AND A SOCIAL CAUSE MODEL OF EXPLANATION WITHIN THE STRONG PROGRAMME

5.3.1 Formulating the Causality Tenet and its Associated Explanatory Schema.

With these preliminaries out of the way we can now formulate the first Causality Tenet (CT). This tenet is central to the understanding of SP; the other tenets will be discussed in the following chapter. Consider some person, or group, x and a belief they have, p. What CT tells us is that x's believing that p is caused by two (sets of) factors, the social, S_x, and the non-social N_x, co-operating together; that is,
$(S_x \ \& \ N_x)$ cause xBp.

There are four quantifiers to be found in the more precise formulation of CT. CT is intended to be general with respect to *all* persons (or groups of persons)[11] and *all* the beliefs they hold; this gives the two universal quantifiers. The next two quantifiers are existential; they select at least one social factor S_x from the class of all types of social factor, and at least one non-social factor N_x from the class of all types of non-social factors. Fully formulated, the Causality Tenet CT says:

CT: For *all* persons x, and for *all* belief contents p such that xBp, *there exists some* social condition S_x of x (from the typology of social factors), and *there exists some* non-social condition N_x of x (from the typology of non-social factors), such that $(S_x$ & $N_x)$ causes xBp.

Though symbols have been used in expressing CT it is clear what they stand for;[12] and they give a clear expression to the purport of CT. Importantly, they highlight the selection of particular social and non-social factors from the open-ended range of the two types of causes, one or more from each which, when all conjoined, cause acts of belief. Though CT is much more explicitly formulated, a strong similarity can be noted between it and both the Marx-Engels Thesis and the various formulations given by Mannheim of his version of SK.

The above suggests a model for the sociological explanation of belief within SP that is in accord with Hempel's Deductive-Nomological (or Covering Law) schema because of its appeal to causal laws. The object to be explained is aBp. The explanation schema which answers the question 'why aBp?' appeals to laws and to particular conditions of believers:

Social Cause Explanation Schema (or Model)
(1) Scientist(s) a stand in given social (cultural, power, etc) conditions S_a and are in certain non-social conditions N_a;
(2) There is a law-like connection such that, for all x, whenever $(S_x$ & $N_x)$ then xBp;
(3) ∴ aBp.

Clause (1) of the schema sets out some "initial condition" of the scientists, including an account of their social and non-social circumstance. Clause (2) is a law of SSK linking social circumstance with belief (the problem that there are no such laws for use in the schema is discussed in section 5.8). From (1) and (2) the conclusion, the *explanandum,* aBp, follows. This schema arises directly from the reconstrual of CT as providing a causal explanation of belief rather than knowledge. This has rival models that will be examined subsequently.

5.3.2 What the Causality Tenet and the Social Cause Model Do and Do Not Say

CT, along with its accompanying social cause explanatory schema, both need further exploration to distinguish them from other theses and explanatory models, with which they can be confused. The first point to note is, as formulated, they concern x's intentional attitude of *belief.* Other intentional attitudes to theories and hypotheses might also be considered besides *belief* that p. For example, as writers as diverse as Popper, van Fraassen and Cohen insist, a scientist might *accept* a theory (for various purposes) without believing it.[13] Or a scientist might *hypothesise* that p, or *entertain* the thought that p, or *conjecture* whether p, or *investigate* the claim that p without believing that p. Perhaps SP could be reformulated as theses about acceptance rather than belief; but SP says nothing about the causes of such intentional attitudes of scientists. As will be seen, in some historical case studies under the banner of SP, there is some confusion about what "object", particularly what intentional attitude, is caused when SP is invoked. The belief/knowledge confusion aside, what SP says is caused is acts of belief; it is silent about the causes of other intentional attitudes of scientists such as hypothesising, entertaining, conjecturing, wondering whether, accepting, and so on, or other activities such as discovering, testing, etc. This is not to say that some of these do not have a social cause; but to express this, SP, and CT with its social cause explanatory schema, needs quite different formulations with different "objects" as the *explanandum*.

The second point is that CT is not concerned with the social causes of other matters that might be within the scope of a more general SS but outside the scope of SSK, or the more particular SP (which, taken at its word, is concerned only with *social causes* of *belief*, or as is said, "knowledge"). For example SS might investigate the *discovery* of some hypothesis and uncover the social causes (if any) of the discovery; but the social causes of *discovery* have nothing to do with CT as formulated. Again, SS might investigate the politics involved in the funding of research into whether some hypothesis or theory is true (highly probable, etc), or to some application of a theory. Or SS might investigate the micro-sociology of the allocation of scarce economic resources for experimental investigation into the claims of one theory rather than another, or for the development of some experimental technique. And so on. But the micro-social investigation into political and economic decisions about what science is, or is not, be researched is outside the scope of CT and not part of what it asserts. In these cases the "object" to be causally explained is not *x's belief that p* but rather *that some science p is (is not) to be funded for research*. This is not to say that "objects" such as the discovery that p or the funding of person (group) y, can not be the topic of social investigation; rather they do not fall within the scope of CT as formulated.

Care needs to be taken over these points because they involve the legitimation of the various domains of application of the sociology of science. And some do take some care – but only some. Thus in his important *The Politics of Pure Science,* Greenberg talks of the 'substance of science' saying that this 'is centered in the laboratory' while 'the politics of science is centered in the committee room'. (Greenberg (1967), p. xi). Though he makes it clear that his book is concerned with the latter (see pp. 4-5), his talk of 'the substance of science' still contains possibilities of confusion over the variety of *explananda,* only some of which are dispelled by the distinction he draws between substance and practice: 'If we are looking at the substance and practice of science, we find the common denominator is a dedication to the understanding of the universe through systematic investigation and measurement, through the harnessing of curiosity, training, discipline and instruments. (Some would add intuition and good luck.)' (*ibid.,* p. 5) Clearly much of what Greenberg calls 'practice' is, despite its social character, outside the scope of SP as formulated. However it remains unclear what the remaining aspect called 'substance' comes to. It clearly includes the explanation of scientific phenomena, and this lies outside the scope of SP. But 'substance' could well include the beliefs of scientists, the very contested domain claimed by SP.

5.3.3 The Claims of SP and the Case of Victorian Statistics

Some of these points also emerge in the work of SP's less committed advocates, such as Donald MacKenzie's account of the rise of statistics in Victorian England. He begins by saying: 'No one doubts that there must be *some* relationship between science and the social context in which it develops' (MacKenzie (1981), p. 2). This unspecific claim is divided into a weak and a strong claim. The weak thesis is consistent with traditional SS in that it makes limited claims such as: 'The extent of social support for science influences the pace of scientific advance, and the direction in which this support is channelled may lead to one scientific discipline growing more quickly than another.' (*loc. cit.*). This is not a point about the *causes of belief* but the causes of the *rates* at which the contents of different sciences unfold; the latter may be affected by social context while the very content itself is not affected (but the *amount* of that content may be affected). The weak thesis is about a different "object" within SS, viz., the *rate of development* of a science and what might causally influence it.

The strong claim is: 'The content of "good" science as well as "bad" can be potentially affected by its social context.' (*ibid.,* p. 3). But this is not a re-expression of CT, since it talks of what 'can be *potentially* affected' rather than what is actually caused. Mackenzie gives this claim two readings: 'One (weak) version of this point of view would be that the production of new ideas in science is socially influenced, but that these ideas are then judged according to general, objective criteria. So the social influence on the content of science would be shortlived.' (*loc. cit.*). Here the object to be explained by SSK would be 'the production of new ideas'. But as for Mannheim, there are 'objective criteria' whereby 'new ideas' that have been produced can be assessed for their belief-worthiness that have nothing to do with social causation. Such a position preserves an old distinction between context of discovery, which can have a social explanation, and context of justification which has none. On the second, stronger, reading social factors replace 'objective criteria' as the causes of belief. This can happen in a number of ways, from Kuhn's notion of the assent of a community of scientists leading to belief, to social negotiations and the other kinds of social causes which CT admits.

The central focus of MacKenzie's book is the social interests that shaped theories of statistics proposed by English Victorians such as Francis Galton (1822-1911) and Karl Pearson (1857-1936). Both had a social interest in eugenics, Galton founding a Chair at the University of London in eugenics which Pearson was the first to occupy. Mackenzie makes much of the possible links between an *interest* in eugenics and the development of statistics by these two theorists. Advocates of SP admit *social interests* (such as an interest in eugenics) as one kind of social factor S_x which can cause belief. (We have yet to discuss the 'social interests' version of CT; this occurs in section 5.5. For the time being we can take the interests thesis within SSK to be an instance of CT but it will be argued that it is different from CT.) For these Victorians eugenics was the study of the means by which the physical and mental characteristics of future generations could be "improved" – and then the implementation of these means by agencies of social control. The development of statistics became a necessary prerequisite to the study of eugenics. Thus an interest in eugenics can explain a number of why-questions such as: (i) why does x study statistics? (ii) why does x pursue one line of research rather than another in statistics? (iii) why does x apply statistics in one area rather than another? But can an interest in statistics explain, say, (iv) why some expression is a theorem within probability theory, such as Bayes' Theorem? Note that this is not the same as explaining (v) why x believes that Bayes' result is a theorem of the probability calculus. Though one might have an interest in discovering such theorems, whether some claim, such as Bayes' result, is a theorem or not is wholly a matter for an internalist theory of statistics to determine. Here there is an important difference between the four objects to be explained. Explaining (i) to (iii) is part of the province of SS. But explaining (iv), rather than (v), is not a task that SP should take on. If it does, SP challenges the propriety of explanations which are solely internal to statistics.

Even though Mackenzie locates a number of cases in which interests allegedly play a role in statistical theory, he also claims that there are cases of belief in statistics where non-cognitive interests (in eugenics, or whatever else) play no role. According to Mackenzie, Karl Pearson had both the broader interests of the professional middle class to which he belonged, and particular social interests in eugenics and in developing statistics for application to eugenics. But Mackenzie goes on to say, contrary to the general tenor of SP and the formulation of CT:

> I am not claiming that Pearson's social background, for example, *caused* his ideas. If my analyses of Pearson's writings [including statistics] and of the interests of the professional middle class are accepted, then all we have is an instance of a *'match'* of beliefs and social interests. Explaining why this 'match' came about exactly when it did, and why the particular individual Karl Pearson should have manifested it, is beyond the present capacity of the sociology of knowledge. ... This does not mean that all we can do is to point to this one instance of a 'match'. It is possible to look at the relationship between the historical fate of a system of belief and that of the class to which it is claimed to be appropriate. (*ibid.*, p. 92; italics added.)

Talk of a 'match', or gesturing at a 'possible relationship', between social interests and the content of some theory of statistics is a far cry from the strong causal claims of CT. As is well known, the mere co-presence, or correlation, in Pearson's mind between mental items such as an interest (in statistics) and a belief (in some statistical theory), is not a sufficient ground for the claim of a causal connection between the two, as required by CT. The careful historian in MacKenzie has won out over the dogma of SP that there must always be a causal connection between social interests and beliefs in statistical theory. Though MacKenzie does not say it, the field is left open to internalist explanations of many of Pearson's beliefs in statistics – and this despite the general interest he had in eugenics and an interest in the application of statistics to problems in eugenics.

The issues discussed above arose because of the very different "objects" within SK and SSK that call for explanation. But they have naturally led to the question whether historical research can show that all instances of scientific belief conform to CT. This matter is more fully addressed in sections 5.6 and 5.7 when we turn to case studies that allegedly support CT. In these studies there is a conflation of co-presence with cause, something that MacKenzie avoids. But at least for MacKenzie there are constraints on the generality of CT, as is evident from the above, and from the following episode he recounts. Mackenzie devotes a chapter (*ibid.*, chapter 7) to a controversy between Pearson and his former pupil George Yule over the best way to measure statistical association of data arranged in contingency tables. The technical nature of the controversy need not concern us. Suffice to say that some might adopt an internalist approach and explain the controversy in terms of differences in their systems of belief. And this is something that MacKenzie does in part do (*ibid.*, pp. 161-4). But the sociologist in MacKenzie wins out when, in the long run, consideration has to be given to the role different social factors and/or interests play in explaining the controversy: 'The differing goals manifested in the work of Pearson and Yule on the association were not accidental. They can be related to different objectives in the development of statistical theory, and perhaps ultimately to differing social interests.' (*ibid.*, p. 168), But the different 'objectives' could be cognitive goals internal to science rather than non-cognitive or social goals.

MacKenzie tells us a story about Person's professional class and his interest in eugenics. Let us grant this. What of Yule? Mackenzie candidly admits that the case for Yule is weak indeed: 'It is difficult to identify very specific goals informing this [Yule's] work, and the most one can clearly point to is the *absence* of the crucial eugenics/statistics connection.' (*ibid.*, p. 180). MacKenzie's italicised *absence* results from his research into Yule's interests. He discovered that Yule was opposed to eugenics; but he could find nothing in Yule's Tory social background or in his social interests that could remotely be said to influence Yule's side of the controversy about statistical association. The historian in MacKenzie wins out in his concluding remark about the controversy: 'Until further evidence can be uncovered, we may simply note the possibility that specific social relations sustained the non-eugenic statistics of Yule and his supporters' (*ibid.*, p. 182) The case of Yule, if not Pearson, is surely a counterexample to CT not to be fudged by talk of *absences* or of the *possibilities* of a connection between *some unspecified* social relation and belief. There were no social interests that informed Yule' side of the controversy that could be found. But there is plenty of evidence for theoretical differences between Pearson and Yule that prompted their controversy that turn on considerations internal, and not external, to statistical theory.

Two issues emerge from the above remarks. One has just been mentioned; MacKenzie's work on some Victorian statisticians shows that there may be no available empirical evidence to support the claims of CT. This is a theme pursued more fully concerning other case studies section 5.6 and 5.7. The second concerns the more precise formulation of CT which is not to be confused with other claims about socio-political influences on science and the mental attitudes of scientists to various scientific "objects" (such as discoveries or questions about what gets funding).

5.4 THE CAUSALITY TENET AND THE RATIONAL EXPLANATION OF SCIENTIFIC BELIEFS BY METHODOLOGICAL PRINCIPLES OF SCIENCE

5.4.1 To What Beliefs Does CT Apply?

There is one sense in which SP is strong: it allegedly applies to all beliefs held by any scientist (or lay person). But there is a problem about the individuation of the belief contents with respect to which CT holds. Are the beliefs to be quite specific, such as *that Earth is an oblate sphere*? Or are the beliefs to be more general, e.g., some of the axioms or basic principles of dynamics? Or are they to be clusters of beliefs such as those, for example, that constitute being a reasonably well informed geophysicist when it comes to plate tectonics? Again, nothing significant turns on whether we allow the 'p' in CT to range over specific beliefs or clusters of beliefs (either a precise set of beliefs or a fuzzy set). However what is important is whether CT applies to just one, or a few, of the beliefs in any cluster, or whether it must apply to all of them. If the former (as the case studies of sections 5.6 and 5.7 show), then there would be beliefs in the cluster which did not arise in accordance with the causal story prescribed by CT.

If unrestricted, 'p' in CT ranges widely over any belief whatever, including beliefs of everyday commonsense, moral beliefs, beliefs in mathematics, beliefs in logic (such as the belief in principles of deductive logic such as *Modus Ponens* or *Disjunctive Syllogism*), beliefs in methodological principles within the philosophy of science (such as prescriptions against *ad hoc* hypotheses, or Mill's methods for judging causes), beliefs within philosophy (such as theories of explanation, or theories about the science/non-science distinction), and finally belief in SP and its four tenets. The last of these is in fact what the fourth Reflexivity Tenet highlights; SP must apply to itself otherwise it would arbitrarily restrict the range of p. We will return to the issue of reflexivity in the next chapter. Of significance here is the extension of SP to non-empirical bodies of belief such as logic, philosophy and methodological principles of the philosophy of science.

In one sense the extension of SP to rules of deductive reasoning such as *Modus Ponens*, or to principles of scientific method is innocent. There is no reason why type (c) processes of education, learning and acculturation cannot cause a person's act of belief in *Modus Ponens*, or in some principle of scientific method (see section 5.2.1 on the typology of social causes). This is in fact the way most of us first come to hear of, and then believe, such principles. However what SP cannot show is why *Modus Ponens* is a *valid* rule, or why such-and-such a principle of scientific method is *correct*. Such type (c) social causation of acts of belief has no bearing on the truth-value of the proposition believed. Theories about how one establishes the validity, correctness and rationality of such principles lie outside the scope of CT. However, as will be seen in chapter 7, social causes due to meaning finitism are said to apply in the case of the meaning of logical terms thereby providing the missing social factors involved in determining validity or correctness.

In order to sort out the different strands of involvement of the social, consider the following particular belief-content B: that the Earth orbits the Sun while rotating daily on its axis. This is a belief that is efficiently transmitted through our current society by parents and teachers (since most do pick up the belief in this way and few miss out). In the pre-Copernican era the opposite belief was equally well transmitted to most members of an earlier phase of our culture. There are two social stories to be told, the first about how most in our society since the late sixteenth century have acquired the belief that B, and a second story about how most of those before the late sixteenth century acquired the opposite belief. Both social stories appeal to type (c) social factors, viz., those concerning the cultural transmission and maintenance of belief. While type (c) social factors are causally efficacious in bringing about acts of believing, they do so quite independently of the truth or falsity of the belief contents; so they can not be reliable transmitters of mainly, or only, truths.

But what of the smaller number of scientists who become astronomers and pick up belief B by studying, and independently employing, the reasons for B given by, say, Copernicus, Galileo, Kepler, Descartes and Newton (or any other reason or evidence)? For these astronomers there is an *internalist* story philosophers and scientists like to tell about the *rational* grounds on which the Copernican model of the solar system was preferred over rival non-Copernican models. Such reasons can be causes of belief. That is, there is a causal pathway leading from the reasons, and the evidence each scientist has, for the truth of B (and not some social condition S_x), to the beliefs that thereby become fixed in the minds of the scientists. For both sociologists who prefer an externalist social cause explanation and those philosophers who prefer an internalist story, the *explanandum* is the same, viz., x's belief that B. But they differ over the *explanans*. For the sociological externalist, the explanatory causal factor is S_x. But for the rationalist internalist, the *explanans* must contain reference to some principles of method M which are then applied by scientist x to B and the pre-Copernican rivals to B (call these rivals 'A').

5.4.2 A Rival to Social Cause Explanation: The Rationality Schema and Its Scope

There is a rival to the social cause explanatory schema of section 5.3.1 that appeals, not to some alleged social law, but to the use of methodological principles by scientists. Suppose that by the criteria of methodological principle M, B is a *better* theory that A (M determines that B is better than A because, say, B has greater conformation that A, or B is a more fruitful hypothesis with wider scope than A, and so on). Setting out the explanation deductively leads to the following model:[14]

Rational Explanation Schema (or the Rationality Model)
(1) Scientist x is confronted with a choice between belief in B or belief in A;
(2) On the basis of methodology M, it would be rational to prefer (believe) theory B rather than A;
(3) x is guided by the theory of rationality implicit in methodology M:

(4) If x holds to, and applies, method M to the choice between belief in B or belief in A and M determines that B is a better theory than A, then x *ought* to believe B (rather than believe A);

(5) ∴ x believes that B (rather than believes A).

The explanatory argument depends on four premises. The first and third premise contains reference to an 'initial condition' of a scientist who uses M when confronted with a choice between B and its rival A. The second premise tells us what the norms and principles of methodology M determine about the relative merits of B and its rival A. (It is this premise with its mention of some methodological principle that makes the explanatory schema, or model, rational.) Finally the fourth premise imposes a rationality condition on all scientists x concerning what they *ought* to believe when they employ method M on B and its rivals, and M determines, by its criteria, that B is better than its rivals. The conclusion is a description of an act that x performs, the act of believing that B. As such the argument form is not strictly deductive, but it has the same form as arguments to do with practical action. When used as an explanatory schema the conclusion of a 'practical· syllogism' offers an explanation of our actions, given our beliefs and desires. The rationality model above is similar in that the conclusion also explains a quite specific kind of act that we have performed, viz., our believing B (rather than believing A). In any rational internalist explanation of why x believes that B, something like these premises about our epistemic beliefs and desires must be employed. In fact the rationality model gives content to the very idea of what an *internalist* explanation of scientific belief is like and sets out one account of the *rationality* of belief in science. Note that this model appeals to no social factor S_x. It provides a quite different explanation of belief than that provided by the *externalist* social cause model of SP.

When should one employ one or other of the rival social cause and rationality models? From sociologists such as Mannheim to philosophers such as Lakatos and Laudan, the rationality model has been given priority. Laudan in fact expresses this in the form of an *Arationality Assumption: 'the sociology of knowledge may step in to explain beliefs if and only if those beliefs cannot be explained in terms of their rational merits'* (Laudan (1977), p. 202). But this assumption needs qualification. We need to allow the possibility that the very same act of belief B could be explained by the social cause model for one person and by the rationality model for another, or by the first model at an earlier stage of a person's career and the second model at a later stage. That is, for principles of method there may be a social story to be told about why they are, or are not, believed by various persons within a culture or across cultures. But none of this bears on their validity or correctness as explanations. Reformulating the *Arationality Assumption* less strongly, we may say that, *in so far as the rationality of science is concerned*, where the rationality model applies there are no grounds for seeking out an explanation according to the social cause model. Of course the employment of the rationality model turns on what counts as a viable methodology M. The further matter of legitimation of our methodological principles was broached in section 1.2 and is well canvassed elsewhere.[15]

5.4.3 Reasons as Causes of Belief

Let us assume that scientist x does reason in accord with the rationality model (in section 6.1 an alternative model based more closely on a scientists' actual reasoning processes will be proposed which does differ from the one just set out). With this caveat, explanations in accord with the rationality model are consistent with the idea that x's reasons for belief in B (viz., what M determines when x applies it in x's choice situation) are also causes of x's belief that B. That is, evidential reasons are causes in the sense that if x had not gone through the process of applying M to A and B, then x would not have come to believe B rather than A. (the counterfactual theory of causation on which this point turns is discussed in section 5.9). As such, reasons can become part of the naturalistic casual nexus in which our minds are situated. Strangely Bloor does not see matters this way, saying that there is an asymmetry between correct and incorrect reasoning in that the latter, but not the former, is caused: 'nothing makes people do things that are correct, but something does make, or cause, them to go wrong' (Bloor, (1991), pp. 8-9). The strangeness of this remark is that if literally *nothing* makes people do things that lead to correct belief, then the belief must be uncaused. Bloor goes on to speak of a *teleological*, rather than a causal, model for reason explanations saying 'the rational aspects of science are held to be self-moving and self-explanatory' (*ibid.*, p. 10). In summing up the differences between the so-called 'teleological model' and SP that makes them rival programmes, he says: 'It [the teleological model] relinquishes a thorough-going causal orientation. Causes can only be located for error' (*ibid.*, p. 12). On this characterisation of the rival teleological model, sociological explanations are said to arise only when it has been determined that some belief is erroneous; so epistemological notions like truth or rationality have a priority in determining which of the teleological model or SP is to apply. But the conclusion only follows on the erroneous assumption that reasons cannot be causes of belief and that teleological, not causal, considerations must apply in the domain of rationality. It is as if Davidson's seminal work on reasons and causes has been by-passed.[16] There is no reason why those who advance explanations in accordance with the rationality model should accept Bloor's peculiar non-causal characterisation of their view.

The above erroneous considerations in which reasons are said not to be causes, leads Bloor to his third Symmetry Tenet, part of which denies the asymmetry between teleological explanations for the rational and the true in contrast with social cause explanations for the irrational and false; there is just one kind of explanation for both, this being the social cause model. We will look at the symmetry requirement more closely in the next chapter. In section 3.5.1 we have already examined Bloor's remark: 'The symmetry requirement is meant to stop the intrusion of a non-naturalistic notion of reason into the causal story. It is not designed to exclude an appropriately naturalistic construal of reason, whether this be psychological of sociological' (*ibid.*, p. 177). What is of importance here is the idea that the norms of reason are understood to work non-naturalistically like a *deus ex machina* intruding upon the causal scene and interrupting it to bring about our beliefs. Within Bloor's naturalism there can be no room for norms that intrude from another realm. But there is a role for natural reason due either to our physical features such as our 'instincts', our 'biological natures' or to our socially ingrained features such as our past training, learning and education in a social context.

Does the rationality model require that norms irrupt upon the causal order? It entails nothing of the sort. The first three premises are merely factual reports either about x's choice situation or the methodology they hold, and what the methodology delivers when applied to some situation. The only norm-like premise is (4) which sets out what ought to be x's act of belief on the basis of x's deliberation about his/her choice situation in the light of having applied M. But in the light of the discussion in the final two sections of chapter 3, this is hardly a *deus ex machina* intrusion upon the causal order any more than is any other 'practical syllogism' which sets out deliberations in the premises and what course of action ought to be followed in the conclusion. The explanation in terms of methodological reasons given above does not irrupt on the causal order and can simply be part of it.

5.4.4 Is There Social Causation of Belief in Principles of Reasoning and Methodology?

Any process of belief fixation using method M that accords with the rationality model, transmutes mere belief into rational belief, or even knowledge. No social factors of types (a) to (e) are at work in fixing these beliefs. But many astronomers, either as children or as adults, might have first acquired belief B (= the Earth orbits the Sun and rotates daily on it axis) by acculturation (social factor type(c)). There is a complex social story told by historians of how the European intellectual community resisted or adopted Copernicus' system in the two centuries following his death. If some scientists did acquire their beliefs by acculturation, then later they might come to sustain their belief B in a quite different way that accords with the rationality model. However, even if these astronomers do not acquire B by a process of cultural transmission, could they have acquired their theory of evidence, reason and even their methodological principles, in this way? That is, could type (c) social factors (or some other type) enter not at the level of belief B but at the level of evidential reasons for B?

Barnes and Bloor (who call themselves 'relativists') answer positively:

> For the relativist there is no sense attached to the idea that some standards or beliefs are really rational as distinct from merely locally accepted as such. Because he thinks that there are no context-free or super-cultural norms of rationality he does not see rationally and irrationally held beliefs as making up two distinct and qualitatively different classes of thing. (Hollis and Lukes (eds.) (1982), pp. 27-8)
>
> 'Evidencing reasons', then, are a prime target for sociological inquiry and explanation. (*ibid.*, p. 29)

Here, not only scientific claims like B, but also the very standards or methodological principles by which they are judged, are to be treated on a par as beliefs that fall within the scope of CT, and are grist for the mill of SP. There is no rational grounding of our methodological principles; there is only their local acceptance, or non-acceptance as the case may be. We can agree that some people might acquire their belief in evidential principles by the same means as they acquire belief B; that is, by acquiring such beliefs under hypnotism, by writing them out as a school punishment, by picking them up as part of their cultural inheritance, by obeying commands of the accepted authorities in society, by being influenced by vested interests, and so on for other social grounds that Barnes and Bloor suggest.[17] But such acquisition of beliefs in principles of reasoning and method is innocent. It leaves untouched the procedures for determining the truth, correctness or validity of such principles; these matters are established by other means (some of which are discussed in section 1.2). It is grounds such as these that establish principles of method and reason independently of any social causal factor – grounds denied in the sociologists' talk of local acceptance.

Advocates of SP appeal to other types of social factor to show that the very principles of rationality are themselves social in character; they appeal to the doctrine of meaning finitism to characterise our very logical concepts employed in evidential reasoning. They deny that any of our principles of reasoning and method are norms that can be acquired by non-social means, for example by being founded by purely cognitive, or *a priori*, means. Once this is realised, it can be seen why the range of 'p' in CT can be taken quite broadly to include principles of method and reason, as well as ordinary scientific beliefs; some social factor, whatever it be, is always allegedly present in the causation of our belief in such principles. The themes adumbrated here are taken up in chapter 7.

Let us return to a further consideration of the possible social factors that might be involved in the way in which most people, and the way in which informed scientists and astronomers, acquire belief B. Scientists such as Copernicus, Galileo, Kepler, Descartes and Newton, first gave us reasons and evidence for claim B; and they believed B on this basis of their reasons and evidence. What social factors, if any, impinge on the process whereby they came to believe B? If any, it cannot be *via*, say, type (c) social causes due to acculturation but must be of a quite different type. Social factors could be present in two ways. First, the way just mentioned and set aside for later investigation, viz., evidencing reasons that have an ineliminable social element due to meaning finitism. The second appeals to other social factors, two of which are: (a) the social processes of 'negotiation', rather than rational processes, which are involved in a given scientific community reaching consensus over some theory or scientific result; (b) the social circumstance or the social and/or political interests that might have been involved in a person coming to believe some theory. It is here that the case studies provided by advocates of SP become pertinent. What they must show is that such social factors cause scientists to believe the very theories and hypotheses for which they are renown. We will investigate two case studies which purport to show that social factors allegedly caused physicists in Weimar Germany to adopt acausal principles in physics (section 5.6), and caused the seventeenth century scientist Robert Boyle to adopt views about 'passive' matter on political grounds (section 5.7). In each case it will be argued that the case for SP is very weak indeed.

Finally, two points about the existential quantifier which binds the social factor S_x in the full expression of CT. Given the order of the quantifiers in CT, for the very same belief p there can be, for different groups of people, a quite different kind of social factor as a cause of belief. To illustrate, using the commonly cited case in the literature to be discussed in section 5.6, Forman alleges that, for some German scientists (G), the particular socio-political circumstances of Weimar Germany (S_G) in which they lived led them to believe (adopt? accept? entertain?) acausality in physics (i.e., either the claim that there are no deterministic laws which underpin statistical laws in physics, or that there are some events that lack a cause). However for another person R, say some budding Ernest Rutherford in New Zealand who also believes claims about acausality (say through reading the relevant literature), the social circumstances S_R are quite different from S_G. Do the quite different circumstances S_R cause R's belief in acausality, or are they irrelevant? The casual process whereby R picks up news of acausality, is in part social (R reads the relevant literature). But if R were to perform his own experiments and/or make the requisite theoretical reasoning, then R's New Zealand social context plays no significant role in bring about R 's belief in acausality.

The world-wide acceptance of the same scientific claims raises a problem for CT. Given that belief in acausality was allegedly caused by the peculiar social circumstances of Weimar Germany, why did different people in different historical-social circumstances around the world also come to believe the same claims about acausality in physics? Several German scientists who advocated acausality lived through quite different social circumstances from Weimar Germany to Hitler's Germany and the post-war Capitalist (or Communist) Germany, while maintaining their belief in acausality. And scientists elsewhere, from the Communist world to the Capitalist West, came to believe in acausality. This raises an issue about whether there are general laws linking social factors with belief (more strictly kinds of belief). The paucity of such laws, and the fact that the same scientific belief can arise and be maintained in quite different social circumstances (since scientific belief is a world-wide phenomenon), are often taken to be serious objections to the truth of SP. Whether there are such laws is taken up again at the beginning of section 5.6 and discussed more fully in section 5.8.

The second point about the existential quantifier binding S_x in CT is this. If one specific kind of factor is not a cause, then some other factor within the broad typology, must be a cause, according to SP. Moreover the factor need not be a proximal cause acting in the foreground of causes producing the effect but may be distal and relegated to the background of causes. Such would often be the case for social factors due to meaning finitism; the meanings we give to the terms in our scientific reports would often be distal background causes against the foreground of other alleged social causes. In listing the social causes there is thus a long disjunction of each type of social factor within the broad typology, *modulo* the possible positions and weightings that each might have within any given causal chain. Given this long disjunction of possible social factors, it becomes especially difficult to either confirm or refute CT. If a confirming social factor has not been found, it does not follow that there is not one somewhere in the disjunction which confirms CT. There are more social factors to investigate including, in the long run, the social factors due to the doctrine of meaning finitism. Nor does it follow that if CT lacks confirmation then it has been refuted. Showing that no social factor is a cause would involve eliminating each disjunct of an open-ended disjunction, i.e., each specific social factor within the broad typology. Disproving CT might be very difficult, given the social role played by meaning finitism in expressing our very scientific beliefs. But perhaps advocates of SP are not bothered that CT lacks both clear confirmation and falsification conditions. But by the scientism of SP, they should be.

5.5 SOCIAL AND POLITICAL INTERESTS AS CAUSES OF BELIEF

As formulated CT allows that type (d) factors, the objective political, social, class, national or gender characteristics of a scientist's circumstance, can be a cause of their scientific beliefs. CT does not say that scientists must be *aware* of the objective socio-historical circumstances causing their beliefs. But sometimes they are and pursue their interests in the light of this. In general, if a person has an interest in something then they are aware they have that interest and deliberately act to realise those interests.[18] The interests might be cognitive, or non-cognitive social and political interests. CT is often understood quite differently to be a thesis about how socio-political *interests* can be a cause of scientific belief (an example of this being MacKenzie's discussion of the rise of statistics discussed in section 5.3.2). This section will spell out the main differences between CT, as formulated in the previous sections, and an *interests* thesis often mistakenly thought to be an instance of CT. Before formulating the *interests* thesis, consider the role of cognitive interests.

Scientists have quite specific cognitive interests in the science which they consciously pursue and which, when realised, may lead them to believe some theory. Thus an interest in high confirmation that results from passing several severe tests can be a cause of belief in the theory. Other cognitive interests in theories that arise out of theories of scientific method include: high explanatory power; broad application; high accuracy over a given range; and so on. In the discussion of Kuhn (section 1.4) we have seen that he calls such cognitive interests 'values', examples of which are simplicity, consistency (internal and external), breadth of scope, fertility, plausibility, and so on. Clearly which theories scientists chose to believe, or accept, or work on, may be closely linked to these cognitive interests, or values. But not all interests may be explicitly held. Sometimes they may be less explicit as when Popper claims, on the basis of the way physics developed early in the twentieth century, that scientists show an underlying interest in theories which are susceptible to revision in that they need to be capable of being criticised and, if necessary, of being replaced. To this end he proposed that there was an implicit cognitive interest in theories that were capable of being revised or, as he put it, are highly testable or falsifiable and are not to be modified in *ad hoc* ways (Popper (1959), p 49).

Scientists can also have non-cognitive interests in their science. No one ought deny the obvious role that personal, professional, political and social interests have in determining what scientists research, from the Manhattan atomic bomb project, to artificial insemination or AIDS. They make their choices about what theories they should work on according to whether or not they enhance their career prospects, their standing in their profession, or what they can do for humanity, from adding to its quality of life to enabling a greater exercise of power over society. But as such, none of this has any direct bearing on CT, which is about the causes of *belief* in some theory, and not about what theory scientists chose to work upon. As Newton-Smith says: 'The thesis that at the macro-sociological level interests do play a role will be called Boring Interests Thesis 1 ... . In using this label I do not want to disparage the thesis. It is just that it is obviously true.' (Newton-Smith (1985), p. 60).

Are there cases where non-cognitive interests of a personal, professional, social and political character are causally efficacious in bring about belief? A budding scientist might come to believe his professor's favourite theory because of job scarcity and a desire to get into the professor's leading research team. And many scientists in the Stalinist Soviet Union might have believed (or did they merely accept?) Lysenko's, and Stalin's, views on genetics, not because of the correctness of their arguments and evidence but because it was expedient to do so in such a repressive society. In contrast Descartes, the convinced Copernican, decided to live in more liberal Holland rather than in France where he might risk, due to possible conflict with the Roman Catholic Church, a fate similar to that of Galileo. Descartes chose to change his socio-political circumstance rather than run the risk of having to tailor his beliefs to the socio-political circumstance of Catholic France.

5.5.1 *The* Interests *Thesis and the* Interests *Explanatory Model*

That non-cognitive interests generally cause scientific beliefs yields the 'Interests' version of CT. As Stephen Yearly, a critic of this thesis, says: '... the aim of the interest analyst is to show how beliefs of all kinds, including mathematical and scientific beliefs, originate in contingent cognitive and social interests.' (Yearley (1982), p. 355). Note Yearley's conjunction of cognitive and social interests. Picking out the social interests alone yields the 'Interests' thesis which has a similar form to CT – but it differs in important respects:

Interests Thesis: for any person x and belief p such that xBp, there are non-cognitive (e.g., political, social, religious etc) interests of x in x's own socio-political-cultural circumstance S_x, such that x's *interest* in S_x (perhaps in combination with other non-social factors) causes xBp.

This Newton-Smith dubs BIT2, or 'Boring Interests Thesis 2, boring because in a strong enough form to be interesting, it is obviously false' (Newton-Smith (1985), p. 62) – false because of the clear role cognitive (and so non-social) interests do in fact play in theory choice.

Associated with the *interests* thesis is also a corresponding explanatory schema; we are to explain why a believes p by appealing to a's social (but not cognitive, intellectual, etc.) interests.

Social Interests Explanation Schema (or Model)
(1) For scientist(s) x there is (a) some social (cultural, power, etc) situation S_x (which may or may not be actual), such that x has an *interest* in S_x (or in bringing S_x about); and there is (b) some p such that xBp;
(2) the particulars, x's *interest* in S_x and aBp, are instances of kinds, a kind of interest I and kind of belief B, and there is a law-like link from I to B;
(3) ∴ aBp.

The *interests* schema, or *interests* model, is directly modelled on the social cause schema, and so it turns crucially on the second premise that requires a law-like connection between kinds of interest, I, and kinds of belief, B. Some such schema must be invoked by advocates of the *interests* version of SP if a Hempel-style explanation is intended (though explanatory models are hardly set out explicitly by advocates of SP).

How do the two models differ? The main difference is that, for socio-causal explanations of why x believes that p, x's objective social circumstance S_x plays a major role. In contrast, in the *interests* model, the cause of x's belief that p is x's *interest* in maintaining some aspect of their social circumstance S_x, or bringing about some non-actual S_x. In a nutshell, the two different hypothesised causes are: S_x itself; and x's *interest in* S_x. In the *interests* model, the causal process unproblematically goes from one mental state, an interest, to another mental state, an act of believing. In contrast in the social cause model, the causal relation is much more problematic; there is alleged to be a causal process going from x's objective social circumstance S_x to x's act of belief without there being any intermediary cognitive attitude to S_x. It is hard to see, even if one is a behaviourist, how such an objective state of affairs such as S_x should be so causally efficacious in bring about acts of believing in x's mind without x also having some cognitive attitude to S_x. Thus merely on the grounds of the causal processes at work, the *interests* model is plausible while the social cause model is implausible.

As indicated above, it is boring (because obviously true) that the sycophantic scientist, or the followers of Lysenko, choose theories to believe in the light of their non-cognitive interests. Such true, but sad, facts can be part of the province of either SSK or SS. But the *interests* model viz., that *all* theory choice involves non-cognitive interests – is part of the hubris of SSK and can be shown false on two grounds. The first is a general argument against non-cognitive interests being efficacious in picking out our successful theories. The second involves reinvestigating the case studies in which it is alleged that non-cognitive interests played a significant role in theory choice; this will be done in sections 5.6 and 5.7.

5.5.2 *The Implausibility of the* Interests *Explanatory Model*

Independently of the case studies, the implausibility of the *interests* model can be argued as follows. It does not require much evidence to show that our sciences have increased our ability to predict and manipulate the world, whether it be the natural or the human world. In our time the ratio of success to failure in predictive manipulability has increased markedly over that of our predecessors. Our non-cognitive interests in increased powers of prediction and manipulability have been amply, if not completely, rewarded with the growth in science. For brevity call our success in prediction and manipulation just 'success'. What explains it? There are two hypotheses about our success in choosing scientific hypotheses. And using the argument form, Inference to the Best Explanation (IBE), we ought to choose (believe) that hypothesis which explains, or makes most probable, this success.

The first hypothesis is advanced by philosophers who hold a traditional conception of scientific method. Scientists have used inferential and methodological principles to choose particular scientific theories out of the much wider range of theories that have actually been proposed. And applications of the theories so chosen have in turn yielded success. Moreover, methodologists argue that there is something right, correct, valid, or true about these principles we have used to chose those theories which in turn yield such success. So not only is it plausible to adopt these principles (there is something right about them), but these principles of theory choice also make success in science highly probable, and enter into explanations of such success.

The second hypothesis is advanced by advocates of the *interests* model, that is, certain advocates of SP. According to the *interests* thesis, scientists have not employed the principles advocated by methodologists but have instead chosen theories on the basis of their political and social interests. And it is the application of these scientific theories chosen in this way that has yielded success. But what is it about such social and political interests, which are in a sphere quite disparate from science, that guarantees that scientists will more often pick those very theories that increase our powers of prediction and manipulability rather than those theories which leave such powers at the same level or even decrease them? It is highly improbable that such socio-political interests should be efficacious most of the time in picking theories that yield success; at best getting success would seem to be random. So social and political interests can provide no explanation of success. Of the two rival hypotheses, the hypothesis of the methodologists provides a better explanation of success in science that the hypothesis of the sociologists. The principle Inference to the Best Explanation licenses us to say, of the superior explanatory hypothesis, that there is something right or correct about our principles of method; the same cannot be said of the sociologists' hypothesis.

It can easily be agreed that our political and social interests will have a role in choosing what theories we work on, or fund; this is Boring (because trivially true) Interests Thesis 1. And these interests may or may not be realised in the subsequent development of science. However, what we need is a reason for thinking that political and social interests will, with high probability, lead us to choose *the very theories that will realise these interests rather than choose theories which do not*. We would be very fortunate indeed if something so different from the scientific enterprise could be so lucky in getting us through the vast range of possible scientific theories by choosing, most of the time, theories which yield success, rather than theories which yield no (or minimal) success. This underlines the improbability of social and political interests always being efficacious in scientific theory choice. And it is what Boring Interests Thesis 2 maintains (but not boring to the sociologists!). That such non-cognitive interests could be so efficacious in choosing successful theories may well excite our interest! But alas the frequency of this is low; and so non-cognitive interests provide a bad, or no, explanation of predictive and manipulative success.[19]

Ought we to believe theories because of the non-cognitive interests they serve? That we should runs counter to Mertonian norms for science (section 4.3), and to the widely held view about the evidential grounds on which theories ought to be believed. But as a factual matter, have some scientists believed theories because of their professional, social or political interests? Two controversial case studies will be discussed in the next two sections. Several of the criticisms directed at the case studies below turn on their failure to establish causal connections; at best often what they show is correlation rather than causation. To reiterate, the central causal claim of CT is this: $(S_x \& N_x)$ cause xBp, or in the case of the *interests* thesis, x's *interest* in S_X causes xBp. But to establish any causal claim, causal methodology must be appropriately applied. There are a range of such methods, from Mill's text-book methods for testing for causes (using the methods of agreement, difference, the joint method and concomitant variation) to other techniques for investigating causation in a single case. Despite their decrying of scientific method, advocates of SP need some well justified methodology at least to establish their own causal claims.

Focusing on the social factor in tests for causation means that x's *interest* in S_x must be a common factor in any situation in which xBp also holds. Also the absence of x having an *interest* in S_x is to be accompanied by the absence of xBp; and any significant variation in x's *interest* in S_x is to be accompanied by a change in what x believes. The mere co-presence of x having an *interest* in S_x with xBp, while necessary for a causal connection, is not sufficient. As will be seen in the case studies, though x's *interest* in S_x and xBp are correlated or are co-present, there is little further investigation into which causes which, whether both are due to some third cause, or they are independent of one another and there is some other cause which gives rise to a spurious correlation. If no causal connection is established then neither CT, nor the *interests* thesis, can get any support. A central weakness of the case studies is their causal methodology – but this is to be expected in an account of theory choice that eschews any role for methodological principles and relies exclusively on either social circumstances, or interests in them.

5.6   CASE STUDY I: ACAUSALITY AND WEIMAR PHYSICISTS

Paul Forman's 1971 paper 'Weimar Culture, Causality, and Quantum Theory 1918-27: Adaptation by German Physicists and Mathematicians to a Hostile Intellectual Environment' is often cited by advocates of SP as one of the central illustrations of their programme, and is used by Bloor to introduce the four tenets of SP (Bloor (1991), p. 7). Forman outlines his thesis: 'that extrinsic influences led physicists to ardently hope for, actively search for, and willingly embrace an acausal quantum physics is here demonstrated for, but not only for, the German cultural sphere' (Forman (1971), p. 3). While intentional attitudes such as 'ardently hope for' and 'actively hope for' might guide research and have a social basis, they have nothing to do with either CT or the *interests* thesis as formulated; only 'willingly embrace', which might be taken to be the same as 'believe', fits CT. The "object" to be explained is why x believes in acausality in physics, where the 'x' ranges over certain German physicists in the decade after World War I. What the content of the belief in acausality amounts to is not always clear in Forman and covers many features of physics. We can take it to denote a variety of claims such as not every event has a cause, or that there are objective chances, or that there are statistical laws that are not reducible to non-statistical laws, or that physical systems do not evolve deterministically as classically supposed but rather indeterministically, and so on.

What does the explaining for Forman is the 'extrinsic influences' of 'the German cultural sphere'; this is a type (d) social factor (see section 5.2). Forman says of this: 'it forms the basis of my attempt to provide … an answer to the question – in its general form crucial to all intellectual history – why and how these "currents of thought", evidently of negligible effect upon physicists at the turn of the century, came to exert so strong an influence upon German physicists after 1918' (*loc. cit.*). But Forman is aware of the vagueness in the Mannheimian way in which he expresses the general form of his question, and adds: 'the historian cannot rest content with vague and equivocal expressions like 'prepared the intellectual climate for' … but must insist upon a *causal analysis*, showing the circumstances under which, and the interactions through which, scientific men are swept up by intellectual currents' (*loc. cit.*, italics added). Of the two possible causal analyses, the psychological and the social, Forman opts for the later 'treating present mental posture as socially determined response to the immediate intellectual environment and current experiences'; further, he seeks a 'model in which certain "field variables" ... are regarded as evoking corresponding attitudes' (*ibid.*, pp. 3-4). Setting aside equivocation over the requisite intentional attitude, we have good reason to take Forman's appeal for a 'causal analysis' to be in accord with not only the Marx-Engels Thesis and Mannheim's pronouncements about the causation of belief, but also CT: for some physicists x living in Weimar Germany in the decade 1918-27 some aspect of x's social/cultural circumstance causes x's belief in acausality.

What aspect? Forman's summarises the cultural milieu of the time:

in the aftermath of Germany's defeat the dominant intellectual tendency in the Weimar academic world was a neo-romantic, existentialist 'philosophy of life', revelling in crises and characterised by antagonism toward analytical rationality generally and toward the exact sciences and their technical applications particularly. Implicitly, or explicitly, the scientist was the whipping boy of the incessant exhortations to spiritual renewal, while the concept – or the mere word – 'causality' symbolized all that was odious in the scientific enterprise.' (*ibid.*, p. 4)

Particularly influential was Spengler's book *The Decline of the West* in which 'over and over again Spengler equates causality, conceptual analysis, and physics, and flays them across the stage of world history' (*ibid.*, p. 33).

Let us take as substantially correct Forman's account of the cultural milieu of the physicists of Weimar Germany with its rejection of analysis, science and especially causality (and let ' S_x' stand for x's being in this milieu). We can also accept his account of the Weimar physicists' *awareness* of this milieu and its hostility to science. According to Forman, it is this *awareness* of S_x (and not S_x alone) which allegedly causes x's belief in acausality. If this is so, then Forman's case study is not an illustration of Bloor's CT but its rival *interests* thesis. Already the case for CT is on shaky grounds since Forman's study is not obviously an instance of it.

Setting this aside, how is Forman to establish his 'causal analysis' (something that would have been excoriated in the cultural milieu he investigates)? It is not enough merely to show that in the mind of some physicist x there are both (A) an awareness of the hostile cultural attitudes to science, and (B) a belief in acausality. The co-presence of A and B in the mind of x is necessary but not sufficient to show that there is a causal connection. What must also be shown is that A *causes* B, i.e., that physicist x's *awareness* of x's cultural milieu with its hostility to causality is the cause of x's *belief* in acausality. However if B, belief in acausality, were due to the physicist's beliefs about matters internal to the physics of the day (call this 'P'), then the causal claim would be undermined, despite the co-presence of A with B and P. We can readily agree that Forman establishes the co-presence of A, B and P; what he does not show is that A, not P, is the cause of B. And this failure is entirely due to the improper application of something as simple as Mill's methods for determining causes. Even though awareness of the cultural milieu and belief in acausality are highly correlated across the minds of several Weimar physicists, the correlation is spurious because a third factor, developing views within physics, works independently to produce belief in acausality in the very presence of the physicists' awareness of their cultural milieu's hostility to causality.

One critic of Forman, John Hendry argues just this saying:

> when we come down to the content of physics, we must of necessity take into account internal as well as external considerations. ... Forman has succeeded in demonstrating that physicists and mathematicians were generally aware of the values of the milieu .... But when we come to the crucial claims, that there was widespread rejection of causality in physics, and that there were no internal reasons for this rejection, then the weaknesses in his argument also become crucial. For there were strong internal reasons for the rejection of causality, and when these are taken into account ... it would appear that the reaction of physicists to the causality challenge was far from being accommodation, and that there may even have been a tendency to isolation. (Hendry (1980), p. 160)

As Hendry argues, well before World War I German physicists gave reasons internal to their physics for adopting acausality; they did not abandon causality after the War for the social reasons Forman cites – even though they were aware of their social milieu.

Which physicists does Forman discuss to establish his case? There is no attempt to find a statistically significant and appropriately chosen test group for investigation in the Germany of the time, and a control group within or outside Germany. Without this no inferences can properly be drawn, and general claims about what social factor influenced the German physicists' belief in acausality are flawed. Instead Forman's case studies are based on about ten supposedly converted physicists, presumably chosen for the convenience of investigation. Not all can be discussed here, but some will be mentioned. It appears that Franz Exner was one of the earliest converts to acausality, but for reasons that, on Forman's own account, had little to do with Exner's cultural milieu: 'of the *Lebensphilosophie* and existentialism which will figure so prominently in most of the following conversions to acausality there is scarcely a hint [in Exner]' (Forman (1971), p. 75). So Exner's case is not confirmation for CT. Forman also tells us that one of Exner's pupils, Erwin Schrödinger, 'repudiated causality for social-ethical reasons in 1922-24'; but during 1925 he 'reconverted back to causality for what were most probably personal-political reasons' (*ibid.,* p. 104) . No evidence is cited that any of these changes of mind, before his 1926 work on wave mechanics, had to do with his cultural milieu. In marked contrast Hendry ((1980) pp. 164-5) has an internalist story to tell in the case of Schrödinger, contrary to that of Forman.

Turning to Hermann Weyl, Forman admits that his case is complex because, even though he adopted acausality quite early in the period under investigation, his reasons for doing so were originally not so much tied to quantum mechanics as to more general problems in physics such as the relation of matter to fields, or to his views about intuitionistic mathematics and mathematical continuum. Concerning the continuum, Forman cite remarks of Weyl who describes it 'as *something which is in the act of an inwardly directed unending process of becoming*'. He also envisages that 'the rigid pressure of natural causality relaxes, and there remains ... *room for autonomous decisions, causally absolutely independent of one another*, whose locus I consider to be the elementary quanta of matter. These "decisions" are what is *actually real* in the world' (Forman (1971), p. 78). As strange as talk of decision in nature may seem, Weyl's obviously internalist remarks, based in speculative theory, offer little support for the claim that his belief in acausality arose from his awareness of his cultural milieu rather than developments in physics and mathematics through metaphysical speculation on the continuum.

Only in 1920 did Weyl link his acausality to issues in quantum mechanics, and to what Forman calls a 'crucial existentialist consideration' (*ibid.,* p. 79) to do with our experience of the unidirectionality of time. But it is misleading to characterise Weyl's concerns with the direction of time, or his talk of 'autonomous decisions in nature', as 'existentialist', the word being used with connotations of a social movement. To so use it is to anachronistically apply it to post-World War I Germany before its full flowering as a movement in post-World War II Europe. Forman's account of Weyl's overall views take us nowhere near the romantic and supposedly "existentialist" features of the cultural milieu which would support Forman's supposition that it produced Weyl's belief in causality. For the physicists mentioned so far, no proper application of causal methodology to their beliefs about their cultural milieu or their physics (as gleaned from their writings cited by Forman), supports the view that it was either the cultural milieu, or their awareness of it, that actually caused their belief in acausality in physics and not their beliefs arising from physics.

Finally, the best case for Forman's thesis is given by the physicist Richard von Mises. But once again, what is presented tells us more about the suggestive nature of Forman's ruminations rather than the historical research necessary to establish causal connections between belief and awareness of cultural milieu. Forman uses von Mises to present an example of the 'suddenness' with which conversions to acausality could take place, and 'its [von Mises' conversion] essential independence of the difficulties encountered in atomic physics' and how it 'provides *prima facie* evidence of a direct connection between the repudiation of causality by a scion of Austrian positivism and his capitulation to the *Weltschmerz* of Spengler's *Decline of the West*' (*ibid.*, pp. 80-81). At last, one would think, here is an instance of Forman's thesis given von Mises' captivation with Spengler and the alleged 'essential independence' of the conversion to acausality from matters to do with physics. But a careful reading of Forman's two page analysis (*ibid.*, pp. 80-82) provides no evidence of a direct *causal* connection between von Mises' Spenglerianism and his new belief in acausality in physics. Why Forman should think the independence 'essential' is unclear; but what Forman does not do is either actually demonstrate the independence of von Mises' belief in acausality from his physics, or demonstrate its causal dependence upon his Spenglerian beliefs.

In February 1920 von Mises gave an address that took causality in physics for granted. Forman goes on to say: 'But when one turns to the thoroughly Spenglerian appendix which von Mises added in September 1921 to the republication of this lecture, one finds his attitude to causality ... entirely transformed'. (*ibid.*, p. 81) The transformation, according to the words of von Mises that Forman cites, is from a system of physics in which causal laws do the explaining, to a quite transformed way of thinking in which, due to the influence of quantum mechanics, statistical laws do the explaining. So far there is nothing Spenglerian involved in this transformation! Further, contrary to his supposed casual analysis, Forman cites from von Mises' writings some evidence from physics for his transformed view. In a long passage from another lecture given that September, von Mises sets out the classical Newtonian deterministic view of mechanics and then says of its limitations in what it can explain: 'All that I want to try to show here is that the accumulated facts which we possess today make it evident that it is highly improbable that this goal of classical mechanics [i.e., to do all the explaining] could ever be attained, and that other, perfectly definite and no longer unfamiliar, considerations are destined to relieve or to supplement the rigid causal structure of the classical theory'. Von Misses gives his reasons for this: 'there are phenomena of motion and equilibrium which will forever escape an explanation on the basis of the differential equations of mechanics' (*ibid.*, p. 82). Thus far, in September 1921, there is a co-presence in von Mises' mind of his awareness of the Spenglerian cultural milieu (suppose Forman is right on this) and a belief in acausality (along with much other belief about physics). Moreover there is an explicit declaration that it is evidence from physics that lead to the adoption of acausality, and not anything to do with his adoption of Spenglerianism.

After citing the above passages from von Mises' writings, Forman says of his change to belief in acausality: 'Admittedly, von Mises has invoked the quantum theory as the occasion for the repudiation of causality' (*ibid.*, p. 81). Not to admit this and to invoke other reasons or rationalisations based on

Spenglerian considerations would be to attribute a massive amount of self-delusion to von Mises, even mendacity given his avowed reasons based in physics (cited above from Forman's own paper). None of this supports Forman's contention that von Mises belief in acausality was 'independent of difficulties encountered in atomic physics' or that it was due to his capitulation to the *Weltschmerz* of Spengler's book, even if he might have independently so capitulated for other reasons. Forman shows von Mises to be aware of his cultural milieu, and to have suddenly embraced acausality. But Forman fails to show that the former caused the later. The explanation of von Mises' change in belief to acausality is the old, familiar internalist explanation based on von Mises' reappraisal of the limited explanatory power of classical physics and the evidence against classical mechanics and for quantum mechanics.

Even if we pass over the problem of whether it is CT or the *interests* thesis that the Forman case studies are meant to establish, the case of von Mises fails to support even the causal claims of either CT or the *interests* thesis, and thus fails to support SP. What shocks rationalists about such a study is the claim that evidence and reason played no role in the adoption of beliefs about acausality, contrary to the received internalist view about what did happen in the physics of the time and the role that evidence and reason is said to have played. But then in reviewing such a case study, what rationalists find shocking is the failure to employ proper causal methodology to establishing causal claims – but one can hardly expect more than this from denigrators of the role of principles of methodology in science.

On any account of what it takes to establish causal connections rather than mere correlations, there are no grounds for supposing that CT gets any support from Forman's study, thereby 'mak[ing] a mockery of Forman's attempt at a "causal" analysis of the phenomena with which he is concerned' (Hendry (1980), p. 170). Hendry's own position is a multi-factorial one in which both social factors and, importantly, factors internal to the sciences of the time play a role. Hendry makes a distinction between social and internal factors; but this is not the same as Bloor's distinction between social and non-social factors that allegedly co-operate in bringing about belief. Rather Hendry's internal factors concern the very content of the science under consideration and its methods of evaluation – but such internalist factors get no mention in SP. Many follow Hendry in agreeing that the prime error of Forman's study is not merely that it fails to be an instance of CT, and thus SP, but that it supposes that externalist considerations can always entirely replace internalist considerations.

5.7   CASE STUDY II: BLOOR ON THE SOCIAL CAUSES OF BOYLE'S BELIEFS ABOUT MATTER

Both Forman's study of Weimar physicists above, and Bloor's study of Robert Boyle's beliefs about the non-active character of matter, the topic of this section, appeal to type (c) and (d) social factors (see section 5.2), or to an interest in such social factors, in explaining scientific belief. Ancient philosophers, such as Lucretius, had attributed to eternal atoms the active power of self-motion. In contrast the seventeenth century scientist, Robert Boyle, is said by Bloor to have favoured a passive, non-active account of the corpuscles that made up the matter of the natural world, and to have rejected the view that they possessed active powers of self-motion. (It is noteworthy that advocates of SP do not offer a social cause explanation of Lucretius' belief about matter while they do of Boyle's.) Bloor summarises his 1982 study of Boyle's passivist view of matter, saying that it 'showed how ancient atomism (in which matter was self-moving and self-organising) was taken over by Robert Boyle and modified by his insistence that matter was passive and that only force was active ... the modification was made to further an identifiable interest of a political kind' (Bloor (1991), p. 166). Talk of 'an interest of a political kind' suggests that it might not be Bloor's own CT that the episode of Boyle's belief in passive matter is meant to illustrate, but rather the *interests* thesis.

In the 1982 study itself[20] the question for which Bloor seeks an answer is: 'How are we to understand the preference that developed in certain quarters, rather suddenly, for an inert and passive, rather than an active and self-moving, matter?' (Bloor (1982), p. 285). Bloor continues: 'In order to explain the change, historians have found it necessary to look at the social context.' (*loc. cit.*) Is no explanation available from a perspective internal to science? For advocates of SP it would appear that we are not to look to either Boyle's own science or to the science of others, such as Descartes' physics which was also based on a non-activist view of matter. Nor are we to look at Boyle's views on religion exclusively. However in a number of works, particularly his *Free Inquiry into he Vulgarly Received Views of Nature*,[21] a close "link", but not a causal link, is to be found between Boyle's views on religion and the natural world. In brief, his view is that the natural world is not an active agent and does not *do* anything, and that the world's matter is uncomprehending, unthinking and inert. However God has active powers and is the ultimate source of all agency in the world. It is God who imparted motion to the corpuscles of the natural world which they subsequently maintain; and he also makes the laws which they obey. However this does not show that matter is totally devoid of all power, a point to which we will return at the end. A further closely related thesis that Boyle also endorsed is that God is distinct from the world and not immanent in it, as some of the Protestant sectarians of the time claimed. For them there is a unity of God and the world (from which it can be inferred that the world has active powers). It was this close unity of the world and God that Boyle, in part, set out to confute in his *Free Inquiry*. God may have created the world but he remains distinct from it and not united with it in any way.

For convenience let us denote Boyle's view on the nature and role of God 'G', and his more scientific view of the material natural world as being made of passive, inactive corpuscles 'PM' (for passive matter). Boyle, of course, entertained both belief in PM and in G at the same time. But did one belief cause the other (or 'influence' or 'shape' etc, these terms merely being more causal talk)? Or were Boyle's beliefs about G and PM causally independent and neither influenced the other? Some historians express the view that Boyle's beliefs in G casually influenced his belief in PM (rather than a more activist account of corpuscles). This might well be the view in Shapin and Schaffer ((1985), pp. 202-3). The thesis would then be that general metaphysical considerations about God and the world causally influenced Boyle's belief in PM. Putting this in another way, for Boyle only those theories of the natural world are to be admitted which are consistent with an overall view of God and His relation to the world. If this is so, the causal claims of CT and the *interests* thesis have been undermined since it is the causal efficacy of Boyle's religious and metaphysical commitments, and not his political interests, that bring about his belief in PM.

Turning to CT we need to ask: is it Boyle's political situation (as CT requires), or his *interests* in (some aspect of) his political situation as the *interests* thesis requires, that allegedly causally bring about and sustain Boyle's belief in PM? Bloor sets Boyle in the political context of his time, which included the English Civil War of the 1640s and the period of the Restoration during which Boyle remained opposed to the sectarians with their views about a close connection between God and Nature. He also tells us of Boyle's commitment to 'latitudinarianism', a view that favoured the restoration of the Anglican Church, but with only limited accommodation to the Dissenters, the non-Anglican Protestants. Surprisingly Bloor tells us of Boyle's opposition to the idea that there is a close unity between God and Nature: "In the place of this animated intelligent universe Boyle put the mechanical philosophy, with its inanimate and irrational matter. This was then used to bolster up the social and political policies that he and his circle advocated. It was called 'latitudinarianism'." (Bloor (1982), pp. 286-7) The remark is surprising; it says that some combination of Boyle's beliefs about G and PM 'bolster' (are causally relevant to?) his social and political interests. This supports neither CT nor the *interests* thesis, in which the causal influence of his political interests is meant to be in the other direction, i.e., political interests cause, and are not an effect of, scientific beliefs in PM.

Continuing the story of Boyle's social context, the Dissenters were radicals wishing to overthrow the hierarchy of the established Anglican church and to organise themselves more freely not only inside the church but, more threateningly, in society as a whole. In the light of this Bloor tells us: 'To say that matter could organise itself carried the message that men could organise themselves. By contrast, to say that matter was inert and depended on non-material active principles, was to make nature carry the opposite message' (*ibid.*, p. 287). The overall context of Bloor's paper is the Durkheim and Mauss thesis that 'the classification of things reproduces the classification of men' (*ibid.*, p. 267). But Bloor gives this thesis an "interests" interpretation in which it is an *interest* that some have in men being organised in a certain way which is reproduced in (i.e., causes) the classification of things. In the case of Boyle's science, what Bloor alleges is that it is his interests in supporting only a limited form of latitudinarianism, and thus supporting only a limited extent to which men are free to organise themselves within society, which leads him to believe that nature's corpuscles are passive rather than active.

For convenience let us call the collection of Boyle's political interests 'I'. Let us also grant that Boyle's life was sufficiently complex for him to have concurrently in his head a number of beliefs and interests. Amongst these will be (1) the set of beliefs in G, (2) the set of beliefs in PM, and (3) interests I. The mere co-presence of any pair of these three items, though necessary for any causal claim between any pair, is not sufficient. In Bloor's study no causal methodology is employed to go beyond mere co-presence of I and PM to show that it is Boyle's political interests I that *cause* belief in PM rather than the other way round. In addition, he needs to show that it is not Boyle's belief in G that cause one, or the other, or both, of belief in PM and interests I. The case is simply unproven. As in Forman's study of physicist's belief in acausality in Weimar Germany, the case for SP made in Bloor's study of Boyle suffers from lack of proper application of causal methodology.

Bloor follows his discussion of Boyle with a similar account of Newton, but he intriguingly adds:

> Of course, neither Boyle or Newton, nor their free-thinking opponents, will be found saying that they believe what they do just because of its political implications, though they were deeply concerned with these. Both sides will believe what they do because experience, or reason, or the Bible makes it plain to them. ...
>
> Both groups were arranging the fundamental laws and classifications of their natural knowledge in a way that artfully aligned them with their social goals. (*ibid.*, p. 290)

That neither scientist can be 'found saying' that they believe their science because of their political implications makes it hard, to say the least, to find direct evidence for either CT or, more likely, the *interests* thesis. Rather their overt appeals are to experience, reason or the Bible. But without some such evidence it is equally as hard to see how the claim that they are '*artful* aligners' could also be founded. Rather the artfulness seems to lie with advocates of SP who have to keep either CT or the *interests* thesis as live options, given that they not only fail to properly apply any causal methodology but also deny that some of necessary evidence can be found – this absence of any evidence being explained away as a form of false consciousness. The absence of explicit declarations in what the two scientists can be 'found saying' can be readily explained in another way; they simply never believed that they held scientific beliefs as a result of their political interests.

So, on what grounds did they believe their science? There are other kinds of evidence, which the scientists can be 'found saying', which has to do with matters such as observations, experiments, theorising, inferences involving these, and so on amongst a host of other such matters. In other words something like the old standard internalist story about how their scientific beliefs were arrived at explains why they cannot be 'found saying that they believe what they do just because of its political implications'. If we are not to accept the account these scientists do give of how they arrived at their scientific beliefs, then we must either attribute a massive amount of self-deception to them, or claim that they have deliberately misled us as to what were the real causes of their scientific beliefs. But again there is no evidence for either such self-deception or mendacity. Rather what this shows is that there is a story, other than the one given by Bloor, about why Boyle held the scientific beliefs he did. As we will see, more recent historiography spells out such a different story using different historical considerations than those upon which Bloor constructed his account.

As is usual in such studies, there is copious citation of historical investigations into the cultural context of science. Focus on just one of Bloor's citations (Bloor 1982, p. 284 footnote 40) of a 1972 paper by J. R. Jacob with the promising title 'The Ideological Origins of Robert Boyle's Natural Philosophy'. J. R. Jacob is often said to have established that Boyle's political interests influenced his scientific belief in passive matter, thereby establishing an instance of Bloor's CT, or of the *interests* thesis. Let us suppose that such claims can be found in Jacob's work. But there is one pressing difficulty; one can also find quite different claims in Jacob's writings about the causes of Boyle's scientific beliefs that do not support Bloor's contentions, or the claims he makes about Jacob's own writings. Here is one example taken from Jacob's 1972 paper. Jacob tells us that contrary to two other prevailing views in Boyle studies he will offer a third: 'I shall claim that Boyle laid the foundations of his natural religion before he developed his corpuscular philosophy, his distinctive contribution to seventeenth century science, and that this natural religion both helped to shape and determine the significance of his full-blown philosophy of nature' (Jacob (1972), pp. 1-2).

Jacob claims that not only are Boyle's views of natural religion merely earlier than his views on nature, but they also caused the latter – talk of 'shaping' and 'determining' being simply more causal talk. If Jacob is correct then, on a sufficiently fine individuation of Boyle's religious beliefs about God from his corpuscularian beliefs (the passive matter hypothesis) and his political interests, it is his religious beliefs that are said to causally influence his scientific beliefs and not his political interests. Unless, of course, his beliefs are overdetermined by both causes emanating from Boyle's religious and political beliefs; but this is nowhere claimed. Jacob continues saying that, from the point of view of Boyle's motivation, 'his piety did indeed influence the development of his scientific or, more properly, his natural philosophy'. In turn Boyle's piety had 'particular social and ideological roots' (*loc. cit.*) that Jacob hopes to trace. Even granted these further causal influences, they take us nowhere near Boyle's political interests before and after the Restoration.

Margaret Jacob writings on the same period tells us of Boyle, as well as a number of other writers: 'All used the new mechanical philosophy, that is, their vision of the natural world, to support a political world where private interest would enhance the stability of the public weal and Anglican hegemony would rest secure.' (M. Jacob (1976), p. 22). Assuming this can be construed to be about passive matter, here is one historian claiming that advocates of the new mechanical philosophy used it to support their political position rather than, contrary to Bloor's claim, their political interests causing belief in the new mechanics. M. Jacob continues in this vein: 'For the latitudinarians, as we shall see, the new mechanical philosophy served as the foundation for a social ideology with a dual purpose: to secure and legitimise church and state against the threats posed by radicals, enthusiasts, and atheists and to reform this established order' (*op. cit.*, p. 25). Again there is an alleged causal influence, in a direction opposite to that supposed by Bloor, of scientific beliefs upon political and religious interests.[22]

But according to M. Jacob the nexus of connections, real and alleged, between religious, political and scientific beliefs are even more complex. It would appear that each protagonist in the struggles of the times is alleged either to have used religious beliefs and/or political interests to support their views on natural science, or conversely to have used their views on natural science to support their religious beliefs and/or political interests. M. Jacob says of this:

> Throughout the seventeenth century, thinkers of whatever philosophical or political persuasion assumed that some sort of relationship, of varying degrees of causality or simply of intimacy, existed between the world natural and the moral and social relations prevailing or desired in the "world politik". Operating, therefore, with basically the same set of assumptions as those of their disparate protagonists about the relevance of the natural order to the political order, churchmen assumed that to

accept the radical's vision or Hobbes' vision of the political order meant to accept
their vision of the natural order, and vice versa. (Jacob, M. (1976), p. 24)

Talk of 'degrees of causality' and 'intimacy' is unclear. But the picture Jacob
presents is that people made *assumptions* about what linkages there were
between political interests and beliefs about nature, and in which direction
the link went – and then employed these assumptions in their ideological
debates. This is different from discovering what actual causal connections, if
any, that might have held between political interests and scientific beliefs
independently of any *assumptions* protagonists of the time might have made
about what the links were. And it is such *actual* causal connections, not
assumptions about them, that need to be subject to test using proper causal
methodology if either CT or the *interests* thesis is to be established. However
this does not clarify the position of Boyle since M. Jacob tells us nothing
about what kind of connections Boyle might have thought there were (if he
did at all), and in what direction they might have gone which would be
sufficient to support either CT or the *interests* thesis in relation to Boyle's
views on passive matter.

The scene portrayed by historians is a very confused and hazy one. What
consolation may there be for philosophers who go to historians for well
established claims about the causes of belief, but come away with a plethora
of competing claims with little evidence of a proper causal investigation
having been carried out? One consolation is that the writing of history has its
fashions, including the writing of the history of science, the historiography of
Boyle being no exception.[23] One of these fashions is said to be the dominance
of contextualism in Boyle studies; talk of socio-cultural context is required by
SSK, in its various guises as CT or the *interests* thesis. Judging by recent
historiography, it would appear that this fashion is now over in most studies
of Boyle written in the 1990s. Thus in a brief review of Boyle historiography,
Michael Hunter says of J. Jacob's work upon which Bloor has relied for his
case, that 'Jacob's views have not stood up to scrutiny in the light of
subsequent research' (Hunter (1994), p. 3). Concerning what connection there
might be between Boyle's science and his contemporary political milieu,
Hunter continues in his introduction to his collection of essays on Boyle:

Jacob's image of Boyle is in many ways rather partial and schematic, based on a
selective reading of Boyle's writings, and often imputing ill-evidenced motives to
him in his controversial works. ... 'his [Boyle's] position was more complicated than
Jacob implied. This is illustrated in relation to the politics of the Civil War period by
Malcolm Oster's essay ...., while the same complexity in Boyle's attitudes is in
evidence in other episodes of which Jacob's account presumed that certain motives
were predominant, when in fact this is hard to substantiate; the need for a total
revision of the view of Boyle's relationship with his milieu given by Jacob is acute'
(*loc. cit*).

In the cited essay we find Oster saying: 'Jacob's treatment of both Boyle
and the Royal Society was open to the charge of being excessively

monocausal' (Hunter (1994), p. 19), monocausality being one of the vices of CT with its insistence on omni-present social causal factors. Oster gives a nuanced account of Boyle's connection with, and attitude towards, the political events of the time, showing that Boyle was much less of a committed political figure than he has been presented, and that Boyle gave qualified support to both Republicanism and the restored Monarchy. With such a more muted account of his politics, there is not the strong support for either CT or the *interests* thesis upon which some have insisted. On the science/interests nexus Oster comments: 'This [Boyle's mechanical philosophy] was very carefully arrived at and was not accepted by Boyle because it could, as the Jacobs have argued, be used as the [Royal] Society's ideological stance against occultist sectaries and radicals who saw science as a "powerful tool for promoting religious, political and social revolution".' (*ibid.*, p. 32). The double-quoted remark at the end of the last citation come from a work of the Jacobs; contrary to them Oster says that science was not used to support political and religious belief. In this way one generation of historians casts serious doubt upon very interconnection between scientific and politico-religious belief that an earlier generation of historians, such as the Jacobs, had claimed. Of course Oster's remark is not a direct attack on either CT or the *interests* thesis, since it concerns the way science was used to bolster political stances. But it does suggest that the more radical and disturbing claim that political stances caused scientific belief must be much more carefully examined.

This brief review of Boyle historiography might well dismay philosophers because of the vacillation of historians between the extremes of "intellectualism" and "contextualism" (i.e., viewing science as either largely internalist or largely externalist). But historians of the last decade take themselves to be in the process of arriving at a more balanced view of Boyle's politics, religion and science. If they are right, then this does not bode well for the extreme socially contextualised accounts of Boyle's scientific beliefs that the Jacobs argue for and upon which Bloor's account of Boyle relies. The upshot is that philosophers, historians and sociologists of science should either suspend belief in, or positively disbelieve, the account of the causes of Boyle's belief in passive matter championed by advocates of SP.[24]

Finally, the above turns on the assumption that, as Bloor supposes, there is such a thing as a totally passive view of matter to be found in Boyle. Recent historians have even undermined this claim. Unless Bloor severely qualifies what he takes his passive matter hypothesis to be in Boyle, he is left with a non-existent "object", Boyle's passive view of matter, for which he has proffered a political explanation as to why Boyle believed in it. Peter Anstey argues that at best Boyle held the view that matter was insentient, that it did not think and that it did not have appetites and the like. This is a far cry from claiming that matter is inert and does not entail that matter is only passive. In fact Anstey finds that Boyle attributed a number of powers to matter such as the power to impart and receive motion, the power to remain in motion in the absence of external forces, and the power to change the direction of motion. And overall he argues for a harmonisation of the powers of God and matter and not a division such that God is the only embodiment of powers while matter embodied none.[25] Anstey's view is that Boyle is what he calls a 'nomic occasionalist'. This is the view that matter has causal powers but that God is responsible, in Boyle's view, for the law-like behaviour of matter. This is a far cry from the position presented to us by Bloor. If Anstey, one of the new wave of historians of Boyle, is right then Bloor, along with the Jacobs, not only gives us an unsubstantiated political explanation of why Boyle believed in passive matter. They also leads us down the garden path in getting us to think that there is something, Boyle's belief in passive matter, to be explained in the first place. We need a much more finely individuated account of what passive matter is supposed to be; but in Anstey's view this is not to be found.

5.8 SOCIOLOGICAL LAWS AND THE CAUSALITY TENET

At the end of section 5.4 it was noted that, according to CT, the same belief held in one social context, for example Forman's claim about the belief in acausality held by physicists in Weimar Germany, could also be held in a quite different social circumstance, e.g., physicists in New Zealand who believe in acausality. This makes it difficult to believe that there are causal laws governing the link between social circumstance and belief, as CT and its social cause explanatory model require. But supposing there are such laws then, using Forman's account as an example, they would have to be of the following form, assuming a Humean account of laws as regularities: there is a type of historical, social and cultural context C (of which the Weimar Germany of the 1920's is but one instance and New Zealand at the same time another instance), and a type of scientific theory, hypothesis or belief B (of which claims about acausality are but one instance), such that all Cs cause Bs (or Cs are followed by Bs with a high frequency). But this says nothing of the non-causal factors N_x that must enter into the law formulation. Taking this into account, a particular joint cause, $(S_x \& N_x)$, must be an instance of social and non-social types – call them 'C' and 'N'. And the particular effect, xBp, is also and instance of a belief type – call it 'B'. The Humean law-like regularity underpinning the causal claim in CT is then of the form: all (C&N)s are Bs. Are such type-type laws to be found in SSK providing support for CT?

At best the laws remain schematic since no example of a law has been proposed by advocates of SSK or SP which picks out the required type of social circumstance and belief in such a way that a descriptively adequate law can be spelled out. In reviewing the attempts to find such laws in the case of interests in the socio-political programme of eugenics and early work in statistics, the sociologist of science Ben-David says of theories of SSK such as SP: 'No success can be claimed for the new Marxian-Mannheimian attempts to find a systematic (that is, a permanent and regular, not just an occasional) relationship among macro-social location, ideology, and scientific theory. Indeed there is little reason to suggest that there should be such relationships.' (Ben-David (1991), p. 462) While he thinks that there might in some cases be a role for social and ideological factors at 'the original stage' of scientific development, they do not persist. Ben-David suggests that even though it might have been influenced by the ideologically charged atmosphere in which it begun, 'Newtonian physics was subsequently adopted by physicists of many religious and political persuasions' (*loc. cit.*) The objection to SP is that it has discovered no significant descriptive causal laws linking socio-historico-cultural types (to be discerned in, say, Germany and New Zealand during the Weimar years – or anywhere else at any time) with types of belief (assuming there are any such significant typologies). Thus there are no laws to back the causal claims central to the formulation of CT, and no laws available for explanation according to social cause models.

Bloor addresses this objection, amongst many others, in the 'Afterword' to the second edition of his book. While recognising that 'the general point seems right' he suggests an alternative way in which laws might arise:

> We do not find, for example, that field theories in physics are associated exclusively with organic social forms, or atomic theories with individualistic societies. Such general connections would break down if only because theories created by one group are taken over by other groups as inherited cultural resources. This is not, however, fatal to the sociology of knowledge. It rules out one simple and implausible definition of the exercise, but leaves others intact. The lack of 'systematic relationships' between 'social location' and 'types of theory' – to use Ben-David's terms – may depend on how broadly 'type' is used. Ben-David's argument overlooked the possibility that sociologists may yet explain why an inherited body of ideas is modified in the way it is, even if the resulting theory is of the same general type. For example, one of the studies Ben-David cited showed how ancient atomism (in which matter was self-moving and self-organising) was taken over by Robert Boyle and modified by his insistence that matter was passive and that only force was active (See Jacob (1976) and Bloor (1982)). Even though the modification was made to further an identifiable interest of a political kind, the fact that the theory was still of the same type (viz., an atomic theory) means that on Ben-David's perspective the covariance and causality passes unnoticed. (Bloor (1991), pp. 165-6)

We need not agree that theories created by one group are taken over by others always as 'cultural resources'. Newtonian mechanics had some of its applications and evidential grounds initially developed in England. But it was taken over by Frenchmen such as Lagrange and Laplace who discovered new or better formulations and applications and more new evidence to support it, as have others across the world in a variety of cultures. Nor has Bloor caught up with the recent historiography on Boyle mentioned in the previous section, as is evidenced by his mention of a 1978 paper by J. Jacob.

In one respect Bloor acknowledges Ben-David's point that, depending on how kinds are individuated, there are no laws of SSK linking types of social circumstance with types of belief. But it is unclear what Bloor thinks Ben-David has overlooked. Returning to Bloor's own account of ancient atomists and Boyle (outlined in the previous section), both believed in the same type of theory, atomism, even though they held modified versions of it; the ancients allowed matter to be self-moving while Boyle (let us suppose) believed that it was passive and not capable of self-movement. Taking 'type' broadly, Bloor appears to concede that there is no law linking (i) some broad type of social circumstance, which the lives of both ancient atomists and the seventeenth century Boyle instantiate, and (ii) the broad type of belief which atomic theory instantiates.

But Bloor also suggests ways in which 'sociologists might yet explain why an inherited body of ideas is modified in the way it is'. Assuming the standard law-like model of explanation, the sociologists' explanation would have to appeal to a law linking (a) a social type, of which Boyle's 'identifiable interest of a political kind' is an instance, and (b) the belief type, of which the modification Boyle made to ancient atomism (viz., abandoning the view that matter can be self-moving) is an instance. To explain the modifications made, the sociologists would have to invoke causal laws linking some type of social circumstance with the type of belief modification made. But even laws linking a type of social circumstance with a type of *modification* of a theory are as hard to come by as the other style of law which links type of social circumstance to type of theory (and not merely type of modification). Bloor admits the latter kind of law is hardly available; but the former kind of modified law is equally unavailable and thus fails to meet Ben-David's objection, also voiced by others.

Bloor accounts for the absence of laws in the following way: 'such laws will exist, not on the surface of phenomena, but interwoven into a complex reality. In this respect they will be no different from the laws of physics.' (*ibid.*, p 167). The Humean account of type-type laws given above allows that the laws not only be about surface phenomena but also hold at a deeper non-phenomenal level. It also allows that there be a number of such deeper laws operating together. No such restriction was imposed on the way the alleged sociological-belief laws required by CT were to be understood. But the situation in SP does differ from physics. In physics the more deep laws are intended to explain surface phenomena, even if they are idealisations about what goes on at the non-phenomenal level.[26] Physics has at least come up with some type-type qualitative and quantitative laws, even though they be idealisations. In contrast SP has not even come up with any ideal laws – this being another aspect of the objection made by critics of SP. Bloor does suggest that laws linking 'cosmological style to social structure ... are a start' (*loc. cit.*) but once again critics will not be satisfied with such a vague law schema.

Finally Bloor appeals to the doctrine of meaning finitism as a source of laws for SP: '*all* concept application is contestable and negotiable, and *all* accepted applications have the character of social institutions' (*loc. cit.*) But these are generalisations about concept application for which an argument is needed to show that they also satisfy further conditions for being laws and not merely the regularities of social institutions. Bloor is correct when he continues: 'Such laws are not what critics expect in answer to their challenge'. But discussion comes to an end with the *ad hominem* response: 'but perhaps that reflects more on them [the critics] than on the sociology of knowledge'. The critics correctly ask for evidence for *laws* rather than mere *regularities*. They would not be satisfied with the thin gruel of weak generalisations that go with concepts or social institutions.

Can the talk of causal laws backing CT be avoided? Bloor seems to be wedded to an account of explanation which presupposes such laws (despite the fact that even their associated regularities are difficult to establish). Carl Hempel is famous for devising and defending the deductive-nomological (or covering-law) model of explanation in which laws, and in particular causal laws, play an essential part in any explanation of any event, including a person's believing something. Bloor seems to assume this model in which laws are essential for the explanation of why someone believes something, viz., aBp. In the absence of any laws we are left with only an explanatory sketch in which the laws invoked are no better than promissory notes to be filled in later.

5.9   CAUSALITY, CAUSAL DEPENDENCE, EXPLANATION AND A REFORMULATION OF THE CAUSALITY TENET

The Hempelian model of explanation is not without its difficulties; and there are rivals to it that do not depend on laws.[27] One such rival view, defended by David Lewis in a paper 'Causal Explanation', turns on the following broad thesis: '*to explain and event is to provide some information about its causal history*' (Lewis (1986), p. 217). Call this the *historical causal explanatory schema (model)* to differentiate it from the *social cause* explanatory schema of SP. What is to be explained is an event occurring at, or over, some time; what does the explaining is the earlier causal structure of events, i.e., the causal history of the event. We can imagine cases in which the causal history is short; but in the world we know causal histories can be lengthy and open-ended with respect to a past history of causal structure. On this model of explanation we do not have to invoke laws but only salient pieces of the causal history (and not the full historical causal structure which in most cases is not available to us). In picking a single event out of the causal history we cite *a*, not *the*, cause. A sufficiently large chunk of the causal history might be designated *the* cause, or the *whole* cause for certain purposes or because of its salience; but strictly speaking the whole cause is the whole of the causal history.

Though Lewis' account of explanation can adopt most theories of causation, he also develops his own distinctive theory which is highly relevant to the formulation of CT. Hume gave us two accounts of the supposedly one definition of 'cause': 'we may define a cause to be *an object, followed by another, and where all the objects similar to the first, are followed by objects similar to the second. Or in other words, where, if the first object had not been, the second never had existed'.* (Hume, *Enquiries*, Section VII, Part II). We may take Hume's talk of 'objects' broadly to include events and states of affairs. His first account appeals to the regularity theory of laws and is none other than the type-type account already used in the previous section. But Hume's 'other words' yield a quite different account of causation that Lewis has developed as the counterfactual theory of causation.

In what follows let *'c'* and *'e'* be particular events. Then Hume's 'other words' say: if *c* were not to have occurred, then *e* would not have occurred. That is, there is a relation of counterfactual dependence between propositions *'c* occurs' and *'e* occurs'. Lewis defines causal dependence between actually occurring events *c* and *e* as follows: '*e* depends causally on *c* if and only if if *c* had not been, *e* never had existed' (Lewis, (1986), p. 167). Causal dependence, so defined in terms of counterfactual dependence, is not yet causation since it is only necessary for causation but not sufficient; causal dependence may not always be a transitive relation while causation is. In the light of this Lewis' defines causation thus:

> Let *c*, *d*, *e*, ... be a finite sequence of actual particular events such that *d* depends causally on *c*, *e* on *d*, and so on throughout. Then this sequence is a *causal chain*. Finally, one event is a *cause* of another iff there exists a causal chain leading from the first to the second' (*loc. cit.*).

Granted this conception of causal explanation, we can see that it is not necessary to invoke laws to explain (but that does not mean that laws are not present for the sequence of events). In the case of CT what we need to causally explain is aBp, i.e., person a's act of believing that p. Such an explanation will appeal to an historical causal structure at the later temporal end of which is the event aBp. The rest of the causal structure preceding the last event, aBp, will contain earlier events some of which will be the social and non-social factors pertaining to a's existence, S_a and N_a. Each of these social and non-social factors will be particular instances of types of social and non-social factors, as outlined in section 5.2. The instances of each type may occur at different temporal points within the causal structure; thus an instance of a type (c) social factor to do with the social transmission of belief might have an earlier place in the causal chain while an instance of a type (d) social factor, such as a particular political interest, might occur much later in the historical structure and closer to the time at which the belief first occurred. The same can be said for non-social factors; they can be spread throughout the causal structure with, say, a particular experience being slightly earlier than the brain processes which comprise an act of believing based on the experience. In contrast one's acquiring the requisite mental capacities in the course of human development is much earlier than both.

Every event in the causal history preceding a's act of belief that p is *a*, not *the*, cause of that act. If some event e earlier than aBp is such that if e were not to have occurred yet aBp does occur, then e may be part of history but it is not part of *the causal* history of aBp. On the counterfactual theory, the Big Bang, a's being born or the constant presence of oxygen around a, are causes of aBp; if any one of these three events had not occurred then aBp would not have occurred (because a would not have existed to believe anything). But such causes are uninteresting and do not tell us, say, why it was *p* that a believed rather than some other q, or why it was that a *believed*, rather than *entertained* or *hypothesised*, that p, or why it was *person a* that did the believing and not some other *person b*. Such contrasting *explananda* do need explanations which cite causes which will explain why just one of these pair of contrasts occurred and not the other.

In another context Lewis proposes a distinction between sensitive and insensitive causation that can be applied here (*op. cit.*, pp. 184-8).[28] When some causes occur their effect is quite insensitive to any of the later intervening events in the causal history. Thus the effect of pulling the trigger of a well-working gun pointed at a person at very close range, viz., their being injured, is quite insensitive to any of the other events in the causal chain (such as a strongly blowing cross-wind between gun and victim). In contrast events such as the Big Bang, a person's birth or the presence of oxygen, even though they are a cause of a's believing that p, are quite sensitive to other intervening events in the causal chain. In the case of events such as the shooting and the injuring, the causal chain runs through a quite stable structure of events. But in the case of the Big Bang, or the presence of oxygen, or a's birth, and a's believing that p, the causal chain runs through a highly unstable structure of which Lewis says the following:

> When an effect depends counterfactually on a cause, in general it will depend on much else as well. If the cause had occurred but other circumstances [in the causal history] had been different, the effect would not have occurred. To the extent that this is so, the dependence is sensitive. ... Sensitivity is a matter of degree.' (*op. cit.*, p. 186).

Using this notion then, from the stance of any point in the causal history, if the chain to the effect is insensitive then the outcome will be highly probable and so foreseeable; but if the chain is sensitive then the outcome is neither probable nor foreseeable. Thus if our interests are in the causes of aBp, then what we need to look for is not any one of the myriad of causes in the full causal history but only those causes leading to aBp which are part of a stable causal sub-structure and which are fairly insensitive to other causes in the structure.

This account of causation and explanation leads us to view CT in a quite new light, and to recast it in many respects. There is no need to appeal to laws to explain why aBp; we need appeal only to causes in a stable causal structure. Accepting Lewis' historical causal explanatory model bypasses objections to SP and its social cause model due to the absence of any significant laws linking types of social circumstance and types of belief. In section 5.3 we understood CT as a claim about particular acts of belief and their causal antecedents in particular social and non-social factors, S_a and N_a. These are now social and non-social events in the causal structure which culminates in aBp. However what is to be made of the requirement in CT that social and non-social factors must *co-operate together* to produce belief? In a loose sense this is correct; within any causal history of aBp there will be both social and non-social factors to be found, but at different temporal points in the causal history. Providing the sensitivity requirement is met, invoking one or the other (but not necessarily both together) will provide a cause, and thus an explanation, of why aBp. But in a more strong sense of 'co-operation' in which any N_x must act in conjunction with some S_x and cannot be without it, such a requirement is misleading and unnecessary. The account of causation adopted here does *not* require that wherever a non-social factor occurs in the causal history there must be a social factor conjoined to it at the same time and place. Some causes might be a complex of both, but not all need be.

One way in which the Strong Programme is *strong* is the conjunction requirement of CT, that is, always $(S_x \ \& \ N_x)$ causes xBp, but never N_x alone causes xBp. But it can now be readily seen on Lewis' historical causal model of explanation, that this requirement is wrong-headed. On that model we can pick out insensitive historical causal sub-structures that do have social factors within them, and citing these do give us a social cause of aBp. But also on that model we might pick out insensitive historical causal sub-structures that have no social factors in them. Or alternatively, any sub-structure with a social factor in it might be quite causally sensitive to the intervention of non-social factors. In such cases as these, citing causes can explain aBp without invoking any social factor at all. Such is the case in the exercise of our cognitive capacities only in thinking from premises to conclusion (to be discussed in section 7.1).

On the counterfactual theory of causation and its accompanying causal account of explanation, CT becomes a quite harmless doctrine that no one would deny. The causal history of any person's act of belief will contain both social and non-social factors. In finding insensitive causal histories to explain why a person believes that p, we might have to invoke the social factors only, the non-social factors only, or a mix of both. There can be no *a priori* insistence that social factors must always be present in every insensitive causal sub-structure. In this way the strength of the Strong Programme is removed and what is revealed is the plausible core of the doctrine that was hidden in its more radical, overblown and misleading formulations. There is a further boon: we overcome the embarrassment for SP that it require causal laws to do some explaining – yet no empirically satisfactory causal laws linking type of social circumstance with type of belief (or type of modification of belief) can be found.

5.10   AN UNNATURAL NATURALISATION

There is one sense in which CT is naturalistic. Any act of belief, xBp, is said to have both social cause S_x and non-social causes N_x. These causal items are the objects and properties that are the subject matter of the various sciences, from sociology to neurophysiology. And only these items are efficacious in bring about another naturalistically construed state, viz., aBp. But from this can there also arise a naturalistic account of knowledge as well as belief? Not at all. If CT were to be construed as part of a naturalistic theory of *knowledge*, then which theory of knowledge would apply to it best? This would not be the justified true belief theory of section 2.2 since there is no mention of justification in CT; rather there is a complete loss of the normativity required by justification, and thus of knowledge. A better candidate might be the reliabilist theory of knowledge discussed in section 2.3. But that theory requires that the belief forming processes specified in the causes of xBp, viz., S_x and N_x, be reliable for the truth of p, over a range of beliefs. But no reliability condition is imposed. In fact it cannot be since the other tenets of SP explicitly require that both true and false beliefs p, and rationally and irrationally held beliefs p, be instances of CT. Thus CT can have no high truth ratio, or reliability criterion, associated with it. So no claim can be made that CT expresses a naturalistic theory of *knowledge*, as opposed to a naturalistic account of the causes of some beliefs (if we considerably narrow the scope of CT to the smaller range of beliefs to which it might evidently apply).

There are other grounds already adumbrated in section 5.1 in which CT, and thus SP, can be understood to be an unnatural naturalisation. Belief is a non-normative descriptive notion unlike knowledge which is normative. If belief and knowledge are construed as similar, as is the want of most sociologists, then all normativity is stripped from knowledge and naturalisation is too cheaply realised. In addition, the normativity contained in any explanation of why we know, as opposed to believe, is also lost. But there is the world of difference between the naturalistic explanations of some beliefs, of which CT might give us an account, and any naturalised theory of knowledge and an explanation of what we know. It is this that quite different models of explanation that appeal to methodological and other principles of rationality aims to provide.

Some theories of knowledge require that the proposition known, viz., the truth p, be justified (as in the case of the model set out in section 2.2). Or if one adopts a causal theory of knowledge, they require that the state of affairs known, viz., p, be somehow responsible for producing the belief that p. Or that the processes producing p be reliable for the production of p. We have already seen that in SP there is no reliability condition imposed on the belief forming processes; SP allows, and even makes a virtue of, unreliably produced true or false beliefs. Nor is their any justificatory link to the claim that p. Nor is there any link between the fact that p and the belief that p, so that any enquirer may be said to reliably track the truth. The only thing, within SP, that is connected to the belief that p is something quite extraneous to the fact p that is supposed to be tracked. The extraneous item is some social, cultural or political condition of a person (or group), or their interest in this condition, that stands in a causal relation to acts of believing. And these items have little to do with the fact that p. So it is hard to see, on these grounds, that SP has anything to do with knowledge. That the belief that p has anything to do with the socio-politico-cultural condition (or interest in this) that gives rise to it, would be purely fortuitous. To overcome this problem, advocates of SP talk of beliefs "mirroring", or being a "reflection" of, the socio-politico-cultural condition that gives rise to them. The very title of Bloor's book talks of "Social Imagery" in this context. But we are lost in unanalysed metaphor in the attempt to overcome the lacuna between the conditions that give rise to the belief and the belief, or act of believing, itself. On these grounds alone, SP is a quite unnatural naturalisation of knowledge. And this central feature of SP, along with its eschewal of normativity, undermines any pretensions it might have to say anything about knowledge.

NOTES

[1]   See for example Machlup (1980), Parts I and II. In his several volumes on the economics of knowledge Machlup feels the need to have a preliminary discussion of the different kinds of knowledge there might be in various disciplines, and even the differing epistemic terms and notions in different societies and cultures. Though he is far more sensitive to the notion of knowledge than Bloor and most other sociologists, he fails to recognise its normative character and is often quizzical about philosophical issues when he says, for example, that he is 'not joining the fussy JTB fanatics – 'Justified true Belief'' zealots' (ibid., p. 38). It is a shame that

Machlup could not get as far as even Plato's initial theory and the reasons for it. Machlup even struggles with the truth condition on knowledge (*ibid.*, pp. 113-7), often confusing something being true with our having tested for its truth. Machlup's 1980 book is seventeen years later than Gettier's 1963 paper which showed that knowledge could not be justified true belief. Russell had raised related objections a half a century before and, of course Agrippa, two thousand years before (see section 2.2). Machlup also fails to note any of the subsequent theories developed in the epistemological literature once Gettier's problems had been raised. However much else that Machlup says is of interest and can be taken on board if one gives up on talk of knowledge and stays with belief, or belief with whatever evidence one may have.

[2]   For a long list of case studies in the history of science which allegedly support the theses of SP, see Barnes' and Bloor's paper 'Relativism, Rationalism and the Sociology of Knowledge', footnote 7 pages 23-5, in Hollis and Lukes (eds.) (1982), pp. 21-47.

[3]   The phrase is from Laudan (1981), p.174. The two papers Laudan (1981) and (1982) are a critique of, and a reply to a response by, Bloor. They are reprinted as chapter 10 of Laudan (1996), the cited remark being on p. 184. Laudan's papers provide one of the more vigorous criticisms of SP in science studies that have been given by philosophers.

[4]   See section 2.1.1 where the notion of a belief is discussed. Not all our beliefs are overtly present to our mind. Thus most of us believe that oranges do not grow in Antarctica; but this is not always before our minds. Such a belief is dispositional in the sense that if each of us is questioned about this belief we would tend to say 'yes', if we want oranges we would not travel to Antarctica, and so on. We may suppose that this disposition is embedded in a brain state.

[5]   Frege thought that propositional contents, or senses, did enter into the causal nexus of the world (Frege (1977), p. 28). But one's view on this does not necessarily impinge on the matters raised here. Note also that Frege insists on an act/content distinction (*ibid.*, p. 28-9).

[6]   There is an important issue in the philosophy of mind and semantics that goes under the name of 'the problem of content'. Very roughly the issue is how the mental states, or brain states in our heads, come to have the propositional content they do have and be about the things they are alleged to be about. For an account of the problem see Braddon-Mitchell and Jackson (1996), Part III. It is not obvious that this problem of content is being addressed by SP; so that will be set aside as an area of concern for SP.

[7]   One sociologist of knowledge does take note of a version of the act/content distinction and that SP applies only to the former and not the later, See Mackenzie (1981), p. 225. His distinction is cast in terms of the Habermas' distinction between the German *Erkenntis* (the act or process of knowing) and *Wissen* (the content known).

[8]   The theory ladenness of observation was argued for in various ways in the 1960s by Kuhn and Feyerabend. A devastating critique, not always fully acknowledged, of the excesses of the theory-ladenness doctrine was published in the same decade in Dretske (1969). Once one is philosophically sensitised to Dretske's distinctions between epistemic and non-epistemic seeing, and primary and secondary seeing, much of Kuhn's discussion of the theory ladenness of observation in Kuhn (1970), especially chapter 10, becomes unreadable because of the way in which it slides over important distinctions. Advocates of SP do not do much better in their adoption of the ladenness doctrine. Another useful criticism of the doctrine can be found in Fodor (1984). There is an attempted reply to Fodor (which does shift somewhat in Fodor's direction), but no reply to Dretske, in Barnes, Bloor and Henry (1996), chapter 1.

[9]   The role negotiation in making determinate scientific belief that evidence would otherwise leave underdetermined is not discussed here. For a critique see Laudan (1996), chapter 1, especially §4, and Brown (1989), chapter 3, especially pp. 54-6. There is a reply to Brown's argument in Bloor (1991), pp. 171-2. There is an excellent critique of attempts by sociologists to use social factors to overcome underdetermination in Okasha (2000).

[10] See Barnes (1982), sections 2.2 and 2.3, for his views on meaning finitism as far as our classificatory terms are concerned. Barnes takes his cue from Kuhn's 1974 paper 'Second Thoughts on Paradigms' first published in 1974 and reprinted in Kuhn (1977). For Bloor's most recent account of finitism based on his interpretation of Wittgenstein on rule-following, see Bloor (1997), chapter 2, especially pp. 19-20, and chapter 3. See also Barnes, Bloor and Henry (1996), chapter 3, for a further discussion, and especially pp. 66-9 where the authors try to come to terms with the idea that there are natural kinds.

[11] The generality of CT with respect to persons reflects the wide scope of SP as a meta-thesis within the sociology of belief generally. CT could, however, be restricted to people within specific communities. Considering scientists, CT could be understood to apply to the scientific community as a whole, or to subgroups reflecting specialisms within science (e.g., particle physicists, meteorologists, soil analysts, etc.). Or CT could be thought of as holding, not only within, but also outside, such subgroups thereby including other scientists not working in some specialism, the intelligent lay person, and those who are not practising scientists. Nothing much turns on the range of x in CT for our purposes.

[12] Note that 'p' in CT occurs as a variable ranging over propositional contents while in contexts elsewhere it occurs as a constant naming a particular propositional content.

[13] If T is a theory, then while a person who *believes* that T also *takes* T to be true, or *holds* that T is true, a person who *accepts* that T may neither take nor hold T to be true and thus not believe that T. Also a person might accept that T for the purposes of working on T, or constructing experiments according to T, and so on, without believing that T. For more on the important belief/acceptance distinction see Popper (1972), chapters 3 and 4, B. van Fraassen (1980), chapter 2, or L. J. Cohen (1992).

[14] The suggestion here is an adaptation of the model of explanation in Currie (1980, p. 462). Note however that this is not the model that Currie endorses; this is set out in section 5.1.

[15] For an assessment of the virtues of rival theories of scientific method see Lakatos (1978), chapter 2, and the papers in Howson (1976); for an assessment of the virtues of Bayesianism as opposed to its rivals see Earman (1992) or Howson and Urbach (1993); for a survey of a number of rival theories of method see the papers in the volume by Nola and Sankey (eds.) (2000), pp. 1-65, their paper 'A Selective Survey of Theories of Scientific Method').

[16] Davidson's theory of mind as part of the causal nexus of the world, and his arguments that reasons can be causes, can be found in essays 1, 11, 12 and 13 of Davidson (1980).

[17] Some of the social factors just listed are taken from the paper by Barnes and Bloor in Hollis and Lukes (eds.) (1982), p 23. Other social factors they list as causes of such beliefs include: established institutions of socialisation; accepted agencies of social control; the furthering of political and technical goals; the practical consequences arising from such beliefs; and so on. What the authors overlook in the case of the last item is that there are theories of rationality, such as rational expected utility theory, that enable us to judge which is the best of the possible consequences of a range of actions we might perform.

[18] What is claimed here is that if x has an interest in (its being the case) that p then it follows that x is aware that they have the interest that p. To deny this is to claim: x has an interest but x is not aware that they have the interest. Of course it can be said that p is in x's best interests even though x does not have p as one of their interests in the sense that they are aware it is one of their interests and they pursue it. But p being in x's (best) interest and x having an interest that p are quite distinct.

[19] A similar argument based on inference to the best explanation is advanced by Newton-Smith (1985), p. 62, and Laudan (1990), pp. 102-3 and pp. 154-5.

[20] Bloor gives a similar account of the causes of Boyle's scientific beliefs in Bloor (1983), pp. 152-5.

[21]   The book was written in the mid-1660s, some revisions were made to it in the early 1680s but it was not published until 1686, but with a Preface dated 1682. The book indicates a continuity in Boyle's thinking about the relationship between God, the world and nature over a long period.

[22]   The historian Michael Hunter reads both Jacobs this way saying: 'The Jacobs lay great stress on the English Revolution and its impact: ... the new science of Boyle and Newton was self-consciously used by its originators to bolster the social and political status quo against the challenge to which it had been exposed in the Civil War and its revolutionary aftermath ....' (Hunter (1990), p. 438). Such a reading of the Jacobs' does not establish Bloor' contention that Boyle's political context was a cause of his belief in the passive matter doctrine within mechanics. The causation cannot go both ways.

[23]   For an account of Boyle historiography of the past half century which begins with the more purely intellectualist (roughly what I have called internalist) approaches to Boyle of the 1950s and 1960s, which is then followed by more contextualist (roughly what I have called externalist) accounts of advocates of SSK and SP, which is followed in turn by more recent accounts that are not so exclusively either, see Sargent (1995) 'Introduction' especially pp. 1-14. Sargent tells us of the rejection by some sociologists (Lynch, Latour, Woolgar) of Bloor's approach: 'In the 1990s a number of sociologists have become more vocal in their criticism of the explanatory project of traditional SSK studies' (p. 6). They reject CT and its accompanying explanatory model, as well as the *interests* explanatory model, in favour of other models (e.g., ethnomethodological, or a descriptivist approach, which eschew explanation, and so on). Whether these 'new wave' sociologists do any better with the historical material is a moot point.

[24]   It should be noted that Margaret Jacob, even though she flirted with a contextualist position in earlier writings, has distanced herself from it. M. Jacob (1995) is an attack on Bruno Latour's postmodernist interpretation of Boyle and his period which also contains side swipes at Shapin's and Schaffer's 1985 *Leviathan and the Air-Pump*. Old intellectual enemies in one cause may become allies in another. See also her critical essay 'Reflections on Bruno Latour's Version of the Seventeenth Century' in Koertge (1998), pp. 240-54.

[25]   See Anstey (2000), in particular chapter 7 section 2 entitled 'The causal efficacy of matter'. Section 3 of the same chapter deals with God and nomic occasionalism. On this view one bit matter has the power to, say, transmit its motion to another bit of matter; but God intervenes with laws which determine, say, how much motion is to be passed from one bit to the other, or in what path, rectilinear or circular or something else, the second bit of matter moves.

[26]   In discussing physics Nancy Cartwright persuasively draws a distinction between non-explanatory phenomenological laws of physics and fundamental theoretical laws which purport to explain but which are idealisations, and are thus in some respects false. This, along with the role played by *ceteris paribus* clauses in such explanations, is discussed in Cartwright (1983) chapters 1 to 3.

[27] For an account of the defects of Hempelian law-like explanation and an account of other models of explanation see Salmon (1990) and Salmon (1998).

[28]   The issues of sensitive and insensitive causation are revisited in Lewis (2000) entitled 'Causation as Influence'.

CHAPTER 6

THE EDINBURGH CONNECTION II:
STRONG AND WRONG[1]

In this chapter the remaining three tenets of the Strong Programme (SP) will be discussed – Impartiality in section 6.2, Symmetry in section 6.3 and Reflexivity in section 6.4. There is also some unfinished business from the previous chapter, discussed in section 6.5, to do with whether or not SP commits us to relativism. The chapter begins with a review of some of the rival models employed by philosophers and sociologists of science to explain belief. What will be argued is that advocates of SP have overlooked one important model of explanation for false, irrational and unsuccessful beliefs that undercuts much of the opposition they falsely set up between philosopher's rational models for scientific belief and their own causal models.

6.1 RIVAL MODELS FOR THE EXPLANATION OF SCIENTIFIC BELIEF

In the previous chapter the causal character of SP was spelled out in the form of the Causality Tenet (CT) with its accompanying *social cause explanatory schema* for scientific belief. Alternatively one could adopt the different *interests* thesis with its accompanying *social interests explanatory schema* for the causal explanation of belief in terms of political, social and cultural interests. These models require law-like connections between social circumstance, or social interests, and beliefs; but as was argued in section 5.8 there are hardly any such laws available for use in such explanations. As Hempel recognised, his deductive nomological model would be seriously handicapped by the absence of any formulatable laws that were not also confirmed to some degree. So he proposed a modification of his model in which at best only an *explanatory sketch* could be offered: 'Such a sketch consists of a more or less vague indication of the laws and initial conditions considered as relevant, and it needs "filling out" in order to be turned into a full-fledged explanation' (Hempel (1965), p. 238). Explanations in accord with both models are at best sketches with little hope of being filled out with well-confirmed laws, or even specific 'initial conditions'.

Is talk of an explanation sketch an acceptable modification of Hempel's original deductive-nomological model? Or does it reveal a basic weakness in the model itself? Many have taken the later view and have proposed models of explanation that abandon laws as an essential component of explanation while allowing a role for causation. Such is one motive behind Lewis' model of causal explanation set out in section 5.9. This provides an *historical causal explanatory schema* which is widely applicable and is much more suited to the social context than the Hempelian model with its commitment to laws. It also has the additional boon that when applied to CT it undercuts the strength of the "Strong" Programme in providing a much more nuanced counterfactual account of the interplay between social and non-social factors as causes of belief.[2]

6.1.1   Rivals Within SSK to the Explanatory Models of SP

Many advocates of an SSK approach to scientific belief have not been happy with the specifics of SP, especially its commitment to causal laws. Some have rejected the extreme scientism implicit in SP due to its causal formulation and its fourth Reflexivity Tenet. However they are happy to adopt the general stance of the other two Tenets of Impartiality and Symmetry. As Lynch puts the matter: 'Bloor's causalist assumptions are not widely accepted in SSK, but his recommendations about impartiality and symmetry are advocated .... Even many of those who do not agree with Bloor's empiricist assumptions and social interest explanations share his skeptical posture towards scientists' and mathematicians' truth claims'. And in explanation of what he means by 'skeptical posture' Lynch says: '"Symmetry" and "impartiality" only require that all theories, proofs, or facts be treated as "beliefs" to be explained by social causes. Bloor's skepticist approach is primarily methodological, as it aims to neutralise the explanatory power of "internalist" accounts ...' (Pickering (ed.) (1992), p. 220). Lynch's understanding of SP's account of both scientific beliefs and their *social causes* accords with that presented here. But there is a difficulty in Lynch's position because he tries to both distance the Impartiality and Symmetry Tenets of SP from its Causality Tenet, CT, and at the same time continue to talk of the explanation of beliefs by *social causes*. But this is a very difficult path to follow since CT sets out the very requirement of social cause explanation that is needed in order to understand what the other two tenets are about. Here Bloor is on stronger ground than his opponents who denude themselves of any social cause explanatory model.

The only way out of this difficulty for Bloor's opponents within the SSK camp is to find some other model of explanation of beliefs, while still retaining some role for social factors as causes of belief. To this end sociologists of science have culled a number of alternative models from miscellaneous sources:[3] philosophical movements such as hermeneutics and phenomenology; particular philosophers such as Wittgenstein (though as will be seen in the chapter 7, different sociologists take different messages from him); models of empathetic understanding (*Verstehen*); models from structuralism, post-structuralism and postmodernism; actor-network models; and so on. However when one examines some of these alternatives to SP it is not clear whether there remains a concern for SP's original problematic, viz., providing *explanations* of scientific beliefs. In so far as they concern themselves with "objects" in the broader SS rather than SSK, then they are not relevant to the issues under investigation. Unless they focus on the explanation of acts of belief, they simply bypass the concerns of SP and do not provide an alternative to it.

This is particularly the case for studies of science based in ethno-methodology and its offshoot discourse analysis. Do ethnomethodological studies focus upon the explanation of why someone believes that p (i.e., the *explanandum* xBp)? Or is their concern more with the adequate observation and reporting of everyday occurrences in the laboratory, or the ongoing practice of science? There is a difference here between an explanation which attempts to answer the question '*why* xBp?', and a description which attempts to determine whether or not it is the case *that* xBp (or some other features of the act of believing, including the very words used to express their beliefs). Garfinkel, the founder of ethnomethodological studies, tells us that there is a strong emphasis on the 'accountable', viz., the observing and reporting of ongoing practices, and says: 'Ethnomethodological studies analyze every-day activities as members' methods for making those same activities visibly-rational-and-reportable-for-all-practical-purposes, i.e., "accountable" as organisations of commonplace everyday activities.' (Garfinkel, (1967), p. vii) Though there is much ambiguity in what Garfinkel says about the nature of ethnomethodology in his differing accounts of it,[4] it seems safe to say that it sets out a programme in SS that is quite different from that of SP; but it by-passes Bloor's main concern of socially explaining why scientists hold the beliefs they do. Issues about SP versus ethnomethodology will arise in the next chapter.

There is one further model for the explanation of beliefs that is worth highlighting, not least because it allows for the explanation of a person mistakenly holding a belief. As will be seen, both the Impartiality and Symmetry Tenets mention the possibility of mistaken beliefs, mistaken in the sense that they are either true or false, rational or irrational and successful or unsuccessful. For brevity call the three contrasting pairs *epistemic* properties of beliefs (which concern truth value, rationality and pragmatic value). We are interested in the explanation of belief p where any of the above six epistemic properties can hold. Advocates of SP often make the assumption that models of rational explanation will deal only with the true, rational and successful while social cause models will deal with the false, irrational and unsuccessful. This is not only a misleading division of labour – it is also quite false, as will be shown.

6.1.2   Further Development of the Rationality Explanation Schema

Consider the model of empathetic understanding which invites us to take the 'inside' track of the mental processes of individual scientists as an explanation of why they believe what they do. In Collingwood's phrase we can take the 'thought-side' and 're-inact', or 're-think', a person's actual thought processes that lead to belief (Collingwood (1994), Part V section 4). In one sense the rational explanation schema of section 5.4 can be understood to do just this, providing the actual thought processes of scientist x are in accord with the principles that a particular theory of method prescribes in the situation in which x applies the method. But x's processes of thinking need not always be in accord with methodological prescriptions even though x believes that they are. To apply the rationality model in such a situation would be misleading since, even though it leads to the *explanandum* xBp, its *explanans* does not contain a correct account of what has gone on in x's mind. What is required is a rationality model for the explanation of belief that tracks the actual thought processes, as correct or as faulty as these may be, that a person goes through in coming to their beliefs.

In such an explanatory model the *explanandum* is an *act* of belief. Talk of an act suggests that a more general model for the explanation of human action, based on practical reasoning, can be adapted to the narrower sub-category of acts of belief. In a practical inference the premises would at least contain information about why a person so acted: i.e., they refer to some want, desire or aim of the actor; and they contain some beliefs the actor holds, usually about the actor's situation and/or about the means for realising wants, desires or aims in their believed situation. The conclusion of a practical inference is sometimes said to be an action, A, or something expressed by a sentence such as the command 'Do A!', or the imperative 'I ought to do A'. Finally, there is an important issue about the logical link that must hold between premises and conclusion. There is much literature that analyses this link; it can be a variety of validity, or of strong inductive support.[5] The practical syllogism can also be adapted to *explain* action, such as someone's act of believing. The premises would then serve as *explanans*, and the *explanandum* would be the report of the occurrence of some act, in this case an act of belief. Here the practical syllogism will be used as an explanatory model.

Such a model of practical action importantly allows for the actor's beliefs to be either true or false; even when mistaken the belief will be part of the reasons for action. For example, a person may believe that, when they spill salt at the table, the appropriate thing to do is to throw a pinch of the salt over their left shoulder with their right hand. The person's aim is to avoid evil or bad luck. The person may, in the presence of salt they have spilled, go through an act of practical reasoning involving these beliefs and desires – and then so act. Or they might spontaneously so act, without going through such thought processes, in the light of their disposition to hold such beliefs and desires. As illustrated, the model of practical reasoning allows for mistaken belief. There may be little evidence for the superstitious belief linking spilled salt with bad luck; and even the belief that salt has been spilled can be false (the spilled stuff may be, say, fine-grained sugar not easily distinguished from salt). In other situations a person might have reasons for acting but fail to act. Again, sufficient reasons may be present in a person's mind but not be what is causally efficacious in bringing the action about.[6] Reasons explain action only when they are causally efficacious in bring about the action.

Focusing only on actions that are acts of belief, the simple practical action schema just outlined sets out the thought processes that a person might go through that lead to an act of belief. The reasons, and the reasoning, tell them what to do; and when they believe, reasons are the causes of the act of belief. The model above can be extended to a Bayesian theory of rational action involving the following: a range of possible acts of belief; a range of possible states of the world; a number of outcomes of those actions, given some state of the world, which can be assigned a probability and to which a value, or "utility", can be attached (i.e., the actor has a "utility function"). Finally there is some principle for decision, such as the principle of maximum expected utility, which tells the actor which act they should perform. Such practical action schema not only tells us what to do, given a pair of beliefs and desires (or the person's "utility function"); when given some act of belief, the beliefs and desires (or the person's "utility function") can also be part of a model for an explanation of an act already performed. In the case of the simple practical reasoning schema, the item to be explained, the *explanandum*, is given by the conclusion of the reasoning schema, viz., x's act of belief that p; and the *explanans* is given by the premises of the practical reasoning schema, viz., the beliefs and wants, or aims or desires (or the person's "utility function").[7]

Let us adapt the above practical reasoning explanation schema in terms of beliefs and aims to the case of the rational explanation of acts of belief by scientists concerning their theories or hypotheses. Instances of the model will have a common *explanandum*, viz., x's (act of) belief that p; but their *explanans* will differ depending on the content of the other beliefs and aims. An example pertinent to explanations of scientific belief is the following. Suppose scientist x adopts some putative methodological principle M which x believes can adjudicate between any pair of rival theories. (Note that x's belief might be false because of unknown limitations on the application of M; or M might contain a faulty principle of reasoning such as the Gambler's Fallacy and x employs this fallacy; or, in the days before the institution of double- or triple-blind experiments, x might have conducted only single-blind experiments; or before Fisher's account of how to properly design experiments, x conducted improperly randomised trials; or x's belief is hopelessly false when method M is taken to be coin tossing; and so on.) Suppose also that x wishes to know of a pair of rival theories T and T*, which ought to be believed. (Suppose x's choice situation is mistaken. There might be another theory T' of which x is aware, and it would fare much better on M than either T or T*; but either x gives T' low initial credibility, or discounts it due to, say, prejudice against its advocates, or simple inattention to all the theories x entertains, etc.) Suppose that x also applies M in the choice context to T and T* and believes that M favours T*. (x might have misapplied M which, in fact, favours T.) Suppose also that x has the general aim of believing the deliverances of what x takes to be scientific methodology. Then any explanation of why x believes T would have to appeal to the actual beliefs and aims of x. Any of x's failings along the pathway to the belief that T (as suggested in various ways above) are of an intellectual character and are not necessarily due to social causes. It will be argued that the above fits a model of explanation that rivals both the social cause and the *social interests* models championed by advocates of SP; it also provides explanations of beliefs when they are false, irrational or unsuccessful.

A model that captures the idea of a 're-enactment', or a tracking, of the actual thought processes of a scientist is the following explanatory schema. It will be referred to as a 'Rationality* Explanatory Schema (Model)', because of its close affinity to the rationality model already set out in 5.4.2 These two closely related models will be collectively called *rationality explanations, or models* (because both appeal to putative methodological principles).

Rationality Explanatory Schema (Model)*
(1) Scientist x believes that he/she is confronted with a choice between belief in A, or belief in B;[8]
(2) Scientist x believes that methodological principle M is the best for choosing between theories in the given choice situation;
(3) Scientist x applies M and believes that M yields that B is a better theory than A;
(4) Scientist x aims to acquire the best belief in the choice situation;
(5) ∴ x believes B.

The schema clearly has strong affinity with a practical inference schema with conclusion 'believe B!' But here we are taking such schema as a model for *explaining* why x believes B, and not a model for deliberation about action. The premises of the schema contain not only beliefs about what epistemic ends one should aim for, but also beliefs about one's choice situation, about the efficacy of method M and about the application of M in the choice situation. The conclusion is the item to be explained, viz., a report of an act of belief. The inference pattern will generally be strong inductive, but can also be construed deductively.

The rationality* model differs from the former rationality model of section 5.4.2 in one important respect. The rationality* model has been explicitly designed to offer explanations of belief which track the actual thought processes of a person while the former model might not necessarily do this (unless the person correctly employs a rational methodology). This allows that the two rationality models can yield the same, or slightly different, explanations of the same *explanandum*, xBp. The main difference is in the second premise in each model. In the earlier model (of section 5.4.2) premise (2) makes a claim about the rationality of some methodological principle. In the later model just set out above, premise (2) is about a *belief*, on the part of x, that some principle of method is a rational principle to adopt. Of course, x may be wrong in this; but the rationality* model of explanation will still track x's processes of belief while the former model cannot. But if x is right, the two models differ only slightly. For this reason it is not inappropriate to refer to both models collectively as 'rationality models'; however there is an important difference to note when x believes falsely that some principle of method is rational.

Some philosophers have explicitly relied in their methodology on the difference between the two models, for example, Lakatos. He claims that rationality explanations are part of an *internalist* rational explanation of scientific belief, while beliefs that cannot be so explained are to be relegated to *external* history. For the externalist leftovers, explanations might be given in terms of the social cause model; so SP does have a secondary role to play according to Lakatos. But Lakatos overlooks explanations of beliefs relegated to external history which do track the actual thought processes of scientists as set out in the rationality* model, but which do not conform to either social cause explanations or explanations in terms of his own more simple rationality model. However, where the actual thought processes of a scientist do accord with methodological principles, then there will be a measure of accord between the two rationality models, even though their *explanans* may not be quite the same.[9]

In the context of SSK, what is important about the rationality* model is that it provides explanations of why x believes that p, where p itself is false or unsuccessful, or x's belief in p turns on something irrational, this being indicated in the *explanans* of the model. (Note that the explanation of why x believes that p (where p can be false or true) will be different from the explanation of why the belief content p is false, or why the content p is true.) This shows that social cause models, including the *interests* model, do not have a monopoly on explanations of belief, even when the belief is false, unsuccessful or irrationally based. Nor does the rationality* model require a prior epistemic evaluation of belief p as to its truth value, pragmatic success or rationality status before it can be applied; the explanation it provides can proceed quite happily without any prior epistemic evaluation. In respect of the last point the rationality* model is not only a rival to both kinds of social cause model; it also undermines one of the reasons often given as to why social cause models have priority over rational models of explanation. This point will become important when we come to consider the Symmetry Tenet.

One final point about the rational* explanatory model. Advocates of SP are quite wrong on other grounds about its domain of application. If rational explanation models are to be applied, they are alleged to apply only in the case of true, rational and successful belief. In the next section we will show that this is wrong. The methodological principles used in the rational explanation models do not always deliver true belief, since many of them are designed not to pick out true hypotheses. Even where the true hypothesis is not available, they enable us to discriminate amongst some set of hypotheses, all of which might be false, for, say, the best confirmed by evidence. Other methodological principles pick out one hypothesis from a set on the basis of greatest simplicity, lack of *ad hoc*-ness, fruitfulness, and so on; and none of these require that the hypotheses be true. There is an underlying assumption about a division of explanatory labour by advocates of SP that does not apply to the rationality models they dismiss.

6.2   THE IMPARTIALITY TENET

The Causality Tenet (CT) of SP tells us that our acts of believing are (i) caused, and (ii) what the causes allegedly are. Concerning (i), acts of belief are part of a causal nexus (even allowing for indeterministic causation for bodily and mind/body connections). This is a view that not only materialists about the mind/body problem would endorse but also most dualists who suppose not only that there is a mind/body separation, but also that mental acts of believing have causes and are not uncaused happenings. Bloor's characterisation of a rival to SP, the Teleological Model, is deeply misleading in this respect because acts of believing are said to be uncaused: 'It [the Teleological Model] relinquishes a thorough-going causal orientation. Causes can only be located for error' (Bloor (1991), p. 12). We have canvassed what is wrong with this view in section 5.4. The teleological model allows that some beliefs arise on rational grounds, and these grounds can be causes of belief; it is not as if such beliefs are without any cause and arise in some indeterministic fashion. And we have also agued that the various rationality models are quite consistent with, and even require on a thoroughgoing naturalism, that reasons for belief be operative as causes of belief. Opponents of SP need not accept the imposition of the peculiar view that reasons are outside the causal order. Since the case for reasons being causes has been well made out (see Davidson (1980), Essay 1), any act of belief due to reasons will be part of the causal order.

The formulation of the Causality Tenet adopted in the last chapter was:

CT: For all persons x, and for all belief contents p such that xBp, there exists some social condition S_x of x, and there exists some non-social condition N_x of x, such that $(S_x \& N_x)$ causes xBp.

The Impartiality and Reflexivity tenets can be seen as special cases of CT. The Impartiality Tenet (IT) says that SP 'would be impartial with respect to truth and falsity, rationality or irrationality, success or failure. Both sides of these dichotomies will require explanation.' (Bloor 1991, p. 7)

IT is an instance of CT for the following reasons. CT is neutral about whether the propositional content of x's belief, that p, is true or false; it is neutral about whether x's belief in p is rationally or irrationally based; it is neutral about whether p is a successful or unsuccessful belief for x to have. Since in CT p, or the belief in p, can have any of these epistemic properties, then IT needs no independent statement from CT, except perhaps for emphasis. As an illustration of IT, consider the case of the Muslim belief that soldiers who die in a holy war go straight to heaven. Success in battle may occur for those who believe this, no matter whether it is true or false, or whether it is held rationally (i.e., for good reasons to be found within Muslim theology) or irrationally (as is maintained by many non-Muslims); and Muslims may believe this whether it is successful for them (they win) or unsuccessful (they lose). IT makes explicit the wide scope of beliefs in CT as a thesis within the sociology of belief in general or the sociology of religious belief in particular.

6.2.1 The Failings of the Impartiality Tenet

Unfortunately IT is not well formulated for three reasons. First, it is not the truth of some proposition p that is to be explained; nor is it p's falsity, rationality, irrationality, success or failure (call these, collectively, the 'epistemic status' of p). Rather, to bring IT in accord with CT the *explanandum* must be x's *act of belief* that p. Then it is the content of the belief, *that p*, which can be true or false, or can lead to success or failure (in, say, its applications or predictions). Or it is x's act of believing that p which is rational or irrational, if we understand these terms to concern the evidence and reasoning upon which x believes that p. In addition, success or failure could be predicated not only of the content, that p, but also x's act of belief that p from which further action by x arises which may be successful or not. Second, concerning SP's conflation of knowledge with belief, IT can not concern the explanation of some person's *knowing* that p since one of the conditions for knowledge, viz., that p be true, is explicitly ruled out. Third, IT introduces talk of explanation whereas CT speaks of causation. However we can minimise this difference by assuming that, in its explanations, IT appeals to the causes set out in CT.

What IT emphasises about SP is that an explanation can be called for any scientific belief that p, independently of the epistemic status of p, or prior to any determination of what that epistemic status is. IT highlights a difference advocates of SP like to draw between their programme and a 'Weak Programme', WP, within SSK. (WP is what Bloor calls the 'teleological model'.) WP is weak in the sense that social explanations of belief will be required only in cases where p is false, irrationally held or unsuccessful. In order to employ WP there must be, first, a prior epistemic evaluation of beliefs as to their epistemic status. Once this is settled, then there is a division of labour in which rational models of explanation internal to science are said to apply only to belief in the true, rational and successful while Weak Programme causal explanations are to apply only to belief in the false, irrational and unsuccessful. In this respect WP is reminiscent of Mannheim's approach to SSK. Apart from its restricted character, why is Bloor concerned to explicitly rule out WP? Only because of the misleading aspects of his characterisation of the "teleological model" of explanation accompanying WP in which only deviant beliefs are caused, or in which the explanation of belief only arises in the case of error. But there is no reason to accept the teleological model and its restricted domain of application; abandoning it allows that all acts of believing are caused and are apt for some kind of explanation, regardless of the epistemic status of p. And this is precisely what the rationality models discussed in the previous section allow.

More fundamentally, is the division of labour concerning explanation suggested by advocates of SP correct? There seems to be an underlying assumption that the rational explanation models can apply to only one side of the epistemic divide, viz., the true, rational and successful. But this is quite wrong and reveals a serious misunderstanding on the part of advocates of SP about the nature of scientific methodologies and their principles.

Consider for example, the application of methods due to Bayes' Theorem. This enables us to discover which hypothesis of a given set of hypothesis has the highest confirmation on the basis of changing evidence. The hypotheses we employ need not all be true; they might all be false since some other unknown hypothesis not in the set under consideration might be the true hypothesis. Within Bayesian methodology some hypothesis can be the best supported on given evidence and yet still be false. However there can still be a rational explanation why some person x believes some scientific hypothesis H (which is false). If x is a conscientious Bayesian, then in an explanation as to why x believes that H, we can appeal to x's use of Bayes' Theorem in the *explanans* of the rationality model. Then we have an explanation of xB(H) (where H is a false hypothesis) that is not of the form of the social cause model required by SP.

Again, consider Popper's theory of method. It is a keystone of his methodology that our theories are all false, though some can be more-true-and-less-false than others in the sense of one having greater verisimilitude or truthlikeness than another. This is not to say that in conjecturing a hypothesis we can never hit on the truth; we could do so, without our coming to know this, or being able to establish it. Popper's theory of method, which involves principles about corroboration, falsification *ad hoc* moves and the like, tells us when it is rational to accept (but not necessarily) believe, or reject, a hypothesis. That is, we can accept a hypothesis that has been corroborated but which, for all we know, may well be false (even though it has increased verisimilitude). If x is a conscientious employer of Popper's theory of method, then there is a rational explanation of why x accepts hypothesis H, *even though H is false*. Again no social cause explanation is needed.

Perhaps the methodologist who most loudly proclaimed that his methodology applies to the false is Lakatos with his methodology of Scientific Research Programmes (SRP). Lakatos liked to emphasise that progress could be made within an SRP even though its 'hard core' could be false; whatever its truth-value the hard core of a SRP is to be conventionally held true or made irrefutable by a decision of its advocates (Lakatos (1978), p. 48). To illustrate, Lakatos spells out the hard core of Prout's programme in chemistry which was later discovered to be false, or Bohr's programme concerning his early theories of the atom in which the hard core was not merely false but inconsistent (*ibid.*, pp. 52-68). Lakatos also liked to show how a SRP could be progressive in that it anticipates many unknown facts while it is engulfed in an 'ocean of anomalies' (*ibid.*, pp. 53-5), some of the anomalies perhaps coming to be understood as refutations of the SRP in the light of a new even more progressive rival SRP.

Finally x might adopt Kuhn's theory of weighted values in which theories are appraised according as they are internally and externally consistent, simple, wide scope, fruitful, solve puzzles, have accurate quantitative predictions rather than qualitative predictions, and so on. None of these criteria entail that the theory chosen on the basis of these values is true rather than false; in fact Kuhn rules out the very possibility that the unobservable aspects of these theories can be known to be true. In his discussion of the possibility of our ever obtaining theoretical truth he expresses deep scepticism: 'the notion of a match between the ontology of a theory and its "real" counterpart in nature now seems to me illusive in principle. Besides, as an historian, I am impressed with the implausibility of the view (Kuhn, (1970), pp. 206). However Kuhn's values do require that the theory be pragmatically successful, providing the value of predictive accuracy is understood to spell out what it means to be successful. That is, there is truth to be found but only at the level of the observable. What is important about the Kuhn's model of weighted values is that truth at the theoretical level plays no role, and that the values he does espouse can be understood without such a notion.

It is surprising that advocates of SP could overlook this feature of nearly all methodologies, viz., that they can deal not only with beliefs in science that are true, but the false as well. The assumption that there must be a division of labour in explanations of why x believes hypothesis H (xB(H)) along the lines of advocates of SP is faulty. Rational models for explaining xB(H) that explicitly appeal to methodologies, can explain why x believes H regardless of the epistemic status of H. There is no need to always invoke social cause models. But this is not to deny that these models have no role at all.

Concerning the pragmatic value of (belief in) an hypothesis, the same models can also apply regardless of whether H is successful or not. Bayesian, Popperian, Lakatosian, Kuhnian and any other methodology, when employed in a rational explanation model, can pick out hypotheses which are successful up to a given time; but such success cannot be always guaranteed by such methodologies unless something like the problem of induction is shown to have a sure-fire solution. In addition Popper argues (the contested point) that an appraisal of an hypothesis according to its degree of corroboration does not entail anything about its future success in predictions; corroboration appraisals are entirely backwards-looking and merely record a hypothesis' success in passing tests up until now.[10] Thus an explanation of why xB(H) can be in accord with rational models even though H is in the long run unsuccessful. Thus not only are advocates of SP wrong about the claim that rationality models can not apply to false and unsuccessful beliefs; they are also wrong about their claim that only social cause models can explain the false and the unsuccessful.

The final case to consider in IT is that of rationality. Rationality is a many-splendoured thing that is left unspecified. Here we will take rationality to be the application of some theory of method to adjudicate between our scientific beliefs. In this case rationality models can explain x's belief by appeal, in its *explanans*, to the application of some theory of method containing correct principles. But the simple rationality model of section 5.4.2 cannot explain x's belief that p when it is irrationally held. Must we now have recourse to social cause explanatory models to explain x's irrationally held belief? No; we still have the rationality* model which can explain x's belief regardless of the epistemic status of the belief.

IT concerns the impartiality of explanatory models with respect to the epistemic status of the *explanandum,* x's belief that p. However if one views IT as an instance of CT, then IT can permit only one kind of explanation, viz., the social cause model. All the other models canvassed above have been ruled out. But there is a more liberal reading of IT, and the impartiality it requires of explanations, along the following lines: for any x and any belief p that x has, then for any epistemic status that p might have, there is an explanation of the belief with that status. This allows that there be quite different kinds of explanation for acts of belief of the sort canvassed in the above; none are ruled out. The less liberal reading of IT switches the order of the quantifiers in the above formulation and says: for any x and any belief p x has, there exists (some one) explanation of xBp, whatever the epistemic status of p. That is, one and only one explanatory model will do the job of explaining why x believes that p no matter what epistemic status p has. The illiberal reading aligns IT not only with CT but also with the requirements of the Symmetry Tenet of SP, to be discussed in section 6.3.

6.2.2 *Impartiality in Rival Programmes Within SSK: the Case of TRASP.*

We have already noted that the Impartiality Tenet is adopted by other practitioners of SSK who reject other aspects of SP, especially CT with its commitment to social cause explanations, and the Reflexivity Tenet which underlines the scientism of SP. This is the case with Harry Collins' version of SSK which he calls 'The Radical Programme', and defines it by Bloor's second and third tenets only. For our purposes it is Collins' rejection of what he calls 'TRASP' that is important: '... there are things that cannot form part of an explanation belonging to the radical programme. Knowledge cannot be explained by reference to what is true, rational, successful or progressive (hereafter TRASP)' (Collins (1981), p. 217) Now of course the banned TRASP cannot be about the explanation of *knowledge* but of acts of belief. What Collins' Radical Programme rules out are explanations of x's belief in p that are TRASP, i.e., it rules out any *explanans* that appeals to the truth of p, or the rationality of x's belief, or success or progressiveness. And in so doing it adopts the illiberal reading of IT. But it is clear that TRASP, like the illiberal version of IT, lives off a narrow diet of mono-causal explanatory models. It ignores any rational explanation model, and endorses only social cause models. In this respect the Radical Programme is as flawed as the Strong Programme it aims to replace. But it is even more flawed that SP if it proposes to reject CT, the very tenet that SP requires to sets out the grounds for adopting any social cause model.

Finally, IT gains a little plausibility through its connection to a methodological prescription about how investigators in the field of SSK are to approach their subject matter, viz., what people believe. In their descriptive investigations, investigators are required to 'bracket off' such matters as whether the beliefs under scrutiny are true or false, are rational or irrational, are successful or a failure, and so on. The field of investigation into belief is wide open and is not to be restricted as to the epistemic status of the belief. There is nothing wrong with such 'bracketing off' requirements – except that some authors refer to this as adopting relativism about belief because of a suspension of epistemic matters. Some sociologists, as we have noted in the case of Shapin in section 1.3.2, have elevated the suspension of all epistemic matters to do with truth to a maxim of method. But to focus on belief and never on truth, is a bad case of truth phobia. Advocates of TRASP are also infected with truth phobia. This is evident when the 'bracketing off' is extended to any explanation of why a person holds the beliefs they do; matters to do with the beliefs' epistemic status are not allowed to enter into any explanation as to why a person believes that p. But this is an unjustifiable restriction of admissible explanations. A methodology M might determine that it is rational to believe that p because, say, of its comparatively high degree of confirmation – and this might be the actual the reason why x believes that p. There are no good grounds for 'bracketing off' such explanations, providing they are correct. TRASP would in effect simply rule out some correct explanations.

In sum the impartiality requirement, that is, the requirement that models of explanation must be independent of matters to do with the epistemic status of the belief or the act of believing, are unnecessary and are often due to truth phobia. And where they might be deemed necessary rationality models can do as well, and even much better, than social cause models.

6.3   THE SYMMETRY TENET

All the above criticisms of SP lay the groundwork for undermining its central distinctive dogma embodied in its Symmetry Tenet. While the Impartiality Tenet and the Reflexivity Tenet (to be discussed in section 6.4) are instances of the more general CT, the Symmetry Tenet (ST) is independent, adding some new content to SP about the exclusivity of the one and only one model of explanation to be adopted. ST says: 'it [SP] would be symmetrical in its style of explanation. The *same types of cause* would explain, say, true and false beliefs' (Bloor (1991), p. 7, italics added).

In science there are particular things we would like explained. Thus we might like to know why, say, the perihelion of Mercury precesses at an annual rate of about 574 seconds of an arc per century. All of this is explained by a combination of Newton's theory concerning the perturbatory effects of other planets (this accounts for 531 arc seconds per century), and a small relativist effect revealed by the general theory of relativity (this accounts for the remaining 43 arc seconds per century)[11]. A quite different matter to be explained is why person x *believes* that Mercury's perihelion precesses at about 574 seconds of an arc per century. Though not as exciting as explaining the first matter, a number of explanations could be given of why x believes this, from x's having read it in a book, or having dreamed it, to x being an astronomer who has worked out this feature of the orbit of Mercury from observations and/or other matters. SP offers us nothing in the way of explanations of the features of the orbit of Mercury itself. Rather it is confined to the second, often duller, matter of explaining why a person believes something (it is not dull in the case of why, say, Einstein believed it). The extraordinary requirement of ST within SP is that there is just one type of explanation to be given of why x believes this. Are explanations such as reading about it, dreaming about it, or working it out, of the same type or not? Hard to say. They look quite different. Perhaps 'just take your pick!'. But if they are not of the required same type (whatever this be) then they, according to ST, cannot serve as explanations of why x believes this. This in a nutshell is the initially incredible symmetry requirement of SP, and the many advocates of SSK who are not devotees of SP but who hold to some version of the doctrine of symmetry. It should also be borne in mind that many SSKers advocate different claims under the banner of symmetry, for example Latour ((1987), chapter 5); here only symmetry as formulated within SP will be discussed.[12]

Let 'p' = 'the perihelion of Mercury precesses at an annual rate of about 574 seconds of an arc per century'. Then p has a truth value (and to the best of our knowledge and within a small degree of error, it is true). The truth value of 'x believes that p' does not depend on the truth value of the contained 'p'. And it is evident that any explanation of 'why p?' will be different from 'why does x believe that p?'. Sociologists of "knowledge" can go about their investigations of such belief claims without becoming astronomers and acquiring knowledge of whether or not p. And often this harmless claim is all that is intended by the symmetry requirement. As was pointed out in section 1.3.2, the intensionality of belief claims makes the world safe from the investigatory ruminations of sociologists whose task ought to be to investigate just one bit of it, viz., who believes what. They do not have to address a further matter to do with the rest of the world, viz., what the truth value of p is itself. But as we saw in 1.3.2, some sociologists make much stronger claims than this and display confusion about the connection between truth and intensional claims like x *believes* that p that are symptomatic of a bad dose of truth phobia. As will be seen from what follows below, the alleged need for ST is a further symptom of truth phobia.

Let us agree that Newtonian mechanics, N, explains all of the precession of the perihelion of Mercury except for the missing 43 arcseconds per century and that Einstein's General Theory of Relativity, GTR, explains the missing 43 arcseconds. Why should one come to believe GTR rather than N? There are good methodological principles that tell us which to believe. In its simple form such a principle might say: in choosing which of two theories, A and B, to believe (accept), then if there is an outstanding anomaly for A while B explains that anomaly (with no loss in what A otherwise explains) then accept B rather than A. In explaining why a person accepts GTR rather than N, one could appeal to such a methodological principle in any rationality explanation. That is, reasons and justifications can be the explainers of belief (acceptance). It was appeal to some such principle that lead Einstein to see that he was on the right track in following up the consequences of GTR, such as its account of the anomaly of the precession of the perihelion of Mercury. As Pais tells us: 'That was the highpoint of his scientific life. He was so excited that for three days he could not work'.[13]

Such explanations that appeal to principles of method can be said to be of a single type – the type that appeals to the (believed) correct application of principles of scientific rationality and method. As will quickly realised, this is not the type advocated with SP. Given both IT and ST, *only* explanations of scientific belief which are of the same type are going to be admitted, the approved type being social cause explanations only. 'Bad luck Einstein! You should not have got so excited about GTR on the basis of the methodological principles that back your reasons for accepting it. You should have looked to your socio-politico-cultural context, and/or your interests in this, as the proper explanation of why you came to believe in GTR.' But had he done so, he would have found nothing in it to recommend his belief in GTR. Yet it is just this that ST advocates for Einstein and the rest of us. If this is not sufficient reason for rejecting ST and its restriction on admissible explanations, then further considerations are offered in the following.

6.3.1 Problems With 'Same Type of Cause'

There is a basic unclarity in the very formulation of ST about what counts as the *same type* of cause since no typology of causes of belief accompanies the statement of ST. In section 5.2 a elementary typology of five types of social causes was suggested, along with meaning finitism as a sixth type. Type (d) causes, such as political circumstance or interest, and type (e) causes to do with negotiation and consensus formation, feature in most of SP's attempts to explain scientist's beliefs (see the studies of Forman in section 5.6 and Bloor in section 5.7). Type (c) causes due to acculturation also feature for other sorts of beliefs. So here are three types of social cause for explaining belief and not just one as insisted by SP. And even within type (d) causes we have distinguished those causes due to social circumstance S_x from those causes that are an *interest* in S_x. So how many types of cause are there? And which is to be employed in the social cause explanatory models? No answer is forthcoming. To say that there is just one type of cause here is to raise a host of unresolved problems of the individuation of types of cause.

In this and the proceeding chapter a typology of explanatory models with distinctive *explananda* for the same *explanandum* viz., x's belief that p, has been built up which includes schema such as the various social cause models, the rationality models, and Lewis' historical causal model. These appeal to a wide variety of items in their respective *explanans*. Advocates of SP appear to bid us adopt only social cause models of explanation. But they are victims of mono-causal mania since they ignore other quite appropriate rationality models.

Most critics of SP have found its commitment to ST as surprisingly dogmatic. So, why the commitment? One argument commonly advanced by SP's advocates is that one needs to have a prior assessment, before applying any model of explanation, of the epistemic status of x's belief that p to determine whether p is true or false, successful or unsuccessful, or x's belief that p is rational or irrational. But we have seen that this is quite wrong. No prior determination of epistemic status is needed before rational explanatory models can be applied. The methodological principles appealed to in these models can have explanatory force even when p is false, irrationally held or unsuccessful.

The last point seems to be the real issue behind talk of 'types of cause' that is misleadingly bound to the doctrine of relativism: 'It [SP] adopts what may be called 'methodological relativism', a position summarised in the symmetry and reflexivity requirements .... All beliefs are to be explained in the same general way regardless of how they are evaluated' (Bloor (1991), p. 158). Setting aside the irrelevant entanglement with relativism, ST comes down to the claim that, quite independently of the epistemic status of x's belief that p, all (acts of) belief are to be explained in the same way. But as has been asked, what *same* way? Once more ST brings us back to the social cause model only.

Perhaps the best critical evaluation of ST is given by Laudan.[14] He distinguishes between three kinds of symmetry according to the epistemic status of x's belief that p. The first kind of symmetry is in respect of p's truth-value, viz., adopt the same kind of explanation of x's belief that p regardless of the truth-value of p. Now as has already been argued, methodologies very rarely decisively show that p is true, or false, though they may tell us about comparative degrees of conformation, simplicity, or whatever else. Even the Popperian claim that we can decisively show our theories to be false but not show them true, is challenged by the Quine-Duhem thesis. As already argued, rationality models which employ methodological claims in their *explanans,* can explain x's belief in p – regardless of the truth-value of p. In the light of this, the symmetry requirement points to rationality models as the one type which best explains x's belief that p, where p is either true or false.

The same considerations apply to the symmetry of explanations independently of the pragmatic success or failure of p (where this is understood methodologically as p's success or failure in predictions, success in passing or failing tests, and so on). Rationality models, particularly the rationality* model which tracks a person's actual methodological inferences, will provide more explanations of such beliefs while the social cause models will provide very few. Once more there is no case for the mono-causal dominance of SP in this area. In fact there is no need for the symmetry requirement at all. Let whichever model for the explanation of x's belief that p do its work, providing it yields the *correct* explanation and not an explanation in accord with some dogma.

The third kind of symmetry Laudan distinguishes is that of rational symmetry, the requirement that the same kind of explanation apply independently of the rationality or irrationality of x's belief that p. We can also include under this heading pragmatic symmetry in respect of the pragmatic success of x's belief that p (and not just p alone, as in the above paragraph). Now if one's focus is on the simple rationality model, then this will only provide an explanation of x's belief that p in terms of the correct employment of scientific method. But if x believes that p on irrational grounds, then this model cannot explain x's belief. But are we thereby forced to adopt social cause models? And if we do then, by the symmetry requirement, are we to drop all rational explanatory models, even in the case of rational belief? This is the conclusion advocates of SP draw given their narrow diet of mono-causal explanatory models. But what they overlook is the availability of a the rationality* model which allows us to track the actual cognitive path that leads to x's belief that p. As we have seen that path can contain errors, falsehoods and even irrationalities (such as using the toss of a coin, the Gambler's Fallacy or some faulty methodological principle not known to be so). But if this is what is actually in x's cognitive path that lead to x's belief that p, then this is what must go into the explanation of x's belief. Once again, rationality models win out over other models. Even further, this gives no grounds for adopting sameness of explanations all round. Once again, a variety of models, depending on how the belief is acquired, is the best approach to finding the correct explanation. Contrary to the dogma of the ST, let a hundred flowers bloom. Let whichever model for the explanation of x's belief that p do its work, providing it yields the *correct* explanation and not an explanation in accord with some dogma.

6.3.2 Bloor's Typology of Symmetries: Logical, Psychological and Methodological

In his 'Afterword: Attacks on the Strong Programme' (Bloor (1991)), Bloor does not take up the challenge made in Laudan's criticism based on the three-fold distinction of ST into the epistemic, rational and pragmatic. But in reply to other critics Bloor draws a three-fold distinction of his own protean ST requirement, viz., psychological, logical and methodological symmetry. Methodological symmetry we have met; it is the requirement that the same explanation of x's belief that p be given independently of the epistemic status of either p or x's belief that p. As we have seen, this does not require us to adopt a social cause model; rationality models can do just as well, or even better, since they, too, can be used independently of the epistemic status of the belief to be explained.

Psychological *asymmetry* arises in the case where, say, an anthropologist uses the 'actor's own categories' to explain action; that is, we are to invoke in explanations only notions that the actors themselves would understand and accept. In the case of witchcraft belief, the explanation of why x believes witchcraft doctrine p will be in terms of x's own reasons q; and this is quite independent of either how the anthropologist would appraise p to be true or false, or whether the reasons q for the belief that p are those the anthropologist could also adopt. Such an 'actor's categories' model of explanation is nothing other than a version of a Collingwood style model in which the anthropologists attempts to re-think, or get the inside track of, the actual thought processes of witchcraft believers. The reasons or methods that the actor's use need not be those that the anthropologist endorses, yet the anthropologists can use them in an explanatory model. Of this Bloor says: 'It is consistent with methodological *symmetry* because the character of the desired explanation is independent of the evaluation' (Bloor (1991), p. 176). True: but note that it is not only social cause models to which this remark can apply. It applies even better to the very different rationality models.

Logical asymmetry arises as follows: 'Members of a witchcraft culture will say they believe in witches because they encounter witches. An anthropologist might say it is because they are symbolising their social experience of living in a small disorganised group prone to scapegoating' (*loc. cit.*). Here the anthropologist offers a quite different explanation that the actors themselves would neither offer nor recognise. Such is also the case when Freudian analysts offer explanations of a person's behaviour in terms of unconscious desires that the person would not themselves accept (there even being a Freudian explanation of this non-acceptance), or structuralist followers of Levi Strauss who offer explanations of myths that the myth believers do not entertain, and so on.

Bloor continues: 'The anthropological theory will logically imply that the witchcraft beliefs (taken at their face value) are false' (*loc. cit.*) But this need not necessarily be so; witchcraft theory and the anthropologists' theory might simply be logically independent of one another. Moreover the actor's and the anthropologists' explanations have the same *explanandum* 'x *believes* that p' (where the truth value of p is irrelevant to the truth of the *explanandum*). So even if it is unclear as to why the 'logical asymmetry' is *logical*, there is at least *asymmetry* in that the anthropologist's explanation need not be one that the actors themselves endorse. But again none of this is grist to the mill of social cause explanation. These models need not feature in either the actor's explanations or the anthropologists'. In sum, Bloor allows two kinds of asymmetry, the psychological and the logical; but these are not to be confused with (they even leave untouched) the methodological asymmetry requirement set out in ST. But, significantly, drawing such distinctions does nothing to establish the dogma that is ST, viz., the requirement that only one type of social causation model is to be used in the explanation of why x believes p.

6.3.3   Rationality Explanations as Intrusions Upon the Causal Order

After explaining the above different versions of asymmetry, Bloor offers a new reason as to why ST should be adopted:

> The problem running throughout most exchanges over the status of the symmetry requirement lies in the clash between a naturalistic and a non-naturalistic perspective. The symmetry requirement is meant to stop the intrusion of a non-naturalistic notion of reason into the causal story. It is not designed to exclude an appropriately naturalistic construal of reason, whether this be psychological or sociological. (*ibid.*, p. 177)

This remark has been already discussed in section 3.5.1. It assumes that the norms of method and reason, if construed objectively, must be akin to a weird kind of normative fact that can intrude, like some *deus ex machina* whenever someone employs them, upon the naturalistic causal nexus in which believers exist (including their social and psychological contexts). The symmetry requirement ST is meant to stop any appeal to such intrusive norms. Much of chapter 3 was devoted to setting out a theory of norms in which they are not a kind of intrusive normative fact but, rather, can be understood in such a way that they fit quite happily with naturalism. So understood they can enter into the causal nexus and be used in explanations of belief. If keeping us free from intrusive norms is the sanitising role that ST is to play, then we do not need ST. We can understand norms quite differently, leaving ourselves free to adopt explanatory models of scientific belief that contain norms.

The freedom from the constraints of ST just suggested is not something that Bloor accepts. He objects as follows:

> Such composite positions, however, are incoherent. They are trying to meet an impossible condition: making reason both a part of nature and also not a part of

nature. If they don't put it outside nature, they lose their grip on its privileged and
normative character; but if they do, they deny its natural status. They cannot have it
both ways. (Bloor. (1991), p. 178)

But section 3.4 shows what has gone wrong here concerning the status of
norms. And sections 3.7 and 3.8 show how to have it both ways through the
supervenience of the normative on the non-normative and through a Ramsey-
Lewis style definition of normative notions. Bloor continues his discussion of
norms of reasoning in relation to a naturalistic defence of rationality based on
the theory of evolution (advocated by Newton-Smith). He says of evidence
for a theory (including the theory of evolution):

... there is still the task of justifying our belief in this theory and saying how we
know it is true. To do this we must suppose that we can intuit evidential relations and
some logical truths. So even here we need access to a realm of epistemological facts,
that is: 'abstract non-physical facts'. ... This abstract non-physical realm must exist
over and above the flux of biological and cultural change if it is to be used to explain
and justify it. If it were grounded in evolution it would have no more probative force
than any other disposition or natural tendency. (*loc. cit.*)

We can agree with the last sentence. But what the above account of norms
shows with its talk of 'intuiting' and of 'abstract non-physical facts' is that it
commits the theory of norms to a thoroughly Platonist construal. But this, as
sections 3.4 and 3.5 show, is just one of several position to adopt about the
status of norms. We are not obliged to adopt the Platonist account above, nor
to adopt an account of norms which strips them of their normativity in order
to naturalise them (as sections 3.6, 3.7 and 3.8 show).

Finally, in order to escape norms understood as a weird kind of intrusive
fact, Bloor appeals to a 'naturalistic construal of reason, whether this be
psychological or sociological'. But if this is an appeal to how we do in fact
reason as a matter of our psychology or our socialisation, it ignores all the
literature in cognitive psychology which shows that as a matter of fact we can
be quite bad reasoners across a wide spectrum of reasoning from deductive to
non-deductive, decision theoretic, etc. Naturalising norms in this way bleeds
them of their normativity, with the consequence that the idea of violations of
the good canons of reason gets lost. But perhaps there is another notion of
naturalisation of the normative that Bloor has in mind; this we must leave to
the next chapter where the theme of meaning finitism is taken up again.

6.3.4 A Further Objection to the Symmetry Requirement

This sub-section develops an independent reason for rejecting ST. The task of
SP within SSK is to provide a causal account of all beliefs. But we have
shown that other non-social rationality models can do this, and do it better.
And they do it while preserving the idea that reasons can be normative and
can be causes of belief. But given Bloor's odd account of the normativity of
reasons, reasons are not possible candidates for the explanation of belief. As
a consequence the only types of cause sociologists are to invoke in their

explanatory models are social (in conjunction with a range of non-social naturalistic factors). Granted a thoroughly scientific (not to say scientistic) programme for explaining belief, Bloor says of theories within SSK:

> If these theories are to satisfy the requirement of maximum generality they will have to apply to both true and false beliefs, and as far as possible the same type of explanation will have to apply in both cases. The aim of physiology is to explain the organism in health and disease; the aim of mechanics is to understand machines which work and those that fail; bridges which stand as well as fall. Similarly the sociologist seeks theories that explain the beliefs that are in fact found, regardless of how the investigator evaluates them. (Bloor 1991, p, 5)

Here we get more of a clue as to what *same type* of explanation might mean in ST. In his commentary on what this might mean Brown ((1989), pp. 38-41) suggests that what is invoked as the *same type* is the following: the same principles (of physiology, mechanics, sociology, etc) are to be used to explain both why observable phenomena P occurs (the bridge remains up) or not-P occurs (the bridge collapses). And in his 'Afterword' to the 1991 edition Bloor makes a similar point by way of an analogy about his methodological symmetry requirement: 'The case is treated by analogy with visual perception. [Person] A simply sees what is there because the relevant perceptual processes are operating properly. [Person] B's vision, by contrast, is 'clouded', or 'occluded' by some interfering factor' (*ibid.*, p. 178). Let us follow up this example of visual perception when a person looks at the Müller-Lyer diagram; they report, of a pair of parallel lines that were first seen to be equal in length, that, when the arrow-heads and tails are added, the pair are then seen to be unequal. This is an example that Bloor uses elsewhere in defence of his required symmetry (Bloor (1981), p. 208).

The two parallel lines by themselves create no illusion when we look at them in normal conditions; our visual belief that the lines have the same length can be accounted for, in part, by the causal mechanisms as set out in our best theory of visual perception, V, concerning the standard viewing of objects. But in the context of the Müller-Lyer illusion we acquire a different visually based belief, viz., that the lines are of unequal length (even if we assume that we are quite familiar with the diagram). What accounts for this? As Bloor suggests, it is the same type of causes, as set out in the theory of perception V, that will be appealed to in both cases of veridical and illusionary belief; but the viewing conditions will be different. In the conditions of viewing the lines without the illusion-making arrow heads and tails, particular conditions C_1 of viewing will apply and the conjunction of $V \& C_1$ will explain why we see that the lines are of equal length. In viewing the lines in the context of the illusion created by the Müller-Lyer diagram, different conditions C_2 apply which include the way in which the causal mechanisms of perception are tricked into working in the way they do in these conditions. The conjunction $V \& C_2$ will explain why we see that the lines are of different length. The same theory of vision V is employed in both cases but in different circumstances, C_1 and C_2. The very same causal mechanisms of perception are not at work in both cases; rather, the causal mechanisms are of the same type (as specified in theory V), but with some different causal factors in the two cases due to the different conditions of viewing, viz., the normal C_1 and the trick played on the perceptual mechanisms in C_2. It is the difference between C_1 and C_2 that explains the two different visually acquired beliefs, one of which is veridical, the other illusory.

Whether our visually acquired belief is veridical or illusory, the same *type* of cause, but not necessarily the same cause, will explain our visually acquired belief about the length of the lines. Analogously for ST, whatever the epistemic status of our beliefs in science, the same types of social (and non-social) causes, but not necessarily the very same social cause, will explain belief. There is allegedly no need to invoke the norms of reason to explain our beliefs – thus the extreme form of naturalism that SP imposes through ST in which all evaluation is eliminated. But does the analogy really work? In the Müller-Lyer example there is tension between (a) accepting the deliverances of belief provided by our visual system, and (b) making *judgements* about what is and what is not veridical. Similarly there will be some tension between (a) accepting the deliverances of belief in accordance with the Causality Tenet, CT, and the Symmetry Tenet, ST, and (2) making judgements about what is and what is not true, or rational.

In the case of perception we do draw a distinction between veridicality and illusion. We do not naively accept the deliverances of belief by our perceptual mechanisms; there are ways of evaluating them so that we can say in one case perception is veridical and in the other illusory. If we did naively accept them, we could not make the distinction that some of our visually acquired beliefs are true and others false. These judgements appear to be independent of the deliverances of the perceptual modules in our brains, even though we may use these modules in the process of making these judgements, e.g., by measuring the lengths of the lines in the Müller-Lyer diagram. Philosophers from Descartes onwards have been at pains to draw a distinction between veridical and illusory perception, including cases such as dreaming. Here is one suggestion as to how we might make the veridical/illusory distinction in the Müller-Lyer case. We note a contradiction between the naive deliverances of belief by our perceptual system when (i) merely looking at the diagram and when (ii) measuring the lines in the diagram. (In terms of the analogy under investigation, think of the deliverances of belief by our social circumstances in accord with CT.) Which of these inconsistent beliefs is correct?

By employing some version of the best overall explanation of our visually acquired beliefs (as Descartes does at the end of the sixth of his *Meditations* concerning the awake/dreaming distinction), we are able to classify those relating to the Müller-Lyer diagram as illusory rather than veridical. We quickly come to the conclusion that the unequal lines viewed in the context of the Müller-Lyer diagram produces the illusory belief. Despite our seeing the lines as unequal, the best overall fit with all our other beliefs is that the lines are really equal in length. But to make this distinction is to employ some form of reasoning (not admitted by CT and ST) as an explanation of why we believe what we do, even in the restricted case of the Müller-Lyer diagram. Thus evaluation of our beliefs must be possible if we are to make the veridical/illusory distinction. And *pace* ST, we must use these evaluations in explaining why we believe what we do. If we do not make the veridical/illusory distinction, and thus do not make evaluations of belief, we would be poor creatures buffeted by what our belief-producing mechanisms turn up (unless we are paraconsistent logicians – but this will not help the case for ST!).

Let us now extend this to CT as applied in the case of scientific belief in which social and non-social, but not normative, factors are at work as causes of belief. We are not obliged to accept the deliverances of our belief-forming processes since they will indifferently produce true or false belief, rational or irrational belief and even inconsistent pairs of beliefs. As in the case of the Müller-Lyer diagram, so in society and in science. Situations will arise in which we need to ask of the beliefs delivered by our belief-forming processes which are true or false, which are rational or irrational. And once we filter our beliefs according to normative/evaluative criteria to do with truth or falsity, rationality or irrationality, we can then override the deliverances of our belief-forming processes using norms of reasoning.

If such overriding takes place, then the overriding by principles of reasoning must be part of an explanation of why we believe what we do. We cannot naively accept the normatively unfiltered beliefs described by SP as the only admissible causes of belief, even though such causal processes may be at work in forming belief. We might give up, as some have done, any normative appraisal of belief, thereby abandoning truth, rationality and other such notions beloved by philosophers. If we do, then just as butterflies are buffeted by the wind, so our minds are buffeted by the beliefs that, according to CT, crowd into our brains. But then one cannot employ any kindred notions such as veridical/illusory, true/false, rational/irrational, etc in with respect to the beliefs that flood into our brains. If we do draw such distinctions then we need to abandon ST which requires the same explanatory treatment of all beliefs. What this brings to the fore is the total inability of SP to provide evaluation of belief. We are just passive acceptors of what of social processes deliver to our brains. And this is what is disturbing about SP.

6.4   THE REFLEXIVITY TENET

Issues of reflexivity have already been encountered in the case of Marx; and Mannheim envisaged that his sociology of knowledge did not arise accidentally because there 'is a specific social situation which has impelled us to reflect about the social roots of our knowledge' (Mannheim (1936), p. 5). Bloor's Reflexivity Tenet (RT) spells out the scientism so far implicit in SP: 'its patterns of explanation would have to be applicable to sociology itself. ... It is an obvious requirement of principle because otherwise sociology would be a standing refutation of its own theories. (Bloor (1991), p. 7) Not to include the claims of sociology, and SP itself, within the scope of CT would be to restrict the scope of 'p' in CT an arbitrary manner and to locate the grounds for belief in these claims in quite other unspecified conditions than those CT requires for all other belief.

In section 5.4 we investigated CT not only in the case where 'p' ranged over ordinary scientific beliefs but also principles of deductive and inductive logic, principles of scientific method, theories of concepts within the philosophy of science such as explanation, observation, and so on. All of these were seen to be grist to the mill of SP which can now be extended to all theories within sociology, any special principles of method that might apply in sociology, and any concepts that might be peculiar to its meta-theory such as the doctrine of *verstehen*, actor-network models, discourse analysis, ethnomethodology, and so on. Nothing is to be left out. But now CT is to be extended to the very four tenets of SP itself; and this gives us the Reflexivity Tenet RT. After all, SP is simply more in the way of belief and as such falls within the scope of CT. That is, there are particular conditions in which the believers in SP exist (e.g., particular conditions in Edinburgh which prevail around members of the Science Studies Unit) and which are such that each is caused to believe SP. Presumably when these conditions do not obtain they fail to believe SP. To express one version of RT, consider the particular instance of CT where for 'p' we substitute the four tenets of the Strong Programme itself (called 'CT(SP)'):

CT(SP)   For all persons x, and for the belief content of the four tenets of SP such that $xB(SP)$, there exists some social condition S_x of x, and there exists some non-social condition N_x of x, such that $(S_x \& N_x)$ causes $xB(SP)$.

It follows from this that CT applies to itself, i.e., there are causal conditions for belief in CT since it is part of SP. This is also a consequence of CT alone since CT is a perfectly general claim about beliefs and so applies to itself.

Can a case be made for restricting the range of CT to avoid reflexivity? It was said earlier that SP was a meta-sociological manifesto which sets out the form that any worthwhile theory within the sociology of scientific belief should take. Thus the four tents of SP are not at the same level as particular accounts of the causation and explanation of beliefs within any of the special sciences (e.g., the belief that the Earth orbits the Sun, etc). This difference might be used as a ground to exempt SP from applying to itself. But if we exempt SP from falling within the scope of CT, then any account of how anyone can come to believe SP in the first place would, if one were not a dogmatist, have to appeal to a theory of belief formation and acceptance outside the scope of CT. What might this be like?

It seems highly likely that we need to appeal to evidential considerations of the sort advocated by traditional methodologists of science with their theories of scientific rationality but rejected by adherents to SP. Barnes and Bloor do in fact present evidence in favour of SP by citing case studies which allegedly support it inductively (see Hollis and Lukes (ed.) (1982), pp. 23-5, especially footnote 7). Thus there is an implicit appeal to the methodology of inductivism in which there is an enumerative inductive inference from the alleged success of case studies to the correctness of the claims of SP. In a complaint about the fact that one of his critics 'may have overlooked the role that actual case studies played in the formulation of [his] programme' Bloor goes on to say that the critic 'has failed to see that I am an inductivist' (Bloor (1981), p. 206).

But is such inductivism outside the scope of SP and part of some theory of scientific rationality? If so, a faulty inductivist method has been employed since it fails to consider the many case studies that do not conform to SP. From a statistical point of view, Barnes' and Bloor's procedure is not sound since no randomly chosen set of historical case studies is used; they restrict themselves only to those studies which have come out of the SP camp. If inductivism is to fall within the scope of SP then how are we to consider inference to the next case? In the doctrine of meaning finitism one of the factors constraining choice of the next case will be the pressure of what others say about the next case, particularly one's mates who are advocates of SP. So inductivism viewed as a case of meaning finitism does not have the rational punch that many of its philosophical advocates as a theory of method thought it should have (e.g., Reichenbach (1949), chapter 11 'Induction').

Inductive evidence based in case studies is fairly weak support for theses as strong as those found in SP. Moreover advocates of SP never, in the context of ordinary sciences, cite inductive evidential grounds as a cause of belief; to do this would be to go against the whole tenor of SP. So, why should SP, a special hypothesis within the sociology of science, be supported differently (using inductive methodology) from other hypotheses in science as far as the causes of its credibility are concerned?

Let us return to CT as applied in the case of SP and explore the difficulties that face thorough-going reflexivity. SP tells us that any belief in science has its social causal conditions S_x (we can set aside the N_x factors in the following considerations). So for any belief p within any science, say in particle physics, we have:

(1) S_x causes xBp

This is no longer a claim in particle physics but an instance of CT, and so a claim within sociology, specifically SP. On the face of it, and in accordance with Bloor's scientism, instances of CT will be as objective a causal claim as any other causal claim in science. Like any such claim (1) is open to test by, say, Mill's methods or some other theory for testing causal claims. Thus we may say that (1) is an objective testable causal claim.

However there is a duality to (1). Not only is it an objective testable causal claim, but it is also a belief within the sociology of science and like any such belief it, too, will have its social causal conditions of credence. That is, for some sociologist of science y (who may or may not be the same as x) there are social conditions S_y which cause y to believe (1):

(2) S_y causes $yB[S_x$ causes $xBp]$.

This is now a "higher order" causal claim about what causes sociologists of science to believe some "lower level" causal claim. Paralleling the argument above we may also show that (2) is an objective testable causal claim. But, again, it is also a belief within SP and must have social conditions of belief in accordance with CT. That is, some sociologist z (who may or may not be the same as x, or y) is also caused to believe (2) by something about z's social circumstance S_z:

(3) S_z causes $zB[(2)]$.

We now have an infinite regress for the causes of belief. While this does not obviously entail relativism it might be thought to entail something akin to it – but this is not so. What the above argument shows is: either SP stops any threatening regress and yields at some point an objective testable causal claim; or there is an infinite regress for causes of belief within the sociology of science. Since there are no grounds for thinking that SP, and CT in particular, are restricted in any way and that there are beliefs outside the scope of application of CT, then CT is committed to an infinite regress of causes of believings. This is an unpalatable consequence. Sociologists of science can never rest content with any belief unless they uncover its causal conditions; so they are forever restlessly entertaining ever-new beliefs because of the infinite regress of causes of belief.

Initially we chose p as a belief within particle physics. But we could have chosen any belief within sociology or, particularly the sociology of science;

or we could have started with thesis CT(SP). In the case of belief in SP itself, we have the following sequence of causes:

(1') S_x causes $xB[SP]$
(2') S_y causes $yB(S_x$ causes $xB[SP])$
(3') S_z causes $zB[S_y$ causes $yB(S_x$ causes $xB[SP])]$
        etc.

We arrive at the same pair of alternatives as before: either the sequence of claims within (a broadly construed) SP stop at some objective testable causal claim about the causes of belief in SP; or there is an infinite regress of belief within SP, so understood.

How might a sociologist of science stop the regress? Stopping at (3) or (3'), or (2) or (2'), seems to lack any sufficient reason apart from avoiding the regress. Stopping at (1) or (1') might be given an independent rationale. Both (1) and (1') lack iteration in the causes of belief and express causal conditions of belief in what one might call first-order sciences, e.g., particle physics, chemistry, biology, ... and finally SP. This, perhaps, is the limited but legitimate scope of application of CT that has been obscured by its over-general formulation. CT is to investigate the causal conditions of belief in first-order sciences only. However this suggestion remains obscure when applied to sociological investigations into the causes of belief. There is no problem where the sociological theories under consideration have as their subject matter, say, class structures or the sociology of the family. But where the sociological theories under consideration have as their subject matter people's beliefs about class structures or the sociology of the family, matters are not so clear. And they become quite muddied where the sociological theories under consideration concern beliefs about beliefs within sociology itself, particularly SP. This is one point at which the above difficulties about the infinite regress can arise. There seems to be no clear way of stopping the regress; restricting the scope of application of CT to first-order sciences seems to leave out a number of beliefs relevant to any sociology of belief.

Nothing yet has been said about the truth or falsity of claims (1) or (1'). But these are the general form of the very claims made in studies such as Forman's investigations into belief in acausality on the part of physicists of Weimar Germany, or Bloor's own study of the belief in passive matter on the part of Robert Boyle. But as was seen, these two studies do not establish their causal claims. So retreat to (1) or (1') might lead us to unsubstantiated causal claims, or even false causal claims. Thus the problem posed by reflexivity is that either there is an infinite regress of beliefs, or if one stops at the lowest level of claim, (1) or (1'), then a good many of these are either false, or unsubstantiated claims, about causation.

6.5   RELATIVISM AND THE STRONG PROGRAMME

On the face of it CT, the central tenet of SP, is thoroughly realist. It concerns a causal connection between, on the one hand x's social (and non-social) circumstance (or interests in this), and on the other x's belief that p. And these causal claims can be tested for their truth or falsity. Though SP prescinds from any involvement with whatever truth-value p has, it does lay claim to the truth of the host of causal claims which instantiate CT. Even though one might speak of the 'social construction of belief', the instances of CT are about a real causal relation, and its causal claims have a truth value.

Bloor's commitment to a version of relativism emerges when he says with Barnes:

> The simple starting-point of relativist doctrines is (i) the observation that beliefs on a certain topic vary, and (ii) the conviction that which of those beliefs is found in a given context depends on, or is relative to, the circumstances of the users. (Hollis and Lukes (eds.) (1982), p. 22)

But (i) by itself is neither necessary nor sufficient for relativism. Varying beliefs on a topic is not sufficient for relativism since the undeniable fact that beliefs vary on a topic can be embraced by both relativists and non-relativists alike within the sociology of belief. Nor is it necessary. If perchance we were all to believe the same thing on a topic then a sociologist of belief would still have the task of discovering why we all believed the same thing and not some imagined alternative. Variety in belief may be an impetus to sociological theorising, but it is not necessary for it.

Clause (ii) is simply a version of CT, expressing the causal dependence of x's believing p upon some social circumstance, S_x, of x. Since causation is relational, it is easy to fall into saying that what x believes is relative, rather than related, to x's circumstances – and Barnes and Bloor do just this. But not everything relational necessarily involves a thesis of relativism. Within SP beliefs are pretty much the epiphenomenal froth on top of underlying social causes, and they come and go as the social causes change. As such, beliefs could be said to be *relative* to social circumstance. Once again, the relational character of causation is not a sufficient ground for any doctrine of relativism. Weak programmers will also admit that some believings have social causes but they are not thereby committed to any relativism other than that of the causal dependence of false or irrational belief on social circumstance.

So far, a relational doctrine like CT does not support a robust doctrine of relativism, except in the loose sense in which advocates of SP might say that x's belief that p is relative to x's social circumstance S_x. Barnes and Bloor continue to advocate relativism when they say: 'But there is always a third feature of relativism. It requires what may be called a 'symmetry' or an 'equivalence' postulate.' (*loc. cit.*) They make it clear that relativism in their sense does not involve a commitment to the claim that all beliefs p are somehow made true by a relativisation move of the form 'p is true-relative-to-x'. Rather it is a commitment to ST in which 'all beliefs are on a par with one another with respect to the causes of their credibility' – and this is so 'regardless of truth and falsity' (*ibid*, p. 23). What these remarks suggest is that the third sort of relativism is nothing other than the Symmetry Tenet ST. But as we have seen there is nothing obviously relativistic about the symmetry requirement. Merely suspending the epistemic evaluation of beliefs in explanations of why x believes p is not sufficient for a robust relativism. Nor is recommending just one explanatory model at the expense of all others necessarily relativistic. In the last section enough doubt was cast on the dogma of ST without also tarring it with the brush of relativism.

Barnes and Bloor also claim that they are committed to an epistemological relativism about reasons in science, and about the so-called 'observational basis' of science which they allege is theory-laden and not basic in any significant sense. The first kind of relativism will be left for the discussion of Bloor's meaning finitism in the next chapter. But perhaps a case can be made for a version of epistemological relativism concerning the theory-ladenness of observations and reports of them in contrast to more foundational views of our observational knowledge. However this is a topic which cannot be pursued here.[15]

Barnes and Bloor are also careful to point out that they are not relativists about truth either. But this may be a subterfuge because they are at pains to reject the claim that truth relativism is self-refuting, and appeal to an argument developed by Mary Hesse[16] which they think establishes this. She claims that a common form of argument against SP would go as follows:

> Let P be the proposition 'All criteria of truth are relative to a local culture; hence nothing can be known to be true except in senses of "knowledge" and "truth" that are also relative to that culture'. Now if P is asserted as true, it must itself be true only in the sense of 'true' relative to a local culture (in this case ours). Hence there are no grounds for asserting P (or, incidentally, for asserting its contrary). (Hesse (1980), p. 42)

For a start it is not obvious what 'P' stands for. Is it the whole sentence, in which case 'P' is an argument? Or is it the first clause up to the semi-colon? The latter will be assumed here. Nor is it obvious what P has to do with any of the tenets of SP, especially CT which makes no mention of truth at all. Perhaps it is the very criteria of truth themselves that are the beliefs which are to be substituted in CT (understood very broadly in which there are no restrictions on the range of 'p'). This suggestion yields as an instance of CT (where 'TT' stands for a theory, i.e., criterion, of truth such as the correspondence, coherence or any other theory of truth): for all persons x of a given community and the theory of truth TT believed in that community, there are social conditions S_x which cause belief in TT, i.e.,

CT[TT]: S_x causes $xB(TT)$

Because of the particular case of the causal dependence of belief on social context alleged here, we may speak of the relativity of x's belief in certain criteria of truth to x's social circumstance, i.e., we may speak of the relativity of truth to social circumstance or culture. CT(TT) understood in this way does yield something close to the first clause of Hesse's proposition P. Thus Barnes and Bloor can not avoid truth relativism, as applied to our very criteria of truth, as they had hoped since it flows from CT understood quite generally. The only way to avoid commitment to the above relativity of our criteria of truth is to restrict CT.

CT(TT) is not a claim about what truths there are; it is a claim about our very theories or criteria of truth. That this is Hesse's view becomes clear in her attempt to show that P is not self-refuting. She continues the above quotation by saying:

> This easy self-refutation is fallacious, for it depends on an equivocation in the cognitive terminology 'knowledge', 'truth' and 'grounds'. If a redefinition of cognitive terminology as relative to a local culture is presupposed in asserting P, then P must also be judged according to this redefinition. That is to say, it is fallacious to ask for 'grounds' for P in some absolute sense: if P is asserted, it is asserted relative to the truth criteria of a local culture, and if that culture is one in which the strong thesis [i.e., SP] is accepted, then P is true relative to that culture. (*loc. Cit.*)

Not only is our theory of truth (TT) to be relativised to social context, but also the very notions of knowledge and grounds for knowledge. The alleged relativism is quite thoroughgoing; it applies to all our epistemic concepts. Moreover it is a relativism to which Barnes and Bloor are committed in virtue of a broadly understood CT. However for Hesse the above argument is not conclusive for accepting SP. What she suggests the argument does is to invite us to '*shift* our concept of "knowledge" so that the alleged refutation becomes an equivocation. This shift is the essence of the strong thesis [viz., SP]; knowledge is now taken to be what is accepted as such in our culture' (*loc. Cit.*) But if this unacceptable consequence arises from the above considerations then there is something seriously amiss. So what is wrong with Hesse's defence of SP?

Let 'p' be some proposition which x holds to be true. But it cannot be true *simpliciter*. It must be a theory of truth that prevails in x's social circumstance. To indicate this let us write not 'true' but 'true-in-S_x', or alternatively 'true-relative-to-S_x'. The hyphenated expression captures the idea that it is not truth *simpliciter* that is under consideration but rather the theory of truth that prevails in x's social circumstance, S_x. Finally let us abbreviate as follows (where 'A_S' is subscripted to the particular social circumstance S in which person x exists):

(A_S): p is true-relative-to-S_x.

Now we have all the ingredients of the classic refutation on Plato's *Theaetetus* of Protagorean relativism as set out in Burnyeat (1976). What Burnyeat argues is that while p may be true in a relative manner, the whole expression A_S is not; that is the whole expression 'p is true-relative-to-S_x' is itself not a truth relative to anything, i.e., an absolute truth. And this truth can be grasped by anyone irrespective of their cultural and social circumstance S* with its own criterion of truth which can be different from that which prevails in S.

There is another way to see what is wrong with Hesse's argument against the self-refuting character of relativist notions of truth. Consider any other person y in different social circumstances $S*_y$ who does not believe p, i.e.,

(B_{S*}): p is false- [not-p is true-] relative-to-$S*_y$.

Now consider the particular case in which for y 'p' is (A_S). Does y hold (A_S) to be a truth about x's system of beliefs or not? Suppose the latter. Then in (B_{S*}) substitute (A_S):

(B_{S*}/A_S): A_S is false-relative-to-$S*_y$; or more fully

(B_{S*}/A_S): {p is true-relative-to-S_x} is false-relative-to- $S*_y$.

If x's truth is not the same as y's truth about what x holds true, then what x holds as his or her system of beliefs becomes opaque and unknowable to y. Conversely does x hold (B_{S*}) to be a truth about what is in y's system of beliefs? If not, then y's system of beliefs becomes opaque and unknowable to x. Thus claims like (A_S) and (B_{S*}) are truths which transcend one's circumscribed set of socially given beliefs. Otherwise neither x nor y can hold correct beliefs about what the other holds as true for each in their own social circumstance. Each is locked into their own set of beliefs relative to themselves, including beliefs about what the other holds as true for themselves, and can never get an objective grip on what are the other's relative truths. Consequently neither can state the doctrine of relativism as something that is generally applicable, or even assert it as applying in their own case. The total opacity of one social group to another is the consequence of a thoroughgoing relativism.

Finally, Burnyeat's point comes to haunt those who would adopt any of (A_S), (B_{S*}) or (B_{S*}/A_S). These are not relative truths. They are all truths that are asserted in an absolute sense. Otherwise there would be an infinite iteration of the truth-relativised predicate. In the light of this, Hesse has not correctly asserted what the self-refutation argument comes to (as set out by Burnyeat). And her argument for its fallacious character misses the vital points made in the previous paragraph and Burnyeat's points about making assertions. In sum, despite Barnes and Bloor's advocacy of a number of doctrines which they call 'relativism' (as set out at the beginning of this section), the symmetry requirement ST is not obviously relativistic. Barnes and Bloor also deny that they are committed to truth relativism by adopting

Hesse's arguments against it. But Hesse's argument is not satisfactory, as was argued above. However there is a commitment to truth-relativism which arises from the broad version of CT in which even our very theory of truth is held relative to social circumstance. Change the social circumstance and the theory of truth to be adopted will, most likely, be different.

NOTES

[1] A similar title is used in Niiniluoto (1999) as the heading for section 9.1 of his book. I wish to acknowledge my borrowing of this apt title.

[2] For applications see, for example, Hawthorn (1991) who takes up the Lewis causal explanatory model and applies it to a number of episodes in past and more recent history, to social science issues and to the understanding of painting.

[3] Many of the wide variety of accounts of explanation adopted in SSK can be found in Lynch (1993), his preferred model being that taken from ethnomethodology. But in so far as ethnomethodology eschews explanations and adopts a more descriptivist account, it is hard to see what it can tell us about the explanation of scientific belief, the very thing that Bloor's SP sets out as its "object" of explanation. But then we have already noted the great variety of explanatory "objects" that are invoked when one speaks vaguely and uncritically of 'explaining science'.

[4] For somewhat differing definitions of 'ethnomethodology' see, for a start, Garfinkel (1967) pp. vii-viii and p. 11. The best overall account of ethnomethodological approaches in SS can be found in Lynch (1993), chapters 1 and 7. Amusingly Lynch tells us of the reception of Garfinkel's various views that 'ethnomethodology has been sustained through communal misreadings of its central text; a virtual consensus constituted by deep misunderstandings of a common set of slogans' (Lynch (1993), p. 13). There is a field day to be had by the reflexive application of ethnomethodology to its own practitioners. For Bloor's criticism that ethnomethodology is allegedly involved in a deep contradiction if it attempts to align itself with Wittgenstein's philosophy, see Bloor's paper 'Left and Right Wittgensteinians' pp. 266-82, a response to a paper by Lynch 'Extending Wittgenstein' pp. 215-65 and followed by Lynch's reply, all in Pickering (ed.) (1992).

[5] The logical relation in practical reasoning is discussed in the various papers in Davidson (1980). See also Lennon (1990) chapter 1. Even though Lennon talks of conceptual links in many cases, this does support the view that the link of 'giving rational support to' can also be inductive in character. For a clear account of practical reasoning which holds that the reasoning schema can be deductive, see von Wright (1983) chapters 1 and 2. Nothing in the text turns on controversies about the nature of practical reasoning. Acts of believing are merely understood to be just one kind of act that follow, in some sense, from the premises of a bit of practical reasoning. And these problems become less urgent when the practical inference schema is understood to be an explanatory schema.

[6] See Davidson (1980), pp. 264-5. What is claimed here is that the act of salt throwing might be performed but not caused by the reasons given, even though the reasons are co-present. In the case of the spilled salt it is conceivable that even though one has gone through the little piece of practical reasoning, this does not cause the act of throwing the salt; what in fact causes it is, say, an involuntary twitch over which the person has no control.

[7] Such a model is akin to the common Humean model of the explanation of actions in terms of desires and belief. And it is the model advocated by Davidson (1980) chapters 1 and 11 to

14, who also argues that reasons are causes of action. Davidson also shows how such a model, which concerns particular beliefs, aims and actions, is reconcilable with the view that there are causal and/or law-like connections at work but that these need not, and do not, feature in the belief-desire explanation given for action. Turning to more complex models, there are many works on Bayesian decision theory which set out models concerning what action we should perform, and in the narrower case on which the above focuses, what act of belief we ought to have; see for example Jeffrey (1983).

8   The schema can be expanded to include degrees of belief in rivals and a choice between three or more theories. Or in the situation in which only one theory A is available, the schema can be altered to deliver either belief in A, disbelief in A, suspension of belief in A, or rational belief in A to a given degree, when methodology M is employed.

9   The kinds of explanatory models that Lakatos advocates for internal and external science (see Lakatos (1978), pp. 118-121), have been criticised in Currie (1980) and Musgrave (1999), chapter 13, section 4. Currie argues for explanations along the line of the rationality* model and against the simple rationality model for explaining scientific belief. The latter model provides what Lakatos calls a 'rational reconstruction' of science in which the explanations do not accord with the actual thought processes of scientists. Lakatos is also left with a problem of accounting for how his rationality explanations might be applied to scientists who have never entertained any of the matters contain in their *explanans*. Thus Lakatos is forced to make such odd remarks as the following; 'The Proutian programme is not Prout's programme' (Lakatos (1978), p. 119). That is, the nineteenth century chemist Prout did not entertain either the rational reconstruction of his theory that Lakatos provides; nor did he entertain the methods of appraisal that Lakatos' methodology advocates and which are embodied in the Lakatosian rationality model. Such models can deviate considerably from actual history while the explanations provided by the rationality* model do not. But in other cases they might, as suggested in the text above, be in accord as when the actual thought processes of the scientist follow exactly the principles of some methodology.

10   On Popper on his corroboration measure and his claim that it is a report of past and not future success in passing tests, see Popper (1972) chapter 1 sections 8 and 9. In the reprints since 1978 of Popper (1972) can be found '*Appendix 2: Supplementary Remarks (1978)*, pp. 363-75, which continues these themes in replies to some critics of his claim to have disposed of the problem of induction and that corroboration is non-inductive. Whatever the outcome of the debate over corroboration versus induction, both Popper and his critics will allow that methodologies can pick hypotheses that are unsuccessful in the sense that either they have predictions that fail in the future or fail future tests, even though they have been successful up until now in either manner.

11   See Will (1993), chapter 5, for a non-technical account which also discusses the challenge that the Brans-Dicke theory might have brought to Einstein's theory by accounting for 3 of the 43 arcseconds per century. This would have rendered Einstein's theory less successful in accounting for all of the discrepancy between what is observed of the precession of the perihelion of Mercury and what Newtonian theory can account for. However the estimates of the solar oblateness of the Sun which would have accounted for the 3 arcseconds per century are now believed to be too small to have a significant effect.

12   Though both Bloor and Latour adopt a symmetry tenet they mean quite different things by it, as is evident in the exchange between Bloor and Latour which begins with Bloor (1999); in the same volume there is a rejoinder by Latour, pp. 113-29 and a reply by Bloor, pp. 131-6. Bloor's position remains clear (though in my view wrong) while Latour's remains obscure.

13   See Pais (1982), p. 20. Also see pp. 253-6 for an account of other effects upon Einstein when the agreement of his GTR with hitherto unexplained astronomical observations became evident to him. As Pais also points out, it was important for Einstein that GTR alone, without

the need for any other special hypothesis, could explain the anomaly. Hitherto physicists had tried to explain the anomaly by appeal to special hypotheses, such as the existence of the intramercurial planet Vulcan. Here there is appeal to another methodological principle about the unity of theoretical explanations, and/or to a principle of simplicity.

[14]   See Laudan (1981) and (1982); these are reprinted in Laudan (1996), chapter 10.

[15]   In Dretske (1969) chapter V, a case is made for a version of perceptual relativity even though he does not endorse other claims about theory-ladenness. See also Shapere (1982) section IV.

[16]   See Barnes and Bloor, in Hollis and Lukes (eds.) (1982), pp. 22-3, for their rejection of truth relativism and footnote 6 on p. 23 for their claim that Mary Hesse has 'thoroughly discussed and thoroughly demolished' the claim that relativism is self-refuting.

CHAPTER 7

THE WITTGENSTEIN CONNECTION:
THE SOCIAL AND THE RATIONAL

In chapter 5 it was noted that the Causality Tenet, CT, is quite general; it applies to any beliefs whatever, such as the causes of belief in the Strong Programme (SP) itself (section 6.4) and the causes of belief in theories of truth (section 6.5). In this chapter we will look at what the general CT has to say about the causes of our belief in principles of inference and the conclusions we draw on their basis. Important here is the influence of Wittgenstein on SP. Advocates of SP claim that logical relations are really social relations of constraint. This is not the truism that social relations of constraint often *accompany* logical relations (as when we correct one another), but a much stronger claim: they allegedly *constitute* logical relations. If this is correct, it is a surprising revision of our notion of rationality and what underpins it. Sociologists claim that support for this view comes from Wittgenstein; in contrast it will be argued there is little support for it in Wittgenstein (given rival interpretations). Barnes and Bloor tell us that there are no universal or absolute justificatory inferences: '... justifications will stop at some principle or alleged matter of fact that only has local credibility ... For the relativist there is no sense attached to the idea that some standards or beliefs are really rational as distinct from merely locally accepted as such' (Hollis and Lukes (eds.) (1982), p. 27). Several consequences of the view that rationality is to be downplayed in favour of mere local acceptance will be examined in this chapter.

The chapter begins with a discussion of whether there are counterexamples to SP based in ordinary inference making (section 7.1). The next section examines some of the rival logical systems to classical logic and asks what explains the adoption of one system rather than another. For SP the differences must be found in social factors or interests. Interests there are. But as the case of C. I. Lewis shows, his objections to Russell's notion of material implication and his investigation into other kinds of implication such as strict implication, reveal that his interests are *intellectual* and *philosophical* and not socio-politico-cultural. The remaining sections consider the influence of Wittgenstein on SP. Section 7.3 investigates the claim of SP that Wittgenstein's account of the 'hardness of the logical *must'* shows that logical relations are really social relations of constraint. This highly implausible doctrine not only ignores the role of semantic model-theoretic considerations that underpin our notion of valid inference, but also turns on a misleading interpretation of some remarks of Wittgenstein. So one attack on traditional ideas of rationality by sociologists is shown not to succeed. The next section investigates Bloor's distinction between inferential practices, codifications and interests in the light of some of Wittgenstein's other comments on practices and form of life. But none of this comes anywhere near showing that logical relations are really social relations.

Section 7.5 takes a broader view and contrasts the naturalistic social causal theory of SP with Wittgenstein's own rejection of explanations in philosophy. In so far as SP is a broad doctrine that covers all our beliefs, then the underlying project of SP differs markedly from Wittgenstein's more quietistic approach in which the task of philosophy is to 'leave everything as it is' and merely to describe our grammar, an activity that will hopefully free us from the bewitchment of language. Does an appeal to Wittgenstein's account of rule following show that social relations underpin logical relations and rationality more generally? This question, explored in section 7.6, is one of the main themes of SP and is here answered negatively. Advocates of SP take seriously scepticism about rule following of the sort developed by Kripke, and adopt a communitarian approach to the solution of the sceptical problem. However it is argued that the communitarian approach is flawed since it must help itself to an account of the meaning of sameness, contrary to what rule scepticism claims, in order to account for how judgements of the community are to be determined to be concordant, i.e., the same. Other respects in which the communitarian response fails are also mentioned. Section 7.7 goes ontological; it investigates the extreme nominalism associated with SP and considers what account SP can give of kinds as opposed to the individuals that make up kinds. It can in fact give no account of kinds independent of our classificatory activities. The final section re-emphasises the loss of normativity entailed by a central aspect of SP, viz., any explanation of our beliefs must be in terms naturalistic items such as social causes and not in terms of norms.

7.1   ORDINARY INFERENCE AS INDIVIDUAL CAPACITY OR SOCIAL RELATION? A REFUTATION OF THE CAUSALITY TENET

Not only do humans reason, but animals also. The Ancient Greek logician Chrysippus is reported to have noticed that a dog, when chasing an animal down a path which divided into three, sniffed down two of the paths without picking up the animal scent, and then ran down the third path without stopping to test it by sniffing. The dog, says Chrysippus, used a version of Disjunctive Syllogism which says: either the quarry went down the first, the second or the third path; but not the first; and not the second; so, the third.[1] Bloor is also aware of Chrysippus' dog and says of its inferential abilities: 'Since Wittgenstein makes appeal to our instincts this would seem to give the disjunctive syllogism a place in his theory' (Bloor (1983), p. 127). From whence the instinctual ability to make such inferences? Dogs are social animals but none of their socialisation appears to prepare them for inference making. They do not learn inferences from their peers; nor are they corrected by them when they go wrong. Nor does the dog community have rules that comprise a dog 'form of life' for inference making. The best hypothesis appears to be that inference making is an individual capacity possessed by each dog to some small degree. And we might suppose that evolution has played a role in producing each animal with the ability to make such inferences. Those who have the capacity to make inferences have a greater chance of survival than others who do not make them. The time Chrysippus' dog might have taken to test the third path by sniffing around might be just the time needed for the quarry to escape. Inference making enhances survival in contexts such as these. Of course dogs need only enough inferring abilities to get by and, like us, may either fail to make inferences in some cases or make the wrong ones in others, as the literature in cognitive psychology amply shows in the case of humans.

Primates and *Homo Sapiens* bear enough evolutionary similarity to dogs; so we can form the unproblematic conjecture that each of us humans has the innate capacity (Wittgenstein's instinct?) to makes inferences, however imperfectly. Since there are no social factors involved here, then the employment of our innate cognitive capacity to form new beliefs which are the conclusions drawn from other beliefs as premises, would appear to be a counterexample to the Causality Tenet CT, the central tenet of SP. Let 'p' be a belief we form as a result of reasoning from other beliefs as premises. Then CT at least says: always there is a conjunction $(S_x$ & $N_x)$ which causes x's belief that p. Now we strike an unresolved matter in the formulation of CT. Does the requirement that S_x and N_x be conjoined mean that they must always be combined *at the same time*, so that for any N_x occurring at time t, there is always an accompanying S_x occurring at the same time t? If so, that would rule out the possibility that some N_x occurring at time t *without* any conjoined S_x occurring at the same time t could, by itself, cause x's belief that p. Understood this way CT does not allow purely non-social causes of belief.

But just this possibility emerged in section 5.9 where causation was understood as a chain of events linked by relations of counterfactual dependence, and an effect was the terminus of a structure of so related events with a longish history. In the case under consideration the terminal effect would be x's belief that p and the long historical structure of events leading up to it would include both social and non-social events. This view of causation does not require, as SP does, that every non-social factor occurring at some time in the historical causal structure be conjoined to a social factor occurring at the same time. In addition it was envisaged that a causal sub-structure leading to some effect be sensitive and thus unstable, or insensitive and thus stable, with respect to other events occurring within the sub-structure. Thus in the firing of a projectile which hits a target, the firing and the projectile hitting the target would be relatively insensitive (though the final event is not always unaffected by) intervening events such as a strong crosswind. But with a sufficiently strong wind as an intervening event, the structure might be sufficiently sensitive to the crosswind so that the effect, hitting the target, does not occur. Again, x's coming to believe that p is sensitive to the many intervening causes that lead back to x's being born (or to the Big Bang, or x's breathing air). If x had not been born (or if the Big Bang, or x's breathing of air, had not occurred), then x would not believe p (at later time t). Thus being born (or the Big Bang, or x's breathing air) is *a* cause of x's believing that p. But causal structures linking these events are highly sensitive to intervening causes. Matters need be only every so slightly different in the history of the world and the effect, x's believing that p, would not occur.

In order to construct a counterexample to CT, expressed strictly as above, we need only find some effect, x's belief that p, which is the terminus of a causal sub-structure which is stable and contains only non-social factors. Such would appear to be the case for Chrysippus' dog. Supposing dogs do have beliefs about the quarry they chase, then that the quarry has gone down the third path is a belief inferred from other beliefs such as: 'it went down one of the three paths but it did not go down the first or the second'. Such beliefs as premises come directly from the perceptual capacities of the dog. The dog detects by sight the division of the path into three other paths; so we may attribute to the dog, on the basis of such visual discrimination, the belief that the path divides into three paths. And it discovers by sniffing the absence of the quarry's scent on the first path, and on the second path; so we may attribute to the dog, on the basis of olfactory discrimination, the beliefs that the quarry did not go down the first path, and did not go down the second. Given these directly perceptually acquired beliefs, and a capacity to make inferences of the type Disjunctive Syllogism (we need not suppose that the dog is logician enough to know Disjunctive Syllogism and its logical conditions of application), then we have a case of belief acquisition by inference that does not causally depend on any obvious social factor.

The causal structure leading to the effect, the dog's belief that the quarry went down the third path, contains the following (illustrated by something like a modularity theory about our cognitive functioning, though this is not the only account that could be used). First visual sensations are input into a "visual module" in the dog's brain that has as outputs beliefs about the visual scene. Similarly there are olfactory sensations as input into an "olfactory module" which has as outputs beliefs about the dog's olfactory surrounds. These in turn are inputs into what can be called an *inference processor* (the "instinctual" or innate inferential capacity) that has as output the belief that is the terminus of the causal structure, such as the quarry went down the third path. There are no evident social events occurring amongst all the cognitive events in the causal structure which begins with sensory input and, via sensorially produced beliefs, ends with an output of a further belief as conclusion.[2]

The same can equally be said of humans. We acquire new beliefs from premises often based on direct perceptual awareness of the world and the capacity to make inferences (either deductively or inductively) from them. And there need be nothing evidently social in the process. We too can infer, from seeing the tracks of our quarry on a muddy path that then divides into three, that if there are no tracks to be seen on two of the paths then the quarry went down the third. Here our new belief as conclusion is the terminus of a stable causal structure which begins with sensory input into perceptual modules in the brain that have as output beliefs about our visual situation. These in turn go through an inferential processor (this could be our "instinctual" or innate inferential capacity); and this processor, in making inferences, has as output the new belief as conclusion.[3] And there is no evident social factor in this causal structure. This, it will be argued, can still be the case even if the central processor earlier acquired socially (through learning) the ability to make inferences, and then much later exercised this ability in coming to believe that p.

The above are examples of relatively isolated stable causal structures in the brain which are linked to the external world through perceptual inputs at one end of the causal structure followed by perceptually acquired beliefs and, then, inferences from them. Conclusions drawn by our inference-making capacities are new beliefs that are the terminus of the causal structure. The events at the "nodal points" of the structure are all non-social cognitive events within our brains from sensory input to belief output. The absence of any social factor in the causal structure shows that there are causes of belief that have only non-social factors as causal events within the stable causal structure. And such examples, of which there are many, are counterexamples to CT with its strong conjunction requirement of social and non-social factors. Of course, social factors can lie outside such relatively stable causal structures. Also social factors can inter benignly even into the beliefs that are premises on the basis of which we make inferences. For example, one can acquire beliefs by a number of social processes, from learning from others to indoctrination. Or the beliefs themselves may be about social artefacts and thus might be claimed to carry a certain social content. In addition the "inferential processor" can, through a social process of learning and not just by instinct or evolution, acquire the capacity to reason. But one can still make inferences from these beliefs in a way in which no further social factor enters into the stable causal structure leading to new beliefs as conclusions.

To illustrate, let us suppose that a person learns at school the axioms of Euclidean Geometry from their mathematics teacher, and learns principles of inference and how to apply them from their logic teacher. Both the beliefs in geometry and the capacity to infer have been acquired socially (and if one has a social constructivist view of geometry the very content of the geometrical propositions can be understood to be about cultural artefacts). Suppose the beliefs and the capacity remain with the person even though their school career is prematurely cut short. Suppose also that in their middle age they renew their interest in studying geometry. Suppose, finally, that on one particular day they devote their time to solving some problems in geometry – and after much effort on that day prove for themselves Pythagoras' Theorem.[4] What counts as the cause of their newly acquired belief on that day? One could consider the much larger causal structure that begins with their school learning and culminates in their belief in the Theorem. But this is a relatively unstable causal structure; events in their later life might be such that they never find the time to devote to geometry, and so they never prove the Theorem (as is the case with most of us). That one gets through life with time and interest enough to think about geometry may depend on a number of quite miscellaneous factors which come together in the causal history for only a few. But the structure is unstable because, if even a few of the miscellaneous factors were not present, or other miscellaneous factors intervene, then one would never have found the time to do more geometry; and so no theorem would have been proved.

A much more stable historical causal structure is that of the day in which the person thought again about the axioms, applied themselves to reasoning about geometry – and then proved the theorem! It is possible for this stable causal structure to contain no social factor. This is not to say that social factors are irrelevant in the story. Rather it is to say that a stable causal sub-structure may exist which one can point to as "the" cause of one's newly acquired belief; and this contains no social factor. But some larger less stable historical causal structure may well contain important social factors. These may well be quite broad social factors that lead to the proving of the Theorem *via* a quite unstable causal structure that contains many miscellaneous, but nonetheless efficacious, events that might easily not have been present. Sure, if they had not been acculturated into geometry in their childhood then the person would not have proved the Theorem. But when focusing on causes the more salient ones are those that bring about their effects through relatively stable causal structures. In addition, even if belief in the premises is socially caused, the social features do not get transmitted down a chain of events in such causal structures. If both belief in the premises of an argument, and belief in a pattern of inference, are socially caused then it does not follow necessarily that any conclusion drawn from the beliefs in accordance with the inference pattern is also socially caused.

For much of the story told so far, the capacity to make inferences can be instinctual, or one bequeathed to us in the course of our evolutionary history. This fits with Quine's account of the evidence relation as a chapter of psychology (see section 3.1). Both Quine's position and the story given above are thoroughly naturalistic. But they hardly preserve the normativity of reason; for on such a naturalism there is no account of the correctness or validity of the inferences, or any reason as to why one *ought* to infer that way, even though one in fact does infer in that way. And unfortunately there is no room left for the empirical possibility that cognitive psychology investigates, viz., how we do deviate from the norms of reason in our actual inference-making. Advocates of SP would also find too limited the account of reasoning just given. Only a psychological, or instinctual or evolutionary story is told of the naturalistic factors which determine how we reason. They also want a large role for a social story to be told as well. But even adding in claims from empirical sociology makes no difference to the problem of normativity.

In emphasising the role of sociology here, Barnes and Bloor reiterate CT when they say: 'The position we shall defend is that the incidence of all beliefs without exception calls for empirical investigation and must be accounted for by finding the specific, local causes of this credibility' (Hollis and Lukes (eds.) (1982), p. 23). Then they add that the sociologist in searching for the causes of credibility must ask:

> if a belief is part of the routine cognitive and technical competences handed down from generation to generation. Is it enjoined by the authorities of the society? Is it transmitted by established institutions of socialization or supported by accepted agencies of social control? Is it bound up with patterns of vested interest? Does it have a role in furthering shared goals, whether political or technical, or both?' (*loc. cit.*).

Just as in the case of theories of psychology, instinct or evolution, sociological investigations of the sort just mentioned can tell us nothing about the correctness or validity of any inferential belief. This is a point the sociologists also go on to make when they add comments about the role of the Impartiality and Symmetry Tenets: 'All of these questions can, and should, be answered without regard to the status of the belief as it is judged and evaluated by the sociologists own standards' (*loc. cit.*). Or anyone else's standards for that matter, including logicians. The point being made here is that our inferential beliefs, and know how, can be acquired by any of the above mentioned socio-politico-cultural means, just as it could be acquired instinctually or by evolutionary processes. That is, we can grant that there is a story to be told in accordance with the tenets of SP of how our reasoning practices sometime get transmitted in society according to the types of cause listed in section 5.2.1. And this is something that anyone opposed to SP on other grounds can grant, just as it was granted in section 5.4.1 that belief in the Copernican and Ptolemaic cosmologies could be social transmitted.

But none of this bears on the following two matters. First, even if our capacity to infer might have developed socially rather than by evolution or instinct, once one has acquired the capacity to infer, its exercise on some much later occasion does not entail that the conclusion believed is socially caused. The person who proved Pythagoras' Theorem might still have their belief at the terminus of a stable insensitive casual structure that contains no social cause. The exercised inferential capacity still remains cognitive, even if the capacity was set up much earlier by a social process. Second, none of this bears on the question of how the correctness or validity of our principles of inference, or of method or reason, is to be established. Is there a non-social story to be told about whether or not some principle of reason R *is correct (valid)*, or some methodological principle M *ought to be followed* (as opposed to *believed* to be correct)? Or can a social story be told here as well? Here we need to avoid the act/content confusion. What we ought to be concerned with here are acts of belief with contents such as: R *is correct*; M *ought to be followed*; and so on. These acts of belief with these contents are also grist for the mill of SP; and sometimes the application of SP might well be correct; but this cannot always be the case by a long shot.

But what of the *contents* of these beliefs? Many philosophers of a rationalistic persuasion have often assumed that an *a priori* story can be told of the justification of these belief contents. Other philosophers have sought a naturalistic account of the grounds of such claims that preserves their normativity and correctness. But the radical turn of advocates of SP is that no such philosophical stories can be told. Rather, any philosophical story that could be told here is to be replaced by a social story. And to show this they adopt a social interpretation of Wittgenstein's account of rule following (the subject of section 7.6). Thus for radical advocates of SP there are social stories to be told all the way down the line and, according to the doctrine of meaning finitism, into the very content of the beliefs themselves.

In opting for a social rather than a philosophical story, some have seen in this an abandonment of any hope of establishing the rationality of our norms of reason and method. In contrast, advocates of SP see this not as abandoning rationality but giving, as they claim, a new kind of account of the grounds on which we should accept principles of rationality. The change is in the account of rationality, but not rationality itself. But such is the difference in the kind of account of rationality given that opponents of SP claim that sociologists have simply changed the subject matter and are talking of something other than rationality; they are not giving a new account of the same old subject, rationality. This last alternative will be argued for here. Bloor appears to opt for something like the former when he says: 'If what he [Wittgenstein] says is true, or anywhere near the truth, the great categories of objectivity and rationality cannot look the same again. ... A social theory of knowledge changes all of this. Objectivity and rationality must be things that we forge for ourselves as we construct a form of collective life' (Bloor (1983), pp. 2-3). But before these issues are engaged, it will be useful to look at the way in which advocates of SP approach the matter of alternative logics.

7.2   THE STRONG PROGRAMME AND THE CAUSES OF BELIEF IN ALTERNATIVE LOGICS

Since advocates of SSK and SP take their task to be the explanation of why people believe what they do, and why there is variability between people concerning their beliefs in some domain, then the explanation of why there are rival systems of logic believed by different philosophers appears to fall within the sociologists' province. And if there is to be accord with SP, any explanation must make reference to socio-politico-cultural circumstance, or an interest in this. But those who hold to classical logic, or one or other of its rivals, cite other grounds for their belief. Their interests are philosophical (or cognitive, epistemic, or intellectual) in character and not socio-politico-cultural. It will be argued in this section that SP, understood strictly according to its letter, fails to give an explanation of why, for many logicians, there are differing beliefs about inference.

7.2.1 Some Rival Logics

The fact that there are logics which rival classical two-valued logic is now commonplace. Many modal, tense, deontic and other logics are not rivals to classical logic because they only add special axioms of modality, tense, deontology, etc. to the axioms of classical logic. But some logics drop some feature of classical logic; in so doing they can be called 'rivals'. Thus there are many-valued logics (which have three or more truth values), intuitionistic logic (which at least has no law of excluded middle), so-called quantum logics which lack associative principles, paraconsistent logics (which have contradictions which are true and which do not contain the principle of Explosion, to be mentioned later); and so on. Of interest here will be the large number of relevance (or relevant) logics that drop a number of other "classical" principles. Also of interest are some modal systems that replace material implication by strict implication.

When Russell and Whitehead produced their axiomatic version of propositional logic, they established what C. I. Lewis regarded as two rather 'startling theorems', also known as 'paradoxes of implication'. Russell and Whitehead used the symbol '⊃' to stand for implication, thereby starting a debate, initiated by the criticisms of Lewis, as to whether or not that symbol in their system really did capture the notion of implication. In fact these results had been noted by medieval logicians; but because there was no linking tradition we can regard the debate as having started anew. The counterintuitive theorems involving implication are as follows.

(i) First "paradox of implication": if not-p, then if p then q. This says: that p is false implies that (p implies anything q). Lewis reads this as 'a false proposition implies any proposition' (Lewis, (1912), p. 522).

(ii) Second "paradox of implication": if p, then if q then p. This says: that p is true implies that (anything q implies p). Lewis reads this as 'a true proposition is implied by any proposition' (*loc. cit.*).

Other seemingly "paradoxical" results also include:

(iii) (p and not-p) implies q. That is, a contradiction implies anything, q. This is also known as the 'Principle of Explosion' because of its "explosive" infinity of consequences.

(iv) q implies (p or not-p). That is, anything, q, implies a tautology.

Lewis and Langford provided a number of proofs of these seemingly startling results in order to show that they turned on well-accepted principles of inference. For example, the proof of (iii) (following Anderson and Belnap (1975), pp. 164-5) can be given using the following rules of inference:

Simplification (i):   from A and B, infer A.
Simplification (ii):   from A and B, infer B.
Addition: from A, infer A or B
Disjunctive Syllogism (DS):   from (A or B) and not-A, infer B.

As a sample, the proof of (iii) is as follows:

(1) Suppose the impossible premise: p and not-p
(2) {(1),Simplification (i)}   p
(3) {(1), Simplification (ii)}   not-p
(4) {(2), Addition}   p or q
(5) {(4), (3), DS}   q   [arbitrary conclusion: QED]

The significance of this little proof is that if one is to give up Explosion, viz., a contradiction implies anything, then a further principle must also be abandoned, either Simplification, Addition or Disjunctive Syllogism (or two of these, or all of them). Earlier relevance logics dropped Disjunctive Syllogism (DS). Why? It was not denied that the inference of 'q' from the contradiction 'p and not-p' is valid in some sense, the sense being that given in classical logic in which inferences are guaranteed to be truth-preserving. For some this is enough. But for others more is required; the conclusion is meant to be *relevant* to the premises. The relevance requirement is like a "work ethic" requirement in that each premise earns its pay by doing some work in bringing about the conclusion; idle bystanders doing no work, get no pay, and are irrelevancies to be avoided. So where are the irrelevantly idle in the above?

In their discussion of the above argument, Anderson and Belnap ((1975), pp. 164-6) focus on DS. Following an argument akin to that in Lewis (1912), they argue that DS should hold only when the 'or' in 'A or B' is intensional and there is some relevance between the two disjuncts. But when the 'or' is extensional, as it is in the Russell and Whitehead system, then the disjuncts can be irrelevant to one another; so, DS ought not to hold. The difference can be illustrated by saying that, in the intentional 'or' in 'A or B', one or the other is the case and there is no other alternative. This is obviously the case in 'either I will stay in or go out' and in 'either it is a bacterial or a viral infection' (supposing that these are the only types of infection). This is not so for the extensional 'or' as in 'either Napoleon was born in Corsica or grass is green'. Moreover, there is an ambiguity in the above argument because Addition holds for both intensional and extensional 'or' while (granting the above) DS holds only for the intensional 'or'; one of these must go on pain of ambiguity. On such grounds of relevance Anderson and Belnap abandoned DS in their new relevant logical calculi. But DS still remains classically valid; but it is not relevantly valid.

Other criteria of relevance are more quantitatively assessable. Thus in the formula 'A implies B', B can be said to be relevant to A in varying degrees such as (a) they share at least one propositional variable, (b) they share all variables, (c) they share at least one variable and one connective, (d) they share at least one variable and all connectives, and so on. Thus *sociative* relevance logics require that A and B share variables, thereby forming a sub-class of relevant logics in which Addition, DS and some other classical principles are not found. Such requirements clearly try to enforce the work ethic for various components that make up the antecedent premises A and consequent conclusion B of the inferences we make. Other relevant logics focus on principles other than DS. Connexive logics are suspicious of Simplification (what work does 'B' do in the inference from 'A and B' to 'A'?) and they may or may not accept Adjunction (given A, and given B, infer (A and B)). And they might also be suspicious of Addition (considering 'from A infer A or B', doesn't B have an air of irrelevance to the premise?). Dropping these, and other such principles, gives a range of relevance logics to investigate. Once relevance logics had been sufficiently explored it was realised that even Aristotle's theory of the syllogism is highly relevant in character. This also proves to be the case for much later Hellenistic logic, and also medieval logic. To relevance theorists the aberration in the history of the logic of 'implies' is the system of Russell and Whitehead which is a very late upstart on the scene, many features of which were well known and criticised by earlier logicians.[5] And this is marked by calling the Russell and Whitehead connective '⊃' 'material implication', thus leaving other senses of 'implication' and 'entailment' to investigation within other systems.

7.2.2 *Logical Intuitions, Logical Codifications and Interests*

The above is a very brief and uncontroversial account of a few aspects of recent developments of logic that is accepted by both Bloor and his critics.[6] Where they differ is over what to make of the differences between classical and non-classical (especially relevance) logics. In order to discuss these matters Bloor proposes a useful three-part model illustrated below.

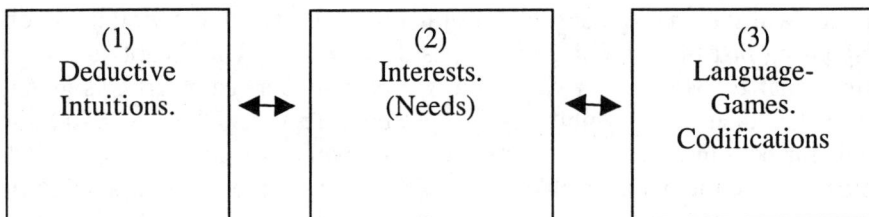

(1) Deductive Intuitions.	⬌	(2) Interests. (Needs)	⬌	(3) Language-Games. Codifications

Box (1) labelled 'deductive intuitions' concerns our belief that our particular inferential practices are correct. Box (3) called, following Wittgenstein, 'languages games', concerns the codification of systems of reasoning. In between is box (2) labelled 'interests' or 'needs' (Bloor, (1983), chapter 6.5, p. 135). (1) and (3) are admitted by most logicians who investigate the connections between, on the one hand, our logical intuitions and, on the other, our logical principles codified according to some theory of syntax and their presentation in axiomatic systems (and one should add their associated models and semantics). Bloor expresses the difference in the following way: 'Box (1) is non-verbal, preverbal; box (3) is verbal' (*op. cit.*, p. 136). More specifically:

> The investigation of box (1) belongs to the psychologist. It refers to our natural reasoning propensities and the patterns that are built up by contact with simple features of the empirical world; .... any claims we make about them [i.e., intuitions in (1)], or any verbal principles that we formulate in order to capture them, belong in another place in the scheme. We are now in the realm of language games, and these I have put in box (3). (*op. cit.*, pp. 135-6)

This much is familiar, being in large part, the contrast between inferential intuitions and their codification in deductive systems.

The important feature that Bloor adds, but which most logicians pass over, is (2), viz., 'the interests that relate to the exploitation of these two things [(1) and (3)]' (*op. cit.*, p. 133). Building on this theme, Bloor claims that 'the structure of the debate over the Lewis theorem [i.e., the proof, given above, that from a contradiction anything follows] is, in fact, a social one' (*op. cit.*, p. 134). And this culminates in the characterisation of (2) as follows:

> This is where the items referred to by Wittgenstein as 'needs' enter the scene. They are represented in box (2). In the light of the discussion of the Lewis theorems, ... I

am right in grouping these under the heading of social interests. My claim is that a pursuit of the causes that make us deploy our intuitions one way rather than another, leads straight to social variables of this kind.' (*op. cit.*, p. 136)

Here no claims are being made about the worth or correctness of one logical system over another, such as the classical versus the many relevance systems. In line with the Impartiality and Symmetry Tenets these matters are being set aside. Rather, what is being addressed is why a person adopts one system rather than another. Bloor makes this important point in highlighting (2) in which interests and needs play a significant role. But the error in his position is that he talks only of social interests. Cannot other kinds of non-social philosophical or intellectual interests also play a role here? For advocates of SP these are played down completely. But as will be seen, even if some logicians might cite social interests as a reason for their belief in some system (and yet others will find such reasons unworthy), there is still a case to be made for non-social philosophical, cognitive or epistemic reasons. But first, to Bloor's case for the role of social interests.

In line with (1) and (3), philosophers can fall into one or other of two camps which might rival one another; call these, respectively, 'intuitionists' (not to be confused with Brouwerian intuitionism) and 'codifiers'. According to Bloor the interests of intuitionists might be such that they disvalue codification because they put emphasis on 'an anti-scientific spirit' in the attempt to 'minimise the achievements of formalisation' and 'to increase respect for other kinds of knowledge' (*op. cit.*, p. 133). These comments are in line with Wittgenstein's attack on Russell's conception of logic and its relation of mathematics scattered through *his Remarks on the Foundations of Mathematics*, one sample of which says: "'Mathematical logic' has completely deformed the thinking of mathematicians and of philosophers, by setting up a superficial interpretation of the forms of our everyday language …. Of course in this it has only continued to build on the Aristotelian logic' (Wittgenstein (1978), p. 300). For another, quite exaggerated, remark, note the following: 'There is no religious denomination in which the misuse of metaphysical expressions has been responsible for so much sin as it has in mathematics' (Wittgenstein (1980), p. 1). But on comparative amounts of sin in connection with religion and mathematics and their social role this is surely false. However it appears that strong views about mathematics such as these are said to lead one to be an advocate of intuitionism at the expense of codification.

Even those who are strong advocates of codification will find that, when they wish to adopt a new codification, they have to appeal to some intuitions, such as particular inferential practices in certain contexts, or to intuitions about the nuances of ordinary language (as do Lewis, and Anderson and Belnap, when they distinguish between kinds of 'or'). But this is said not to be without a social aspect. It leads to more work in formal logic since 'previous workers have not solved all the problems. There are still prizes to be won and reputations to be made' (Bloor (1983), p. 134). It is for reasons such as these, and others, that Bloor alleges that the debate over the Lewis theorem is a sociological one. But does a closer examination of the reasons some have had for entering into the "Lewis debate" support the case that only *social* interests prevail? Might not the above appeal to unsolved problems, or to the new problems that arise, be non-social philosophical grounds for entering into the debate? This will be argued next.

7.2.3 C. I. Lewis' Non-Social Interests in Rejecting Material Implication

What reason does Lewis offer for his criticisms of Russell's and Whitehead's attempt to axiomatically capture, and provide theorems concerning, our notion of implication? In part Lewis' objection is based on an appeal to intuitions about the meaning of 'implies' and his distinction between various kinds of disjunction (such as the intensional and extensional 'or') that could be used in the definition of 'implies'. But this is not mediated by any social interest that Lewis explicitly mentions; rather it is a scrutiny of meaning in relation to what an axiomatic system purports to be about. The logicians' interest in capturing the nuances of the meaning of 'or' hardly betrays a social interest (even though meanings might be said to be social). Rather it is a cognitive and intellectual interest in preserving important distinctions overlook in previous logical systems, despite the fact that there might be reputations to be made, and prizes to be won (or an interest in these). One may simply have an intellectual interest, as does Lewis, in exploring two logical systems, one with an extensional and one with an intensional 'or', and then investigating which captures best our pre-logical notion of implication.

Another of Lewis' objections is of a quite different character and concerns the failure of the logical system's 'implies' to be useable in the context of hypothetico-deductive confirmation. This would be a serious failure since the hypothetico-deductive methodology uses the notion of implication while being a cornerstone of much of science. Lewis' objection is:

> One of the important practical uses of implication is the testing of hypotheses whose truth or falsity is problematic. ... If the hypothesis happens to be false, it implies anything you please. If one finds facts x, y, z, otherwise unexpected but suggested by the hypothesis, the truth of these facts is implied by one's hypothesis, whether that hypothesis be true or not – since any true proposition is implied by all others. In other words, no proposition could be verified by its logical consequences. If the proposition be false, it has these "consequences" anyway. (Lewis, (1912), p. 529)

Lewis is raising a problem for our understanding of hypothetico-deductive methodology in the light of one of the "paradoxes of implication", that a false proposition implies any proposition. We need not enter into a debate about what Lewis says here; assume he is right to be concerned that true consequences can not "verify" a hypothesis since these consequences will follow from a false hypothesis anyway. Granted this, then either we should reject the claim that all aspects of implication have been captured in Russell's system concerning 'implies', or that hypothetico-deductivism is in serious trouble as a methodology in science which centrally employs the notion of implication (as the word 'deduction' suggests). The former has been the course most commonly taken with the discovery of the difference between material implication and other kinds of "implication" such as logical consequence. And what interests are served here? Not social but intellectual interests in resolving the difficulty for hypothetico-deductivism that Lewis raises.

Hypothetico-deductivism has other problems associated with it, such as that of irrelevant tacking on of other hypotheses to the one under test. This is one reason that Bayesians' cite as a superior feature of Bayesian confirmation; it is not subject to the irrelevant conjunction problem (Earman (1992), p. 64). The problem is this. If H implies E, then E is commonly said to confirm H. But conjoin any irrelevant X to H and still (H and X) implies E; and so E confirms (H and X). Thus X free-rides for some confirmation. This turns on another classical principle: (H implies E) implies ((H and X) implies E). This is an instance of the principle of monotonicity in which the conclusions of H remain unaffected by the addition of any further premises such as X to H. Dropping such a principle and adopting non-monotonicity, in which adding premises does make a difference to what can be inferred, has become one of the cornerstones of work in Artificial Intelligence and in Default Logics under the heading of 'non-monotonic logic'. Some practitioners might abandon hypothetico-deductive confirmation for Bayesianism, or investigate non-monotonic rather than monotonic logics, for Bloorian socio-political reasons including the reason that 'there are still prizes to be won and reputations to be made'. But such quite external reasons appear lame, irrelevant and even dishonest to those who believe what they do because of the problems that can be solved, this being a time-honoured methodological principle of choice between rival point so view, including logical systems.[7]

In introducing a category of interests and needs (box (2)), Bloor claims that 'the structure of the debate over the Lewis theorem is, in fact, a sociological one' (Bloor, (1983), p. 134). But the social factors are not locatable on either Lewis' or Russell's side of the debate about implication. We have looked at an early consideration of Lewis'; let us now focus on a later view (also held by his co-author Langford). Lewis is famous for introducing the notion of strict implication as a rival to Russell's account of implication. This led to systems of strict implication which are now labelled within modal logic as systems S1 to S5 depending on the axioms adopted. But some of these systems, Lewis recognised, also had unintuitive consequences; they also give rise to paradoxes of strict implication which parallel the paradoxes of material implication.

Lewis and Langford devote a whole chapter of their joint book to the topic of the variety of implication relations (Lewis and Langford (1959), chapter VIII 'Implication and Deducibility'). Such a variety would not be evident to an intuitionist, a devotee of box (1) only. It becomes evident only when codifiers, devotees of box (3), do their syntactical and axiomatic work (with, of course, an eye on intuitions of box (1)). The authors tell us: 'Thus there are an indefinitely large number of 'logics', or possible canons of inference, every one of which is true throughout and states true laws of inference. If our *intellectual habits and interests* were slightly different, we might chose some other than the logic of traditional deduction to be our guide' (*ibid.*, p. 260, italics added). With reference to these different intellectual habits and interests, the authors go on to mention other possible logics which they might investigate. One such is the three-valued system of Lukasiewicz where the values are 'known to be true', 'known to be false' and 'doubtful'. Other logics in which they might have an interest are those that ignore the Law of Excluded Middle. And so on. But in no case is their appeal to social interests; rather there is instead an appeal to the 'intellectual habits and interests' of other systems of logic with their notion of implication.

Lewis distinguishes between those varieties of implication that turn on only the truth values of the premises and conclusion and those non-truth value implications that require more than this (such as his own systems of strict implication). In this connection it is worth citing part of a passage from Lewis and Langford in which they set out how intellectual interests, not social interests, play an important role here:

> The real defect which all truth-value logics have, in use, is pragmatic. ... What a proposition implies, in any truth-value meaning of the word, is different, if the proposition is true, from what it implies if false. And this dependence of what implies what upon truth or falsity defeats one of the most important *interests* which we have in our observation of logical relationships. This characteristic *interest* moves us to note, primarily if not exclusively, only those relations which remain invariant whether the terms of them are true or false. One notable example of the *interest* in question is the desire to be able to draw the consequence of an hypothesis whose truth or falsity is undetermined. (*ibid.*, p. 261; italics added)

We need not argue the merits of Lewis' case, except to note that it reiterates a point (mentioned above) that Lewis had made in 1912 about an *intellectual* interest, and not a social interest, in problems that Russellian implication raise, for example, for the hypothetico-deductive method, or the need for a notion of implication that is truth-value independent (as suggested above). In Lewis' own account of why he advocates his own theory of implication both intuitions and their connection with codifications play a role. And interests also play a role – but only intellectual interests are mentioned. Bloor's leap from talk of interests to sociological interests is just that; it is a leap over a chasm of interests that are not social at all.

7.3   IS THE HARDNESS OF THE LOGICAL *MUST* REALLY THE SOFTNESS OF A SOCIAL RELATION?

In the previous section we considered a number of different logical systems with different accounts of what 'implies' might mean. Setting aside the now obviously misnamed conditional of material implication, there is Tarski's model-theoretic notion of logical consequence (in which Ψ is a logical consequence of Φ just when Ψ is true in every model in which is Φ true).[8] And there are other attempts to capture the notion of implication through Lewis' notion of strict implication in various modal systems, the various notions of entailment in the various relevance logics, and the accounts of implication (and validity) in a host of other systems. Belief in any one of these, according to advocates of SP, is to be accounted for in terms of 'sociological variables' or interests in these. But as was argued, non-social interests which are philosophical, cognitive or epistemic in character, play a much more significant role in explaining why we believe (accept? work on?) one or other of the notions of implication embedded in their associated logical systems. In this section we will focus on other matters. These are not whether our belief in a given concept of implication (whatever it be) is socially caused, but whether the very concept of implication itself is social in some manner. And if it is social, we need to investigate what impact this has on the account of rationality of our norms of inference making.

In order to establish the social character of the concept of implication advocates of SP adopt, and adapt, many of the views of Wittgenstein on mathematics and rule following. Here the focus will not be so much on the texts of Wittgenstein of which there is copious rival commentary and interpretation. Rather, is will be on the understanding of his views as filtered through sociologists of scientific knowledge, such as Bloor and others, who advocate the doctrine of meaning finitism. In this section we will look at what Wittgenstein calls 'the hardness of the logical *must*'(Wittgenstein (1978), Part I §121). But both occurrences of 'the' are somewhat unfortunate in that there is no one logical must, as the above list of kinds of implication indicates; and so there is no one hardness to consider but rather a family of them of varying degrees of "hardness". For example, whatever logical hardness, or compulsion, Disjunctive Syllogism has for the classical logician, these are features the relevance logician does not acknowledge. Moreover, Wittgenstein's focus was largely on the axiomatic system of logic developed by Russell and Whitehead; his focus was not on the model theoretical approach being developed in the 1930s and 1940's, the time of his work on logic in his later philosophy. And he seems to have hardly focused on the contemporaneous work of Lewis on implication. Nor does Wittgenstein discuss the 'hardness of the logical *must'* in relation to probabilistic reason, Bayesian or otherwise; this is a matter that gets little attention in Wittgenstein's writings or those of his sociological followers.[9] However he could not have been aware of the flowering of rivals to classical logic in the second half of the twentieth century thereby bringing to logic a richness not known since medieval times. What reasons do advocates of SP have for thinking that logical relations are social relations?

Kusch is one advocate of SP who puts the sociality of our logical concepts quite bluntly. He says approvingly of arguments against a dualism of the rational and the social mounted by followers of Durkheim and Wittgenstein: 'Their strategy in this arena is to argue that rationality, meaning and objectivity are tied to normativity, and that normativity, right and wrong, demands consensus and thus social collectivity' (Kusch (1996), p. 86). If we include principles of implication under the heading of rationality then we can raise the *Euthyphro* question. Is a principle of implication rational (and thus normative) because the social consensus says so? Or is there social consensus that a principle of implication is rational because the principle is rational? The last supposes the independence of the rational from the social and allows, on the supposition that there is a community of *good* reasoners, that there can be social consensus about the rational because the collective comes to acknowledge its independent rational character. The former supposes no such independence. In this respect it differs little from Bloor's own social consensus definition of "knowledge" discussed in section 5.1.1 in which any individual difference from what the collective knows is idiosyncratic and merely "belief". Similarly, any individual difference from the collective determination of what is rational could be viewed as idiosyncratic and irrational.[10] Such a view makes the empirical investigation into the frequency with which people endorse good or bad implications difficult indeed, for collective endorsement can never be wrong.[11]

Kusch also claims 'that the rational supervenes on the social' and that this 'rules out the following possibility: it cannot happen that a scientific culture shifts from, say, antipsychologism to psychologism without a change in social institutions, social arrangements, or social interests' (*ibid.*, p. 92). As an historian of ideas, Kusch has investigated the acceptance, and then rejection, of psychologism in late nineteenth and early twentieth Western European society (especially Germany), and he has employed the theses of SP to this end. But his explication of his claim does not show that it is the *rationality* of some principle of implication that is supervenient on the social. Rather what is supervenient on the social is *some social group's acceptance* of some doctrine, that of psychologism, and later *their acceptance* of a rival doctrine, that of antipsychologism. What is required for the supervenience of the rational on the social is that some very principle of implication R (say) supervene on the social. To show this it would not be enough to show that *some group's acceptance of such a principles of implication R* supervenes on the social. In section 3.8 it was argued that our principles of method and reason do supervene. But the supervenience basis was not social; rather it was the descriptive properties of our beliefs upon which the normativity of principles supervened.

However Kusch's claim does show that SP does make a supervenience claim, of sorts, through its Causality Tenet CT which says: for all x and p (including principle R as an instance), $(N_x \& S_x)$ cause xBp. This might be rendered alternatively as follows in the case of principle R: there is no change in x's belief that R unless there is change in the supervenience base $(N_x \& S_x)$. But note that what supervenes is x's *belief* that R, and not R itself. Thus R and its rationality are not the supervenient items when SP is construed this way. And the supervenience base, *pace* Kusch, is not just the social but the combination $(N_x \& S_x)$. Kusch's formulation would rule out rationality in animals while Bloor's would permit it (with suitably "bracketed off" social factors).

Bloor's own view of implication and other logical relations is this: 'What, in the realm of language and ideas, we refer to as logical relations, and logical constraints, are really the constraints imposed upon us by other people. Logical necessity is a moral and social relation. Wittgenstein is quite explicit on the point' (Bloor, (1983), p. 121). We will return shortly to the grounds on which Bloor thinks that Wittgenstein is explicit on this point. In claiming that logical necessity is a moral relation, we can assume that it is at least normative in that we *ought* to follow any rules of logical necessity. And one might perhaps add that there is an ethic of correct inference making, and of belief, of the sort Descartes set out in the fourth of his *Meditations* concerning error and our wills. But this is an "add on" rather than part of what logical necessity might be.

What is of central concern is the claim 'logical necessity *is* a social relation (of constraints imposed by others)'. The word 'is' does not appear to be the 'is' of predication; rather it is the 'is' of identity, a reading made more clear by Bloor's use of the words 'are really'. This is a far cry from what logicians would say, for example, that logical necessity is to be cashed out semantically in terms of truth in all possible worlds, implication relations being a sub-class of logical necessities. If we grant that the 'is' of identity is the correct reading, it is unclear what sort of identity claim is intended. Is it some kind of Kripkean *a posteriori* necessary identity? Just as we discover empirically the necessity that water is collection of H_2O molecules, so in parallel we discover, through empirical sociological investigation, the necessary identity that the logical is the social. Alternatively is it a reductive relation? Or some kind of supervenience relation?

Which of these is intended is unclear. If it is not some 'is' of identity that is intended, then we are left with a rather weak claim that logical relations are often accompanied by relations of social constraint, a claim to which we need not object. Alternatively the claim might not be one of identity since talk of 'really are' can plausibly be understood as eliminativist. We might have thought that there were really such things as logical relations; but there are no such things. Logical relations go the way of phlogiston, Zeus and Santa Claus. All that really exists are the social relations of constraint between people. Talk of the existence of logical relations sound too Platonistic; and Platonism, according to most commentators on Wittgenstein, is to be rejected. Why rejected in the case of Bloor is not always clear; but one reason might be the fear, set out in section 3.4.2, of non-naturalistic items intruding upon the natural. So there are in reality no logical relations, but there are social relations of constraint that play a similar role.

Whatever account of identity we attribute to Bloor, what reasons does he give for his view? He continues the above, interspersing his comments with remarks from Wittgenstein:

> Of course the laws of logic can be said to compel us, he says, 'in the same sense, that is to say, as other laws in society'. A clerk who made idiosyncratic calculations and inferences would be punished. If we don't draw the same conclusions as other people, then we get into conflict, 'e.g., with society, and also with other practical consequences' (*ibid.*, pp. 121-2, with quotations from Wittgenstein (1978), Part I, §116)

Shapin and Schaffer also sloganise the cited remark of Wittgenstein under the heading of chapter IV of their book (Shapin and Schaffer (1985), p. 111); '... the laws of inference can be said to compel us; in the same sense, that is to say, as other laws in society'. But in an excellent critique of the reading of Wittgenstein by advocates of SP as an overly scientistic supporter of their sociological naturalism, Friedman gets us to rethink what is actually being said here (Friedman (1998), p. 259 footnote 45). The hapless clerk referred to is told, according to a regulation, to allocate people to certain places according to height. One clerk reads out the men's names and heights. Another clerk allocates them to sections, but has to do this by making an inference from the information supplied by the first clerk. As a sample inference we are given the following (see Wittgenstein (1978), Part I §17):

> For example: a regulation says 'all who are taller than five foot six inches are to join the ... section'. A clerk reads out the men's names and heights. Another allots them to such-and-such sections. – " NN is five foot nine". "So, NN to the ... section". That is inference.

This is Wittgenstein's sample inference, though it is enthymatic. The first clerk reads out the height of NN saying "NN is five foot nine". The second clerk has to supply some missing mathematical premises in order to infer to

"NN is taller than 5 foot 6 inches". Given this interim conclusion as a further premise, and the regulation, the second clerk can now infer "So, NN to the ... section". Later Wittgenstein adds to this story (in Part I §116) the following: 'The clerk who infers as in (17) *must* do it like that; he would be punished if he inferred differently. If you draw different conclusions you do indeed get into conflict, e.g., with society; and also with other practical consequences'. (Why punished remains unclear, and contrasts with Socrates' more gentle dialectical questioning of the Slave Boy to correct his errors.)

Does this story show that the logical relation involved in the inference is really a social relation? The inference considered by itself can be drawn independently of other aspects of the story. But then the added aspects tell us that the clerk is performing the task of allocation as the result of a regulation; and the regulatory powers (let us suppose) require correct allocation to section on pain of punishment. But none of the added story about the regulatory powers adds to the inference itself; it merely tells about one way in which the clerk can fail to meet the requirements of the regulatory powers. He can do this by making incorrect inferences. Or he could do this by mishearing what the other clerk said. But nothing in this supports the strong claim that the inference relation is really one of social constraints upon the clerk. At best they are consequences visited upon the clerk as a contingency of his situation.

We could well envisage the clerk making similar inferences in the course of a party game, or as a way of filling in time. In this case inferences can be correctly made in contexts that have no sanctions imposed for getting the conclusion wrong; or the only sanction is the ribbing the clerk might get for making mistakes that his more rational companions recognise. Alternatively the clerk could play the game by himself and allocate tin soldiers to places on a board according to height – and then, perhaps, by sight see that he might have some in the wrong place, in which case he would check not just his measurements and placings but also his inferences. Unless he was excessively puritanical, the clerk would hardly punish himself for error; but he would check himself. The rule of implication underlying the inference made is one thing; the sanctions of the regulatory powers are another. And the later do not constitute the former. But they may accompany the former in some cases.

A question can be raised here as to whether Wittgenstein is talking about inference making, as when we pass from one proposition we believe to another, and then come to believe the second proposition; or whether he is talking about a logical relation between propositions. Most logic textbooks draw a distinction between inference making, which they often characterise as psychologistic, and logical relations that are not psychologistic at all. Wittgenstein makes it clear in one passage (*ibid.*, Part I, §8) that he is talking of inference-making in the psychologistic sense, and that he is not talking of any logical relation. If this is the case, then Bloor has mistaken the nature of the topic under discussion by Wittgenstein in these passages. It is not a logical relation, but rather a psychologistic relation of inference, that is up for consideration. If this is so, then neither of these can be said to be social relations at all.

The same passages that Bloor cites from Wittgenstein are cited to the same end by him in a later book. Of Wittgenstein's talk of punishing the clerk, Bloor says not only of the clerk's inference but more generally of rule following:

> The argument so far may be summarised like this: in following a rule we move automatically from case to case, guided by our instinctive (but socially educated) sense of 'sameness'. Such a sense does not itself suffice to create a standard of right and wrong. It is necessary to introduce a sociological element into the account to explain normativity. Normative standards come from the consensus generated by a number of interacting rule followers, and it is maintained by collective monitoring, controlling and sanctioning their individual tendencies. Consensus makes norms objective, that is, a source of external and impersonal constraint on the individual. It gives substance to the distinction between rule followers thinking they have got it right, and their really having got it right. (Bloor (1997), p. 17).

But the last phrase 'really having got it right' cannot mean 'really' but rather only 'coming into accord with the consensus'. Here the *Euthyphro* problem raises its head again. We may say either of the following: x (e.g., the clerk's inference) is really right because there is consensus about x; or, there is consensus about x because x is really right. Clearly Bloor opts for the former. And in so doing Bloor opts for an account of objectivity and rationality that is merely that due to social consensus, and nothing more. In so opting many would say that Bloor is not merely giving a new account of our old notion of objectivity and rationality; he is really changing the subject and talking about something else, viz., what we collectively endorse.

Finally we are to see in this the claim that inferences really are social constraints, such as those upon the clerk. And these constraints are not merely contingencies of the social circumstances of the clerk; they are allegedly constitutive of his very activity of inferring. And this latter strongly social reductivist account of logical relations is got by appeal to, and interpretation of, the texts of Wittgenstein as authoritative on the matter. Setting aside the appeal to authority, can the interpretation be sustained? So far, 'no' as suggested above. So what diagnosis can be offered?

What has gone wrong here for those like Bloor, Shapin and Schaffer who not only see in this example the claim that the laws of inference can compel in the same sense as laws of society, but also claim that the laws of inference are really just social relations? In part, as Friedman suggests, it is an over-hasty generalisation from the particular example of the clerk. But the fault can also lie with Wittgenstein's comparison of the compulsion of the laws of logic with the laws of human society. He offers no argument for this and merely claims that the comparison holds. But there are differences. The laws of society are backed by a police, judiciary and a regime of punishments according to the kind of law breaking. If the laws of inference compel us, then it is not exactly in the same way. There are no court actions, visitations from the inference police or regimes of punishment for making faulty inferences. The analogy breaks down in this respect. But of course there are consequences for us if we infer as others do not, but not those such as a visit from the inference police or a court trial. Others might sanction us in various ways, from joking to avoidance. But who is to say that the others with whom we come into conflict have right on their side? We might all draw wrong inferences, as many do in the case of the Gambler's Fallacy, and the host of other common fallacies uncovered in investigations into our cognitive psychology. The few can infer (rightly) P while the many infer (wrongly) Q. Once more social constraints are no guarantee of rightness of inference.

These matters aside, what is not part of Wittgenstein's view is that, because the laws of inference do compel us in allegedly the same way as laws of society do, the laws of inference really are social relations (just as the laws of society are). Contrary to Bloor, Wittgenstein is *not* explicit on the point that logical necessity is a social relation. He is silent on it. He is silent, one can charitably suppose, because he does not make the *non-sequitur* that Bloor appears to make in inferring as follows: laws which are logical compel, and laws which are social also compel (in the same way), so logical laws are really the same kind of law as social laws (though they will differ in content). It is hard to see how Bloor and his sociological followers can get to the strongly reductive conclusion that logical inferences are really social relations of constraint from what Wittgenstein says about similar compulsions unless some mistake like this has been made. But this is a diagnostic point only. We have no reasons in Wittgenstein, or from elsewhere, for the identity or eliminativist thesis: logical laws really are social relations of constraint. The two are distinct. Social constraints might accompany logical laws some of the time, but they do not constitute logical laws.

7.4 WITTGENSTEIN ON LOGICAL RELATIONS, PRACTICES, CODIFICATIONS AND FORM OF LIFE

What account of logical relations does Wittgenstein give? This is a large matter that we can only partially enter into here. Bloor, like other commentators, shows that Wittgenstein did not hold with the Platonist view that the correctness of implications lies in their being about some non-natural realm of normative fact. Nor are they somehow products of our understanding. Nor are they true in virtue of a theory of 'analytic validity', i.e., the theory which says that the logical terms 'and', 'or', 'if ... then', 'all', and so on, have a fixed antecedent meaning. On this last view the justification of our principles of implication is based in a doctrine about the meaning of logical terms, and the theory of meaning is itself meant to have some firm *a priori* foundation. But for Wittgenstein, not only is the correctness of our theory of implication up for question, but also the very theory of the meaning of logical terms on which it supposedly depends. There is no foundational justification for our laws of inference by an appeal to the meaning of logical terms, though an interesting analytic link may be indicated here. Nor is a justification to be sought in the codification of our inferential practices as axiomatic systems (though it is unclear in Wittgenstein whether, along with this purely syntactical proof-theoretic objection, there is also intended a semantic model theoretic objection as well). As interesting as such codifications may be, they do not answer the justificatory question.

7.4.1 The Bed-Rock of Practice and Usage

Along with most other commentators, Bloor understands Wittgenstein to be adopting a version of conventionalism. This is not the conventionalism that might be found in the conventions of meaning, this in turn being understood, as above, in terms of some theory of analytic validity that appeals to the meaning (perhaps holistically understood) of the logical terms. According to advocates of SP, this is a violation of Wittgenstein's methodological prescription encapsulated in the remark 'the meaning of a word is its use'.[12] The meaning of expressions is to be accounted for by their use, and not use in terms of meaning. That is, Wittgensteinian conventionalism in which use is fundamental and meaning derivative is of a piece with his view that justification stops in our *practices*, including our linguistic and inferential[13] practices:

> If I have exhausted the justifications I have reached bedrock, and my spade is turned. Then I am inclined to say: "This is simply what I do". (Wittgenstein (1967), Part I §217)
>
> I *go through* the proof and then accept its result – I mean: this is simply what we *do*. This is use and custom among us, or a fact of our natural history. (Wittgenstein (1978), Part I §63).
>
> The danger here, I believe, is one of giving a justification of our procedure here [of checking a calculation several times] where there is no such thing as a justification · and we ought simply to have said: *that's how we do it.* (*ibid.*, Part III §74)

In respect of logic, logical form and inference making, the kind of justification envisaged here contrasts markedly with the view that Wittgenstein had held in the *Tractatus*. Of his former view Wittgenstein says 'no empirical cloudiness or uncertainty can be allowed to affect it – it must rather be of the purest crystal' (Wittgenstein (1967), Part I §97). Or 'the strict and clear rules of the logical structure of propositions appear to us as something in the background – hidden in the medium of the understanding' (*ibid.*, §102). Here what was formerly hidden in the medium of the understanding with the purity of a crystal, is replaced by the more earthy spade hitting a rock in the relative openness of practice. There are many issues to discuss concerning this appeal to practice. We will focus on only the following important Wittgensteinian point, viz., that justifications stop, and they stop in our practices and usages, and not in meanings or whatever else. This also suggests a link to the tripartite account of inference making of the previous section where intuitions, codifications and interests were distinguished. Now we can place inferential practices in this schema; they fall into box (1) labelled 'intuitions', this term in this context being understood broadly enough to include our inferential practice as well.

Concerning our inferential practices, there are presumably a potential infinity of token occurrences of such practices. These might be reduced somewhat by recognising that some *token* inferential practices are all of the same *type*, that is, there are potentially many instances of, say, the type of inference Disjunctive Syllogism. But in order to establish types of inference we must embark on the first fundamental steps towards codification. This involves the very important issue of how types of inferences are to be recognised by means of a theory of logical (syntactical) form. Let us, for the sake of the argument, simply suppose that we have a way of classifying our ordinary language remarks in one or other of the several recognised logical forms. For example, Wittgenstein's sample inference of section 7.3, setting aside its enthymatic character, is of the simple form: All ϕ are ψ; NN. is ϕ; so NN. is ψ.

Once types of inference have been identified, then it is possible to move on to the next step of codification, that of allocation to different axiomatic systems *via* their proof in that system. (Here semantic modelling also plays an important role; however we will consider these matters, as Wittgenstein and advocates of SP appear to do, only at the level of syntax and proof theory.) Thus the axiomatic system of Russell and Whitehead will capture many inferential forms. But it will not capture modal inferences (such as Lewis' systems of strict implication) since it does not make modal terms like 'necessary', contingent' etc. explicit in its syntax and axioms. It will also include some inference patters that intuitionistic logicians will not want to accept. It will also include the inference patters that Lewis found startling (see the previous section) and that relevance logicians would not want to accept. What this suggests is that our ordinary inferential practices are a motley that goes relatively unnoticed until codification takes place. But once codification does take place the differences in our inferential practices become evident. It turns out that there is not one codification of our practices but a family of codifications many of which rival one another (classical, relevance, intuitionistic, etc). In his later writings on logic Wittgenstein has some harsh things to say about the whole idea of seeking foundations in logic and mathematics in the sense of the programme of logicism announced by Russell. But what Wittgenstein overlooks, and perhaps could not have been aware (even though the Lewis systems were available to him), is that codification of our inferential practices has had one salutary effect. Our growing theories of logic have revealed that there can be rival codifications, and thus our token intuitive practices, in so far as they instance these rivals, are not unified in any way. The practices are a motley which admit no single but many rival codifications.

Thus consider the inferential practices of two people: C, whose inferential practices are classical, and R, whose inferential practices are those of relevance logic. When it comes to making an inference of the type Disjunctive Syllogism, C will give the Wittgensteinian response to a query from R about how his practices are to be justified: 'This is how I do it!' But R will object saying: 'No! That is not how it is done! *We* do not do that, and would never do it!' Again classical logicians will accept principles of inference involving the law of excluded middle and non-constructive proofs that and intuitionist would not.[14] Intuitionistic mathematics was in fact a going enterprise of practise even before its codification took place in intuitionistic logic. And its rejection of certain principles of inference was fairly explicit even though its founder, Brouwer, did not think that any codification would be adequate to his intentions. Here we have examples of clash of practice, perhaps embodying little understood rival intuitions, made clear only with some degree of codification, or with a specification of the intellectual purposes they serve (not necessarily social).

7.4.2 Insane Alternatives But Sane Rivals

What does Wittgenstein say about such clashes? In a number of places Wittgenstein gets us to think about what it would like to do things differently. There is a sequence of seemingly weird suggestions in *Remarks on the Foundations of Mathematics* which are linked to a quotation from Frege which Wittgenstein queries: '"... here we have a hitherto unknown kind of insanity" – but he [Frege] never said what this 'insanity' would be like' (Wittgenstein (1978), Part I §152). We might initially say, as Wittgenstein does: "The laws of logic are indeed the expression of 'thinking habits' but also the habit of *thinking*. That is to say they can be said to shew: how humans beings think, and also *what* human beings call 'thinking'" (*ibid.*, §131). But there is an air of contingency about our habits of thinking that is brought to the fore in a number of cases of practice which seem weirdly "insane" to us, but are nonetheless possible.

For example, there are those who might measure by shrinking rulers (*ibid.*, §140). There are those who might measure timber not in terms of weight or volume but area when spread out on the ground (*ibid.*, §148-50). There are those who might hand over coins in a purchase according to the amount that pleases the buyer (*ibid.*, §153). Finally Wittgenstein envisages a case where a person might draw, from the same premises, one conclusion on one day, another the next day, a third on the third day (or the same conclusion as on the first day), and so on. (*ibid.*, §155). After this last possibility he asks: 'Are our laws of inference eternal and immutable?' But in a way this is the wrong question to ask. Rather, cannot there be rival practices that we could indulge in? And this brings us closer to the matter of rival logics where the proper question is not to ask about whether there is just one logic that is eternal and immutable but whether there is a pluralism of logics, the differences between them having to do with matters of intellectual interest and purpose (not necessarily social interests as advocates of SP would insist).

What do all these seemingly 'insane' alternatives illustrate? There is a useful discussion of them by Stroud that bears on issues to do with inference. After discussing the above examples and others like them, Stroud says:

> The point of Wittgenstein's examples of people who do not "play our game" is only to show that our having the concepts and practices we have is dependent upon certain facts which might not have obtained. They show only that "the formation of concepts different from the usual ones" is intelligible to us; but it does not follow from this that those concepts themselves are intelligible to us. (Stroud (1966), p. 493)

We will return to the theme of the contingency of many of our practices; they are part of a human natural history that might have been different. At this point sociology might find a role for itself in at least describing the practices found in our actual human natural history. And if there is variability within actual human natural history, then a causal explanatory SP might have a further role in explaining such variability. But given our actual practices, there is a certain sense in which they are constitutive of what we understand by words such as 'counting', 'inferring', 'measuring' (as in the case of wood), 'purchasing', etc. Stroud's point is that we understand what it means to say that our practices might have been different and that we might use shrinking rulers, or measure wood by the ground area it occupies, or infer different conclusions on different days, or pay according to the pleasure of the buyer, and so on. But Stroud's extra point is that from this we cannot draw the conclusion that such ways of measuring, inferring, purchasing, etc are intelligible to us. We have given up on the, albeit contingent, practices that are constitutive of our concepts of inferring, measuring, purchasing, etc. What diminishes in intelligibility is that our concept of measurement allows that it might be done by using shrinking rulers, or by investigating the area wood occupies; or that inferring allows variable conclusions to be drawn. Here we are over the horizon of what is intelligible to us. We seem to employ our concepts but have abandoned some of their constitutive elements.

Granted this, just how close an analogy is there between these seemingly 'insane' cases and the notions of implication that differ from the classical notion of implication? There is no strong analogy at all. Even though different inferring practices are involved, it is not the case that the rival concepts of implication are unintelligible (but they might be highly contested). Thus even though Wittgenstein does allow that there might be different practices in a number of domains, only some of these become unintelligible to us; others, such as rival inferential practices, still remain intelligible rivals.

Wittgenstein says about logical inference (setting aside the over-psychologistic character of the first clause):

> ... so long as one thinks it can't be otherwise, one draws logical conclusions. This presumably means: so long *as such-and-such is not brought in question at all*.
> The steps which are not brought in question are logical inferences. But the reason why they are not brought in question is not that they 'certainly correspond to the truth' – or something of the sort – no, it is just this that is called 'thinking', 'speaking', 'inferring', 'arguing' (Wittgenstein (1978), Part I §156)

But it is a rather conservative response to claim that our actual practices in drawing inferences are somehow strongly linked to what we call 'inferring', etc. To take another leaf out of Wittgenstein's book it would be better to say that there are a family of practices, some of which rival one another, that are to be called by the generic term 'inferring' etc. The recent systematic investigation of logic has shown us that that there is a plurality of possible practices and codifications of inference; and none of these show that any of these logics are temporally fleeting and mutable despite their rivalry. Thus intuitionists have an intellectual purpose in adopting restricted modes of inference that a classical logician would not necessarily share; and relevance logicians have interests and purpose not shared by the classical logician. What needs to be bought out is the variety of such practices and the intellectual ends to which they may be used. Wittgenstein even envisages such rival practices himself when in the *Philosophical Investigations* he envisages kinds of negation other than that in classical logic, for example negation which does not iterate, or double negations which do not become affirmative but which increase the strength of the negative (Wittgenstein (1967), pp. 147-9). Such rivals to our classical practices are both intelligible, and some could even be given a formal representation. They do not strain at the horizon of the intelligibility of what we understand by negation.

Let us allow that for Wittgenstein there can be practices within inference making that rival those of our natural history, even though he says little about them. And in the little that is said, there is next to nothing that an advocate of SP can find that supports the idea that logical relations are social relations, or shows that differences due to rival practices arise from socio-politico-cultural causes. Wittgenstein also says little about why we might change our practices, a topic addressed in the next section. And, as will be argued in the next section, in so far as advocates of SP do attempt a scientific causal explanation, they depart from much of the Wittgenstein they would otherwise co-opt to their cause.

Feyerabend offers a few quite different suggestions about how we might criticise and change our methodological practices in science (see section 1.5.2). He draws a contrast between (i) our actual folk scientific practices in the case of scientists making methodological decisions about particular moves in the game of science, and (ii) the attempts of Popper, Lakatos and others to establish general binding principles of method, or of rationality (see section 3.6). Here the sphere of methodology parallels the sphere of inference making in which there are, on the one hand, our inferential practices and, on the other, our codifications of such practices in rules of syntax and axiomatic systems. Feyerabend offers criticisms of a one-sided devotion to either our codifications of rationality or to our practices, seeking what he ultimately calls a "dialectical interaction" between the two. Practices can fail for a number of reasons; they might be popular for not very good reasons, or they might deteriorate, or time worn old practices might be inadequate in new circumstances requiring new practices,[15] and so on. Codifications might also fail for a number of reasons: they might be so abstract that their conditions of application in particular cases are unclear; they may simply fail to apply in the real world; if they are taken as the dominant item in the practice/codification relationship they might suppress good practice, and so on (Feyerabend (1978), pp. 16-27).

These points can apply to the field of inference making as well. But other ways in which one might assess practices and their codification have already emerged. Thus intuitionistic logicians often make a good point to the effect that some existence proofs in mathematics are just that – they prove only the bare existence of some mathematical property or object. They do not offer a proof that shows how one might construct or find such an object or property. And if this is one's goal, then new modes of proof are called for. And again, relevance logicians might wish to add to the goal of truth preservation the further goal of relevance of conclusion to premises, thereby changing what is to count as an acceptable inferential practice or codification of practice (it might help us avoid some paradoxical inference procedure). In both cases it would be an exaggeration to infer from these differences that there is a different 'form of life' in which some mathematicians are indulging. But there is a significant difference in what is taken to be an inferential practice or rule.

We have briefly looked at a number of features of Wittgenstein's ideas on inferential practice and changes in such practice. As significant as these may be, none has any bearing on the claim of advocates of SP that logical relations are social relations. The descriptive enterprise of spelling out our practices, the enterprise of codification and the notion that there can be different practices and codifications (axiomatisations) are significant matters which a descriptive, but not a causal explanatory, SSK might take as one of its tasks. But such a descriptive enterprise can say nothing about matters to do with the correctness or validity of the inferences we make. No social story emerges that would support SP in this domain. Rather we need to fall back on what logicians tell us in semantics and model theory about what constitute notions of valid inference.

7.5   THE SCIENTISM OF THE STRONG PROGRAMME AND WITTGENSTEIN'S ANTI-SCIENTISM IN PHILOSOPHY[16]

As noted, Wittgenstein has little to say about how we change from one practice to another. Often practices and the rules which govern them are linked by him to form(s) of life. And with some changes in practice it is alleged that there must be a change in form of life. As he puts the matter:

> new types of language, new language-games, as we may say, come into existence, and other become obsolete and forgotten. (We can get a *rough picture* of this from the changes in mathematics.)
> Here the term "language-*game*" is meant to bring into prominence the fact that the *speaking* of a language is part of an activity, or form of life. (Wittgenstein (1967), Part I §23)

Within these activities we can include the practice of inferring. As Wittgenstein admits, these practices can come and go over time. But what he must also allow is that there can be co-existing rival inferential practices, or a plurality of inferential practices. Depending on how closely or loosely particular inferential practices are used to individuate the notoriously vague expression 'form of life', we can say either that there is a change in a form of life, or that the form of life remains the same despite some changes in inferential practice. Here we will opt for the latter, taking a leaf out of Wittgenstein's book in saying of those contexts where he talks of form of life, "we often do not know how the expression 'form of life' is being used here". It does seem exaggerated to say that because classical, intuitionistic and relevance logicians indulge in different inferential practices that each must have a different form of life. Rather, within the one form of life there can be plurality of rival practices. And such is the pluralism of the current scene within logic and the theory of inference (though not a pluralism readily acknowledged or indulged in by all).[17]

According to Wittgenstein our practices are part of a contingent natural human history which we could set about describing: 'What we are supplying are really remarks on the natural history of human beings; we are not contributing curiosities however, but observations which no one has doubted, but which have escaped remark only because they are always before our eyes' (Wittgenstein (1967), part I §415). Many of the examples of different practice that Wittgenstein cites, such as the "insane" ones of the previous section, are not part of our actual natural history but are imagined cases which could form a part of a possible human history. The point of the imagined cases is to illuminate, and make a stark contrast with, our actual practices that we do not very readily recognise.

7.5.1 Scientific Explanation versus the Primacy of the Description of Language Games

It is an important part of Wittgenstein's view of us that some of our practices and associated form(s) of life are detrimental to us, both intellectually and otherwise. Concerning this, philosophy might provide some therapy: 'Philosophy is a battle against the bewitchment of our intelligence by the means of language' (*ibid.*, §109); and 'What is your aim in philosophy? – To shew the fly the way out of the fly-bottle' (*ibid.*, §309); and like remarks. Suppose after much buzzing we do get out of the fly bottle. It remains unclear whether we merely abandon a bit of language that has bewitched us, or indulge in some new language game free from the earlier defects. Wittgenstein appears to adopt both views. The latter position emerges in the citation two paragraphs above from the *Philosophical Investigations* when Wittgenstein speaks of 'new language games coming into existence'. But the more quietist Wittgenstein emerges when he says: 'What we are destroying is nothing but houses of cards and we are clearing up the ground of language on which they stand' (*ibid.*, §118); and, 'The results of philosophy are the uncovering of one or another piece of plain nonsense and of bumps that the understanding has got by running its head up against the limits of understanding' (*ibid.*, §119); and finally the self-imposed question which is given no answer 'I must speak the language of every day ... *Then how is it that another one is to be constructed?*' (*ibid.*, §120). These remarks at best suggest a purgation of our old practices and not a replacement of them by new practices.

Whichever the case, it is part of Wittgenstein's pessimism that we might never make sufficient effort to free ourselves of bewitchment. And even if we could, according to him little, if anything, can be said about how changes in practices and language-games lead to a change in form of life, and what account we might give of these changes:

> Who knows the laws according to which society develops? I am quite sure they are a closed book even to the cleverest of men. If you fight, you fight. If you hope, you hope.
> You can fight, hope and even believe without believing *scientifically*. (Wittgenstein (1980), p. 60).

So we cannot expect from Wittgenstein any explanation, even a scientific explanation appealing to laws or causes, of why our practices and forms of life do change. As we will see, the best we can do is provide only descriptions, and not explanations, of our practice and associated form of life from within our prevailing form of life.

But some "clever" men have emerged amongst us who claim some insight into the laws governing social changes and what we believe, and who see it as their task to offer an explanation of such change. What they offer is, contrary to what Wittgenstein holds, a scientific account of why such changes occur. One such account, amongst several, is the naturalistic empirical causal science of sociology as set out in SP. Its whole task is to explain socially, and in accordance with the general character of a causal explanatory science, why we believe what we do and why beliefs change in the way they do. Here is one way in which the advocates of SP differ from Wittgenstein – they advocate a naturalistic science of belief and its change while for Wittgenstein this remains a closed book even to the cleverest.

Wittgenstein is well known for his rejection of the idea that philosophy is to be understood as on a par with an explanatory science:

> It is true that our considerations could not be scientific ones. ... We must do away with all *explanation*, and description alone must take its place. And this description gets its light, that is to say its purpose, from the philosophical problems. They are, of course, not empirical problems; they are solved, rather, through an insight into the working of our language ... *in despite of* an urge to misunderstand them. (Wittgenstein (1967), Part I §109)
>
> Philosophy may in no way interfere with the actual use of language; it can in the end only describe it. (*ibid.*, §124)
>
> Philosophy simply puts everything before us, and neither explains or deduces anything. – Since everything lies open to view there is nothing to explain (*ibid.*, §126)
>
> Our mistake is to look for an explanation where we ought to look at what happens as a 'proto-phenomenon'. That is, where we ought to have said: *this language game is played*. (*ibid.*, §654)
>
> The question is not one of explaining a language game by means of our experiences, but of noting a language game. ... Look on the language game as the *primary* thing. (*ibid.*, § 655-6).

Many other such remarks can be cited from Wittgenstein in support of his quietism. What they all show is his anti-scientism, his opposition to the view that philosophy is to be understood and practised along the lines of any science. It is also quite clear that in philosophy the explanatory task of science is to be eschewed in favour of the description of the workings of our language and the language games the embody. Importantly there is a *primacy* to language games such that we can only note and describe them. The primacy is quite strong in that there can be no further explanation as to why we adopt one game rather than another, or change from one game to another. Any such attempt would be wrongly called an 'explanation'; at best it is just another game, another 'proto-phenomenon', that we can play and which we can only note and describe.

Given the above, it is already clear that the agenda of SP is quite different from the agenda that Wittgenstein prescribes for philosophy. SP is to be understood as a naturalistic science with a large role assigned to sociology. SP prescribes in its central tenets not description but causal explanation. And finally, it offers nothing in the way of descriptions of the character of our language-games but, on the contrary, purports to offer much in the way of description of our socio-politico-cultural context (or our interest in this) and how this allegedly causally explains our beliefs, including our beliefs about language, logical inference, and the rest.

Bloor is aware of Wittgenstein's anti-scientism and counters by drawing a distinction between the domain of science where explanations can flourish and the domain of philosophy which, according to Wittgenstein, ought to offer no explanations but only descriptions: 'The 'disease' of 'wanting to explain' ... refers here to philosophical not scientific explanation' (Bloor (1997), p. 133). This might suffice if it were not for the generality of SP. The goal of SP is at least to explain (scientifically according to the tenets of SP) specific scientific beliefs. The social-cause models of explanation that accompany SP explicitly illustrate this. But SP's target *explananda* are not just scientific beliefs, but all beliefs including ordinary classificatory beliefs such as 'this is a swan', our philosophical beliefs about explanation itself, causality itself, truth itself (see section 6.5), our logical beliefs about inference itself, and the like. Given this very broad target of kinds of belief to be explained, SP strays from the territory of explaining only scientific beliefs into the territory of explaining beliefs which are about meaning or beliefs which are philosophical in character. In so straying SP comes into conflict with Wittgenstein's conception of the task of philosophy. For certain kinds of philosophical belief Wittgenstein is adamant that only a descriptive account can be given; no explanatory account, whether naturalistic-social-causal or otherwise, can be given. Given Bloor's own explanatory project he raises the question: 'Did he [Wittgenstein] not denounce the search for causes and the construction of explanatory theories?' And he replies on behalf of his own endeavours: 'I will be going against certain of Wittgenstein's stated preferences, his chosen method, and perhaps his deepest prejudices' (Bloor (1983), p. 5). It is important to have these significantly different agendas out in the open.

7.5.2 Wittgenstein and the Claims of Ethnomethodologists

Wittgenstein's remarks have their legion of interpreters.[18] For some there is the one true interpretation to be found and defended; for others, understood in certain ways they are a source for further development. The latter is often the case for advocates of SSK broadly understood. Both the causal explanatory programme of Bloor's SP within SSK, and Michael Lynch's descriptive ethnomethodology within SSK, pay allegiance to the authoritative character of Wittgenstein's writings – but they take his remarks in divergent directions. Lynch puts one of the differences thus: 'Rather than trying to explain a practice in terms of underlying dispositions, abstract norms, or interests [the first and the last being part of SP], a task for sociology would be to describe the ensemble of actions that constitute the practice. This is precisely what ethnomethodology seeks to do' (Pickering (1992), p. 290). In this respect the descriptive programme of ethnomethodology would appear to come closer to Wittgenstein than does the casual explanatory programme of SP. Here the sociological heirs to the subject which used to be called philosophy divide into their camps, including Strong Programmers and ethnomethodologists, who contest their respective Wittgensteinian inheritances.

In section 6.1.1 we encountered Garfinkel's ethnomethodology as a rival within SSK to the causal explanatory models of SP. In the introduction to a collection of his papers, Garfinkel tells us that his ethnomethodological studies seek to treat even 'the most commonplace activities of daily life' as 'topics for empirical study' (Garfinkel (1967), p. 1). But studied in what way? For Garfinkel they are at least to be 'account-able': 'When I speak of accountable my interests are directed to such matters as the following. I mean observable-and-reportable, i.e., available to members as situated practices of looking-and-telling' (*loc.cit.*). Is this observing and reporting, or looking and telling, the same as what Wittgenstein means by 'noting' our language games, or by 'describing' the natural history of our practices, as opposed to explaining them? Garfinkel hardly comments on the work of Wittgenstein; and in any case many of his comments on his own enterprise are obscure (for example, the copious use of dashes to no clear purpose). But the sociologist of science and advocate of ethnomethodology, Michael Lynch, does his best to take the fundamental ideas of ethnomethodology in the direction of the Wittgensteinian description of our natural history. [19]

Lynch regards Wittgenstein as a useful source for ethnomethodology, since, as he tells us, 'ethnomethodology has become an increasingly incoherent discipline despite incessant efforts by reviewers and textbook writers to define its theoretical and methodological program' (Pickering (1992), p. 232). Taking Lynch's remarks on the state of ethnomethodology as authoritative, how can Wittgensteinian philosophy boost its prospects? Lynch's contention that 'a task for sociology would be to describe the ensemble of actions that constitute practice' makes a direct link to the primacy Wittgenstein accords to the language games that comprise our natural history and the claim that they can only be noted or described and not explained. But as the examples of "insane" cases described in section 7.4.2 show, Wittgenstein also gives us an imaginary ethnography, and not just an ethnography of our actual practices. The point of the imaginary cases is often philosophical and conceptual, viz., to tell us, by way of contrast, something about the nature of our concepts tied to actual practices. So even if there is room for Lynchian ethnomethodologists to extend Wittgenstein's programme by giving a much fuller description of our actual practices, there is no similar role for ethnomethodology with respect to imagined practices.

Even though Lynchian ethnomethodologists reject explanation in favour of description, this only brings them a little closer to Wittgenstein's conception of the task of philosophy than advocates of SP; for what they count as description, and what is to be described, differs markedly from what Wittgenstein intends. Lynchian ethnomethodologists' descriptions of scientific activities[20] often has a different "object" from Wittgenstein's. An example to be discussed shortly concerns the activities of several astronomers on the night they discovered an optical pulsar. But the description of such broadly construed activities hardly focuses on the "object" the description of which Wittgenstein took to be central, viz., our grammar. Wittgenstein says of his philosophical task: 'Our investigation is therefore a grammatical one. Such an investigation sheds light on our problem by clearing misunderstandings away. Misunderstandings concerning the use of words ..." (Wittgenstein, (1967), Part I §90). And 'It [grammar] describes and in no way explains the use of signs' (*ibid.*, §496). What Wittgenstein means by grammar and the descriptive enterprise which has as its "object" grammar, is not a straightforward matter to grasp.[21] But we are familiar enough with the kind of practice that it involves within his philosophy. Some even see a larger philosophical enterprise in the description of our grammar. Thus Friedman ((1988), p. 261 footnote 47) compares Wittgenstein's grammatical investigations to Kant's transcendental/empirical distinction and makes the useful point that Lynch's ethnomethodology 'misses the Wittgensteinian connection, ..., between 'grammatical' explorations of the 'essential' features of our language and the exhibition, from within our language, of the normativity of such features' (*ibid.*, p. 266). Such is the case, for example when Wittgenstein uses his "insane" cases of an imaginary human natural history in order to display some feature of our current concepts.

So, how close does the Lynchian ethnomethodological idea of non-explanatory description come to Wittgensteinian description within philosophy? The answer is 'not close at all'; in fact their very enterprise presupposes something like Wittgenstein's descriptive grammar. This can be easily illustrated by a few examples from an ethnomethodological study of an episode within science, the paper by Garfinkel, Lynch and Livingstone (1981) 'The Work of a Discovering Science Constructed with Materials from the Optically Discovered Pulsar'. Once pulsars were postulated in 1967, the hunt was on amongst astronomers to optically identify the first pulsar. This was done by a group of astronomers on the night of January 16, 1969. A tape recording which had been left running that night recorded the astronomers' excited remarks. The tape and a log the astronomers made, constituted the materials for ethnomethodological analysis.

In the ethnomethodologists' 27 page paper, 15 pages are used to set out some of this material and the astronomers' final paper (with a little running commentary); the remaining 12 pages are devoted to their account. One might think that the material in the last 15 pages is sufficient for its own description; but it is not. What one finds in Garfinkel's, Lynch's and Livingstone's 12 page article (hereafter called 'GLL') is a substantial body of presuppositions bought to this material that is not descriptive of the situation but quite theoretical in character. In many cases the theory is highly contested, and the descriptive terminology is hardly illuminating. For example, there is an issue of individuating and describing what is to count as the astronomers' 'night's work' in their observatory, evidence for which appears on the tape and in the log. On this GLL say: "So we think of their night's work as this: their unavoidably 'situated' practices become progressively witnessable-and-discourse-able as 'the-exhibitable-astronnomical-analyzability-of-the-pulsar-again'" (GLL (1981), p. 135). This remark is peppered with Garfinkel dashes (as in Garfinkel's deliberate 'account-able' cited above). Even what is to count as "the night's work" is laden with the perspective of a Garfinkel ethnomethodology, the subject matter of which in this case has little to do with Wittgensteinian grammar.

Like all practitioners of SSK who eschew any realism, GLL set aside any matters to do with the actually existing pulsar that was optically identified: 'Soon after we examined the tape we decided to set aside as an inadequate rendering of their discovery the disengaged transcendental pulsar of a Galilean science' (*ibid.*, p. 136). The reason they give is dubbed 'the coroner's problem'. Coroners end up with corpses and not the life the body lived. Similarly in the case of the astronomers, talk of the actual optically identified pulsar is like being left with a corpse that excludes the astronomers' processes of search and discovery. We will return later to the obvious point that there are various ways in which one can describe the activity leading up to an alleged discovery without begging any questions in the description of that process as to the existence of what is discovered. Instead, let us focus on the way GLL go about describing the situation using terms that come from philosophical theories ('transcendental', 'Galilean', etc) which bear little relationship to pure ethnographic description. Moreover they say: 'We construed the IGP [independent Galilean pulsar] as a "cultural object" and in that status incorporated it into their night's work as a feature of their work's "natural accountability"' (*ibid.*, pp. 137) Why Galilean? There is a footnote reference to Husserl's discussion of 'a science in the Galilean mode', whatever this might be. And why is the IGP regarded as a cultural object to be *incorporated* into the nights' work? If one does not wish to suppose that there is an independently existing object, a pulsar, English has the resources to indicate this without accepting aspects of Husserl's account of science as Galilean or indulging in talk of the incorporation of cultural objects.

GLL attempt to explain how the astronomers got from the pulses on their oscilloscope and the other instruments with which they worked, to talk of the IGP by using an analogy of a potter shaping up clay into a pot. They say: 'The analogy of the potter's object is suggestive on a point that we need more than anything else to make, but find it difficult to put in a few conventional words: the intertwining of worldly objects and embodied practices.' (*ibid.*, p. 137). It is hard to know what practices without bodies would be like; so the term 'embodied' is pleonastic. And a footnote tells us that the word 'intertwining' has been borrowed from Merleau-Ponty. Moreover, in reading their jargon-laden paper one can feel acutely GLL's observation about the crucial point they are trying to make, but which they 'find difficult to put in a few conventional words'. But finally, one might be pleased to see talk of 'worldly objects', and see their link to the astronomers' practices, as a gesture in the direction of Wittgenstein (supposing these practices to be in part linguistic). Pulsars, too, can be worldly objects, along with the telescopes, oscilloscopes and other instruments with which the astronomers interact, and not cultural objects (as are their instruments). But at best this is an analogy that breaks down; the potter's finished pot is an object, and a cultural one at that, but the optically identified pulsar is an object *tout court* and not anything cultural.

As a final example for comment, consider GLL's remark:

> From the local historicity of the embodied [sic] night's work they [the astronomers] extract a cultural object, the independent Galilean pulsar. The IGP retains the material contents of astronomical *things* in their entirety. Nevertheless, the IGP, the potters object, is a *cultural* object, not a 'physical' or a 'natural' object' (*ibid.*, p. 141).

Much unnecessary mystery is created by wanting to say that the IGP retains its material as an astronomical thing in the same way as, one may suppose, the potter's pot retains it clay yet remains a cultural object. But this is a very misleading analogy concerning the way the individuation and naming of new objects works. In the above passage the emphasis is on the claim that the IGP is not the independently existing pulsar that most, the astronomers' included, would suppose was the object identified on the night of January 16, 1969. The overwhelming impression from this remark is that the astronomers did not discover any actual pulsar at all on that night; rather they made a cultural object in much the same way as a potter makes a pot. It is not as if the pulsar optically identified is an independently existing thing somewhere in space; rather it is a product of the astronomers night's work in their observatory and laboratory. That this is the stance of GLL is evident in further remarks they make: 'Our insistence that the IGP is a cultural object insists upon this about it: astronomically detailed specifications exhibit as the pulsar's Galilean independence the local production properties of the [astronomers'] work'. And a little further on GLL describe the optically discovered pulsar 'as an endogenously produced identifying detail of [the astronomers' night work]'. The night's work is in fact so efficacious that it provides 'for [the IGP's] transcendental orderliness'. (*ibid.*, p. 141) Given GLL's language, the astronomers are not mere discovers of stars; rather, like God, they are star makers!

There is much that can be criticised in the account that GLL give of the work of the astronomers who detected the first optically viewable pulsar (see, for example, Fox (1988)). The line of criticism suggested above is that GLL's anti-realism about the pulsar is unwarranted.[22] One may wish, as GLL do, to describe the work of the astronomers on the night of January 16, 1969 without supposing in their description that there was an actual object discovered. We can grant the need for practitioners of SSK not to make question-begging assumptions in their description of searches about what does, or does not, exist. De Soto searched for the Holy Grail that did not exist; however we can describe his activities in a way that is independent of whether or not the Holy Grail exists. And we can do much the same for the activities of the astronomers. In the second half of the nineteenth century many searched for Vulcan, the alleged inter-Mercurial planet; and we can describe their search without begging any questions as to the existence or non-existence of Vulcan. However there is a difference in those cases of search where the object of the search does exist and a point is reached where success in the search is recognised. We call these 'discoveries'; and the very use of the world 'discover' is a success word. It would be a serious lacuna in any ethnomethodological account of a discovery that it can only cope with its search phase, but can provide no account of when a search culminates in success or failure. What is missing is a philosophically informed account of the nature of a search for some object, its discovery, the individuation of the object, how language is used to refer to that object, and a sensitivity to those cases where the object of the search either does, or does not, exist which does not leave all objects of discovery as intensional objects which lack actuality.

The anti-realism of GLL is of a piece with the anti-realism of the account in Latour and Woolgar (1986) concerning the discovery of chemical substances; and it suffers from the same deficiencies. But their anti-realist stance is not the main issue under discussion. We embarked on the detour into an examination of an often-cited paper concerning an ethnomethodological approach in science studies to see if the kinds of descriptions it offered were akin to those offered by Wittgenstein of our natural human practices. There is clearly little in common. First, the "object" is different in that grammar and language games are not being described but, rather, quite general activities, such as those of astronomers discovering something new in the heavens. The astronomers will, of course, employ language and indulge in language games; but these are not the object of study in the paper. Second, there is an appeal to, and use of, a range of philosophers such as Husserl and Merleau-Ponty. But these appeals are to specific doctrines in philosophy that have little to do with the analysis of language games. Finally, there is a general presupposition of strong anti-realism that is unwarranted. In fact the anti-realism is so strong that it leaves the ethnomethodologist incapable of giving an account of when investigators actually encounter real objects, such as optically evident pulsars, or chemical substances. At best scientists encounter only constructs of their own making. The anti-realist ethnomethodology of GLL is philosophically loaded to such an extent that what they provide can not be regarded as merely a description.

As suggested, there is no need for GLL to put the matter of how the real pulsar was discovered in the way they do. There are alternative ways that concern criteria of identification and how words are introduced to refer to observable items. This is not always a straightforward matter; far from it. But this is to suggest that elements of philosophy other than that of Husserl, such as theories of identifying reference, might be of better use here. More recent so-called 'analytic' philosophy of language is replete with theories that ethnomethodologists could well employ about how to individuate objects and introduce names. And they do not beg questions about realism.

In using philosophical theories to get their so-called 'descriptive' ethnomethodological enterprise going, ethnomethodologists cannot be adopting Wittgenstein's account of the description of grammar. Rather they are presupposing something like the grammatical investigation that Wittgenstein recommends. For example, just how do we attach referents to words? In part this is something that Wittgensteinian grammatical investigations will tell us about. And it is something that ethnomethodologists might employ in their account of the practices of those astronomers during their night's work. But these very ethnomethodological descriptions, even if their "object" of study is the very language used by scientists, will presuppose the very grammatical story that Wittgenstein thinks is the proper province of philosophy. Whether one agrees with Wittgenstein that this is *all* there is to the business of philosophy is one matter; but it can legitimately be *part* of the business of philosophy. From this it follows that the enterprise of grammatical description will be presupposed even in a discipline like ethnomethodology. So it will not take over as one of the heirs, or perhaps the only heir, of the subject called 'philosophy'. Philosophy is still a going enterprise not to be supplanted by sociology, cultural studies or whatever else.

Bloor's SP and Lynch's ethnomethodology also differ strongly over the issue of how we are to understand Wittgenstein's account of rule following. Lynch takes the non-sceptical interpretation while Bloor follows Kripke's sceptical account and his positive communitarian resolution of the sceptical problem. This is an issue which not only divides practitioners of SSK but also philosophers who attempt to find some consistent stance in Wittgenstein's scattered remarks on rule following. In what follows we will look at some of the difficulties that philosophers have found in the Kripkean communitarian interpretation that have a bearing on the doctrines of SP and especially its meaning finitism.

7.6 COMMUNITARIANISM, MEANING FINITISM AND THE STRONG PROGRAMME

7.6.1   Scepticism About Rule Following

Bloor's meaning finitism has already been introduced in section 5.2.2. Here we continue the discussion within the context of Wittgenstein's remarks on rule following, and in particular the communitarian interpretation that advocates of SP given it in the light of Kripke's influential interpretation. Briefly put, the sceptical problem arises as follows. Given a rule such as 'Add 2' a person might be trained to proceed as follows (beginning with 2): 2, 4, 6, 8, and so on up to 1000. But when asked to continue beyond 1000 they proceed as follows: 1004, 1008, 1012, and so on. If queried the person could say that there is nothing given in their previous instruction 'Add 2', and in the examples from 2 to 1000, that tells them to go on 1004, 1008, etc. rather than 1002, 1004, 1006, etc. In fact a wide range of alternative ways of proceeding can be imagined. The difficulty being expressed here is the following: there is nothing in the instruction given 'Add 2', and the examples initially produced, that determine which way to proceed since both ways of proceeding (as well as many others) are consistent with the instruction and the examples. The problem raised here concerns what it is to understand a rule and to keep applying it in the same way. There seems to be a paradox here, which Wittgenstein expresses as follows: 'This was our paradox: no course of action could be determined by a rule, because every course of action can be made out to accord with the rule. The answer was: if everything can be made out to accord with the rule, then it can also be made to conflict with it. And so there would be neither accord nor conflict here' (Wittgenstein (1967), Part I §201).

In commenting on this, Kripke understands Wittgenstein to be giving a sceptical argument which is of a kind with Quine's arguments for indeterminacy of translation and inscrutability of reference, and Goodman's arguments about the predicate 'grue'. The sceptical position about meaning of any expression, including 'Add 2', is starkly presented: 'The sceptical argument, then, remains unanswered. There can be no such thing as meaning anything by any word. Each new application we make is a leap in the dark; any present intention could be interpreted so as to accord with anything we may chose to do. So there can be neither accord, not conflict' (Kripke, (1982), p. 55). The scepticism expressed here opens a gap between a rule ('Add 2') and the extension of the rule (the infinitude of numbers). The extension *accords* with the rule. But the problem is in the very relationship of accordance between a rule and its extension. The relation is indeterminate because it yields no account of why one set of numbers rather than another is the rule's extension.

There are those who reject Kripke's sceptical interpretation of Wittgenstein on rule following. The focus of their objection is on an alleged misinterpretation of Wittgenstein in claiming that he endorses the idea that there is a sceptical gap between rule and extension. Rather, they read Wittgenstein as an arch anti-sceptic whose considerations about rule-following are meant to be a *reductio* of there being a sceptical gap at all. According to one diagnosis of what has gone wrong, Baker and Hacker say that the locus of the different interpretations is to be found in the sceptic's *externalist* account of the connection between a rule and what is in accord with it, which allows for a sceptical gap, and an *internalist* account, which does not. 'To understand a rule is to grasp an internal relation between the rule and its (potential) extension. ... Correctness and incorrectness are determined by the internal relation between the rule and what counts as accord with it. It is not a *discovery* that '1002' follows '1000' in the sequence of even integers. Rather, getting this result is a *criterion* for following the rule of this series' (Backer and Hacker (1984), pp.76-7). We will not follow up the various interpretations of Wittgenstein on rule following, on whether or not a sceptical reading is appropriate, in particular, on what might be meant by the distinction between external and internal relations between a rule and its extension. Rather, we will accept the sceptical interpretation, and consider Kripke's resolution of its difficulties by appeal to a community. It is this stance and resolution that advocates of SP adopt.

The solution to the problem of the sceptical gap addressed here is by appeal to what the community might say and do in a given circumstance. Thus the person who was taught to follow the rule 'Add 2' in the cases from 2 to 1000 (and was corrected or applauded when they did it in accordance with their community of teachers) would be corrected by the community when they continued '1004', 1008' etc. In line with a long-standing communitarian interpretation of Wittgenstein, Kripke tells us:

> ... if one person is considered in isolation, the notion of a rule as guiding the person who adopts it can have *no* substantive content. There are ... no truth conditions or facts in virtue of which it can be the case that he accords with his past intentions or not. As long as we regard him as following a rule 'privately', so that we pay attention to *his* justification conditions alone, all we can say is that he is licensed to follow the rule as it strikes him. ...
>
> The situation is very different if we widen our gaze from considerations of the rule follower alone and allow ourselves to consider him as interacting with a wider community. Others then will have justification conditions for attributing correct or

incorrect rule following to the subject, and these will *not* be simply that the subject's own authority is unconditionally to be accepted. (Kripke (1982), p. 89)

Any individual who claims to have mastered the concept of addition will be judged by the community to have done so if his particular responses agree with those of the community in enough cases, especially the simple ones … (*ibid.*, pp. 91-2)

Clearly Kripke's solution to the sceptical gap he detects puts strong emphasis on the social aspects of rule following. The solution also tends to make the social aspects constitutive of rule following, rather than the uncontroversial claim that they are a constant accompaniment, in the shift from talk of truth conditions to community-wide assertability (justification) conditions. Bloor's view, however, is broader than just community conformity, since the formulation of SP has both social and non-social factors as causes of belief, even in the case where the belief is about the outcome of the application of a rule. But of course the social factors are the dominant ones. Bloor also highlights an aspect of the Kripkean communitarian view which he calls 'meaning finitism'. This is largely the view that the future uses of a term are not in some manner laid out in advance but rather 'meaning is generated in a step-by-step fashion as we go along' (Bloor (1997), p. 136).

Bloor's social communitarian response to rule scepticism and his meaning finitism come together in the following extended passage. It also reflects the causal tenets of SP and shows how SP offers a scientific social causal account of both belief and meaning of the sort contrasted with the anti-scientism of Wittgenstein in the precious section.

According to meaning finitism, we create meaning as we move from case to case. We *could* take our rules and concepts anywhere, in any direction, and count anything as a new member of an old class, or of the same kind as some existing finite set of past cases. We are not prevented by 'logic' or by 'meanings' from doing this, if by these words we have in mind something other than the down-to-earth contingencies surrounding each particular act of concept application. … The real sources of constraint preventing our going anywhere or everywhere, as we move from case to case, are the local circumstances impinging on us: our instincts, our biological nature, our sense experience, our interactions with other people, our immediate purposes, our training, our anticipation of and response to sanctions, and so on through the gamut of causes, starting with the psychological and ending up with the sociological. That is the message of Wittgenstein's meaning finitism. (Bloor (1997), pp. 19-20)

The above applies not only in the case of continuing to follow the rule 'Add 2' after 1000, but also saying of any new item in front of one 'it is a swan' (and not a goose or duck), or making an inference to some conclusion. No rule, nor any of the past cases that were in accord with some rule, determine whether or not some next item is in accord with a rule. What prevents us from classifying anything as the next case is not an appeal to non-natural *abstracta* such as meanings or a Platonistic realm of norms and rules. What does take us to the next case is the total range of naturalistic causes which includes non-social factors from biology to our sensory experience and, importantly, social factors including our past training and socially acquired dispositions, the community consensus and our interaction with the community. All these social and non-social factors come together to causally bring about our response, and the response of others, when we go on to say of the next case '1002' (rather than '1004'), or 'it is a swan', or make a logical inference.

What has happened to the idea of a rule in this account of meaning finitism? It remains unclear whether there *exist*, in some sense, rules (of meaning, or of adding, etc.) to be followed, or that there are no rules to be followed and that rules really are (in some reductive or eliminative sense discussed in section 7.3) the social relations that arise in one's interaction with one's community when the public language is used. Some version of the latter seems to be the case. In fact Bloor adopts the distinction made by Backer and Hacker between external and internal connections between a rule and its extension, claiming that the very internal relation, which holds between a rule and what is in accord with it, is constitutively social.[23]

There is no denying that many aspects of language are social. But the focus here is on a particular claim about the sociality of language. This is the *communitarian* view of the sort found in Kripke, or in Bloor, which allegedly gets us from the sceptical predicament that 'there can be no such thing as meaning anything by any word' to some account of how meaning is attributed to the words of our language. And it is here that, in the case of SP, the focus of attention falls entirely on social factors and hardly on the accompanying non-social factors. In what follows other aspects of Bloor's meaning finitism will be set aside. A number of points, already found in the philosophical literature, will be mentioned that undermine the communitarian position. It is not the aim of this section to provide a positive theory (though many non-communitarian responses can be found in the literature, including those which deny the very rule-scepticism which gives rise to the problem in the first place).

7.6.2 Trouble for the Communitarian Response of Rule Scepticism – the Problem of Sameness

For those who appeal to one's community to overcome meaning scepticism, what can be said of Robinson Crusoe isolated on his island? Does he follow rules, or not, if there is no community to correct him? Kripke's view is that Robinson Crusoe isolated on his island can follow rules, correctly and incorrectly. But Crusoe's physical isolation is to be distinguished from the case of a person who is *considered in isolation*: 'an individual, *considered in isolation* (whether or not he is physically isolated) cannot be said to do so [follow rules]' (Kripke, (1982), p. 110). In contrast, Backer and Hacker provide evidence that Wittgenstein himself did not find the case of Crusoe problematic; and Bloor argues to the same end.[24] So the case of Robinson Crusoe does not count against the communitarian response.

Quite different considerations are raised by Boghossian ((1989), pp. 521-2) in his extensive review of the large number of responses to Kripke's interpretation of Wittgenstein that bears on the community versus individual aspects of rule-following. One problem Boghossian discusses was first raised by Goldfarb ((1985), §II). Consider some Crusoe *in isolation* (neither the historical Alexander Selkirk nor the fictional Robinson Crusoe). Kripke alleges that a rule can have no substantive content because there is no fact of the matter that could provide truth conditions or facts for the meanings of the words of isolated Crusoe, or how he continues on in accord with his past intentions. As far as the assertability (or justification) conditions are concerned, they will not be able to distinguish between following a rule and *thinking* that one is following a rule, thereby undermining the idea of following a rule "privately". Let us accept this since our purpose is not to evaluate the "private language" argument[25] but to consider the communitarian response.

Turning to persons in a wider community does yield assertability conditions, but not truth conditions. Others now have assertability conditions for when a person is or is not following a rule, other than the person's say so, that go well beyond an individual considered in isolation. To illustrate Kripke says: "*Smith* will judge Jones to mean addition by 'plus' only if he judges that Jones' answers to particular addition problems agree with those *he* is inclined to give. ... If Jones consistently fails to give responses in agreement ... with Smith's, Smith will judge that he does not mean addition by 'plus'" (Kripke (1982), p. 91). The example must be qualified to allow for some smoothing of irrelevant disagreements between Jones and Smith, obvious errors, or a weighting for easy or obvious cases over difficult one's, and so on. (Perhaps other conditions need to be added such that Smith and Jones are not intoxicated, on drugs, approaching senility, and so on.) In what follows these qualifications will be acknowledged by an appropriate 'most'. The same will apply not only to Smith but also to other members of the Jonesian community such as the assessment of Jones by Brown, Black, etc. That is, the assertability condition for Jones' adding (cases of smoothing some disagreement aside), is something like the following schema:

(A) x is warranted in asserting that Jones means addition by 'plus' provided Jones uses 'plus' in the same way (in most cases) as x is inclined to use 'plus'.

In the schema we can replace 'x' by the Smith of Kripke's example. And other members of the community such as Brown, Black, and so on, are also instances of x.

It is useful to contrast (A) with another assertability condition:

(B) x is warranted in asserting that Jones means addition by 'plus' provided Jones responds with the *sum* to most arithmetical queries that he encounters.

While Kripke's sceptical argument is understood to show that there is no fact of the matter as to what the words of our language mean, there is no similar Kripkean scepticism as to what numbers sum to what. Unlike meaning, for Kripke there are realms of fact to appeal to, such as arithmetical facts and a host of other commonsense and naturalistic facts.[26] In the very formulation of (B) we use notions like addition in appealing to what is the actual *sum* of various pairs of numbers that Jones encounters. And in using such notions we must rely on how we are to correctly continue using the rules of addition. Also in (B) there is no reference to matters to do with a community and their agreement or disagreement. (B) applies to the individual in isolation without reference to any other individual. Even Jones can stand in for x. So Jones can be right or wrong about the sum in the absence of a community to applaud or criticise him. (B) might be said not to represent the intentions of those who adopt the communitarian response to rule scepticism. However there is an important point that arises from further consideration of (B).

Turning to Goldfarb's critique, the first step in his argument is that if assertability conditions like those in (B) are employed then one of Kripke's main conclusions cannot be drawn, viz., there are no assertability conditions for a language used by a person in isolation. There are facts of the matter about what sums to what that can be used to distinguish between a person, in isolation, following a rule correctly and *thinking* that they are following a rule correctly. But this is the very distinction that, on Kripke's view, is needed if the very idea of a 'private' language is to be rejected. So if the claim that *no* distinction between a person using a rule correctly and *thinking* they are using it correctly can be made on the basis of some given assertability conditions, then they must be ruled out. It is clear that (B) can not be ruled out on these grounds.

The next step in the argument is as follows. One of the crucial features of (B) is that to state the assertability conditions the notion of a *sum* is used. To make the assertability condition useable at all, the users must already know how to continue using rules of addition correctly. That is, users need to know that 'plus' means addition. But of course this is one of the very matters that is up for consideration, viz., what it means to follow a rule for addition – or a rule for anything else for that matter. Rule scepticism is a global thesis about all the terms in our language which does not allow that some like 'plus' get excluded from its scope. Suppose, then, that we rule out any assertability condition that depends on any notion that requires knowledge of how we are to continue correctly according to any rule, including the rule for 'plus'. Now we come to the heart of Goldfarb's objection. We have simply ruled out too much. And there seems to be no intermediate position in which we reject (B) but do not rule out too much. It is hard to see how any assertability condition we could envisage would not contain a notion that required us to know how to continue correctly using it according to some rule.

The point just made applies to the very assertability condition (A) that Kripke suggests and that communitarians adopt. In particular (A) requires that Smith, in making the judgement that Jones' response is the *same*, or *different*, from the one that Smith is inclined to give, requires a notion of *sameness*. The eighteenth word in (A) is 'same'. And Smith's employment of this notion, like any other, requires that Smith know how to go on correctly according to a rule for sameness. Thus if Jones is following the rule 'Add 2' and is to do so from the number 1000, and Jones goes on to 1002, then Smith must make the judgement that this very number is the *same* as the one he is inclined to give in adding 2 to 1000. But if Jones went on to '1004' then Smith must make the judgement that this number is *different* from the one he is inclined to give. And so on, for all the other cases in which Smith has to judge sameness with respect to any other case of Jones performing addition.

But more can be said. Not only must Smith be able to go on judging sameness in respect of what Jones comes up with, but (most of) the other members of the community, Black, Brown, etc. must also be able to judge sameness in respect of what Jones comes up with. But not only Jones. In order to discover what the community agrees to, each member must judge of every other member that the answers they are inclined to give are the *same*. However we are required to rule out correct knowledge of how to go on using the word 'same', as we did for the knowledge of addition required to go on adding in the case of (B). But to do this is to render unusable the very assertability conditions that the communitarians require to state their case. If they do not rule it out, then they must appeal to facts of the matter as to what is same or different. This cannot also be something which is up for grabs across the community in which Smith, Black, Brown etc. do in fact go on in different ways alleging that this counts as the same. This is the covert assumption of communitarians that is really denied to them by the global character of the meaning scepticism they otherwise endorse, viz., no term on our language has a determinate meaning. They help themselves to the notion of sameness of response across the community that is denied to them by the very thesis of rule scepticism to which their communitarianism is supposed to provide an answer.

This is a matter of which Wittgenstein appears to be aware when he said: 'If you have to have an intuition in order to develop the series 1 2 3 4 . . . you must also have one to develop the series 2 2 2 2 . . . .. But isn't *the same* at least the same?' (Wittgenstein (1967), Part I §214 & 215). 'The word "agreement" and the word "rule" are *related* to one another; they are cousins. If I teach anyone the use of the one word, he learns the use of the other with it. The use of the word "rule" and the use of the word "same" are interwoven.' (*ibid.*, §224 & 225). So when two or more members of a community follow any rule such as 'Add 2' there are judgements of sameness and difference to be made each step of the way, and across the entire community. Using the example above when Jones goes on to number 1002, we need to know that the various tokens of '1002' in the mouths of all the members of the community means 1002. This is the very matter about which total global rule scepticism can provide no guarantee. But it is the very matter that communitarians must help themselves to if they are to state their assertability conditions. That is, there is one rule about which there needs to be a fact of the matter, if the community is to say of Jones that he has got it right or wrong when it comes to adding. And this is following a rule about Jones' having gone on in the *same* way as each member of the community (or most of them). Not to allow this is to have an assertability condition that is unusable.

7.6.3 *More Problems for the Communitarian Response*

Perhaps the SP advocate of communitarianism can respond by saying that none of the above gives an adequate account of what they intend. The following tries to accommodate this objection, but in so doing goes beyond what Kripke intends. The suggested repair is due to Blackburn ((1984), pp. 85-6) and can be expressed as follows. What the communitarian requires is not a judgement that 'plus' means addition (Blackburn's example is 'red' means red). Rather they require a more elaborate judgement of the following sort: 'this is what most in my community would call 'plus'. The assertability condition might then be something of the following sort:

(C) x is warranted to assert that Jones means addition by 'plus' provided that Jones' uses 'plus' in the same way (in most cases) as most in x's community are inclined to use 'plus'.

In putting matters this way we need not only be able to use the word 'same' to refer to sameness (as discussed above), but we also have a new predicate "what most in x's community would call 'plus'". What this predicate means is as much up for grabs is was the original notion of 'plus', according to rule-scepticism; there is no fact of the matter as to its meaning. Just as there can be our "normal rule" as to what any predicate means, there can be what Blackburn calls 'bent rules' of the sort that the rule sceptic envisages along the lines of alternatives to our "normal rule" 'Add 2'. And this applies to any predicate, including "what most in x's community would call 'plus'".

To resolve this we could consider a new predicate: "this is want most members of x's community would call 'what most members of x's group would call 'plus'"? But this leads us to a regress in which each more complex predicate (such as "this is what most in my community would call 'plus'") is not even synonymous with the base predicate from which it derives (in this case 'plus'). As Blackburn comments: 'The standard for correctness in description seems to have shifted, as it were, from conformity with how things are, to conformity with each other. If I live in a community which calls the Earth flat, is not this just one more element of the dance, on which I ought not to get out of step?' (*ibid.*, pp. 86-7). But the situation is even worse than Blackburn describes. Not merely have we shifted from how things are to merely being in conformity with one another. There is no fact of the matter as to whether we are in conformity with the community since this requires a notion of sameness.

What the above shows is that, for those who take the communitarian way out of rule scepticism, attempts to express the assertability conditions for a person to mean addition by 'plus', they must employ, in the assertability condition, the notion of being the same. But if rule scepticism applies, as it should, to uses of 'same' then the communitarian cannot express their assertability conditions. Alternatively, if the assertability conditions are expressed in an different way to accommodate a possible objection from the communitarian rule-sceptic, then notions are introduced into the assertability conditions which, by their own rule-scepticism, are problematic; but resolving this leads to a regress.

Blackburn raises further objections to the communitarian response that are of significance for advocates of SP.[27] This is the objection that the very same problems that confront the individual can also confront the community as a whole. An individual person can be considered as a continuous series of "person time-slices" over a longish period of time. It is then possible to draw an analogy between the person time-slices and the various persons of a community. The rule-following sceptic gets a later time-slice of a person to wonder whether they are using a rule in the same way as their earlier time-slice. Similarly the rule sceptic can get one member of a community to wonder whether they are using a rule in the same way as other members of the community. On the communitarian view, the community is to escape the rule sceptic by claiming that they each (near enough) use a rule in the same way so as to make the same judgement about the performance of Jones, and of one another. But then so can the individual Jones use a rule in the same way by appealing to the sameness of judgement about the current Jones' time-slice use of the rule and an earlier Jones' time-slices use of the rule. By the same token if the rule sceptic is to win against the individual then they can also win against the community. Just as the rule sceptic undermines the sameness which is required of the individual's use of a rule, so they also undermine the sameness required of the community's judgement not only of Jones' performance but their own collective performance as well. Thus the community is no better off than the individual in the face of a challenge from the rule sceptic. If the individual succumbs to the rule sceptic's challenge then a similar argument can be constructed for the community. Conversely, if the community can be let off the hook, then a similar argument can be constructed which also lets the individual off the hook.

In what follows a different kind of objection, due to Boghossian ((1989), pp. 534-6), is raised in which it is envisaged that the community might be quite wrong. One common objection to the communitarian response is that (most or all of) the community can be wrong; and so the hapless Jones who might have been out of step with the community, might not have been wrong after all. (Not everyone was a flat Earther, especially those who tried to change the views of flat Earthers). To use Blackburn's excellent comparison (see previous footnote for references) a violinist in an orchestra might play a wrong note in the sense of being a discordant one amongst the concordant many. But, for the communitarian, to ask whether the concordant many are wrong does not have an answer. A Wittgensteinian response would be to say 'This is just what we are playing!' An appeal to what a score says runs up against the rule sceptic who might say, for example, that there is no fact of the matter as to whether the symbol for C-sharp in a score means C-sharp or not. However the analogy of the orchestra does not apply fully in the case of addition, as has been just argued for its assertability conditions. As Blackburn puts it: 'My community may all suddenly start saying that $57 + 68 = 5$, but this does not make me wrong when I continue to assert that it is 125. I am correct today in saying that the sun is shining and that daffodils are yellow, regardless what the rest of the world says' (Blackburn (1993), p. 223).

But there is another way in which the community can be wrong, as Boghossian describes (assuming that they do not collude with one another – this being understood from now on). Once again problems for any individual can reappear for the community as a whole. But this is just the kind of problem, as in the case of the orchestra as a whole, that the communitarian must rule out. But can they?

The word 'horse' means horse, and not cow. Let us suppose that Jones has a disposition to apply 'horse' to horses. However Jones might look at a horse in unpropitious viewing conditions (e.g., on a misty night), or while he is not his true self (e.g., drunk, drugged, tired), and may be disposed under these conditions to call some cow 'horse'. His all-wise community might then correct him. But why can't the conditions which afflict Jones when he calls the cow 'horse' also afflict Smith, Black, Brown and the rest of the members of Jones' community? Since they are all rather like Jones in respect of physical and cognitive make-up, they can equally be disposed, in the same non-standard conditions of viewing, to call the cow 'horse'. But then the community cannot call these 'mistakes' because that is what the community is disposed to say. There is no further court of appeal. Similarly Jones might mis-call some horse 'cow' in some unpropitious conditions of viewing; and the community does likewise because of the shared dispositions in circumstances of viewing. However none of these can be called mistakes because that is what the community is disposed to say. So in the idiolect of the community each word 'horse' and 'cow' will have in their extensions both horses and cows, and each will omit, respectively, some horses and some cows.

Extending the above, a whole range of items might get into the extension of the terms 'horse' so that it is a disjunct of miscellaneous items such as (cow, zebra, camel, etc) depending on what the community determines under the conditions of viewing that prevail for them. The problem does not apply only to the word 'horse' but also to the words 'cow', 'zebra', 'camel', and to a host of quite different words. So the terms of the community's language do not have the bearers we take them to have but some wild disjunct of miscellaneous items. This must be reckoned to be a failure for the communitarian view since many of our terms do not mean, or have as their bearer (or extension), some long and miscellaneous disjunction of items.

The same can be said of the consensus theory of truth (a view that many advocates of SP are inclined to adopt). We may suppose that the consensus theory says something like the following: for proposition p, p is true iff there is consensus in the community that p. But there can be consensus about the false. As just envisaged the community can agree that it is a cow when it is in fact a horse. And they can agree that $57 + 68$ adds to 5 when it really adds to 125. It ought to be no part of the truth-conditions of a proposition that the community agree about it. Of course it is open to communitarians to reply that their view is a revisionary one that is not to be understood along the lines of the ordinary or classical accounts of truth. But then it might be doubted that consensus truth has anything to do with truth, despite the occurrence of the word 'truth' in the name of their doctrine.

These objections seem to strike at the heart of the communitarian answer to how it is that we follow a rule and mean what we say. This is not to say that there are not other answers to the problems which arise from assuming that there is something to be said on behalf of those views of meaning which proceed on the assumption that the rule sceptic needs to be answered (as opposed to those who think we have been led up the garden path by rule sceptics). Nor is it to deny that there are important social and community aspects which might contingently accompany either our learning to follow a rule or our use of a rule to mean what we do. But it is to question the claim of SP that the social communitarian view is *constitutive* of what we mean by 'plus' or 'infer' or any of the other terms of our language. In particular Bloor's claim that logical relations are really social relations of constraint has not been established by the communitarian view of rule-following. We are not to be saved from the rule sceptic, and thereby give an account of meaning, simply by calling in community social services.

7.7   NATURAL KINDS AND MEANING FINITISM

In the above there was a supposition that most of our words have an extension, viz., there are terms that mean, or are about, or refer to, actual items. Thus there are proper names for certain items such as persons, rivers and towns, and there are names for kinds of thing (colours, substances, species). This is a view not denied by Wittgenstein. He crucially reminds us of this just after telling us that for many, but not all, cases of tokens of 'meaning', that the meaning of a word is its use in the language. He continues immediately adding 'And the *meaning* of a name is sometimes explained by pointing to its *bearer'* (Wittgenstein (1967), Part I §43). Moreover language-games involve two items, a linguistic item used by a person and a non-linguistic item, an *activity*. One such activity is using a word to name a bearer, but as Wittgenstein is at great pains to point out, we must not wrongly narrow the idea of language games to just naming. So there are two kinds of meaning for a term, its use and its bearer. Wittgenstein mentions the proper name 'Excalibur', and the names of colour kinds such as red, green, white, blue, sepia, etc. There is no reason not to suppose that he would not think that 'horse' and 'water' would not also name (respectively) an animal and a substance kind; and he has a lot to say about the term 'game' which is akin to a generic social kind. Further, we should not suppose that the instances which fall within a kind are sharply demarcated. Some kinds can have determinate extensions but other kinds, such as games, can have fuzzy boundaries (see Putnam (1975), p. 217). Kinds, like being bald, are fuzzy in that some individual items definitely fall under the kind, and others outside, while there is a penumbra in which other items seems to be neither one nor the other but may vary in their probability as to whether they are in or out.

An important philosophical matter that Wittgenstein, Kripke and Putnam all address is how a name gets attached to its bearer, no matter whether it be a proper name for a single item or a kind name. In the case of Kripke there is a discussion of how proper names such as 'Moses' or 'Nixon' get attached to their bearers, and how kind terms like 'tiger' and 'water' get attached to their bearers. Here we will say little of the different philosophical theories about how names are connected to their bearers, from Wittgenstein's seemingly 'descriptivist' account to the Kripke and Putnam non-descriptivist account which often relies on bare causal relations. In what follows we will simply assume that there are names, and that there are, on the whole, bearers for those names (problems about fictional an other cases being set aside); that is, on the whole, names have an extension (whether it be a single item, a number of items, a kind, and so on).[28]

The above is an attempt to counter the common view that Wittgenstein was only interested in how language was used, the language games of which word usage is part, and the rules we follow (call these 'horizontal links'). He was not interested in how names relate to their bearers (call these 'vertical' language-item or language-world links). But this is wrong. He carefully mentions both meaning as use and meaning as bearer in the same passage cited at the beginning from *Philosophical Investigations* §43. Both horizontal and vertical aspects need to be taken into account. The book by Hintikka and Hintikka ((1986), chapter 9) is devoted to countering the 'received view' (as they call it) in which only the horizontal features of use, language games and community or individual rule-following is countenanced in commentaries on Wittgenstein, while the vertical language-world relations are either played down, ignored or even denied. Do advocates of SP follow the received view and play down or deny that names have bearers (or to use a technical term of logic, have an extension, whether it be a kind or a class with individual items as members)? It would seem that they do in the light of what Bloor says. In this section we will follow this up for the case of kinds and ask what SP's communitarian view of rule-following does for our talk of natural kinds (such as colours, tigers, water, etc). The focus is on natural kinds and not on our talk of social kinds (money, marriage, games, etc.), artefactual kinds (such as screwdrivers or computers) or even miscellaneous classes of thing that we might name (such as call 'Charlie' all those observable things within 10 metres of you, or that you currently own, etc.).

On the matter of whether the kind words of our language have an extension (i.e., bearers), Bloor seems not just agnostic but definitely atheistic; there are none. He is a nominalist who admits that there are individual items, such as this cat, that cat, and so on. Presumably we can introduce proper names for each; or use descriptive phrases such as 'this cat' (accompanied by pointing) to pick out an individual item, and so on. But he is an extreme nominalist since he claims that there are no natural kinds of thing such as cats, colours, etc. (We have already seen that for SP there are no abstracta, no normative facts corresponding to rules, and no logical facts.) Bloor, we might say, is a class nominalist.[29] There are only individual items that we classify in various ways by putting them in to classes. We in fact construct or make kinds; they are not there objectively and independently of our classificatory activities. The classifications are highly variable in that they change over time and locality in the very individual items that fall within a classification. That we classify in the way we do, and that such classifications change, is all grist to the mill of SP which attempts to find socio-politico-cultural causes (or interests in these) which lead us to classify in the way we do, or change our classifications. Change our socio-politico-cultural circumstance (or interests) and our classifications change. There is no room for fixed classifications, or natural kinds that are socially and politically and culturally invariant and independent.

The above seems to be the purport of the following remarks of Bloor ((1997), pp. 23-4):

> One of the most basic concepts used in philosophical semantics is that of the 'extension' of a term. The extension of 'swan' is the class of all things, past, present and future, correctly called swans. The extension of 'water' consists of everything that might have been, or might be, truly classified as water. ...
>
> For a finitist there is no such thing as the 'extension' of a term or concept, or, if the word 'extension' is used, it radically changes its significance. ... Without (closed) extensions of the traditional kind, however, we no longer have propositions with a determinate content .... There is no class of things existing in advance of the application of a label. Here and now, there is no determinate class of things which will, or could, truly be called swans. The content of that class depends on decisions which have yet to be taken, and so does not yet exist. Particular things, or individual objects, exist in advance, but not classes of things. The claim is that 'extensions', as philosophers have characterised them, don't exist. They are simply fictions generated by a philosophical theory.

Let us set aside matters to do with whether or not our propositions have a determinate content and just focus on the world. We have admitted that some kinds are fuzzy; but fuzziness is not Bloor's concern in the above. Rather the traditional idea of a class (even a fuzzy one) of all past, present and future swans is simply not available for someone who is both a meaning finitist and a class nominalist. Perhaps we should not take Bloor at his word when he adopts the strong position that extensions, understood to be classes, do not exist for our kind terms. Perhaps we should say they do exist, but only temporarily because they undergo change with change in our classificatory practices and beliefs. On this reading our kind terms do have an extension but it is extension-at-a-time-relative-to-our-classifications; there are no kinds with an invariant, fixed extension independent of our classificatory activities at any time. The latter is part of realism about kinds such as swans, or water, eschewed by the finitism and kind-constructivism of SP.

Bloor is at least a realist about individual items. We do not have to do anything in order for them to exist; they can exist without us. Presumably individual rocks, grains of sand, drops of water, and so on, exist independently of us, and our perceptions, our beliefs our classificatory systems and our language. (The realists' sense of independence is given by a counterfactual: x could exist even if people were not to exist, or people were not to hold beliefs or theories about x, or have a language in which they talk of x.) But Bloor is not a realist about kinds of thing; what kinds (which he takes to be classes of thing) there are depends entirely on our classificatory practices, beliefs and language. Since these can vary over time, place, culture, epoch, class, etc, then our classificatory systems, and changes in them, are grist to the mill of the tenets of SP. In this respect Bloor's SP, even though it is realist about individual objects, it is non-realist and constructivist about kinds.

Scientists tend to be (uncritical) realists about kinds. They hold that there are a range of objectively similar kinds of sub-atomic particles such as electrons, protons, and the like. And there are kinds of elements, even kinds of isotopes of elements. There are the potentially infinite number of kinds of chemical substances such as water, common salt, etc. And there are a wide variety of living species, i.e., kinds of living thing.[30] However we do construct cultural artefacts; but these can co-exist with natural kinds. More strongly, the former require the latter as, for example, there can be no screwdrivers unless they are embodied in some bit of metal.[31] Granted this, we should allow that physicists, chemists and biologists make classifications which they hope will correspond more and more to actually existing natural kinds such as elementary particles, atoms, molecules, compounds, cellular and multi-cellular creatures and species. The more objectivist view that there are natural kinds is well expressed by the metaphor derived from Plato about nature being 'carved at its joints'.[32] In contrast the non-objectivist view of SP is that the only joints are the ones we ourselves carve; they are not there before the carving that makes them.

The above illustrates two rival views about kinds, and how we distinguish kinds. The first view is that there are natural kinds that 'follow nature's articulations' as Plato strictly puts it. These kinds are not only independent of our human classificatory activities. They also share many properties and their shared constitutive properties enter into explanations of the common behaviour of their many instances.[33] Such is the case for the many kinds found in science, as mentioned above. Further, the kinds themselves play a role, in some theories of meaning, in determining how kind terms get their reference. Such a role is given prominence in the theory of kinds and kind terms developed by Kripke ((1980), Lecture 3) and Putnam ((1975), chapter 12). In their theory of the reference fixing of terms like 'cow', 'swan' or 'water', the bearers, the kinds cow, swan, water, respectively, play a definite role. Briefly on the Kripke-Putnam reference fixing theory for kind terms, what fixes a term to its bearer is an initial term introduction in the presence of samples of the kind, the term being used to refer to the kind exemplified. Each term is then transmitted through a community, each person using the term with the intention to refer to the same kind as those from whom they picked it up. Moreover, instances of the kind play a definite role in determining how we go on to say that the next item is a cow, swan or water. Much of our common-sense and growing scientific knowledge plays a role in helping us pick out kinds by specifying a number of stereotypical properties of the kind (for example, we readily pick our water by its stereotype of colourlessness, liquidity, thirst quenching, the fact that it often falls out of the skies, and so on). The sciences might even tell us something about the constitution or essence of a kind. But such knowledge is not necessary in order to refer to the kind cow, swan or water. Humankind got on quite well, and still gets on just as well, even though most have never heard of the chemistry that tells us that water is a collection of H_2O molecules (give or take a small admixture of impurities).

This contrasts with the view that arises from the doctrine of meaning finitism. Advocates of meaning finitism do not admit that there are natural kinds independent of our classificatory activities; so they are not available for Kripke-Putnam style reference-fixing. Instead they focus mainly on our classificatory practices and our rule following which, on the communitarian view, is simply the community getting all it members to reach an overall consensus about what individuals might fall under what rule for classifying. And in this process the actual cows, or swans, may play no role at all. What plays a role is the socio-politico-cultural factors that bring about consensus. Also playing a role are the descriptive properties that are used to pick out kinds, the very descriptive properties which Kripke and Putnam were at pains to show could not play the reference-fixing role assigned to them. Within SP· the descriptive properties associated with kinds can change over time; so there are no kinds invariant with respect to the properties we use to classify. But the price for this position is high. As shown, the communitarian view is unworkable as a theory of rule-following; and at best only proper names can have bearers, not kind terms.

The above also quickly gives rise to an incommensurability of kinds, as the following illustrates. The Greeks had a term for the kind *fire* ('pur' or 'πυρ'). And Aristotle had a theory of fire in which, give or take some earthy impurities, was a pure substance. We too have a term 'fire' which refers to the very same thing that the Greeks referred to, viz., fire. Both we and the Greeks knew how to produced it, and what to do with it such as cook, warm ourselves, and so on. Such commonplace stereotypical properties help us pick out samples of fire. But we no longer hold Aristotle's theory; rather fire is not even a substance but is radiation in the heat and light wave-bands. To build too much of Aristotle's theory into the way in which we are to refer to fire would be to introduce an unnecessary incommensurability of reference between the Greek word 'πυρ' and our 'fire'. But this is what advocates of SP are inclined to do, especially since they hold to a strong version of the theory-ladenness of our scientific terms. And this is bolstered by their communitarianism about meaning and rule following, and their view that our kind terms really do not have fixed extensions. For them it is unproblematic that the terms 'πυρ' and 'fire' have incommensurable reference to different "kinds" constructed variously by us according to the different theories we hold. But much of this is highly counterintuitive.

What can be said of Wittgenstein's doctrine of *family* resemblances in this context? Wittgenstein's example of games is one of *family,* and not species or genus, resemblances. Taking talk of a family to be at a higher taxonomic level, there may well be something like a cluster of defining properties for families of kinds. Here a cluster of properties can be a large disjunction of conjunctions of properties where the properties in each conjunction contain a minimum number of properties (greater than a half?) from some selected class of properties (which itself might have fuzzy boundaries). Thus the cat family, which ranges from lions to domestic cats, is a broad family which admits of a cluster definition. But each cat species has a more specific subset of properties from the cluster which individuate it as a species within the same family. Also the family itself is a kind the extension of which is well demarcated and only slightly fuzzy (as is the case of each kind in the family). Using Wittgenstein's example of games, these are a family which, again, has a cluster of properties as its definition. But this does not mean that the many species or kinds of game do not have much tighter definitions in terms of a sub-set of properties selected from the cluster. Thus chess is a well-defined game which allows hardly any variation in its defining rules; and the same is true of rugby (as distinct from league or soccer), and of hopscotch (e.g., the strict defining rule of feet falling on a line means a foul). The doctrine of family resemblances is just that – the resemblances between the various kinds that make up the family. This leaves untouched whether or not the kinds of the family themselves have a much tighter set of defining properties. And when one comes to natural kinds there might even be an essence to discover. So the doctrine of family resemblances can be consistently held with the idea that there are kinds (within the family) which are natural in the sense specified above.

Though they recognise the difference between these two views, advocates of SP do not really come to terms with the idea of natural kinds.[34] As we have seen, most sociologists are strongly nominalistic, being anti-realist, or constructivist, about kinds, as the following illustrates:

> Should we then think of our inherited system of classification as the 'framework', 'constructed by men [sic]' into which things and objects are fitted? This is sensible. How we classify will depend on which 'framework' we inherit and, because different systems of classification are inherited in different cultures, how the members of those cultures classify will be different. (Barnes, Bloor and Henry (1996), p. 48).

This might be viewed as an unexceptional account of the way we do actively classify. Of course, historians of science must allow that we have made different classifications as our sciences have developed. But the explanation of this, like other scientific beliefs, might have nothing to do with the explanatory claims of SP but much to do with the science in which the terms occur. Moreover the extent to which the sciences and different cultures do differ in their classifications can be highly exaggerated by sociologists.[35] We can dub this 'cultural constructivism' of kinds. But this cuts no ice concerning the doctrine of natural kinds. Even if we grant that many of our classifications have change over time, this still tells us nothing about whether or not some kinds exist independent of our classificatory activities. However talk of 'frameworks' that comprise our classificatory systems counts against an account of kinds which are independent of our classificatory systems.

The sociologists continue:

> Is it not the existence of nature which accounts for people with the same framework using it in the same way? Is this not where the 'resemblances' come in: because we 'feel the resemblances' just as everybody else does, we fit things into our given framework just as they do? This is a beguiling conjecture, and it is most important to be aware of the grounds which make it necessary to reject it …. (*loc. cit.*)

Here the sociologists reject any attempt to appeal to any sameness in classificatory systems across cultures in terms of commonly recognised resemblances. We need not go into their reasons (but they turn largely on social aspects of language learning). What we need to note is their rejection of a commonly accepted view of kinds based on similarity relations that we can all recognise whatever our culture. And they make no mention of the role our current classificatory systems play in the success of our science; from this it can be inferred (using inference to the best explanation) that there is something right, or approximately right, about much scientific classification (i.e., it corresponds more and more closely to what natural kinds there are). The sociologists' position also runs against our current science in which there are well-developed theories about how we classify, such as cladistics, of the sort elaborated in Sober (1988) which are then used to discover, not construct, the naturally occurring species. The reasons given above are not enough for the sociologists to scuttle the idea of kinds independent of any framework we posses.

The above illustrates one way in which the communitarian view can run riot about the meaning of our kind terms. Given that a community can, in its rule following, allegedly move from past cases to any future case, then the idea that kinds are class constructs fits well with this view. However we have shown that the communitarian view of rule following is not sustainable. Moreover, SP buys into a particular metaphysical view of the world; it is not neutral on many philosophical doctrines but is strongly partisan. What it needs to show, but has not done, is that the idea of natural kinds is not sustainable. And showing this is quite different from telling us how our classificatory procedures do work in fact. We can grant that we have, in the past, made classifications that are wrong (assuming truth is one of the goals of science in making classifications – it need not be the goal for non-scientific activities). Earlier we were wrong about alloys being compounds when we discovered that they were, on the whole, mixtures; Aristotle was wrong about fire being a substance; the Karam are right about there being a kind *cassowary*, but wrong about not putting it in the family of birds; and so on for countless cases. Not to admit this is to say that, in some magical way, alloys-as-compounds once existed, according to the classificatory system of the then science; but then they went out of existence and in their place there appeared alloys-as-mixtures. Once there was Aristotelian fire but now there is our fire. And so on. This is Kuhnian world-changes with a vengeance.[36]

None this in any way undercuts the idea that there are natural kinds which we hope our classificatory terms in the long run capture. The world has repeated same-kind items scattered through it, such as electrons, drops and puddles of water, and the like. It is not the case that every item is so different that no kinds can exist (though this might have been the case just after the Big Bang it is not the case now). This is a refection of the idea that the classifications we make correspond more and more to the kinds that there are – and that there are good methodological principles in science that we use in making this claim. But within SP this idea can have no force because there are no kinds to which our theories can approximate. At best there is merely a social story to be told about why we classify the way we do at any time.

7.8   'SOCIOLOGY IS A WAY OF SENDING US TO SLEEP'

The Strong Programme with its central Causality Tenet CT is a naturalistic theory of the causes of any belief and any change in belief, and of the causes of why we mean what we do using our language, the causes being drawn from all the sciences from physiology to sociology. There is nothing normative in the story that SP tells us about our beliefs, or belief change. However the very presupposition that there are such causal connections to be found presupposes that there is a methodology for arriving at correct causal connections, a methodology often badly employed in the alleged case studies drawn from the history of science that are meant to illustrate SP's claims. The story SP tells is one in which all normativity is drained from any account of why we ought to believe one thing rather than another. Our beliefs are merely froth on top of the underlying naturalistic causes.

As has been illustrated, SP helps itself to a particular controversial interpretation of Wittgenstein on rule following and meaning. On this view social factors become all pervasive, not only shaping what we believe but even what we mean by our beliefs. But as we have seen, the scientism of SP is not one endorsed by Wittgenstein, a point in part acknowledged by advocates of SP even though they appeal to much else that is in Wittgenstein (or more correctly, their reading of his often obscure claims). For Wittgenstein there are quite general facts about human natural history that shape our practices and the language games with which they are involved. Sometimes Wittgenstein gives us examples of this actual human history, but more often that not he invents fictitious, but possible, examples of human activity and language games to illustrate one or another philosophical point. Advocates of SSK, from ethnomethodologists to SPers, see themselves as advancing Wittgenstein's programme by exploring more of the actual natural history of our human practice, especially in their case studies of the sciences and scientific activity. But in this there is still an absence of any normativity.

In a way Wittgenstein anticipated the scientistic approach to human activity and beliefs embodied in SP. Thus he asks and replies:

> Are the propositions of mathematics anthropological propositions saying how men infer and calculate? – Is a statute book a work of anthropology telling us how the people of this nation deal with a thief etc.? – Could it be said: "The judge looks up a book about anthropology and thereupon sentences the thief to a term of imprisonment?" Well, the judge does not USE the statute book as a manual of anthropology. (Wittgenstein (1978), Part III §65; capitals in original)

The point here is quite evident. What might be said in an anthropological vein, that is, in the vein of an empirical science, about our beliefs (logical and scientific) has no bearing on what we *ought* to believe (any more than reports about practices concerning thieves tell a judge what ought to be done). The point here is simply a correct instance of Hume's "is-ought" principle. Again, we find Wittgenstein saying, in respect of thinking of mathematical statements as empirical claims, proofs as experiments and learning empirically about how calculations come out: "What does it [the mathematical proposition resulting from calculation] say, though? What relation has it to these empirical propositions? The mathematical proposition has the dignity of a rule.' (*ibid.*, Part I §165). In his commentary on these and related passages Friedman ((1998), p. 262) says that the word 'dignity' is, in the original, the German 'würde' which also has the connotation of worth. Thus the mathematical proposition is worthy, and this has normative connotations. Similarly the judges' rule has worth, in the sense of normativity. And so do all rules. Once again the emphasis is on the normativity of our rules of reason, a feature lost in the scientism of SP (but not lost in the account of norms given at the end of chapter 3). And this holds both for the naturalistic causal-explanatory story told in SP about science and for the anti-explanatory descriptivist approach to science practised in ethnomethodology.

In what seems to be an important summing up of his methodological approach Wittgenstein says ((1967), Part II §xii):

> If the formation of concepts can be explained by the facts of nature, should we not be interested, not in grammar, but rather that in nature which is the basis of grammar? – Our interest certainly includes the correspondence between concepts and very general facts of nature. (Such facts as mostly do not strike us because of their generality.) But our interest does not fall back upon these possible causes of the formation of concepts; we are not doing natural science; ....

Wittgenstein's main philosophical interest is in grammar, and the descriptive enterprise associated with it. His interest is not with science, even a science devoted to uncovering and explaining the natural history of human activity. However he recognises a close connection between human activity and, in the long run, grammar. We may not share these interests in just Wittgenstein's way, or share them at all. But what is clear is that the advocates of SP cannot share these interests because what Wittgenstein alleges we must fall back to is the descriptive account of grammar, and not the explanatory pretensions of SP in this area. These explanatory pretensions must themselves rest on a described grammar.

Whatever connections SP might have with Wittgenstein's philosophy, it is clear that it is flawed in many respects, one of its pervasive flaws being the non-normative character of the scientism SP embraces. Even SP's communitarianism is not normative (from whence the ought of community conformity?). What can we say of the self-proclaimed sociological heirs to the subject that used to be called philosophy? We may adapt a remark of Wittgenstein, replacing his generic word 'science' by one of its species 'sociology': 'Man has to awaken to wonder – and so perhaps do peoples. Sociology is a way of sending him to sleep again' (Wittgenstein (1980), p. 5).

NOTES

[1]   Chrysippus' dog features large in Anderson and Belnap (1975), especially pp. 296-7 in §25.1 entitled 'The Dog'. They discuss the rejection of Disjunctive Syllogism in their investigation into relevance (or relevant) logics.

[2]   The example of animal inference is discussed in the final section of Bloor (1991a), pp. 135-9. However he views such cases as boundary conditions and not as counterexamples to SP with its central CT. It remains quite unclear in this case why a counterexample is relegated to a mere boundary conditions for the application of a theory, thereby preserving the theory from refutation. The more central cases which Bloor thinks are paradigmatic for SP are those of human knowledge and inference making which, he maintains, contain central social elements. It is this which is contested in the text following. It is a theme initially developed in Nola (1991) to which Bloor (1991a) is a critical response.

[3]   Talk of a causal structure containing perceptual modules and inferential processors which are not necessarily domain specific, and which perform at least the task of making inferences, is adapted from the theory of modularity proposed in Fodor (1983). The modularity theory is not the only theory that can be adapted to explain what goes on in the case of inference from perceptually acquired beliefs. But it is one of the more plausible theories of what happens that does suit the case at hand.

[4]   This example was used as a refutation of CT by ordinary human inference in Nola (1991). There is a response to the objections made in Bloor (1991a).

[5]   See Hyde and Priest (eds.) (2000) which is a collection of papers by the late Richard Sylvan's work on relevant logics which he called 'sociative'. Chapters 3 and 4 are a useful introduction to various relevance and paraconsistent logics and the different kinds of relevant linkages that have been proposed. Chapter 5 is a preliminary investigation into the history of relevance in logic beginning with Aristotle. Chapter 10 is a paper by Sylvan and Nola which investigates sociative and connexive logics as a way of solving many of the paradoxes that arise in the philosophy of science from the grue and the raven's paradox to problems to do with

confirmation and hypothetico-deductivism that can be resolved within such logics, and a theory of confirmation embedded in connexive and sociative logics.

[6] A similar account can be found in Bloor (1983), chapter 6.4.

[7] A personal report may be in order here. The paper in Hyde and Priest (eds.) (2000) chapter 10 by Sylvan and Nola was written, not with any prize in mind (such a thought never entered our heads) or any other social interest. An interest in relevance logics is very much a minority interest in the field, and not always career enhancing. Our interest was purely philosophical, viz., to attempt to solve some of the problems about combining theories of confirmation within the context of classical inference by changing that context to one of implication within sociative and connexive logics.

[8] The Tarskian account is not without its difficulties; see Etchemendy (1990) for suggested modifications.

[9] In von Wright (1982) pp. 137-62 there is a chapter entitled 'Wittgenstein on Probability' which sets out the limited extent to which Wittgenstein considered the logical theory of probability.

[10] For what it is worth, this is hardly the position that Wittgenstein adopts in respect of truth and falsity (and we may think of the pairs rationality/irrationality and knowledge/not-knowledge in a similar way): '"So you are saying that human agreement decides what is true and what is false?" – It is what human being *say* that is true or false: and they agree in the *language* they use. That is not agreement in opinions but in form of life' (Wittgenstein (1967), Part I §241. Even though this says nothing about a theory of truth, it correctly rejects the idea that human agreement decides what is true or false. Rather agreement lies elsewhere in the language used, or in the form of life (whatever this is), and not at the level of opinions.

[11] As a simple *tu quo que* Kusch (1996), p. 88, offers an example of syllogistic argument of the form: All A are B: No C is B; so, No A is a B. This is used to get readers to accept a particular conclusion. Are we to accept that the conclusion is a correct implication of the premises simply on the grounds of collective endorsement by Kusch's readers (and others)? Or are there independent grounds of the sort Aristotle and others have given which do not turn on collective endorsement for the correctness of the implication in syllogistic reasoning?

[12] It is worth looking at the full context of Wittgenstein's famous remark in the *Philosophical Investigations*. It has the form of a one-way methodological claim in which use is more fundamental than meaning: 'For a large class of cases – though not for all – in which we employ the word "meaning" it can be defined thus: the meaning of a word is its use in the language' (Wittgenstein (1967), Part I §43). In many cases (not all) it is our use of language (i.e., our practice) that is to explicate meaning. The converse does not hold, viz., meaning explicates practice, or use. Wittgenstein also goes on to say: 'And the *meaning* of a name is sometimes explained by pointing to its bearer'. In this case the bearer becomes fundamental and explains what a terms means, as well as use. This is important for all those who would downplay the role of bearers, or what we refer to, in explaining meaning.

[13] It is explicit in Wittgenstein that '"obeying a rule' [e.g., a rule of inference] is a practice", and 'to obey a rule, to make a report, [etc]... are *customs* (uses, institutions)' (Wittgenstein (1967), Part I, §202 and §199).

[14] There are many places in his later writings where Wittgenstein discusses Brouwer's view of mathematics and the nature of existence proofs. For a discussion see, for example, Marion (1998), chapter 6.2 'Excluded Middle and Existence'. However it remains unclear about the extent to which Wittgenstein might have viewed the rivalry between classical and Brouwerian methods of proof, along with intuitionistic logic, as rival practices and codifications leading to a pluralism within logic.

[15] As an illustration in the case of logic, the existence of distributed data bases (for example a person might keep a number of diaries, one in a book, one in a computer, etc) there might be

entries in each that are inconsistent with one another. A classically programmed inference machine (with access to the data bases) would, because of the principle of explosion, keep drawing inferences in this situation. One solution might be to adopt some other logic that does not involve explosion.

[16] Much of this section, as well as the previous and following sections, owe a debt to the fine paper by Friedman (1998) in which the rival agendas of SP and Wittgenstein are set out, especially the non-scientific character of Wittgenstein's conception of the task of philosophy, and its contrasting role with science for both the doctrines of SP and ethnomethodology within SSK.

[17] Such a pluralism is explicitly adopted in the papers by Richard Sylvan on relevant and other logics collected in Hyde and Priest (2000); see pp. 398-9. See also Haack (1978), chapter 12, for an account and defence of pluralism in logic.

[18] It is well known (and it will be discussed in the next section) that Wittgenstein gives a sequence of numbers 2, 4, 6, 8, ..., and then asks 'how does one go on?'. As is also well known, one can go on in any way. Similarly, given a finite sequence of numbered sections of Wittgenstein's later writings on could ask his legion of interpreters 'how does one go on?' And his interpreters do go on in many different ways allegedly faithful to the initial sequence of numbered remarks. Perhaps Wittgenstein's own work, set out as numbered sections, exemplifies the thesis of rule scepticism that many find in his work.

[19] See Lynch (1993) especially chapter 1 but also chapters 4 and 5. See also the papers in Pickering (ed.) (1992) by Lynch 'Extending Wittgenstein: The Pivotal Move from Epistemology to the Sociology of Science' pp. 215-265, a response by Bloor 'Left and Right Wittgensteinians' pp. 266-82, and the reply by Lynch pp.283-300. In this encounter some of the differences between Bloor's and Lynch's approach to SSK emerge starkly.

[20] For an example of the ethnomethodologists' descriptions see Lynch in Pickering (ed.) (1992), pp. 247-56; the case of astronomical discovery mentioned here is discussed in the text below. Though the claims of SP might be exciting but false, the descriptive work of ethnomethodologists on the recordings of scientists' conversations is the converse of exciting.

[21] For an account of what grammar and its description might mean, see Hacker (1972) chapter VI, and Garver (1996).

[22] In his paper in Pickering (ed.) (1992), pp. 247-56, Lynch discusses the GLL paper and makes some remarks that look as if he is attempting to distance himself from the earlier paper's anti-realism: 'The IGP is the pulsar that by night's end is assigned an identity and a set of astronomical properties' (*ibid.*, p. 248). This looks realist enough. However he continues to endorse GLL's talk of the IGP as a "cultural object". But one cannot have it both ways. It is hard to know what confusion is involved here. Is it merely a confusion of (a) a person using a word to name and talk of some object (whether it exists or not), a matter which is clearly cultural, with (b) whether the name a person uses actually refers, which is not a cultural matter?

[23] 'From the sociological standpoint socialization, consensus, and the like, far from being outside the internal relationship, are actually *constitutive of it*' (Bloor, 'Left and Right Wittgensteinians', in Pickering (ed.) (1992), p. 272).

[24] See Bloor (1997), chapter 8, much of which is devoted to the case of Crusoe. See Baker and Hacker (1984), pp. 38-42 and 1985 pp. 171-9 for their account of Wittgenstein on Crusoe.

[25] Commentators often distinguish the idea of a private language in the sense of following a rule privately from that in which one names a sensation. For a suggestion of the latter, see Wittgenstein (1967) Part I §275, which contrasts with §269 in which a private language is one in which only I '*appear to understand*'. Further, §202 makes clear that 'to think one is obeying a rule is not to obey a rule. Hence it is not possible to obey a rule privately'.

[26] Kripke does not tell us what are to count as facts. Clearly there are no meaning facts. But there can still be other kinds of fact such as those of common-sense, those of the various

sciences (from physics to sociology), those about our mental states and those of arithmetic. In criticism of Kripke, Bloor ((1997), chapter 5), attempts to fence off sociological facts as a way of lessening the argument to meaning scepticism through lack of the right kind of fact. But it is clear that Kripke could take naturalistic sociological facts on board and still arrive at his sceptical conclusion.

27   See pp. 221-5 of 'The Individual Strikes Back' in Blackburn (1993). Similar considerations also arise in Blackburn (1984) chapter 3 §3.

28   In what follows talk of kinds, sorts and types will not be distinguished. Moreover the bearers of kind terms are not always classes (with instances as their members) but kinds. The current class of (the now extinct flightless bird) moa is not the bearer of the kind term 'moa'. Nor need it be the class of all moas past, present and future; nor all moas in all possible worlds. Classes are often individuated by their members. But if we envisage that moas might not have gone extinct and that there might have been more moa than there have been, then the class of all moas is not the kind moa. Finally, if one has nominalistic qualms about kinds, types or sorts as being too "abstract", then remember that classes are also abstract objects, perhaps in a way more problematic for nominalists than kinds.

29   Class nominalists are at least committed to one kind of abstracta, classes, as Bloor appears to be. It is rather hard to be a pure nominalist admitting only concrete naturalistic individuals items; if one admits classes, then they are distinct from the individual members that make them up. Classes are quite abstract entities, a point which SP does not face. On class nominalism see Armstrong (1978), chapter 4.

30   So far we have only spoken of substance kinds. But in science we must also admit kinds of event, such as reflection, refraction or attraction, and kinds of process such as catalytic processes, attraction processes, meiosis, metabolism, etc. The focus in the text on substance kinds (whether named by count nouns such as 'tiger' or mass terms such as 'water') should cause no problems.

31   There is an excellent discussion of realism, the social construction of kinds and the connections between these, in Searle (1995), chapter 1.

32   In the *Phaedrus* 265E to 2656A (see the translation and commentary of Hackforth, (1952)), Plato actually talks not of 'carving nature at its joints' but of making divisions that follow 'objective articulations'. As Plato emphasises, it is clumsy classifiers who hack like butchers across these articulations in ignorance of them. Thus it is not so much the objectivists who 'carve' nature at its joints – the carving metaphor has been misplaced. We find the world already "carved" into kinds independently of our thoughts, theories and activities. Rather it is non-objectivists, particularly constructivists, who are liable to hack, in ignorance, across nature's articulations; the objectivist merely follows nature's articulations.

33   A realism about kinds which eschews nominalism and which rejects the view of kinds as an artefact due to our language is spelled out in Ellis (1996), and more fully in Ellis (2001). Ellis also provides a theory about the nature of kinds, their essence and how their essential properties enter into explanations. A less robust but nonetheless realist view of kinds can be found in Quine's essay 'Natural Kinds' (in the book of Quine (1969), pp. 114-38). Quine considers the resemblance relations which can underpin kinds but puts greater emphasis on how science postulates kinds as part of its explanatory framework; such explanatory efficacy is part of our evidence that there are natural kinds. There is nothing in this that would make kinds social constructs.

34   In Barnes, Bloor and Henry (1996), pp. 66-9, the authors provide one attempt to come to terms with meaning finitism's apparent lack of acknowledgement that the world might contain natural kinds. However their discussion remains at the level of how names are attached to natural kinds and how one goes on to the next case of any given kind, two important issues more fully discussed in Kripke (1980), Lecture 3 and Putnam (1975), chapter 12. But this says

nothing about whether or not there are kinds independent of our classificatory practices. As becomes clear in their text, the sociologists doubt that there are objective kinds existing independently of our classificatory practices.

[35] One famous case often cited is Bulmer's (1967) 'Why the Cassowary is not a Bird' thereby indicating the Karam of Papua New Guinea do not classify the cassowary in the way we do. But what is evident is that the Karam do at least recognise a distinct kind *cassowary*, just as we do. It is not as if, between us and the Karam, one misses a kind classification that the other makes. And this is the only point being argued in the above. Where we and the Karam differ is in the higher reaches of classificatory concepts. The kind cassowary does not fall under the broad classificatory term bird but elsewhere in the Karam taxonomy of the cosmos. The late Ralph Bulmer was a colleague of mine. Intrigued, I asked him just how much difference in classification of their flora and fauna there was between us and the Karam. Bulmer's reply was that he was always impressed by the way Karam and other Papua-New Guinea classificatory systems fitted most of the classificatory system of western science, occasionally more perceptively but rarely less.

[36] See Kuhn (1970) chapter 10 for the idea that, with change of paradigm, the incommensurability of successive paradigms is such that the ontology of kinds and other entities postulated in each paradigm changes with change in paradigm, i.e., the very word itself changes. For the realist the world is still there; what has changed is not the world but our beliefs about it. But such change in belief, on the Kripke-Putnam view of reference, does not necessarily entail change of what we talk about. That can remain invariant despite change in our theories. For an account of the fixing of reference for scientific terms that avoids Kuhnian world-changing incommensurability, see Kroon and Nola (2001).

PART III

THE FRENCH CONNECTION:

FOUCAULT

SYNOPSIS OF PART III

The two chapters of Part III deal with some of the views of Michel Foucault, especially his "power/knowledge" doctrine. Foucault's theories are applied by him to all the sciences, but mainly the human sciences. And since they make claims about the non-rational way in which theories succeed one another, his views have a place in this book. Chapter 8 deals with his obscure and difficult doctrine of what a discourse is, and his claim that there are deep ruptures and discontinuities dividing all discourses, especially those of the sciences. This is the topic of Foucault's most abstract 1969 book, *The Archaeology of Knowledge*. Chapter 9 deals with a claim, central to his post-1969 writings, that the causes of the ruptures and discontinuities have to do with the nexus of power in which all discourses, scientific or not, exist.

Section 8.1 begins with Foucault's conception of knowledge. It is quite unlike the conception of knowledge outlined in Part I, and in fact it hardly addresses traditional epistemological concerns. Most of the section is taken up with unravelling the use of the two French terms, *connaisance* and *savoir* (for which the one English word, knowledge, has to do double duty), and showing why what Foucault says has little bearing on traditional epistemology and must presuppose it. Foucault does not always use either term in their traditional sense; often he uses *savoir* to stand for something like the presuppositions that make a discourse possible. So, one of Foucault's archaeological tasks is to uncover the presuppositions that make the various discourses in science possible. Since the notion of a discourse has a prominent role in Foucault's theory, section 8.2 addresses two related questions. What makes one discourse different from another? And what makes the statements of a discourse different from one another? As will be seen, much turns on the somewhat unclear notion of the "rules of formation of a discourse". In turn, this introduces more of Foucault's technical vocabulary concerning the rules for the formation of enunciative modalities (see subsection 8.2.4), and rules for the formation of concepts and strategies (section 8.3). The remainder of the chapter attempts to come to grips with the notion of rules for the formation of objects. Section 8.4 broaches this issue by discussing a case, the construction of the "object" madness. What this reveals is that Foucault's theory of discourse is not so much about madness, or the varieties of madness, but rather about our *discourse*, or *talk* about madness. In fact it turns out that the various psychiatric discourses are not about the real existing varieties of madness (such as depression, schizophrenia, etc.); they are about, not these real objects, but what are called *ersatz* objects, different ersatz objects being constructed in each of our discourses. Section 8.5 sets Foucault's views against the broader issues of realism and antirealism in science, and contrasts the view that there are natural kinds in the various sciences that we set out to discover with Foucault's view that we do not discover anything about an independent reality, but rather construct the ersatz "objects" and kinds of our discourse. Section 8.6 focuses on the strong role that the context of our statements plays for meaning and other matters, especially how contextualism gives rise to Foucault's constructivism concerning ersatz objects. With this background, section 8.7 returns to the issue of how discourses are to be individuated, and the difficulties in his position. The final section raises a paradox for Foucault's constructivism as it applies to his own theory of discourse. This is a serious problem for his theory of discourses that Foucault recognises, but fails to remedy.

Foucault's archaeology is a descriptive enterprise that attempts to uncover the breaks and ruptures that have arisen in our discourses. Chapter 9 deals with Foucault's genealogical enterprise that attempts to say something about the causes of discourse discontinuity. Sections 9.1 and 9.2 trace the development of his view that it is power, or social relations of power, that bring about discourse change. Foucault seems never to have doubted that power might have little to do with knowledge; but what the connection between power and knowledge might be was always an unresolved problem for him. Section 9.3 considers Foucault's doctrine of power and raises some critical points against it, arguing that his notion of power is far too broad. Relations of power are seen everywhere in all contexts. However this is not because power arises where we had not noticed it until Foucault pointed it out; rather, the omni-presence of power is a defect of Foucault's theory. Section 9.4 examines Foucault's doctrine of "power/knowledge" and what it might mean, since "knowledge" is also caught up in the nexus of power. Two of Foucault's central crucial claims about the knowledge-power link are distinguished, viz., the view that power produces knowledge, and the converse that knowledge produces power. The second claim has some plausibility when carefully understood, but hardly advances beyond earlier claims of Francis Bacon. But the first claim shows how closely Foucault's doctrine resembles that of the Strong Programme of Part II, and thereby inherits many of its faults. In section 9.5 six major objections are raised against Foucault's "power/knowledge" doctrine, especially the claim that power produces knowledge. The final section 9.6 briefly considers how the doctrine impinges on the notion of truth that Foucault employs, and what difficulties there are in this. The final upshot is that Foucault's power/knowledge doctrine has as little to recommend it as the Strong Programme it resembles, and in fact even less.

CHAPTER 8

AN ARCHAEOLOGICAL DIG THROUGH FOUCAULT'S TEXTS

The works of Michel Foucault have had an extraordinary influence in sociology, history, literary studies, cultural studies – and in some cases even philosophy. Despite this it will be argued here that one of Foucault's central doctrines exhibits many of the features, and thus the defects, of the sociology of science discussed in Part II. However his terminology and his emphasis on different factors such as power set him apart as a distinctive theoretician. Much of the commentary on Foucault is hagiographical. In part this is due to some quite pervasive obscurities that few commentators attempt to clarify while supporting Foucault's general stance. But it is also in part due to the contagious mistrust of notions like truth and rationality in Foucault. His attack upon such notions has strongly influenced the postmodernist rejection of so-called 'enlightenment' intellectual values by his followers. But much of this, as will be argued, is simply mistaken. The mistake is in part due to an over-inflated account of what these "enlightenment" values might be, along with a poverty-stricken view of what are the current accounts of rationality in science and elsewhere (see chapter 1), and a failure to recognise the normativity of knowledge. In turning off the lights of the "enlightenment", Foucault and his followers manage to undercut the very basis of their own doctrines. We need not turn to any "enlightenment" philosophers for an account of what these values might be; but some idea can be gleaned from the discussions in Part I of knowledge and our epistemic values, particularly chapter 1 where the idea of a critical tradition concerning science is outlined.

For Foucault and his followers the expressions 'knowledge', 'power', 'discourse', 'genealogy', 'power/knowledge' and 'power/discourse', (sometimes with '-' in place of '/') occur with a frequency that is often inversely related to the degree of comprehension they engender. In the case of the last two expressions it is assumed that there must be some significant relation that the sign '/' could stand for amidst the many evident implausible connections. A critical analysis of what '/' signifies needs to move beyond bland characterisations like 'knowledge occurs in, or is influenced by, its social and/or power context'. As for the Strong Programme, finding a significant relation always remained a problem for Foucault. Even in the year before his death, he said in an interview in response to a query about the claim that reason is power:

FOUCAULT You must understand that is part of the destiny common to all problems once they are posed: they degenerate into slogans. Nobody has said, "Reason is power". I do not think anyone has said knowledge is a kind of power.
INTERVIEWER It has been said.
FOUCAULT It has been said but you have to understand that when I read – and I know it has been attributed to me – the thesis "Knowledge is power", or "Power is knowledge", I begin to laugh, since studying their *relation* is precisely my problem. If they were identical, I would not have to study them and I would be spared a lot of fatigue as a result. The very fact that I pose the question of their relation proves clearly that I do not *identify* them. (Kritzman (1988), p. 43)

But Foucault is being disingenuous here because, as will be seen in the next chapter, Foucault does talk of identity. Moreover Foucault seems not to recognise the symmetry of the 'is' of identity in feeling the need to express twice the thesis attributed to him, once as "Knowledge is power", and again as "Power is knowledge'". But as will also be seen, Foucault could be understood to adopt other relations, such as causality or counterfactual dependence. Taken charitably, Foucault clearly insists that some relation, but one not as strong as identity, must hold between power and knowledge. This weaker claim is something that not all his followers accept.[1] However Foucault did not envisage an even weaker and more plausible claim, viz., that only occasional historically contingent connections of causal dependence might hold in some cases; nor did he envisage, as will be argued, that there is no relationship at all in the direction from power to propositional knowledge that ....

The task of the next chapter of Part III will be to explore and critically evaluate Foucault's "power/knowledge" doctrine. It will be argued that Foucauldian power plays the same role in his "power/knowledge" doctrine as do social factors in the Strong Programme. Understood this way, the "power/knowledge" doctrine is heir to exactly the same defects as the Strong Programme. But worse; since Foucauldian power is only one kind of social factor, the "power/knowledge" doctrine turns out to be an even narrower doctrine than the Strong Programme, thereby adding further to its implausibility. It will also be argued that, despite the rhetorical flourishes that surround the "power/knowledge" doctrine, little more can be truthfully claimed for it than what one can find on behalf of the 400 year-old slogan attributed Francis Bacon, viz., 'knowledge is power'.

This chapter will discuss a number of preliminaries necessary for an understanding and critical evaluation of the "power/knowledge" doctrine. Foucault develops a theory of the nature of discourse using a technical language of his own. He uses in an idiosyncratic way a number of terms such as 'knowledge', 'discourse', 'statement', 'archaeology', 'genealogy', and so on. These are not to be taken at their face value, even though the first three have a currency outside the context in which Foucault (or more strictly his translators) has used them. What they mean in Foucault is often quite unclear.

Most of these terms, along with many others, get an extensive, but often obscure, outing in one of Foucault's most dry and abstract works, *The Archaeology of Knowledge* (first published in French in 1969 and translated into English in 1972). The terminology persists in his later more genealogical works in which the "power/knowledge" doctrine flourishes. In *Archaeology* the discussion of these notions moves at a high level of abstraction rarely relieved by illustrative examples that might help the hapless reader. There is not much commentary on this book by his followers, and a commentary will not be given here. In this chapter only a few of Foucault's notions and doctrines will be discussed sufficient to enable one to grasp the "power/knowledge" doctrine which emerged fully in his post-1969 works, while often expressed in the language of the 1969 book.

The focus will be on notions such as knowledge and discourse, the associated ideas of the unity of a discourse and discursive formations, the very broad contextualist theory of meaning and language that Foucault presupposes, and the relation of all this to knowledge and to science. As will be seen, Foucault's claims are a mixture of the correct, the sensible and useful along with the bizarre and downright perverse, the mixture being such that it is often hard to tell which is which. In what follows an attempt will be made to cut through much of this sufficient to put the "power/knowledge" doctrine in an appropriate setting.

Before proceeding further there is one general theme occurring throughout Foucault's work that needs highlighting. Foucault had been impressed, often over-impressed, by the fact that our bodies of belief pertaining to any domain, such as biology, economics and grammar, have undergone profound changes over time. In his view the changes have not been gradual; nor have they been cumulative; nor have there been underlying continuities amid piecemeal adjustments; nor have there been long periods with the same underlying trends with only surface ripples of difference. For Foucault, such a stance characterises much of what he calls the 'history of ideas'. In rejecting such a history of ideas, Foucault produces his own contrasting view, which is central to his "archaeological" approach. For Foucault the changes in the history of science, and thought more generally, are quite cataclysmic and proceed by ruptures, breaks, discontinuities and deep differences in which there are not to be found any underlying continuities and trends. Foucault presents himself as the theoretician of such cataclysmic change. As such he shares some of the more radical pretensions that can be found in Thomas Kuhn's 1962 book, *The Structure of Scientific Revolutions*. Kuhn argues that science proceeds by a series of revolutions characterised as "paradigm changes" which issue in a number of different incommensurabilities. These incommensurable changes are not just in the laws of a pair of successive "paradigms", but also in their experimental or observational practice, and their concepts, models, values, methods, ontology and observational base (given the doctrine of the theory-ladenness of observation).

We saw in section 1.4 something of the radical stance of the 1962 version of Kuhn's book with respect to the paradigm-relative character of our criteria of methodological evaluation. This paradigm relativity is extended to a host of other items, including the ontology of the theories within a paradigm and the concepts employed in the claims made within the paradigm. In these respects both Kuhn and Foucault are typical products of the 1960's with its ethos of radical change, challenge, rejection and difference. In what follows a number of comparisons will be drawn between the Kuhn of paradigm incommensurabilities and the Foucault of ruptures and discontinuities. But there are also differences to be noted as well. For one thing Kuhn is not as implausibly radical as Foucault who pushes the idea of discontinuities to extremes. And later, as we will see, Foucault married his doctrine of change to the "power/knowledge" doctrine, something that Kuhn resisted (as we have seen in section 1.4) in its Strong Programme version. However in so far as their 1960s works are concerned, there are a number of quite fruitful similarities to note amid the differences. The important common feature to note is that, granted that the history of human thought and theorising is cleaved by rupture, discontinuity and incommensurability, both set out to provide what we might call a 'meta-theory' about such radical change. Kuhn provides a meta-theory about the nature of paradigms and Foucault a meta-theory concerning discourses. These meta-theories contain a distinctive battery of concepts and doctrines about radical change. In the case of Foucault, there is a meta-theory about discourse (a term used in a special sense by him), discursive formations and a theory of "discursive regularities" which involves talk of such things as "rules of formation" for items such as objects and concepts. Much of this chapter will be devoted to spelling out what all this means, and just how viable it is as a meta-theory. Unfortunately much of the chapter will be negative. Not only does Foucault exaggerate the discontinuities that have actually occurred, but he also develops a misleading, obscure and tendentious meta-theory that contributes to the exaggeration.

8.1   FOUCAULT ON KNOWLEDGE

If we are to grasp Foucault's "power/knowledge" doctrine (*pouvoir/savoir*) then we need to know not only what relation '/' stands, for but what the two relata are, viz., power and knowledge. We will leave the discussion of the first of the relata, power, to the next chapter; here we will deal with the second relata, knowledge. In chapter 2 a number of features of knowledge were spelled out. (a) The different constructions that 'know' can enter into in English such as 'know how to...', 'know why...', 'know how...', know what...', know that...', and so on; (b) the important distinction between 'know that ...' and 'believe that ...'; (c) a number of different theories of knowledge that ... from Plato's tether account to a version of a foundationalist theory of knowledge, or a reliabilist theory, etc; (d) social aspects of knowing and knowledge; and so on for other aspects of our very rich concept of knowledge. Each of these distinctive aspects has to be treated separately to avoid confusion. In many writings of Foucault it is as if none of these important distinctions exist; hence there is much epistemological confusion. On some occasions Foucault shows awareness of these quite traditional aspects of our notion of knowledge that philosophers have been at pains to set out in the history of philosophy. On other occasions he either appears to be oblivious to these aspects of knowledge, or uses the notion in his own idiosyncratic way – in which case one needs to be aware that his claims then have little to do with traditional concerns about knowledge.

Importantly, one needs to be aware when Foucault is continuing a traditional debate when he uses epistemic notions, and when he turns his back on that debate and conducts another debate that has nothing, or very little, to do with the former debate – though, confusingly, the same vocabulary gets used. If the latter, then in providing an account of 'knowledge' in his idiosyncratic sense, Foucault is not taking part in the traditional debate about knowledge but is simply changing the topic of conversation. In so far as he does not take part, then Foucault cannot be said to even be revising the traditional conceptions of knowledge for which philosophers have argued, and which do get embodied in our "enlightenment values". Such values are those of truth, evidence, rationality, etc; and there are the norms associated with knowledge and methodology more generally (see chapter 3). These are quite central to traditional concerns about knowledge, whatever more particular theory of knowledge one adopts. However matters are not always so simple, as will be seen. Foucault often runs with the hares of rational epistemology and hunts with the hounds of postmodernism panting for the blood of rationality. Here it will be argued that Foucault's pursuit of knowledge, in the sense of *savoir* to be discussed shortly, must move entirely within the traditional framework of knowledge, despite claims to the contrary that his concerns are often not those of traditional epistemology. But alas, what Foucault advocates often fails to meet the requirements of traditional epistemology.

As already indicated, Foucault seeks 'a general theory of discontinuity' (Foucault (1972), p. 12) in our thought and practices and not a theory that emphasises its continuities. It is often claimed that such a theory cannot be located within traditional epistemology. But problems of discontinuity between systems of beliefs have concerned epistemologists as traditional as Descartes whose epistemology is in part a response to the scepticism induced by the scientific revolution of his day, though he did not produce a theory of discontinuity. And it remains a concern in current philosophy of science when philosophers as different as, say, Popper, Kuhn, Feyerabend, Lakatos, and even Bayesians, wrestle in their own way with ontological, semantic and epistemological problems arising from the discontinuities and renovations in the sciences. As already indicated Foucault's views bear a strong resemblance not only to those of Kuhn, but also Feyerabend and Lakatos, and in a different vein, the many sociologists of knowledge who were writing at the same time. The quite different linguistic garb in which he expresses his views, and his different lineage *via* French writers such as Canguilhem, Bachelard and Althusser, should not obscure this fact.

For a book entitled *The Archaeology of Knowledge* little is said about knowledge (the French title includes 'Savoir' and not 'Connaissance' though the English 'knowledge' has to do duty for both French terms, as it does for the German 'wissen' and kennen'). Nor is anything said about the epistemological distinction between knowledge and belief; the French words for belief (the verb *croire*, the noun *croyance*) hardly occur at all in the book and are not contrasted with any kind of knowledge. We can, however, glean a little about his notion of knowledge when Foucault says of his archaeological enterprise in a chapter of *The Archaeology of Knowledge* called 'Science and Knowledge':

> Instead of exploring the consciousness/knowledge (*connaissance*)/science axis (which cannot escape subjectivity), archaeology explores the discursive practice/knowledge (*savoir*)/science axis. And whereas the history of ideas finds the point of balance of its analyses in the element of *connaissance* … archaeology finds the point of balance of its analysis in *savoir* – that is, in a domain in which the subject is necessarily situated and dependent, and can never figure as titular. (Foucault (1972), p. 183)

This is hardly crystal clear. But there is a strong flavour of an epistemology in which, like Popper's theory of objective knowledge, the supposed independence of the individual knowing subject is downplayed in favour of a more dependent role with respect to an objectivised "what is known". That is, talk of an individual subject, as in 'S knowing that p, is downplayed in favour of remarks like 'it is known that p' which refer to no subject of knowledge. Also Foucault is at pains to distinguish his archaeological enterprise (of which more in subsequent sections) from the common notion of a "history of ideas". In large part the difference will turn on the special theory of discourse that Foucault develops as a tool of archaeological analysis that is not found in the history of ideas; indeed it can be used in the very analysis of what the history of ideas itself says as well as what it might investigate.[2] All we need note now is that Foucault uses the terms *connaissance* and *savoir* to mark the difference between what the history of ideas aims to obtain as knowledge, and what archaeology aims to obtain as knowledge. But it is hardly clarificatory to use these two French terms in a non-standard way to mark a distinction that in itself not always easy to grasp.

8.1.1 Knowledge as Connaissance

A footnote especially provided for the English translation of *The Archaeology of Knowledge* throws a bit more light on the two axes:

> The English 'knowledge' translates the French *'connaissance'* and *'savoir'*. *Connaissance* refers to a particular corpus of knowledge, a particular discipline – biology or economics, for example. *Savoir*, which is usually defined as knowledge in general, the totality of *connaissances*, is used by Foucault in an underlying, rather than an overall, way. He has himself offered the following comment on his usage of these terms:
>
> 'By *connaissance* I mean the relation of the subject to the object and the formal rules that govern it. *Savoir* refers to the conditions that are necessary in a particular period for this or that type of object to be given to *connaissance* and for this or that enunciation to be formulated.' (Foucault (1972), p. 15 footnote 2)

Again, this is hardly clear or adequate. For the translator *connaisance* is a body of knowledge (why not belief?) that is demarcated in some way, such as biology, economics, grammar, etc. Nothing is said about rival theories that might be held by different people at the same time, or by the one person at different times. However it is reasonable to assume that, say, the 18th century biology of Lamarck (further refined to some period in his life since he held different theories at different times) is different from the *connaissance* held by Darwin either before or after he developed the theory of evolution. Some in this context also speak of 'scientific *knowledge*' and mean by this the different theories held at different times, including those which have been shown false, or are at least not conclusively true. We could, less controversially, refer to these *connaissances* (indexed to a person and a time) as a "bodies of belief" or "sets of belief", or even theories, held by a person at a time, thereby avoiding issues to do with one of the implications of 'knowledge that...', viz., truth. However the contrast between knowledge and belief, so common in philosophy, is passed over by Foucault. And many of his followers, instead of talking, in the plural, of 'bodies of belief', or sets of beliefs, or theories, use the gross neologism 'knowledges' instead; and this obscures all connection of knowledge to truth and evidence. Despite this, we do have one common use of 'knowledge', viz., that found in those scientific theories for which we have some reason or evidence for their truth (or truthlikeness). And this can, for our purposes here, be deemed to be uncontroversial.

A slightly different emphasis is given by Faubion in his 'Introduction' (Faubion (ed.) (1998), pp.xxvii-xxx). In ordinary French the verb *connaître* functions like the English 'is acquainted with' or 'is familiar with'. But Foucault also uses it in a broader sense including acquaintance with such things as a body of learning, or what one has mastered. Understood in this way, a person can be acquainted with a range of items such as another person, a town, a body of regulations, Beethoven's sonatas, Lamarck's earlier views on evolution, his later views, Darwin's theory of natural selection, and so on. Thus there is continuity from the ordinary French to Foucault's more technical use of *connaissance*, this last perhaps being best understood as acquaintance. And amongst the "objects" with which we are acquainted can be included the various sciences. From knowledge by acquaintance of some "object" x (e.g., natural selection), it is a short step to claim that we have knowledge *of* x; and such knowledge of x in turn issues in a lot of propositional knowledge that ..., which is about x.

8.1.2 *Knowledge as* Savoir

The English 'knows that...' is commonly translated using *'savoir'*. The translator suggests that talk of *savoir* is more general and underlying than *connaissance*. Here we might think that *savoir* has something to do with more general matters, such as the conditions that must apply for any body of belief to be called 'knowledge', as set out in some definition. But this thought is dispelled when the translator goes on to talk of the *totality* of *connaissances*. Here the generality is not thought of in terms of conditions (such as truth and evidence conditions) for any body of *connaissance* to be knowledge. Rather it is the less interesting conjunction of *connaisances*, such as the conjunction of biology, economics, etc (though nothing is said about how one might avoid inconsistencies arising in such conjunctions thereby voiding any pretence to knowledge).

   Foucault's own gloss on the difference neither completely supports what the translator says, nor makes matters much clearer. But at least there is a difference in level of generality. For Foucault *connaisance* pertains to one kind of object and the 'formal rules that govern it'. Talk of rules here concerns Foucault's theory of discourse to be elaborated later. However we can say now that such rules pertain to a particular scientific theory (or discourse) with which we are acquainted; and these rules give rise to the "objects" of that theory (discourse) with which we can also be acquainted. For example, in a biological theory held at a given time, there might be "objects" such as *species*, evolutionary relations between species such as *descent*, and mechanisms of evolution such as *natural* and/or *sexual selection*. As will be seen in Foucault's theory of discourse (see section 8.4), for different scientific discourses at different times there will be different "objects". Here we can follow Foucault and say of a person who has mastered the appropriate sciences that they are *acquainted* with, or *know*, such "objects" as species, sexual selection, etc. (Can some "objects" be the same yet occur in different scientific discourses (which have different "formal rules"), or is there always some radical Kuhnian incommensurability? This is a controversial matter to be addressed later in section 8.5).

Further, there is the special Foucauldian claim that 'formal rules govern objects'. What this means will have to wait until section 8.4. But what Foucault says here does make some contact with the traditional notion of knowledge, as when we claim that we have knowledge *of* such "objects" as species, relations of descent, or specific evolutionary mechanisms, and so on. As suggested above, talk of 'knowledge *of* x' can be understood along the lines of 'being acquainted with x', or being familiar with, or able to recognise, x'. And all of these locutions need to be cashed out in a display of knowledge that ..., viz., propositional knowledge (*savoir*) about such "objects". Some might be unhappy to talk of 'knowledge' in the context of scientific theories that are false. They would be happy to fall back on talk of a "body of belief" (for which there is some evidence providing some degree of confirmation for the existence of such "objects"). But as noted the French for 'belief that ...' (Greek 'doxa') that contrasts with 'knowledge that...' (Greek 'episteme') hardly occurs in Foucault's writings.

The ordinary French *'savoir'* is commonly understood to mean 'know(s) that ...' where the blank can be filled by any propositional content such as that expressed by 'roses are red', 'electrons are negatively charged', etc. Foucault does use *'savoir'* in this sense. But he also uses it in a quite technical sense that has little to do with the propositional knowledge that ... that arises in each science. *Savoir* is also the knowledge (if there be such) that pertains to the postulation, in accordance with Foucauldian formal rules, of this or that "object" in particular *connaissances*, and the conditions that are necessary for this. The difference seems to be that *connaissance* concerns the particular knowledge we have by acquaintance in some given science (which can be cashed out as knowledge that...). In contrast the technical use of *savoir* is intended to provide quite special knowledge of the *conditions* that make *possible* the postulation of this or that "object" in a discourse, changes in the character of the "objects" and the abandonment of one "object" for another as discourses change. Philosophers sometimes use the word 'transcendent' to refer to 'conditions that are necessary for some item to be possible', and a transcendental investigation is an investigation that yields knowledge of such conditions of possibility. But 'transcendent' is too grand a term for the *savoir* that Foucault seeks. It will turn out that the conditions of possibility are themselves highly contingent historical, social conditions that are themselves open to change. So what will be proposed here is that the contingent conditions which are necessary for, but make possible, the postulation of a range of object across a number of theories or Foucauldian discourses be called ' presuppositions of a discourse'. Then Foucauldian *savoir* aims to get a quite particular kind of knowledge that ..., in the quite traditional sense of getting true beliefs along with a justification for their truth , viz., knowledge that the presuppositions of a discourse are such-and-such. Is Foucault successful in providing such special knowledge? The thrust of the rest of the chapter is 'no'.

8.1.3 Savoir *as an Elusive Transcendent Object of Discourse*

One might be inclined to see in Foucault's special use of *'savoir'* similarities to what Collingwood says of the *presuppositions* of any theory, including its ontological assumptions. And we have already notes a similarity with the views of Kuhn in which objects are *postulated* relative a *paradigm*. Again one could note a similarity with Lakatos' talk of a "hard core" and "heuristics" of a "scientific research programme" with their differing ontologies. There is also an affinity with Carnap's talk of framework principles in which we can ask questions *about* the ontology of an entire framework, as distinct from questions about what exists *within* a given framework. (As an illustration we can ask within the number framework 'Is there a prime number between 7 and 12?'; but to ask 'Are there numbers?' is to pose a quite different question about the number framework itself.)[3] Again one may wish to talk of an historical *a priori*, a phrase Foucault uses ((1972), p. 206). But the term *'a priori'* must be stripped of its Kantian connotations in which it has to do with knowledge which can be obtained independently of experience. Here the *a priori* is relative to a given set of conditions that make something possible, and these conditions are themselves highly contingent and can only be known in a non-*a priori* fashion, viz., they can be known empirically (*a posteriori*).

Finally one might see a considerable affinity between Foucault's quest for *savoir*, in the sense of discourse presuppositions, and Wittgenstein's talk of the contingent natural history which underlies our language games and their associated form(s) of life (outlined in sections 7.4 and 7.5). One of Wittgenstein's tasks is to describe (but not explain) what this natural history is like, and sometimes to envisage changes in it that then result in incoherence elsewhere because they are not our current practice within our actual natural history. One difference is that the changes in Wittgensteinian natural history and language games occur only slowly, if at all; in contrast changes in Foucauldian *savoir*, the presuppositions of our discourses, are commonplace in the history of thought and practices. But both kinds of change uncover an alleged "incommensurability" between different discourses or language games. There is also the incoherence that a person engaging in one discourse, or language game, will find in the different language games or discourses of another. And this applies, especially in the case of Foucault, not just to different people but also to different periods in the history of human thought, and different practices in different discourses.

Such parallels are instructive in showing that Foucault's quest for *savoir* (in his non-standard sense) has been a quest elsewhere in philosophy, including "positivists" such as Carnap. But such parallels might not explicate all that Foucault intends, especially when we consider his talk of "formal rules" in this context. Whatever the case, if we grant Foucault's technical uses of *connaissance* and *savoir*, then there is nothing to suggest that Foucault is not engaging in a quest for knowledge that ... in the quite traditional sense, i.e., to believe truths which are supported by evidence for their truth. But there is a difference in the two kinds of "objects" of knowledge being sought. The one is the quite standard knowledge by acquaintance (*connaissance*) that yields knowledge of "objects"; this can ultimately be cashed out as knowledge that p where 'p' ranges over the propositions of some theory within science. For example, 'p' ranges over, in the case of Darwinian biology, claims about what species there are, what characteristics they have and how they differ, what evolutionary relations they stand in, and what mechanisms bring about evolutionary change. But, as suggested, this "knowledge" is best thought of as a body of belief, for there are many cases where our scientific theories fail to meet, or only partially meet, the strict conditions for knowledge set out in chapter 2. As Popper would put it, our theories might not be wholly true or false; instead they have some degree of verisimilitude or truth-likeness.

The second kind of knowledge that p (Foucault's technical *savoir*) is a more special higher-level, or presuppositional, kind of knowledge where the 'p' have something to say about the conditions necessary for the postulation of the objects of a range of theories, these objects being given by formal rules. Whether we can have such a kind of knowledge that ... is a moot point, as will be seen, due to the very obscurity of Foucault's theory about such rules. But at least we have reduced somewhat the mystery that surrounds Foucault's idiosyncratic use of the two French terms that get translated as 'knowledge'.

Finally in a 1968 interview about the time *The Archaeology of Knowledge* was being composed, Foucault puts a different gloss on his enterprise:

> Knowledge [*savoir*] is not the sum of scientific knowledges [*connaissances*], since it should always be possible to say whether the latter are true or false, accurate or not, approximate or definite, contradictory or consistent; none of these distinctions is pertinent in describing knowledge, which is the set of elements (objects, types of formulation, concepts and theoretical choices) formed from one and the same positivity in a field of unitary discursive formation. (Faubion (1998) p. 324)

This makes *connaissance* much less like knowledge by acquaintance and more like knowledge that..., to which can then be applied the principles of methodological appraisal that Foucault lists. This is one of few places where Foucault talks of methodological appraisal; and it is clear that it applies to the propositions of each *connaissance*. But what of *savoir*, of which it is said that these methods of appraisal cannot be applied? Curiously, this kind of knowledge is said to be a set of elements, and this is not propositional. Perhaps this is an error and what should have been said is that it is knowledge *of* the set of elements that is sought; this takes us closer to the knowledge of the presuppositions of discourse(s). But then such knowledge can be methodologically appraised.[4]

In the "power/knowledge" doctrine both kinds of knowledge that ... are allegedly linked in some way to power. But already one can see, in the light of the discussion of the Strong Programme in Part II, that the prospects for any connection between power (understood as a relational social item) and the first kind of knowledge involving such propositional contents, is dim indeed. The second is Foucault's special kind of knowledge that ... which pertains to the presuppositions of our discourses. A power connection with this kind of meta-knowledge seems even more dim. The next chapter investigates just how dim the connection is – in fact it reveals the dark at the end of the Foucauldian archaeological tunnel.

8.2   FOUCAULT ON DISCOURSE AND THE IDENTITY CONDITIONS FOR STATEMENTS AND DISCOURSES

In this section we begin the foray into Foucault's meta-theory about discourses. We need to discover what Foucault means by 'discourse' since he takes over an ordinary term and gives it a technical meaning for his own purposes. Also bound up with this is the question about when one discourse begins and when it ends by being replaced by another discourse, i.e., the unity and disunity of discourses, or their identity conditions. This is important for Foucault because it will set out a way of demarcating one discontinuity or rupture from another, thereby allegedly revealing the multitude of deep fissures that are said to cleave human thought and practice. However his account is entirely a product of the linguistic turn in philosophy. What incomparabilities he finds are entirely at the linguistic level, and what "objects" there are in our theories are, again, determined entirely by rules which are linguistic in character and do not involve, as one might suppose, semantic language-world relationships. This, we shall see, has disastrous consequences for his account of discourse, and of archaeology more generally.

8.2.1   Introducing Foucault's Meta-Theory About Discourse Individuation

The meta-theory of discourse which Foucault develops in Part II of *The Archaeology of Knowledge* is 'neither lucid nor uncontroversial', as Dieter Freundlieb puts it in one of the best critical accounts of its shortcomings (Freundlieb (1994), p. 153). The excessively abstract language, along with vital points that need further elucidation, often disappear into metaphor or highly florid language. And a self-referential paradox arises, as will be seen, when the tenets of Foucault's theory are applied to itself – a problem that Foucault recognises but passes over.

Foucault's central task is to give an account of the identity conditions of any discourse. Though his meta-theory applies quite broadly even to mathematics (though this is often denied, see Foucault (1972), p. 189), the four examples he commonly cites are those of medicine, psychology, economics and grammar (see *ibid.*, p. 31 and p. 46). But how are such broad discourses to be distinguished? This question becomes more urgent when one asks how more specific discourses are to be distinguished, for example, in economics the distinct views of the Physiocrats and the Utilitarians, not to mention many other economists. Again, distinctions need to be drawn between all the sub-discourses in biology such as the earlier discourse of Lamarck, who proposed a single series of animal evolution culminating in humans, that is presumably to be individuated from his later discourses in which the earlier view is modified and then replaced. Further, Lamarck's various discourses are to be individuated from the discourse of Darwin of 1859 (not to mention many other discourses with biology). For Foucault such distinctions are commonplace and may be based on unclear criteria, possibly derived from the "history of ideas" that he rejects. Given a meta-theory of discourses and their identity conditions of the sort Foucault proposes, such common pre-analytic distinctions may not survive and the boundaries might get completely re-drawn leading to a completely new and unrecognisable set of classifications. If this is the case, then Foucault's meta-theory of discourses and their unity cannot be tested using the pre-analytic categorisations we already possess, for these will be radically revised or rejected. It must stand on other grounds. And this is one of its difficulties; though bold in what it attempts, as will be seen it is unclear on what grounds Foucault's meta-theory does stand.

The basic unit of analysis within Foucault's meta-theory is that of a *statement*, or a concatenation of statements which make up a discourse. However for clarity let us begin with the distinction Peirce made between sentence tokens and sentence types. Tokens of the same sentence type can occur in different discourses (as utterances or inscriptions). Using an illustration that will be discussed again later in this chapter, Foucault tells us that according to his view:

> The affirmation that the earth is round or that species evolve does not constitute the same statement before and after Copernicus, before and after Darwin. ... The sentence 'dreams fulfil desires' may have been repeated throughout the centuries; it is not the same statement in Plato and in Freud. (*ibid.*, p. 103)

Part of the point being made here is that there can be different occurrences of different sentence tokens of the same sentence type at different times and places. Examples of sentence type are given by the following printed tokens: 'The Earth is round'; 'species evolve'; and 'dreams fulfil desires'. Different tokens of these types can occur in the same, or in different, discourses such as the discourses of pre- and post-Copernicans, pre-and post-Darwinians, and Plato and Freud.

But Foucault goes beyond this in claiming that different statements are made when these tokens of the same sentence types occur in different discourses. To illustrate, let us suppose three tokens of the sentence type 'dreams fulfil desires' (more strictly the Ancient Greek equivalent type) uttered or written) by Plato on the same day (in, say, 380BC) will result in the same statement being made by him on the three occasions. But any token of 'dreams fulfil desires' (more strictly the German equivalent token) uttered or inscribed by Freud at any time will result in a different statement being made by Freud from the statement made by Plato. (Let us suppose that the different tokenings of the same sentence type by Freud make the same statement; they would not if we could individuate different Freudian discourses in which the tokens occur.) Here there is alleged to be incommensurability between the tokenings of the same sentence type in that they make two different statements, one by Plato and another by Freud. For Foucault the difference in the statements made need not always turn on a difference in meaning; in fact Foucault does not often talk of meaning. What makes the difference in statements has yet to be spelled out in Foucault's meta-theory. What appears to be ruled out is that Plato and Freud make the same statement. From this is seems we are to conclude that they cannot express the same proposition. And one cannot deny what the other asserts.

Note that what one cannot say is that the Plato tokens and the Freud tokens make different statements *because* the statements occur in different discourses. What we have yet to do is individuate discourses. According to Foucault we are to do this by building up from the notion of different tokens of the same sentence type being used to make the same or different statements, and then use the notion of same or different statements to individuate discourses. To appeal to discourses at any point in the building process would destroy the meta-theory on the rock of circularity.

8.2.2 How Are Statements to Be Individuated? Four Suggestions Rejected.

How might such building be done? Foucault envisages four possible ways, all of which he finds unsatisfactory. The first way employs the object, or item, that the statement is about as individuator of the statement. Thus statements about the same object can be grouped together, but those about different objects fall into different groups accordingly. 'Statements different in form, and dispersed in time, form a group if they refer to one and the same object.' (*ibid.*, p. 32) More explicitly, we can say that different tokens of sentences (uttered or inscribed) at different times make the *same* statement if they have the *same* subject matter, i.e., are about the *same* objects. Here we rely on the prior individuation of objects such that if they are the *same* object, then we allegedly have a sufficient condition for the sameness of the statements, viz., they are about the same objects. But this is not adequate. Though 'dreams wake me up' and 'dreams fulfil desires' are both sentences about dreams they cannot make the same statements in English since they say quite different things about dreams. We have to add at least that what is said of dreams, the property ascribed, is also the same. This difficulty can be overcome by taking Foucault's umbrella term 'object' to cover also properties, relations, etc; but confusion over vocabulary then needs to be avoided in broad use of the term 'object'; some objects are also properties and relations.

This aside, Foucault has deep qualms about the identity conditions for such objects, and our ever being able to talk of the same object across different discourses (whether we take the "object" to be what item the sentence is about, or what property or relation it ascribes to the item). For example, there is supposedly the one object, madness, which statements within psychopathology are supposed to be about. But Foucault holds that the object, madness, has changed with discourse change. If we are to avoid circularity we cannot appeal to such a discourse-dependent object to individuate statements within any broad psychiatric discourse. Even more radically he holds that the very object, madness, is made, or constructed, by our discourses and cannot be used to individuate them. The issues involved in these claims about "objects" in the sciences generally (of which so far astronomy, biology, economics, grammar and psychiatry have been mentioned) will be the topic of most of the rest of this chapter. All we need observe now, is that Foucault rejects (*ibid.*, pp. 31-3) the idea of a discourse-independent object (understood broadly) to which we could appeal in individuating sameness and difference of statements, and thus sameness and difference of discourses that they make up.

The second possibility concerning statement individuation appeals to the force, or *style* or *manner* (*ibid.*, p. 33) of the statement made when a sentence token is used. To illustrate the notion of force, the very same sentence 'there is a bull in the field' can be used on different occasions with different forces. Thus on one occasion it may be used to inform, another to warn (when said with rapidity and emphasis), yet another to ask a question (when uttered with a rising inflection), and so on. Foucault also argues that medical statements have a style and manner that can differ over time and that is related to force. So can force, style and manner of a statement individuate (kinds of) statements? Foucault rejects this not because of its unimportance, but because he holds that statements in medicine can be modified or rejected on grounds that have nothing to do with style or manner, and he wishes to preserve this. Foucault could have made his point by saying that sameness of force, style and manner is just one of a set of necessary conditions for grouping statements as the same. And this, as will be seen, is his final view.

Thirdly, could the sameness of concepts employed in statements be either a sufficient and/or necessary condition for the sameness of statements? Foucault envisages this possibility (*ibid.*, p. 34-5). But worries about the individuation of concepts also concern him, especially when a concept at a lower level in a system of classifications reappears in a new system of classification. Thus a whale, once classified under the broad category fish, was later classified within the category of mammal. And to use Foucault's example, some speech disorders were once classified as a kind of madness, but later this was abandoned (*ibid.*, p. 40). Is the lower-level concept of a whale, or speech disorder, the same or not when they are, respectively, when the concept changes classificatory systems? There is an answer available in recent analytic philosophical theories about the semantics of kind terms, in which the kind whale, or kind speech disorder, remains unaffected by changes in higher classifications. Thus Kripke argues that our reference to tigers using the term 'tiger' is unaffected by the discovery that tigers are, contrary to our former belief, not mammals but reptiles, or robots controlled from Mars (Kripke (1980), pp. 115-28). But this answer is not available to Foucault whose whole project turns on ignoring (either deliberately or inadvertently) semantic considerations.

Fourthly and finally, Foucault envisages the possibility that sameness of statement be judged on the basis of what he calls sameness of theme or strategy (Foucault (1972) pp. 35-7). For example, evolution is a theme that runs through biology over several hundred years. But it is also a theme in linguistics (the evolution of language) and it impinges on other more remote discourses from religion to ethics. Though Foucault finds such thematic linkages important, they are too broad to use for the individuation of statements, and thus discourses.

This will suffice for an account of what Foucault says about the individuation of statements by means of their concept sameness and thematic sameness, a topic that will be addressed again in section 8.3.

8.2.3 *Foucault's Hypothesis About Statement, and Discourse, Individuation.*

The previous sub-section reviewed four criteria on the basis of which different statements might be determined to be the same, and so be used to determine the sameness, or difference, of discourses. Though they are not unimportant aspects of statements, they do not do the job that Foucault needs, viz., an account of discourse individuation. Delving further into the meta-theory of Part II of *The Archaeology of Knowledge,* we find that Foucault develops a further notion to articulate the identity conditions of a discourse that builds on the four rejected criteria just mentioned. This is the notion of a *discursive formation* crucial to which is the notion of *rules of formation*. There are four kinds of rules of formation: there are rules for the formation of objects, of enunciate modalities, of concepts and of strategies. What each of these means is often an obscure matter, and will be the subject of separate sections. One problem is that Foucault hardly cites any such rules; he simply assumes that such rules exist. In this respect he is like many other philosophers of language who allege that language is a rule-governed activity – but then they hardly ever give us example of such rules.

Let us go along with the assumption that rules of formation exist (even though they may not be explicitly expressed). Foucault says of these rules that they 'are conditions of existence (but also of coexistence, maintenance, modification, and disappearance) in a discursive division' (*ibid.*, p. 38). This strongly suggests that the rules of formation are the necessary conditions of discourses that govern their objects, enunciative modalities, concepts and strategies, and that make these possible. Such rules are the presuppositions of discourses. And as discussed in section 8.1.3, it is the goal of Foucauldian *savoir* to provide knowledge of such rules. Could this knowledge ever be explicit? If so then a distinctive body of knowledge that ... concerning the rules could be set out. But such rules are rarely ever set out by Foucault; what one gets instead are illustrations of how the rules apply in particular cases, such as the discourse in psychopathology about, say, madness. If this knowledge is rarely explicit, is it then implicit, or tacit? If so, then when we actually indulge in some discourse we need not be reflectively aware of the rules that are its presuppositions. Assuming Foucault's theory, this would have to be the case. A useful analogy might be with our tacit knowledge of the rules of grammar that we, as expert native speakers, obey most of the time but of which we are not, and need not be, reflectively aware as an explicit body of knowledge. *Savoir*, understood as tacit knowledge, need not be easily accessible knowledge that .... It would taken an "archaeological", or some other, investigation to make explicit what is normally not an evident, but a hidden, aspect of our discourses. Put this way, *savoir* might be viewed not only along the lines of knowledge that ... but as knowledge how to ... , thereby involving skills and abilities. Foucault does not envisage this possibility since he does not make the explicit/tacit distinction; and his anti-structuralism eschews any approach which would locate rules as hidden structures underpinning our skills and abilities that make up our knowledge of how to ....

Foucault alleges that the four rules of formation often clump together. When they do, there are regularities that hold between them, thereby yielding a discursive formation. That such regularities come to prevail provides the identity conditions for a discursive formation, and thus a discourse: '... whenever, between objects, types of statement, concepts or thematic choices, one can define a regularity (an order, correlations, positions and functionings, transformations), we will say, for the sake of convenience, that we are dealing with *a discursive formation'* (*ibid.*, p. 38). Foucault is unclear about whether the regularity holds between any two or three of the above items, or it has to hold between all four items. Let us assume the last. Then any regularity will break down when one of the rules of formation for any one of the four items changes.

The identity conditions for any discourse ultimately boil down to a conjunction of the four rules of formation. Given that a regularity holds between the four rules of formation, then any failure in a regularity due to a change in any one formation rule, which yields either changed objects, or changed enunciative modalities, or changed concepts or changed strategies, will bring that discourse to an end. If a new regularity were to arise when one of these changes occurs then one would enter into a different discourse governed by this regularity. For Foucault, each discourse is a unity that has been formed by the bundling together of formation rules that have a regular association with one another. Each discourse will have arisen from a discontinuity which has an incommensurability with, or a rupture from, or a break with, a preceding discourse because of a breakdown of some regular association; and the same cleavages will occur when any succeeding discourse emerges through the formation of new associated regularities.

In the light of this account of the four kinds of formation rules and the individuating role they are to play, it will be clear what are some of the difficulties with Foucault's meta-theory of discourse. One difficulty is that there are breaks and discontinuities at every turn producing an implausible multiplicity of different discourses. Change only the style or manner of presenting a sentence but every thing else remain the same, or change a lower level concept by changing its position in a system of classification but all else remain the same, then any regularity will fail – and that discourse has been abandoned. Foucault's exaggerated conception of what ruptures and discontinuities have actually occurred, is bolstered by the somewhat arbitrary identity conditions he postulates for discursive formations in his meta-theory. Moreover the account of the rules for the formation of objects has highly implausible aspects which lead to a peculiar kind of linguistic idealism, or irrealism. This will become a major issue later in section 8.4.

Much more needs to be said, and will be said in subsequent sections, about the identity conditions for a statement and how this contributes to the identity conditions for discourses. For the remainder of this sub-section let us turn to a few more remarks that Foucault makes about his basic unit of analysis, the statement. The starting point of any Foucauldian archaeological investigation will presumably be what inscriptions, or recorded utterances, of statements still exist. For existing languages, and those no longer spoken or inaccessible to us, 'a language (*langue*) is still a system for possible statements, a finite body of rules that authorises an infinite number of performances' (Foucault (1972), p. 27). This invokes the notion of a generative account of language, familiar to students of syntax, as a finite set of rules (some of which might be construed to fall under the Foucauldian rubric of 'rules of formation'). The rules can produce an infinite number of 'elements' of a discourse – the elements being statements (or more accurately speech acts, as will be shown in the next section).

But Foucault also needs a notion other than that of all the *possible* statements that might be made in a language. As an empirical basis for an active archaeologist, he needs the narrower notion of the *actual* finite set of statements that have been made over a period of time using the same rules of formation. This Foucault calls the 'field of discursive events'. These would be actual surviving inscriptions or utterances, not all of which we may actually have to hand at any one time, but could so over time. It is this actual "field" that is available for investigation by advocates of archaeology (as well as others). Once given such a field (uncovered in documents, inscriptions, etc), the important question for Foucauldian archaeologists of discourses becomes 'according to what rules has a particular statement been made?'. And the important question concerning an actual field of discourse becomes 'how is it that one particular statement appeared rather than another?' (Foucault (1972), p. 27). Here Foucault's meta-theory of rules of formation, as obscure as this remains, must have bite when it becomes part of the very investigation that the archaeologist must carry out as he or she ponders what rules actually hold that turn the very inscriptions and utterances actually encountered in a 'field of discursive events' into statements. No appeal can be made here to the lame 'history of ideas' or ordinary or common understandings; these too must rest on some Foucauldian prior archaeological investigation into what are the rules governing the discourse under question.

So, quite quickly, questions concerning the identity conditions for statements and discourses come to the fore. What gives traditional modes of classification into, say, grammar, biology or economy, their identity and what makes them different? Foucault bypasses traditional classifications, not necessarily because he rejects them, but because he has a quite different meta-theory about what makes the unity of any discourse, thereby exposing the decisive breaks that have allegedly occurred between discourses. In the light of his meta-theory, the traditional classifications of discourses, three of which have just been mentioned, might or might not survive as unitary discourses.

Already we have seen the importance of the distinction between sentence types and tokens and the statements that can be made by their use. We need to say a little more about this and then elaborate on a further important distinction, that between a statement and a proposition. Given this we can say more about enunciative modalities. In the next section 8.3 we will add to the account of rules for the formation for concepts and themes given in this section. Much of the remainder of this chapter will be concerned with rules for the formation of objects, discourse identity and the irrealism that Foucault is committed to by his meta-theory.

8.2.4 Statements Versus Propositions – and Enunciative Modalities

So far a discourse is, at least, a set of statements, and discourse identity turns on statement identity. Foucault sets out a theory of what statements are in Part III of *The Archaeology of Knowledge* which makes much use of Anglo-American analytic philosophy, particular mention being made of the 'speech act referred to by English analysts' (*ibid.*, p. 83). In this passage from his book it is clear that Foucault wishes to distance his notion of a statement from that of a speech act. Foucault's French word *'énoncé'* is commonly translated as 'statement'. But Dreyfus and Rabinow ((1983), pp. 54-9) argue for a strong link between Foucault's account of a statement and Austin's notion of a speech act. They also helpfully report (*ibid.*, pp. 45-7) a 1979 correspondence with the philosopher John Searle in which Foucault admits that what he means by a statement is the same as what John Austin meant by a speech act.

For our purposes consider three aspects of speech acts distinguished by Austin ((1962), p. 108). (i) Locutionary acts: these are utterances, or inscribings, of (tokens of) sentences (of a given type) which already have a given sense and reference. (ii) Illocutionary acts: these are locutionary acts which result in a speech act having a certain conventional force. Thus the illocutionary act might convey information, issue a warning, give an order, ask a question, make an undertaking (as in promising, or during a marriage ceremony), etc. (iii) Perlocutionary acts: these are the effects we usually bring about in a hearer by our utterance, e.g. the hearer is informed, deterred, surprised, etc. In what follows we will be concerned only with the illocutionary act of stating, e.g. 'patient Charlie is going to recover from his operation', which at least has the perlocutionary effect of informing someone. But more than this can be conveyed by an utterance of the sentence. Said by the right person, a doctor, in the right place, his or her consulting rooms, the remark will be taken up not merely as a bit of information but something with the status of knowledge, or at least something with the force of testimony.

Also important is the notion of the Fregean sense, or the propositional content, of what is said, viz. *that* patient Charlie is going to recover from his operation. For Austin the sentences used in locutionary acts already have fixed a definite sense and reference, i.e. meaning. Important here is the proposition asserted when a locutionary act is performed in a given context.

This is to be distinguished from the different aspects of Austinian speech acts, or Foucauldian statements (*énoncé*). Like Austin, Foucault insists on the distinction between a proposition and a statement (though the distinction is not always observed by him or his translators). A Foucauldian statement, understood as an Austinian speech act, is quite different from a statement understood as a proposition. Propositions are not speech acts. They are the bearers of truth and falsity and stand in logical relations. In contrast speech acts are a species of event and as such they enter into causal relationships; they do not directly enter into logical relationships.

Austin emphasises the quite local context-dependence of the occurrence of most speech acts (indexed as they often are to person, time, place, language, etc). In contrast, Foucault is interested in the ways in which the context-dependence of a statement (from now on to be understood as the same as a speech act) can become a general presupposition of our discourse. The context may not be evident to us when our discourse takes place against a background of institutional settings of which we are generally unaware. The apparent independence of our discourse is unmasked when we make a Foucauldian 'archaeological investigation' into the way in which our discourse works within the context of its hidden institutional setting.

For Foucault a statement such as 'patient Charlie is going to recover from his operation' uttered by a doctor, and the statement 'patient Charlie is going to recover from his operation' said by our neighbour are different statements because of the institutional backing of the doctor who, we may suppose, we are more inclined to believe and who has ways of supporting the claim. We are inclined to ask of our neighbour 'How do you know?' but not of the doctor whose authority to make the claim we tacitly accept. Moreover we often take the situation of the doctor as providing testimony about the truth of what is said, and even epistemic credentials thereby providing knowledge. Despite these differences between the speech acts of the doctor and the neighbour, what is said, the proposition, is the same in both cases, viz. *that* patient Charlie is going to recover from his operation.

We are now in a position to set out more fully what Foucault means by rules for the formation of enunciative modalities. Such rules govern the speech act side of an utterance or inscription and not the propositional content expressed. Foucault does not state these rules explicitly. But along with many philosophers of the social sciences, he supposes that rules play a large part in providing a link between the speech acts we perform and the various social facts, social institutions and other settings that give our utterances the force that they have. Foucault puts emphasis on the special style or manner of presentation of the statements of a discourse. His enunciative modalities of a discourse concern such matters as the qualifications of a speaker; or the places such as the journals, lecture halls or laboratory in which a person may announce their views; or the authority or status of the speaker; and so on. And such enunciative modalities clearly involve not only the illocutionary force of a person's utterance but also the perlocutionary effect upon a hearer of the utterance.

The sorts of authority implicit in scientific statements and the perlocutionary effect the statements normally have upon hearers, rely on the quite particular form that a science's institutional setting has taken in its society that goes well beyond anything Austin envisaged concerning the context of speech acts. If we go along with Foucault in saying that there are rules which govern these institutional settings (even though not explicitly mentioned), then these will be the presuppositions of the discourse that give it the kind of enunciative modality that it has. If we suppose that change in institutional settings and changes in rules of formation for enunciative modalities are linked to one another, then any change in these will cause changes in the illocutionary force and perlocutionary effect of any speech act.

Foucault says of medical discourse: 'Medical statements cannot come from anybody; their value, efficacy, even their therapeutic powers, and, generally speaking, their existence as medical statements cannot be dissociated from the statutorily defined person who has the right to make them, and to claim for them the power to overcome suffering and death.' (Foucault (1972), p. 51) Perhaps we can agree with Foucault that some medical discourses can have therapeutic affects; the soothing worlds of a medical specialist might work in much the same way as a placebo. Foucault also argues that such authority of doctors was extended to those who worked in medical research laboratories. During the past century such authority has been further extended to all research scientists whatever their field of investigation. On the whole we take their utterances to be those of experts and their testimony to be expert.

In Foucault's view, the rules of formation of enunciative modalities are necessary, but not sufficient by themselves, to individuate a discourse. But if there is a change in rules governing enunciative modalities then there will be a change in the discourse. If the institutional settings of scientists change then the associated enunciative modalities might also change. And a change in enunciative modality is sufficient for a different scientific discourse to emerge. Thus if doctors were to lose their authority then, on the basis of Foucault's exaggerated conception, medical science would be a different discourse, even though the hypotheses and concepts of the science remain the same. In the light of this the statements of a discourse are definitely not be identified with the propositions of the discourse. Finally, there will be discourse change where we might least expect it, on a pre-analytical understanding of a discourse. Given Foucault's identity conditions for a discourse, discourse change can come easily and readily; merely change the kind of speech act performed while uttering or inscribing the same sentence (and expressing the same proposition).

8.3   RULES FOR THE FORMATION OF CONCEPTS AND STRATEGIES

8.3.1   Rules for Concept Formation

Consider, first, rules for the formation of concepts (see Foucault (1972), Part II chapter 5). Statements within discourses, such as grammar, economics or biology, will contain concepts. Since for Foucault the statement is the basic unit of analysis, concepts are not independent of the statements in which they occur (or conjunction of them within a given discourse). Foucault does not tell us what are the identity conditions for concepts, either taken singly or taken in groups. He sets out no explicit definitions of single concepts (a paradigm of which would be, in a biological context, 'father' = 'male parent of off-spring'), or the more flexible cluster style of definition. Not does he consider how the defining properties of a concept might alter over time. As a result we have no clear criterion for concept sameness and difference. For an example of a single concept, we may wish to know if the concepts *species*, or *evolution*, in biology are the same from pre-Darwinian discourse to Darwinian discourse, or the concepts *dreams* and *desires* are the same from Plato's discourse to Freud's. In the case of groups of concepts, the unit under consideration might be a number of sentences in which there occur a number of concepts which, when taken together, implicitly define one another. For example, we might follow Kuhn ((1970), chapter IX) who asks us to consider the several laws of Newtonian, or Einsteinian, mechanics in which concepts like space, force, acceleration and mass occur, and are implicitly defined in terms of one another in the context of their respective Newtonian or Einsteinian laws. For Kuhn these concepts are incommensurable.

Given Foucault's interest in locating ruptures and discontinuities, he unsurprisingly says that these concepts are not the same from discourse to discourse. As far as the meanings of our concepts are concerned they will exhibit a Kuhnian incommensurability according to the different discourses in which they occur. Note a circularity that needs to be avoided. We wish to know what are the identity conditions for discourses and their component statements; hence an appeal to the notion of the rules of formation for concepts within statements. Even if we lack a clear idea of when a concept is the same or different, we should not solve this problem by appealing to the discourse or the statements in which it appears to give its identity conditions. It is better instead, but still somewhat unsatisfactory, to leave the individuation of concepts an undecided matter.

There are a variety of ways in which there can be conceptual change. As well as a term occurring in two discourses expressing different content, new concepts may be introduced from one discourse to another, or old concepts may be dropped when adopting a new discourse. Also concepts can enter into new relations with one another; they can overlap, or exclude or displace, one another in different ways. New ways of describing or classifying or categorising may also come and go yielding conceptual change. For Foucault it is also important to note not only what concepts include but what they exclude as not falling within their domain; with conceptual change there will be change in not only what is included but also excluded. But this follows automatically; for all significant concepts there will a range of things that fall under them, and a range that do not. Foucault notes all these ways in which concepts can differ, but says little about how they get their meaning and what counts as a conceptual change. There is nothing in the way of a theory of meaning that even compares with what, say, Kuhn says of the these matters in his 1962 *The Structure of Scientific Revolutions*.

Rather than focus on logical relations between, and the meanings expressed by, concepts, Foucault spells out a number of non-logical relations in which sentences, and their concept-expressions, can stand. He tells us, somewhat obscurely:

> Natural History was not simply a form of knowledge that gave a new definition to concepts like 'genus' or 'character', and which introduced new concepts like that of 'natural classification' or 'mammal'; above all, it was a set of rules for arranging statements in series, an obligatory set of schemata of dependence, of order, and of succession, in which the recurrent elements that may have value as concepts were distributed' (Foucault (1972), p. 57).

Here the focus is on the sentences in which 'recurrent elements' are identified as concepts. But what are the non-logical relations between the concept-containing sentences? These are the ways in which statements can be organised into wholes; the ways in which statements get written down, and the order given to them in such writing; the ways such statements are used to report observations; the way they are critically evaluated, accepted or abandoned; and so on. Foucault seems to hold that these non-logical relations in which sentences stand can also carry over to the concepts they contain, and thereby provide some illumination as to what are rules of formation for concepts. But this is a moot point; they do not provide a clear account of the presuppositions that are necessary for, and make possible, the particular concepts of a range of discourses.

Grant that concepts can be related to one another non-logically as well as logically. When discussing what discourse presuppositions hold for concepts, Foucault asks: 'Could a law not be found that would account for the successive or simultaneous emergence of disparate concepts?' (*ibid*., p. 56). Consider, first, closely related (we cannot say "same") concepts occurring in different discourses. Finding a law governing the way in which related concepts might occur in different discourses might be a tall order; but finding a common cause might not be. As an example one may note the emergence of concepts of chance in both early twentieth century physics and biology and look for a common ground for this, even if no "law" can be found. But within Foucault's theory of discourse, we are precluded from saying that it is the same concept of chance that occurs in both physics and biology. And this is unfortunate because there can be a common background of mathematics, and thus concept of chance, that is employed in both physics and biology. Secondly, consider disparate concepts. Foucault seems to bid us consider some "law" governing the emergence of disparate concepts. It is hard to see what law could govern a matter such as the (roughly) simultaneous emergence at the turn of the twentieth century of (a) Thompson's concept of an electron and (b) the rediscovery by a number of biologists of Mendel's concept of a gene (though it was not always called a 'gene'). No such laws seem to govern the emergence of the concepts of electron and gene at all. The absence of any regularity, let alone "law", makes futile the search for discourse presuppositions concerning concept formation of this sort.

Amongst other non-logical connections between disparate concepts, Foucault also bids us to consider their successive or simultaneous disappearance, their continued co-existence and their concomitance. As an example of concomitance, consider one serving as an analogy for another, or providing a model for the other as when, say, in early classifications of animals there is an attempt to mirror the order of the cosmos as a whole. It is hard to see what "law" might be found governing the emergence, continued co-existence or disappearance, of disparate concepts, or even significant connections and regularities. Once more, knowledge of such presuppositions of these discourses, seems unavailable. Perhaps in the case of concomitance, in which analogies are drawn between disparate concepts in different discourses, some such connections might be found. But the concepts cannot be all that disparate, and must have some common elements on the basis of which analogies are drawn. Thus if an analogy is to be made between concepts of animal classification and concepts of the order of the cosmos, then these concepts must have some similarities for any analogy to hold. The upshot of the above is that while something can be said about the logical and non-logical relations in which concepts stand, very little can be said, beyond claims about their bare existence, about the rules of formation for concepts. So, little knowledge (*savoir*) is to be found about the presuppositions of concept formation in discourses. Much the same can also be said about the rules for the formation of themes and strategies (*ibid.*, chapter 6) – to which we now turn.

8.3.2 *Rules for the Formation of Strategies and Themes*

Foucault identifies themes that may survive different but related discourses. Thus within different discourses concerning language studies, there is an underlying theme of a unity of all "Indo-European" languages, the very concept, Indo-European, emerging at a particular point in discourses about languages. Also accompanying this is a further theme that there is an original language (now non-existent) from which all the members of the Indo-European family emerged. In other quite different discourses there is a theme of the evolution of species common to different biological discourses. And such a theme can also come to be a theme within language discourses, if the idea of an evolution of languages takes hold. Foucault also calls these themes 'theories' or 'strategies'. Given these differing themes or strategies, Foucault sees it as his task to (a) discover the different themes that have occurred in the history of different discourses, (b) when they begin and end, (c) if they are related in any way to one another, and if so whether this is a matter of "necessity" or of chance, (d) whether they are analogous to one another (i.e., themes of evolution in biology, geology and language), and so on. Here there is an appeal to the rules governing the formation of themes, which are then part of the discourse presuppositions to be uncovered that make such themes and strategies possible.

Here Foucault is forced to recognise something correct about the theory of ideas that he rejects in adopting his radical rupture model, viz., that some discourses '... form, according to their degree of coherence, rigour, and stability, themes or theories' (*ibid.*, p. 64). Thus to even speak of the same theme of, say, the existence of an original Indo-European language across different discourses in language studies, we need there to be an identifiable set of concepts (such as the concept *Indo-European languages*), and an identifiable set of sentences expressing the theme, across the different discourses. If there are such common elements across discourses what can be said about Foucault's general stance concerning ruptures, discontinuities and similar cleavages? Does this mean that there must be far fewer cleavages than he envisages, and thus far fewer disparate discourses? This would seem to be an appropriate response in the face of the exaggerated picture Foucault insists upon of massive discourse difference. But it is not a response Foucault makes. Rather he accommodates this problem for his meta-theory by introducing more technical terminology. This he speaks of 'possible points of diffraction of a discourse' and distinguishes 'points of incompatibility', 'points of equivalence' and 'points of systematization' all of which are attempts to deal with the problem of how there can be 'discursive sub-groups – those very sub-groups that are usually regarded as being of major importance, as if they were the immediate unity and raw material out of which larger discursive groups ('theories', 'conceptions', 'themes') are formed' (*ibid.*, p. 66). But all of this is mere semantic wiggle. The multiplication of technical language to get over the excesses of his own meta-theory helps obscure the obvious, viz., that there are many continuities of objects, enunciative modalities, concepts and themes (strategies) across the many discourses Foucault envisages. He recognises that there is a problem in talking of common themes, but the exaggerations of his meta-theory prevents him from finding a satisfactory solution to the problem.

In his discussion of themes Foucault also talks of how choices are made between theories. In Parts I and II there was a discussion of the rational grounds on which theory choice can be made using the epistemic and methodological criteria encapsulated in our critical tradition. But mention was also made of other kinds of choices about theories on quite other grounds. These include: choices made by some funding body about financial support for the testing or development of some theory; choices made by some budding research scientist about what theory they should work on for yielding the most fruitful research outcomes; choices about what scientific developments can best influence practices outside that science (e.g., the development of accurate clocks for navigation, and thus trade, purposes); and so on. In his talk of "strategies" Foucault mention examples of the latter kind of choice, such as how the development of theories of grammar influence pedagogy, how theories in economics came to be adopted and their prescriptions followed by governments and businesses, and so on. In this context Foucault speaks (*ibid.*, p. 68) of an 'authority' and of 'rules and processes for the appropriation' of discourse. Whether there can be rules governing the appropriation of discourses is doubtful; they just get appropriated. Nonetheless, it is important to consider ways in which choices about theories can be made which have nothing to do with epistemic grounds for theory choice (but which presumably do depend on such grounds). What this shows is that Foucault has a broad notion of themes and strategies that covers quite diverse aspects of science.

The above acknowledges the various themes and strategies that can occur in different discourses. And it mentions a problem this raises for Foucault's meta-theory of discourses with its emphasis on discontinuity. But there is very little to find in Foucault about the rules of formation for themes and strategies, beyond a supposition of their existence. And it is to such rules that we are supposed to turn to get knowledge (*savoir*) of the necessary presuppositions of a discourse. Once again the transcendental knowledge Foucault seeks remains elusive.

8.4   RULES FOR THE FORMATION OF OBJECTS: THE CASE OF MADNESS.

We now turn to the final rule of formation for objects which, in conjunction with the other three rules for the formation of enunciative modalities, concepts and strategies (themes), give the identity conditions for a discursive formation, and also identity conditions for statements and discourses. In what follows the term 'object' will be used broadly both for what a statement is about (thus in 'dreams fulfil desires' the sentence is about dreams and desires), and for the properties or relations ascribed to what the sentence is about (in this case the relational property of being fulfilling).

We have seen that Foucault rejects the idea that the rules for the formation of objects are sufficient for discourse identity; but they are necessary. He also maintains a second distinctive thesis which will be called 'constructivism' with respect to objects (taken broadly). Quite literally, objects are the constructs of our discourses. What this means and its ramifications will emerge in this and subsequent sections. Foucault does not discuss these claims generally but illustrates them in the case of our talk of objects in the context of psychopathology (in particular, madness). 'The objects with which psychopathology has dealt ... are very numerous, mostly very new, but also very precarious, subject to change, and in some cases, to rapid disappearance' (*ibid.*, p. 40). As examples Foucault cites such objects as motor disturbances, hallucinations, speech disorders, sexual aberrations, criminality, dementia, and so on. On these he comments: 'A variety of objects were named, circumscribed, analysed, then rectified, re-defined, challenged, erased' (*ibid.*, p. 40-1). Thus an important task for any archaeologist is to say how certain (kinds of) objects became the topics of our discourses how they were altered, classified, and then reclassified, and finally how they ceased to be topics of later discourses. Here the focus of interest is not so much the objects actually postulated at some stage in the development of psychopathology; this has to do with what Foucault calls *connaissance*. Rather the interest is in *savoir*, the conditions under which this or that object are postulated, or made possible, how they emerge in some discourse, are then are altered and/or abandoned. This is a much more *transcendental* enterprise, which we have preferred to describe as the necessary conditions or presuppositions of the possibility of objects across a number of discourses.

8.4.1 Foucault's Transcendental Foray into Rules for the Formation of Objects

Foucault gestures towards the required presuppositional knowledge (*savoir*), but it ultimately remains elusive. He reminds us of the quarry when he says:

> The conditions necessary for the appearance of an object of discourse, the historical conditions required if one is to 'say anything' of it, ... the conditions necessary if it is

to exist in relation to other objects, if it is to establish with them relations of
resemblance, proximity, distance, difference, transformation, – as we can see the
conditions are many and imposing. Which means that one cannot speak of anything
at any time' (*ibid.*, p. 44).

Foucault goes on to tell us; 'It [the object] exists under the positive conditions
of a complex group of relations' (*ibid.*, p. 45). What relations might these be?
Foucault distinguishes two sorts. The first are *primary* relations which hold
'between institutions, techniques, social forms, etc' (*loc. cit.*), examples of
which are families or judicial authorities. But also important are the
'secondary relations that are formulated in the discourse itself: what, for
example the psychiatrists of the nineteenth century could say about the
relations between the family and criminality' (*loc. cit.*). The emphasis here
must be on what the psychiatrists *could say*, because Foucault adds that what
they *could say* 'does not reproduce, as we know, the interplay of real
dependencies' (*loc. cit.* emphasis added). Here it is important to separate, as
Foucault appears to do, the *real* relations of dependency which hold and
which might be the topic of some discourse, from *what could be said* of these
real relations at some time by psychiatrists in their discourse. And note that
what *could be said* might have little to do with those *real* dependencies, i.e.,
has little to do with truth. In the examination of discourses we are to focus on
what they say, setting aside completely anything to do with their truth. The
separation of the realist's notion of what *really* holds, from the irrealists what
is *said* to hold (but might not actually hold) will be of considerable
importance later.

Foucault also goes on to also say that what the psychiatrists *could say*
'does not reproduce the interplay of the relations that make possible and
sustain the objects of psychiatric discourse' (*loc. cit.*). That is, what the
psychiatrists say moves at the level of *connaissance*, and not the higher level
of *savoir*. The primary and secondary relations have made possible a certain
discourse for psychiatrists. But the users of some discourse need not be aware
of the conditions (if any) that make their discourse possible; this is something
for the Foucauldian archaeologist to uncover. And what is uncovered is a
highly contingent, set of primary and secondary relations of an historical,
social and discursive nature. For Foucault it is these contingent conditions
that make the formation of objects possible. What this indicates is that a
focus on only the objects of some discourse is misdirected; it might give us
some *connaissance* but not the more transcendental *savoir*. So: 'we are sent
back to a setting-up of relations that characterise discursive practice itself;
and what we discover is ... a group of rules that are immanent in a practice'
(*ibid.*, p. 46).

We have now located the elusive quarry, the presuppositions of discourse,
the transcendental knowledge that Foucault seeks. But it is not a knowledge
that holds for all discourses. If it is available at all, it is a particular, highly

contingent, knowledge of the presuppositions of a particular discourse, in this case the set of rules governing the formation of the "objects" of the discourse. And we are to use such rules to individuate discourses. So, what are some of these rules? Foucault does not state them explicitly but assumes they exist. We can get a hint of aspects of the rules by examining how the objects of a particular discourse are formed; and this means turning to case studies. In what follows the illustrative case will be the "object" madness.

8.4.2 The Illustrative Case of the Object Madness (1): The Being of Madness, and its Rules of Formation.

Talk of rules for the *formation* of objects suggests that the rules themselves actually *form*, *make* or *construct*, objects. Such talk suggests a non-realist, or irrealist, approach to the status of objects, or kinds of object. In contrast for a realist, objects such as electrons, mass, and cats, are said to exist independently of anything human, including our perception, thought, theories, discourses, languages, and even rules of formation. Also various bodily conditions from blood circulation, neurotransmitters and their functioning, to even madness, can also exist in the sense of the realist. Despite some seemingly strong claims to the contrary, Foucault does not deny such independent existence. Rather the existence of such objects is irrelevant to his meta-theory about discourses with its rules of formation for what we might call 'discourse-relative objects'. Let us see how this is so in the case of madness. We will turn to the issues of realism and anti-realism in Foucault in section 8.5.

The case of madness arises when Foucault considers matters of statement identity, especially the following sufficient condition: if two statements are about the same object, then they can be grouped as the same statement (taking 'object' broadly). What bothers Foucault is the conditions under which we can speak of the *same object*. Pre-analytically we may be unclear as to when the different discourses about madness are about the same or different "objects". Foucault says that all the statements in psychotherapy

> ... seem to refer to an object that emerges in various ways in individual or social experience and which may be called madness. But I soon realised that the unity of the object 'madness' does not enable one to individualise a group of statements .... There are two reasons for this. It would certainly be a mistake to try to discover *what could have been said* of madness at a particular time by interrogating the being of madness itself, its secret content, its silent, self-enclosed truth. (Foucault (1972), p. 32, italics added).

In one respect Foucault is right. If our interest is in the range of things that *could have been said* in a discourse then this is a matter for historians of ideas, or "archaeologists", who investigate the use of language, the concepts employed and the statements made; it is not a matter for medical researchers. But do medical researchers, even those who adopt different discourses, at no time engage in some way with 'the being of madness itself', however misdirected their research may be or how false some of their scientific claims may be in early, and even later, phases of their science? What Foucault proposes is that even if scientists are engaged in some way with 'the being of madness itself', their discourses need not be. What the discourses are about, their "object", and the being of madness itself are two distinct things in Foucault.

The case of madness is controversial, as many of the Foucault and non-Foucault inspired recent accounts of its history and our dealings with madness reveal. However Foucault's talk of the 'being of madness' can be misleading in a number of ways. As Freundlieb puts it: 'even from a realist rather than Foucault's nominalist perspective, madness is not considered a natural kind, as it were, but a family of illnesses that often manifest themselves in different forms, depending at least in part on their cultural context' (Freundlieb (1994), p. 157). Because of the motley of things that have been called 'madness', even for a realist there may well not exist 'the being of madness itself'. There is no single kind, genus or family of thing to be spoken of in the case of madness; instead 'madness' names a family of illnesses the boundaries of which are indeterminate and have been variable over time. Even some central members of the family might lack a single 'being', for example, extreme persistent depression or schizophrenia. But for yet other members there might be a 'being' to be investigated; such is the case for Huntington's Disease.

One kind of 'madness' is a form of dementia now known as Huntington's Disease which involves a progressive growth in uncontrollable twitching of the body, especially arms legs and mouth, and a progressive degeneration of cognition and emotion accompanied by violent outbursts. The classic symptomatic description of this variety of 'insanity', as Huntington calls it, is to be found in his classic 1872 paper. Before Huntington it was included under a collection of diseases variously known as, 'chorea' (from the Greek for dance) or 'St Vitus' Dance' or 'the dancing mania'. Paracelsus classified the various kinds of chorea, distinguishing medieval forms of the Dancing Mania which were psycho-social in origin from *chorea naturalis*, i.e., chorea 'coming from the nature' of the sufferer. Included within this category would be suffers of Huntington's Disease (HD) and other kinds of dementia, but not demonic possession. From the 1830's Elliotson and Waters had noted its occurrence in families, as did Huntington in his 1872 paper; but it was not until 1908 that its Mendelian dominant inheritance was properly described. In the 1980s the gene for HD was located towards of the tip of the short arm of chromosome 4; but its precise DNA sequencing was not discovered until the early 1990s. What effect the defective gene has on certain brain cells causing cognitive, emotional and motor degeneration is still a matter of investigation.[5]

In what sense were doctors such as Elliotson, Waters and Huntington, and even Paracelsus, in touch with, to use Foucault's phrase, 'the being of madness itself', in this case the being of HD? On the Putnam-Kripke theory of reference-fixing, it is possible, even at the beginning of an investigation, to be understood to be referring to a genuinely existing kind of disease, viz., HD, while not being able to identify all cases of HD or misidentifying some, while holding some false beliefs about HD and at the same time failing to recognise some of the central truths about 'the being of HD itself' (i.e., its genetic basis and its effects on some but not other cells in the brain). In the case of HD its symptomatic descriptions, including its hereditary character, were sufficient to latch the early investigators onto a single kind of mental disease even though they knew nothing about the nature of the kind or how that kind produces the effects it does. In the light of this example we need to be cautious about talk of the being of madness. Thus there may be varieties of madness which do have a 'nature' which remains unknown to us even when we successfully referring to it. But in other cases there might be words for varieties of madness that pick out no kind with a common nature. At best the way Foucault introduces the issues obfuscates many important matters to do with successful reference.

8.4.3   The Illustrative Case of the Object Madness (2): Rules of Formation for the Object Madness, Constructivism, and Ersatz Objects

Let us now turn to Foucault's positive account of the formation of the object madness, which follows on immediately from the previous citation:

> mental illness was constituted by all that was said in all the statements that named it, divided it up, described it, explained it, traced its developments, indicated its various correlations, judged it, and possibly gave it speech by articulating, in its name, discourses that were to be taken as its own. Moreover this group of statements is far from referring to a single object, formed once and for all .... The object presented as their correlative by medical statements of the seventeenth and eighteenth century is not identical wit the object that emerges in legal sentences or police action; similarly, all the objects of psychopathological discourses were modified from Pinel or Esquirol to Bleuler; it is not the same illnesses that are at issue in each of these cases; we are not dealing with the same madmen.
>
> One might, perhaps one should, conclude from this multiplicity of objects that it is not possible to accept, as a valid unity forming a group of statements, a 'discourse concerning madness'. Perhaps one should confine one's attention to those groups of statements that have one and the same object: the discourses on melancholia, or neurosis for example. But one would soon realise that each of these discourses in turn constituted its object and worked to the point of transforming it altogether. (Foucault (1972), p. 32)

This is Foucault's constructivism in a nutshell, especially expressed in phrases like 'discourses constituting their objects'. The 'being of madness itself', or particular kinds of madness such as HD, drop out of consideration and are left 'secret, silent and self-enclosed'. Instead what we have are our various discourses about madness. And if we try to replace this discourse by further discourse about more specific objects such as melancholia or neurosis, etc, all we have are further discourses with their own discourse-constituted objects. At no point can we tell when we have, in any of our discourses, reached out and touched a bit of reality. Foucault's meta-theory even frees us from the need to find an independently existing object, madness, melancholia, etc, to give unity to our various discourses. Rather, the picture is inverted. There is a dependence of objects on discourse in which the rules of formation for each identifiable discourse make, or construct, the objects of the discourse. Objects come ready-made for the set of sentences of a discourse by the very rules of formation that give the discourse its identity. The sentences of a unified discourse are not satisfied by some independently existing objects; rather objects, fictional or not, are constructed to automatically satisfy the sentences of an identifiable discourse.

Such a profligate postulation of objects is well-known in philosophy. Sometimes such objects are said, following Brentano,[6] to be the 'intentional objects' of our discourse; they are also said to have 'intentional inexistence' rather than genuine actual existence. For Brentano the objects were strictly mental; in contrast Foucault is much less psychologistic in that his 'objects' are the 'subjectless' intentional inexistents postulated in our discourses. Moreover Foucault would not want to restrict 'intentional inexistence' to things only, but extend it to other objects such the kinds, sorts and properties of which we speak in our discourses.

Another philosopher who is often understood (perhaps wrongly) to be a profligate postulator of objects is Meinong. In the 'Meinongian jungle' of objects, it is supposed that for any seemly denoting linguistic expression such as 'the golden mountain',[7] or 'the round square', there is and object to be assigned to it which subsists rather than exists. Foucault's profligate postulation of objects for each of our discourses is similar in this respect to the postulation of intentional or subsistent objects, but for reasons different from those which motivated Brentano or Meinong. Since Foucauldian objects, given by rules of formation, are not quite the same as Brentano's inexistent objects, or Meinong's subsistent objects, then let us call them 'ersatz objects'. Whether any ersatz objects of discourses are to be identified with actually existing objects remains an open question. If emphasis is put on Foucault's anti-realist tendencies then, since all objects are constructed or constituted by discourses, no Foucauldian ersatz object is identical with any actual object (assuming that there is an actual discourse-independent reality). But what is sorely needed, even if one takes on board the important Foucauldian point that the ersatz objects of some discourses (especially medical discourses) clearly undergo change, is some measure of just how far the ersatz object is from the actual object(s) under investigation. But no account of this seems possible.

Continuing the long quotation given above from its last sentence, we find Foucault saying about separate discourses for melancholia, neurosis, etc:

> But one would soon realise that each of these discourses in turn constituted its own object and worked it to the point of transforming it altogether. So that the problem arises of knowing whether the unity of a discourse is based not so much on the permanence and uniqueness of an object as on the space in which various objects emerge and are continuously transformed' (*loc. cit.*).

Following from this Foucault tells us:

> The unity of discourses on madness would not be based upon the object 'madness'.... It would be the interplay of the rules that make possible the appearance of objects during a given period of time' and 'the unity of discourses on madness would be the interplay of the rules that define the transformations of these different objects, their non-identity through time, the break produced in them, the internal discontinuity that suspends their permanence.' (*ibid.*, pp. 32-3)

Here it is not an object, 'madness' that confers unity on a discourse. Rather rules do double work: they confer unity or disunity on discourses, and at the same time constitute the different objects of different discourses. But what are these rules? The example of madness requires that that there be rules, but none are given. What is given are the various discourses about madness, and the statements that make them up. Once again the special kind of knowledge (*savoir*) of the necessary presuppositions that make the discourse(s) about the object madness (and other objects) possible has not come to the surface. We know something (*connaissance*) about the object of a given discourse about madness; but we know little (*savoir*) of the presuppositions that make this possible.

8.5   REALISM AND NOMINALISTIC ANTI-REALISM ABOUT OBJECTS AND KINDS

8.5.1   Realism and Anti-Realism

Already we have found Foucault's talk of 'objects' ambiguous in that it can refer not only to ordinary things but also their properties and relations. But the word 'object' can function as an umbrella term not only for particular things (e.g., the Sun) or events (the Sun's setting today), but also kinds of thing (e.g., stars), and the properties and relations that all these can have. Foucault is a nominalist who claims that the only things that exist are particulars, e.g., events such as the occurrence of speech acts of utterance or writing, or things such as particular people, particular grains of sand, etc. Traditionally for nominalists only actual particulars exist; there are no kinds, sorts, properties or abstracta such as possible objects, numbers, Platonic universals, essences, and the like. Given that we readily talk of these latter kinds of "objects", such as possible speech acts, or numbers, then a task for nominalists is to explain such talk when they alleged that such "objects" do not actually exist. Anti-nominalists, who assert that some of the latter objects exist, come in a number of varieties; they range from mild anti-nominalists who would admit kinds while eschewing universals, to extreme anti-nominalists who advocate a profligate ontology in admitting them all, including Platonic universals. Foucault is a strong nominalist who countenances no variety of anti-nominalism, not even kinds. For Foucault the very concept of a natural classification arose at a particular time in biology (see Foucault (1972), p. 57). But this cannot be right as that idea is at least as old as Plato and Aristotle. That there are natural kinds independently of what concepts we have is a matter upon which a realist about kinds would insist.

It is evident that we do classify particulars into kinds or sorts such as people, sand, speech acts, and the like. One obvious way in which we make classifications is on the basis of the similarity and difference relations particulars bear to one another. Systems of classification of materials based on chemical characteristics, or of biological species based on characteristics ranging from morphology to genomes, represent our most sophisticated ways of employing such relations. But we may ask of the classifications that we make: in the short (or longer) run do they approach what are commonly called natural or objective kinds of thing? Or, to use Plato's phrase, do our classifications 'follow the objective articulations' of nature? (see chapter 7 footnote 32).

Anti-realists answer in the negative. For them no sense is to be attached to the idea that our classifications ever approach any articulations in nature; more strongly they allege there are no such articulations in nature. The kinds we suppose to exist are not based on nature but on the way we classify according to the peculiarities of our biological makeup, our interests, our political arrangements or power relations, or whatever, as we saw in section 7.7 concerning the Strong Programme. Kinds are our constructs and have no independent existence. In contrast, the realist about kinds answers positively, and in doing so rejects nominalism about kinds. But not all the classifications we make need follow nature's objective articulations. Even though we classify some plants as weeds, weeds form no biological species, genus or family. The issue concerning realism versus constructivism about kinds arises not only for our everyday classifications but also for the classifications made in the various sciences, such as electrons (of physics), schizophrenics (of psychiatry) and labourers (of classical political economy).

For realists our scientific classifications are alleged to track (with varying degrees of success) the articulations of the natural and the social world. Part of the explanation of the (partial) success of our sciences in enabling us to deal with the world is the (partial) success with which they have latched onto the way in which the world is articulated with its kinds, properties, relations and so on. For nominalists such an explanation is not available since the classifications into kinds and properties employed in our sciences are merely constructs that we make. Nominalists must perforce remain content with the extent to which our sciences enable us to cope with the world without attempting to explain why we cope as well as we do.

The terms 'realism' and 'anti-realism' get used in different ways by different philosophers; so some explanation is in order. Broadly speaking there are two views about the objects of our discourses. The first is realism, the view that (most of) the objects about which we talk (believe, theorise or discourse) exist independently of our talk (beliefs, theories, discourses).[8] The independence is such that if we were to change our talk (beliefs, theories, discourses) those objects would continue to exist. The opposing non-realist view is that objects are not so independent and that our talk, beliefs, theories or discourses, make or 'constitute' objects. These objects exist, but they have a dependent existence in that if we were not to talk, or believe, theorise or discourse, in the ways we have, or if we were to go out of existence, then the objects would not exist.

For the non-extremist both views can be held, depending on the category of object under consideration. One can be a realist about some kinds of object but not others. Most are realist about scientific kinds such as electric charge, quasars, tectonic plates, viruses and DNA molecules. All these objects continue to exist even if we change our discourses or there were to be no discourses at all. In contrast, we are inclined to be non-realist about objects such as money, stock market crashes (social events), universities (institutions), spoons (artefacts), and so on. In being non-realist we do not want to say that these "objects" do not exist. They do exist, but in a manner dependent on some of our beliefs, practices or intentions (but not others). For example, we intend to use pieces of curiously marked paper as money, and believe of such objects that they are money; but to lose specific intentions or to not have specific beliefs would mean that there is no money.[9] In this respect we can say that there is such a thing as social reality (e.g., money does exist, there is the property of being married) but it is a (social) construct of ours and not an independently existing object in the strong sense of the realist (as above). On either side of this *via media* there are two extremist positions: super-realism in which the only objects which exist are those which are understood on the realist model; or constructivism, in which the only existing objects are those which are understood on the anti-realist model.

There is another issue within realism that has a bearing on Foucault's discourse-relative objects. Realists commonly assume that their current theories about unobservable objects, processes and properties are a good guide to what exists in the world even though we do not have direct perceptual access to them, for example, electrons, pulsars, genes, gravitational attraction, and the like. However realists can be unsettled by an argument which Putnam calls 'the disastrous meta-induction'. It purports to show that we should not have confidence in the existence of the unobservables that our current theories postulate. On the basis of historical investigations into science it can be shown that 'just as no term used in the science of more than fifty (or whatever) years ago referred, so it will turn out that no term used now (except observation terms, if there are such) refers' (Putnam (1978) p. 25). Such example include: terms of astronomical theories referring to celestial spheres or epicycles and deferents; terms of the Cartesian theory of motion that referred to vortices; terms of chemical theories that referred to phlogiston; the theories of heat that supposed it was a flowing substance called 'caloric'; Thompson's term 'electron' versus Bohr's term 'electron' as used in quantum mechanical theories of the electron; and so on. In the light of these historical examples how do we know that what is currently postulated in our theories will be denied to exist by scientists working many years from now? We do not. The conclusion of the pessimistic meta-induction is that we cannot be sure that what exists according to our current theories will not have the same fate as our ancestor's theories.

Is this a good argument against realism? Is it not the case that more terms, in say chemistry, have been kept on than abandoned since the last spectacular case in which we abandoned a term naming a chemical substance, viz., 'phlogiston'? Some have argued that the number of abandoned terms, and their putative referents, in science has been small compared with the number kept on. We cannot argue this case here, but endorse the positive answer (see Lewis (2001)). However Foucault would not be bothered by the pessimistic meta-induction. In fact he would not be pessimistic about it at all! He might even positively embrace it as a consequence of his theory about the ruptures and discontinuities that can be found in our discourses about what exists. The realists' worry that they might be wrong about what exists is not even an open possibility for Foucault. We should just relax and ride along with the host of varying objects that come and go according to the rules of formation for objects that accompany discourses.

Similarly Foucault would embrace Kuhn's doctrine of the incommensurability of objects in different paradigms, a doctrine which goes along with the pessimistic meta-induction. Given that the "objects" of discourses are ersatz, Foucault adopts a view close to Kuhn's concerning the incommensurability of the referents of terms in successive theories. Foucault would embrace Kuhn's claim that 'when paradigms change, the world itself changes with them' and that 'after a revolution scientists are responding to a different world' (Kuhn (1970), p. 111). Kuhn is, of course, bothered by talk of the world literally changing with changes in our beliefs. But he feels there is point to this language, part of which is to underline the radical incommensurability between objects postulated from paradigm to paradigm. It is precisely this which Foucault endorses about the ruptures and discontinuities he alleges occur in the history of our human thought and which are captured in his account of discourse-relative objects – except Foucault finds even more incommensurabilities between objects than does Kuhn. Also, Foucault's views on the truth of theories or discourses is of a piece with Kuhn's remark: 'There is, I think, no theory-independent way to reconstruct phrases like "really there"; the notion of a match between the ontology of a theory and its "real" counterpart in nature now seems to me illusive in principle. Besides, as an historian, I am impressed with the implausibility of the view' (Kuhn (1970), P. 206). Similarly for Foucault; even if there is an independent reality, our scientific discourses are only ever about the incommensurable ersatz objects they construct.

8.5.2 Foucault's Anti-Realism About Kinds

Let us now return to the contrasting views of realists and anti-realists about kinds. Nominalists are, on the whole, realists about only particulars and are anti-realist about kinds, sorts, properties and the like.[10] This is the position of Foucault; so his umbrella term 'object' can at least be taken to stand for particular actual things. But more often than not Foucault uses the word 'object' in an extended way that can be quite confusing. In adopting a strongly nominalist stance towards kinds, sorts or properties, none of these "objects" (as we might say) exist independently of our modes of classification made possible by our various discourses. That is, when it comes to the existence of classifications, kinds, sorts and types of thing which underpin our use of classificatory or predicate expressions of our discourse, Foucault is strongly anti-realist. But Foucault speaks as if the general terms and predicates of our language still refer to kinds, properties and relations; for, as will be seen, he still wishes to speak of the "objects" of our discourses. But these are objects only as a matter of courtesy. It is as if the general terms and predicates of our discourses are fitted out with "objects", but these "objects" are not to be understood in any realist sense as somehow being part of the furniture of the world; instead they are artefacts, or constructs, of our discourse. Such objects we have called 'ersatz'.

No better evidence for his anti-realist nominalism can be found than in the often-cited passage from the 'Preface' to *The Order of Things* in which Foucault discusses Borges' story about the classification of a 'certain Chinese encyclopaedia' in which our normal animal classifications are challenged by a totally different set of classifications. Thus:

animals are divided into; (a) belonging to the Emperor, (b) embalmed, (c) tame, (d) sucking pigs, (e) sirens, (f) fabulous (g) stray dogs, (h) included in the present classification, (i) frenzied, (j) innumerable, (k) drawn with a fine camelhair brush, (l) *et cetera*, (m) having just broken the water pitcher, (n) that from a long way off look like flies. (Foucault (1970), p. xv)

What could be the point behind such a seemingly bizarre classification of animals? Foucault recognises that we use resemblance relations in making 'a considered classification, when we say that a cat and a dog resemble each other less than two greyhounds do' (*ibid.*, p. xix). But he does not pursue the theoretical grounds on which such classifications are made, for example those in texts on cladistics.[11] Rather he doubts the certainty of any such classifications. As an historian of the sciences, his task is to uncover the very different classifications we have allegedly made throughout the history of biology, grammar and economics, the phrase 'order of things' being used to describe the resultant history of such classifications. One of the attractions of Foucault's work is the very different orders of classification he allegedly uncovers in the history of our sciences, and the grounds for them. However the style of investigation that he proposes is taken to undermine the objectivity of our kind classifications and to support a nominalism in which classifications are taken to be our constructions, relative to a given discourse.

It is noteworthy that bizarre classifications of the sort conjectured in Borges' Chinese Encyclopaedia have never been proposed by anyone, the Chinese included; it is a piece of Borges' fiction. Such bizarre classifications have had considerable discussion in recent analytic philosophy. It is well known that quite arbitrary sets of objects can be specified such as the set of objects: {the Pope's nose, the South Pole, the Eiffel tower, the number 3}. Such an arbitrary collection has little interest to anyone, and as a result has no name. But we do have general and predicate terms in our language largely for the convenience of talking about a number of things at once or, with the help of logical quantification, to pick out one of a class of items. As the history of our language shows, we change our general terms over time; 'electron' is a recent addition while 'foolscap' has now dropped out of use with changes in standard paper sizes. And sometimes there is point to a seemingly arbitrary list of items such as: {skinless chicken, fresh fish, fruit, water, long walks, regular sleep}. The point of this classification may not be obvious since it crosses broad categories of animal, vegetable and human activities. However a point emerges when we are told that they are items on a list of dietary and exercise recommendations drawn up by the National Heart Foundation for those with heart conditions.[12] There is no name for all the items on the list, but the classification is not without purpose.

Again analytic philosophers have become acquainted with the infamous words 'grue' and 'bleen', two predicates used by an advocate of nominalism, Nelson Goodman,[13] to stand for 'green before time t or blue after' and 'blue before time t or green after'. One of the issues raised by our standard blue/green talk when contrasted with the grue/bleen talk, is the role played by these terms in our inductive inferences. From a number of emeralds observed before time t to be green, we standardly infer that the next emerald to be observed after t is green; but we could equally as well infer that the next emerald to be observed after t is grue, and thus blue. The same evidence of observed green emeralds leads to contrary predictions about future observations. Inductive inference have been troubled not only by Hume's critique that there is no justification for the conclusions drawn but also by Goodman's 'new riddle of induction'. Those not of a nominalist persuasion have looked to the role that kinds can play in justifying our inferential practices and in overcoming Goodman's riddle.[14]

Finally the bizarre, to us, Foucault/Borges classification cannot be made without supposing that there are relatively stable classifications such as, in (d) and (l), the kinds *pig*, *water* and *pitcher*, or the properties *sucking*, or *being broken*. Such relatively stable classifications are necessary for even expressing the bizarre classifications themselves; without such stability the bizarre classifications could not be described. What this shows is that, given the classificatory expressions we already have in our language, we can make any other classifications in the fashion of Foucault/Borges, no matter how bizarre. And while many of our classificatory expressions do not always pick out robust natural kinds, some do, such as pig or water; other expressions may still pick out kinds, such as artefacts like pitcher, or functional kinds such as being broken.

According to *The Order of Things,* how are our classifications made? In this work (first published in French in 1966) Foucault appeals to the 'fundamental codes of culture – those governing its language, its schemes of perception, its exchanges, its techniques, its values, the hierarchy of its practices' (Foucault (1970), p. xx). Explanations of the classifications the codes make is a task to be left to scientific and philosophical theories. At this point in his thinking of the matter, Foucault's interest is in an 'intermediate region' in which 'codes of culture' impose a new 'grid' on language, or new schemes of perception, etc. In particular, the book discusses the changing kaleidoscope of classifications in language, biology and economics from the sixteenth century onwards. But talk of 'codes of culture' as the determinants of classifications and their changes does not survive in his next book of 1969 *The Archaeology of Knowledge*. In the later book the rules for the formation for objects, including our kind classifications do much of the work. However this is not his last word on the matter of explanation which, as will be seen in the next chapter, turns very much on power. For the purposes of this section we need only note the nominalism which informs Foucault's constructivist account of kinds and the changing classificatory systems we have adopted, whatever be the causes of the classifications and their changes – from 'codes of culture', to 'rules of formation', and finally to power.

8.6   THE CONTEXTUALIST THEORY OF MEANING AND 'ERSATZ' OBJECTS

It would be wrong to understand Foucault as denying the existence of a world of independently existing objects, or even social constructed objects (such as money). And perhaps if pushed he might even concede that the world comes bundled into natural kinds. However we do not have any epistemic access to this world (even though we are part of it and interact with it), Our discourses draw a veil over it. All we have access to is what we say in our discourses. And what we say, the objects we talk about, the concepts we use, are all given by rules of formation for our discourses. We have called this Foucault's anti-realism. But it might also just as well be called his irrealism with respect to objects, since all the objects are ersatz. Even if the ersatz objects (and kinds) of our discourses do bear a relation to the objects (and kinds) that really exist in the world, we could discern nothing of this relation. All we have presented to us are our discourses, the (ersatz) objects they contain, and the properties and relations that our discourses postulate. In this section we will investigate further Foucault's doctrines about our discourses, the claim that they "constitute" their (ersatz) objects and the irrealism with respect to these objects. We will also consider another feature of his views, that of contextualism about sentences, about meaning and about objects.

8.6.1   Constructivism About Objects.

Foucault's constructivism allows that there is a reality independently of our discourses; so he is not fully idealist about the world. But, whatever that reality is like, it plays little role in shaping our discourses. This he expresses emphatically: 'In the descriptions for which I have attempted to find a theory, there can be no question of interpreting a discourse with a view to writing a history of the referent'. And he continues with respect to the object madness:

> ... we are not trying to find out who was mad at a particular period , or in what his madness consisted, or whether his disturbances were identical with those known to us today. ... We are not trying to reconstitute what madness itself might be, in the form in which it first presented itself to some primitive, fundamental, deaf, scarcely articulated experience and in the form in which it was later organised ... by discourses .... Such a history of the referent is no doubt possible; and I have no wish at the outset to exclude any effort to uncover and free these 'prediscursive' experiences from the tyranny of the text'. (Foucault (1972), p. 47)

What is denied here is the common view that there is a discourse-independent object such as madness, and that we have learned more of it as our theories have developed. However our growing grasp of such a discourse-independent madness plays no role in determining what Foucauldian discourses are like, or what objects they are about. So, what determines what the words of our discourses are about?

One answer that the realist would favour is that it is madness itself of which we have obtained growing knowledge (but not necessarily a growth that is without some discontinuities along its path). Or, better, it is each member of the family of illnesses falling under the umbrella term 'madness' with all their properties and relations, that play some role in helping fix the reference of the terms on our discourses to worldly objects (and kinds). Realists could take on board much of what Foucault says about the growth of our theories about aspects of madness, but not necessarily his special theory of discourses in which he clothes his account of their emergence. Our sciences also involve an engagement, no matter how imperfect, primitive, incomplete and misleading it might be, with madness itself (or the members of the family of illnesses which are called 'madness'). It is this engagement with madness that helps determine many aspects, but not necessarily all, of what we say about madness in our sciences. The realist looks in part to the semantics, the word-world connections made in our theories, to help determine why we have the discourses we do. Of course the realist will have to be aware of the impact of ideology, interests and whatever else, that help shape both our theories and our acceptance or rejection of them. But importantly it is the engagement with aspects of the world (in this case madness) that helps explain some (but not all) of the central features our discourses possess. In filling out how this is so, one can appeal to an account of how the proper names and kind terms of our discourses get their reference fixed that has been developed in Kripke ((1980), Lecture III) and in Putnam ((1975), chapter 13).[15] In contrast Foucault's emphasis is entirely upon the way in which the rules of formation "constitute" objects (and kinds). In this semantic word-world relations do not figure; the realist answer just given is not available to Foucault.

In rejecting the priority of objects over discourse, Foucault advocates a priority of discourse over objects in which rules of formation not only give identity to a discourse but they 'make' or 'constitute' the very "objects" the discourse is about. In effect, what we are investigating is no longer purely syntactical aspects, or mixed syntactical/pragmatic aspects, of our discourse, but rather discourse-object relations. But to call this 'Foucault's semantics' would be deeply misleading because Foucault has no semantics of the sort which sets out discourse-world relationships. It is his lack of a semantics that underpins his anti-realism, his evocation of power/knowledge and his resort to genealogical explanation of aspects of our discourse. After all, semantics (i.e., word-world connections), does explain some of the functions of our language. If there is no such semantics to do some explaining, then resort has to be made to something else – power in Foucault's case.

Any vestige of a semantics with a modicum of realism in it, drops out of consideration as Foucault develops his constructivism about the objects of discourse. This becomes evident as the last quotation cited above continues:

> But what we are concerned with here is not to neutralize discourse, to make it the sign of something else, and to pierce through its density in order to reach what remains silently anterior to it, but on the contrary to maintain it in its consistency, to make it emerge in its own complexity. What in short we wish to do is to dispense with 'things'. To 'depresentify' them. ... To substitute for the enigmatic treasure of 'things' anterior to discourse, the regular formation of objects that emerge only in discourse. To define these *objects* without reference to the *ground*, the *foundation of things*, but by relating them to the body of rules that enable them to form as objects a discourse ... . (Foucault (1972), p. 47-8)

In the same vein, Foucault says that discourses 'are not, as one might expect, a mere intersection of things and words; an obscure web of things, and a manifest, visible, coloured chain of words; ... not a slender surface of contact, or confrontation between a reality and a language (*langue*)' (*ibid.*, p. 48). Talk of 'a mere intersection of things and words' might lead one to think that discourses do involve referential relations between words and things. However the referential relations do not hold between words and discourse-independent objects but the ready-built ersatz objects of discourses.

The extreme anti-realism of Foucault's ontology becomes more evident when he goes on to say of discourse that it is 'the loosening of the embrace, apparently so tight, of words and things, and the emergence of a group of rules proper to discursive practice'. Here the prime role is played by 'rules [which] define not the dumb existence of reality ... but the ordering of objects'. Foucault's project becomes 'a task that consists of not ... treating discourses as groups of signs (signifying elements referring to contents or representations) but as practices that systematically form the objects of which they speak' (*ibid.*, p. 49). Foucault's project is put more radically in his 1970 lecture 'Discourse on Language'. There he tells us that his thinking is to be guided by three decisions: 'to question our will to truth, to restore to discourse its character as an event; to abolish the sovereignty of the signifier' (*ibid.*, p. 229). Even when the sovereignty of the signifier is abolished, it might still be kept on in some more plebeian role of keeping track of items of the discourse-independent reality which Foucault does admit exists. However the abolition is total. Discourse glides over reality without making any referential contact at all; what contact it does make is only with its own ready-built ersatz objects, none of which need have any counterpart in reality, or even bear any truthlikeness (once we question even the will to truth).

8.6.2 The Contextualism of Statements

The idea that discourses construct their own ersatz objects is a persistent theme in Foucault and is bolstered by his contextualism about meaning and reference and about statements. The passages above illustrate this in the case of objects. But much the same applies for whole statements; their meaning is so contextualised that it does not survive transportation from discourse to discourse. Foucault says concerning the identity conditions for statements:

> The identity of a statement is subject to a second group of conditions and limits: those that are imposed by all the other statements among which it figures ... . The affirmation that the earth is round or that species evolve does not constitute the same statement before and after Copernicus, before and after Darwin; it is not, for such simple formulations, that the meaning of the words have changed; what changed was the relation of these affirmations to other propositions, their conditions of use and reinvestment, the field of experience, of possible verifications, of problems to be resolved, to which they can be referred. The sentence 'dreams fulfil desires' may have been repeated throughout the centuries; it is not the same statement in Plato and Freud. (*ibid.*, p. 103)

Some of this we have already met; but what the fuller passage makes clear is how the broad context in which a sentence occurs contributes to the statement made, and what meaning is expressed. These depend contextually on other sentences as well as a host of other matters. So, what can this tell us about the identity conditions for statements (taken to be Austinian illocutionary speech acts), and the propositions they express? Foucault tells us on the following page (*ibid.*, p. 104) that in the case of, say, the translation of a scientific text from French to English, there are not two statements, one in French and one in English, but rather one statement but in different linguistic forms. This brings the notion of a statement quite close to the idea of a common propositional content which remains the same despite the differing speech acts involved (one involving an inscription in French the other in English). So, we are to understand that statements can contain a common propositional content of the sort expressed by the English sentence 'the Earth is round' that remains invariant for even non-English speakers. And these speakers might make their utterances before or after Copernicus. But how does the idea that there is a common propositional content, *that the Earth is round*, sit with the idea that much of the total context of a statement determines that content, where the total context includes such items as experience, possible verifications, problems to be resolved, and whatever else? Not very well.

There are important contextual features that determine what a person states, or reveal that two tokens of the same sentence are ambiguous. But none of these should undermine the large number of cases in which there is a common propositional content expressed when tokens of the same sentence type, say 'dreams fulfil desires', are uttered. So how come Freud and Plato cannot at least express the same propositional content? Freud would say things that Plato would not, such as talk of Oedipus complexes or the role of a censor in letting the suppressed out of the subconscious; and Plato would say things that Freud would not, such as elaborate upon his theory of Forms. However from this it does not follow that the sentence 'dreams fulfil desires' has different propositional content when used by Plato and Freud. To claim this would be to adopt a quite grossly inflated contextualism in which everything one says contributes to the meaning of any single utterance.

In the scientific text example, Foucault says that there can be something in common which survives translation from French to English. So, in the Plato/Freud case there can also be something in common which survives not only translation (this time to and from Ancient Greek and German) but also survives transplantation from one context of utterance to another, one body of surrounding beliefs to another, one sequence of experiences to another (Freud and Plato would have had quite distinct sequences of experiences), and one possibility of verification to another possibility. Moreover there must be something in common surviving Plato's agreeing with, or denying, what Freud asserted, viz., the claim that dreams fulfil desires; otherwise each would say something totally different and their statements would be like 'ships passing in the night'. We are lead to the claim that Freud and Plato cannot be saying the same thing only by an over-inflated contextualism which has connections for each sentence branching out into remote areas irrelevant to meaning.

In the case of the 'affirmation that the Earth is round', the very same propositional content is affirmed both by the Ancient Greeks and Copernicus. We should not deny that there are aspects of our speech acts (some of which Foucault mentions) which can differ so that two sentence tokens of the same type can be used to make different statements (an obvious example is 'I have a headache' said by two different people). However what needs to be corrected are excessive claims in which it is insisted that two sentence tokens of the same type must be used to make different statements where the contexts of the two tokenings differ in ways Foucault mentions. In correcting the excesses, a case can then be made for continuity of meaning and reference for many statements, such as 'the Earth is round', made in different contexts. This is one of the boons of the theory of meaning espoused by Kripke and Putnam, amongst others; it rejects an over-inflated contextualism which allegedly determines meaning. But this has been commonplace in philosophy since well before Frege developed his theory of sense to explain it more fully.

8.7   THE INDIVIDUATION OF SENTENCES AND DISCOURSES – ONCE MORE

We are now in a position to say more about Foucault's identity and difference conditions for discourses. When might two discourses be different? One sufficient condition for discourse difference would be: when they are about different objects. Thus a sufficient condition for discourse difference (SDD) is: for two discourses, if there is no common subject matter, then the discourses are different; i.e., different objects, so different discourses. By contraposition, this is equivalent to a necessary condition for discourse identity (NDI): if two discourses are the same, then they have the same subject matter. Though Foucault does not discuss these quite plausible conditions, let us suppose Foucault would agree with them.

Much of Foucault's discussion of discourse identity is devoted to the following condition which he rejects: 'statements different in form, and dispersed in time, form a group if they refer to one and the same object' (Foucault (1972), p. 32). (Note that this assumes that the other rules of formation remain constant.) We may re-express this in Foucauldian terms as a sufficient condition for discourse identity (SDI): if the rules of formation of objects are such that there is one domain of objects (individual things, kinds, properties, relations, etc.) which the rules of formation assign to the component parts of the statements (names, predicates, etc.) then the statements belong to the same discourse. More briefly (SDI) says: same objects, so same rules of formation for objects, and so same discourse. Upon contraposition, this is equivalent to the following necessary condition for discourse difference (NDD): different discourses, so different objects. (Note that the various criteria under discussion presuppose that we have criteria for the identity and difference of objects; this is not a matter we need take up here.) Foucault is emphatic in his rejection of (SDI). But before discussing his reasons, let us investigate an independent reason realists can give for rejecting it.

Contrary to (NDD), can there be different discourses (theories) yet they are about the same objects? The view of Foucault set out above is that, since discourse constitute their own objects, then two different discourses cannot be about the same objects. Whether or not we can refer to the same object across different theories, or across different Foucauldian discourses, has been a matter of considerable debate in the philosophy of science of the last forty years. Kuhn and Feyerabend,[16] amongst others, argued that we cannot refer to the same object across deeply differing theories (paradigms) by adopting some version of the theory-ladenness of meaning and reference, particularly the idea of referential incommensurability. Their view is that, assuming we have criteria for individuating one theory, or paradigm, from another, the meanings of all the (descriptive) terms in the same theory (paradigm) are determined by the total context of the theory (paradigm) in which they occur. A corollary is that where there is a change in theory (paradigm) there is a change not only in the meaning expressed by terms but also a change in the reference of the terms. And with such a change in reference there is alleged to be a change in the ontology of the theory. For convenience let us call this 'the contextual theory of the meaning/reference of terms'. For Kuhn and Feyerabend a version of (NDD) accompanies their contextualism concerning the meaning of most terms in a theory, but especially the theoretical terms.

Putting this in Foucauldian terminology, where there are different scientific discourses (theories, paradigms) employing the vocabularies of the same or different languages, there will be different Foucauldian rules of formation governing the statements of the discourses; assuming a version of the contextual theory of meaning/reference for discourses, the discourses will be about different objects. But this is simply another way of expressing (NDD), 'different discourses, so different objects'. But now note: (NDD) is logically equivalent to the sufficient condition for discourse identity (SDI) which Foucault rejects. Viewing matters this way exposes a serious tension in Foucault's views, a tension which arises from the contextualist theory of meaning/reference which carries (NDD) with it. Foucault does not really circumnavigate this difficulty. This aside, the comparison of Foucault's contextualism for the statements of a discourse with the contextual theory-ladenness of meaning and reference, adopted at one time by Kuhn and Feyerabend, is instructive. Despite the differing verbal garb, the issues with which Foucault wrestles are closely related to issues that were being debated in the philosophy of science in the 1960s when Foucault was writing *The Archaeology of Knowledge*.

Realists about our sciences should also reject criterion (SDI), but for reasons that go against considerations based on the contextualist theory of reference for scientific discourse. As has been pointed out, many of those working in semantics and the theory of reference for the terms of our sciences have shown us how reference can remain invariant across quite different scientific theories, and this can also be applied across Foucauldian discourses. We cannot explore here their account of how the reference of names in a theory are fixed to objects such as particular things, kinds and properties;[17] but it can be illustrated. In discourses as different as those of the Ancient Greeks, Copernicus and Einstein, their reference to the Sun, or to the Moon, remains invariant. In Bohr's early non-quantum theory of the atom and his later quantum theories, the term 'electron' referred to the same, and not to a different, kind of object. Again Aristotle's and our talk of fire remains invariant despite the fact that he thought it was a pure substance while we now think it is most un-substance like in being electromagnetic radiation bought about by chemical combustion. And so on for many other examples where there are different discourse but sameness of objects, contrary to the claims of both (NDD) and (SDI).

Realists maintain that there are discourse-independent objects and that we can hold different, even incompatible, theories, or beliefs about them while successfully referring to them; so they reject (NDD). (NDD) can also be questioned on other grounds, such as the possibility of different axiomatic presentations, and thus different discourses, for the same domain of objects. Realists can agree with Foucault in rejecting (SDI), but for different reasons.

If these criteria are rejected then, as our current theories of reference-fixing show, it becomes possible that reference remains invariant despite the fact that we might hold much false belief about what we refer to. This is a significant point which is ruled out by Foucault's appeals to the contextual theory of meaning and reference. Foucault inherits many of the difficulties philosophers have found in Kuhnian and Feyerabendian incommensurability alleged to hold between theories, or discourses.

8.8   A REFLEXIVE PARADOX IN FOUCAULT'S THEORY OF DISCOURSE

Granted the above account of Foucauldian discourses, a severe difficulty for Foucault's whole project, clearly bought out in Freundlieb (1994), can now be elaborated. In writing *The Archaeology of Knowledge*, Foucault set out to challenge the traditional divisions between our various discourses. Of this project Foucault asks: 'What new domain is one hoping to discover? What hitherto obscure or implicit relations? What transformations that have hitherto remained outside the research of historians? In short, what descriptive efficacy can one accord to these new analyses?' (Foucault (1972), p.71). The surprising words here are 'discover' and 'descriptive efficacy', i.e., the truth, or falsity, of his analyses. In hoping to *discover* a 'new domain' of objects, or *discover* new relations so far not noticed by historians, Foucault is assuming that in his discourse about discourses (i.e., his meta-theoretical discourse) he can make genuine discoveries about the objects of discourses; and he can assess truths and falsities.

But this is the very thing he denies to all the other discourses he investigates, for example our discourses about madness. Even though there might be an independent reality concerning madness, it remains secretly hidden from our discourses. In investigating all other discourses, except Foucauldian meta-discourses, we do not *discover* objects but *construct* ersatz objects out of the statements of the discourse. Though we might discover the objects that we have, in some manner unbeknownst to us, constructed through our individual and social activity, in no way are there discourse-independent objects waiting to be discovered – except those of Foucault's own meta-discourse in *The Archaeology of Knowledge* about all other discourses. But this would be to give Foucault's meta-discourse a privileged status that no other discourse has. Similarly in asking about the descriptive efficacy of his meta-discourse, he supposes that it can be descriptively adequate or inadequate, i.e., descriptively true or false. But this is not a possibility allowed for any other discourse.

The objection can be phrased in another way. All discourses are rule governed, including, one must suppose, Foucault's long meta-discourse in *The Archaeology of Knowledge*. Since rules of formation determine (ersatz) objects for each discourse, then the rules governing his meta-discourse determine (ersatz) objects for it. Two consequences flow from this. The first is that Foucault recognises that there are other traditional ways of dividing discourses that might be backed by a rival discourse about discourses. But both his own, and any rival, meta-discourse will either be the same discourse about the same "objects" (in which case the idea that there a rivals in this area disappears, contrary to the whole *raison d'être* of Foucault's enterprise); or they will be different meta-discourses and thus be about different (ersatz) objects. Since the latter is the case, then the rival meta-discourses are about quite different objects and are therefore not comparable.

The second consequence is this. How is Foucault's meta-discourse to be bought to bear on some particular lower-level discourse, such as the discourses about madness, or about biology, etc? Does he make genuine *discoveries* about the lower-level discourse, as his above remarks expressly hope? Or has he in his meta-discourse about any lower-level discourse, as Freundlieb suggests, 'simply bought yet another set of discourse-specific objects [of each lower-level discourse] into existence that have no identity over time' (Freundlieb (1994), p. 159)? That is, (ersatz) objects bear no relation to the objects (actual or ersatz) of the lower-level discourse? The latter must be the case on his theory of discourse applied to itself. Foucault, instead of making discoveries about our lower-level discourses, has merely constructed what might be called 'counterpart discourses'. Such counterpart discourses are as much a construct of his meta-discourse as the counterpart discourses themselves are of their own ersatz objects.

Such consequences are disastrous for Foucault's project. If his own meta-theory is taken to apply to itself, then it cannot provide a general theory of our actual historical discourses with whatever continuities and discontinuities they might have. Unless Foucault's meta-discourse is privileged in a way no other discourse is, then it produces only artefacts as the objects of lower-level discourse – and these are not the same as the objects of any historical discourses. Foucault recognises this problem in the final chapter of *The Archaeology of Knowledge* where he has a discussion with an imaginary interlocutor who questions the transcendental character of his own theory of discourse. In reply Foucault 'admit[s] that this question embarrasses me' and admits that his 'discourse about discourses' 'is avoiding the ground on which it could find support'. His reply is simply a re-assertion of his position, saying of his meta-discourse that 'its task is to *make* differences: to constitute them as objects, to analyse them and to define their concept' (Foucault (1972), p. 205). As obscure as some of this may be, the italicised 'make' is Foucault's. The differences that are alleged to hold between our discourses are not discovered; they are made; they are even *constituted as objects*. If this is the case then we have no grounds for believing anything Foucault says about the unity or disunity of our actual historical discourses. This gives, along with the role played by ersatz objects, a precise meaning to Foucault's candid confession: 'I am well aware that I have never written anything but fictions' (Gordon (1980), p. 193). Foucault allows his interlocutor the lame response that perhaps his theory of discourses, like all other discourses, is 'a discipline still in its early phases' which 'leaves its future development to others' (Foucault (1972), p.206). But no future development will remove the self-reflexive paradox wrought by some of the extremes of his theory (especially its theory of objects) unless they are completely abandoned.

The above reflexive paradox arises for all the technical notions employed in Foucault's meta-discourse, including his use of terms such as 'statement', 'discourse', 'rules of formation', 'objects', and the like. They are simply terms of a meta-discourse which has its own meta-rules of formation different from the meta-rules of formation of any other meta-discourse. What of the terms beloved of philosophers such as 'truth', 'knowledge' and the like? Or what of Foucault's own use of the term 'power' in his later meta-discourses? If the terms of traditional philosophy such as 'truth' and 'knowledge' have no role in Foucault's own meta-discourse, then at best their role would be within the many lower-level discourses of philosophers that are open to analysis in terms of Foucault's meta-discourse. On his meta-discourse there will be rules of formation governing the use of 'truth' and 'knowledge' in their respective philosophical (and other) discourses and there will be no common conceptions of truth and knowledge to be found across all (lower-level) discourses. Thus the philosopher's privileging of truth and knowledge across diverse discourses is undercut, a consequence that many followers of Foucault endorse in embracing relativistic account of truth and knowledge. Foucault toys with this position, but towards the end of his life he regarded it with some scepticism. However in asking about the 'descriptive efficacy' of his own meta-discourse, Foucault cannot adopt this view about either truth or knowledge. He needs to admit something that is not discourse-relative.

There is much more that could be said about Foucault's meta-discourse on discourses and its serious drawbacks; but this should suffice. His theory is a culmination of his researches in his earlier books into the various breaks and ruptures he alleges have throughout the history sciences he investigates. But those earlier works sometimes lacked the more fully developed meta-theory of discourses that lie at the heart of his descriptive archaeological enterprise; and so Foucault indicates ways in which they might be corrected. However very few have taken up his difficult and obscure theory of discourse with its technical apparatus. After these books came Foucault's own rupture, his genealogical phase in which he attempts to explain why the discourses he discerns, with their breaks, ruptures and discontinuities, changed. To this we turn in the next chapter.

NOTES

[1]   Many commentators suppose that there is a strong connection between knowledge and power. Thus in the 'Afterword', Gordon (1980), pp. 233-7, takes the strong identity reading, as does Rouse (1987), p. 24. In chapter 11 of Kusch, (1991) the connection is said to be an internal-essential relation.

[2]   Foucault (1972) devotes Part IV chapter 1, pp. 135-40, to the distinction between archaeology and history of ideas; see also pp. 4-8.

[3]   See Collingwood (1972) Part I; Lakatos (1978) chapters 1 and 2; Carnap, 'Empiricism, Semantics and Ontology' in Carnap (1956), pp. 205-21.

[4]   This is not all that can be said on these matters. Nor is Foucault always consistent in his account of the difference between *connaissance* and *savoir*. Thus in a 1978 interview reprinted in Faubion (2000), pp.256-7, Foucault says: 'I see *'savoir'* as a process by which the subject undergoes a modification through the very things that one knows [*connaît*]', an example of such *savoir* being knowledge of 'the economy while constituting oneself as a labouring subject'. This puts a different construal on the two kinds of knowledge that might not be wholly reconcilable with the account in the text. And oddly it is the "knower" that undergoes change.

[5]   For more on the historical background to Huntington's Disease see Hayden (1981) or Harper (1991).

[6]   The terms 'intentional object' and 'intentional inexistent' can be found in Brentano, (1973) p. 88; but they have a much earlier origin in medieval philosophy.

[7]   Foucault does discuss empty descriptions such as 'the golden mountain' in Foucault (1972), pp. 88-96. But the discussion is tortuous. We are invited to consider not only the referent of statements but also the correlates of propositions and a statement's *referential*, viz., 'rules of existence for the objects that are named, designated or described within it [the statement], and for the relations that are affirmed or denied in it' (*ibid.*, (p.91). But once again the objects are not given independently but are constructed.

[8]   Here the views of Devitt (1991) chapters 2 ands 13 are followed. However Devitt adopts a nominalism which eschews a strongly realist view of kinds.

[9]   The matters raised here about the construction of social reality are more fully discussed in chapters 1 to 5 of Searle (1995).

[10]   Note that extreme anti-realists who strive to remain nominalists, such as Bishop Berkeley, would not grant even this much about the existence of individual objects.

[11]   For one discussion of how we make classifications in biology see Sober (1988).

[12]   The example is suggested on pp. 254-5 of Windschuttle (1996).

[13]   See Ch. III and IV of Goodman (1955). The definition of 'grue' and 'bleen' given above is one of several variants.

[14]   There are many accounts of how an appeal to kinds can overcome some of Hume's problems about induction. One of the more recent is Kornblith (1993).

[15]   While the accounts of reference fixing developed by Kripke and Putnam show how reference to objects and kinds can be invariant across discourses, other accounts can also do this. There is a Ramsey-Lewis style of definition that can be applied to fix the reference and meaning of the terms of our discourses, especially theoretical or non-observational terms and how they are attached to non-observables. For examples of this, especially the introduction of the term 'cadaverous particle' by Semmelweis, see Kroon and Nola (2001).

[16]   See chapter 10 of Kuhn (1970), and Feyerabend (1962), especially pp. 28-9. Their position is criticised in Kroon and Nola (2001).

¹⁷ For an account of reference-fixing for a variety of theoretical terms see Kroon and Nola (2001).

CHAPTER 9

GENEALOGY, POWER AND KNOWLEDGE

For Foucault archaeology is largely a descriptive enterprise, using the apparatus of his meta-theory of discourses; what is to be described are the various discourses found in human thought and practice, along with all their ruptures, breaks, discontinuities, and the like. In contrast genealogy is an explanatory enterprise in which, according to Foucault, what does most of the explaining is power; what is to be explained are how various discourses are adopted and maintained, and why they are changed and then abandoned. Also to be explained are the various kinds of knowledge, including what Foucault calls *connaissance* and *savoir*.

Foucault acknowledges that Nietzsche has been an important influence on his post-1969 works that deal with genealogy. In a 1971 essay 'Nietzsche, Genealogy, History' (Faubion (1998), pp. 369-91), Foucault lists the different words Nietzsche used to talk about genealogy, especially in the essay *On the Genealogy of Morals*. The word 'genealogy' comes from the Greek *'geneá-logos'*, a tracing of the descent of a person from their ancestors often to establish a person's pedigree. The word is also related to the Greek *'genesis'* which means origin in the sense of a beginning or a creation. Not unrelated are the German terms used by Nietzsche in different contexts such as *'Herkunft'* (closely linked to the idea of descent), *'Ursprung'* (source, causal beginning), *'Entstehung'* (emergence) and *'Geburt'* (birth). Nietzsche does not always carefully distinguish between these terms, but Foucault does, and then narrows his discussion to the difference between *'Herkunft'* and *'Entstehung'*, both often translated as the bland 'origin'. What this suggests is that in the explanatory exercise of genealogy there might be a number of different "objects" as candidates for explanation. However for those acquainted with the idea of explanatory contrasts in the analytic literature on explanation[1] there should be no problem in specifying exactly what is the "object" to be explained. However this may not be enough to dispel some of the lingering obscurity that surrounds what genealogy is meant to be about.[2]

This chapter will be devoted to a critical evaluation of Foucault's "power/knowledge" doctrine. Already at the beginning of the previous chapter the problem of what '/' might stand for was raised. We will see that in the various expressions of the doctrine the second item is not always knowledge; often discourse, or truth occur there as well. In the previous chapter enough was said about what one of the relata might be, viz., knowledge or discourse. In this chapter it remains to spell out what the relation '/' and the other relata, power, might be.

9.1   THE CAUSE OF DISCOURSE DISCONTINUITY

Why are there discontinuities between discourses? In his work up to and including the 1966 *The Order of Things* one can find detailed, even exaggerated, descriptions of alleged discontinuities in a number of fields, particularly the human sciences such as grammar, economics, biology and medicine. The 1969 *The Archaeology of Knowledge* is Foucault's attempt to set out a theory in which discourse discontinuity can be described. However in the 'Foreword' he wrote to the 1970 English translation of *The Order of Things*, Foucault mentions some unresolved problems for his enterprise one of which is 'the problem of causality' of which he says:

> It is not always easy to determine what has caused a specific change in a science. What made such a discovery possible? Why did this new concept appear? Where did this or that theory come from? Questions like these are often highly embarrassing because there are no definite methodological principles on which to base such an analysis. (Foucault (1970), p. xii-xiii).

Foucault suggests a few agents of change such as the development of new instruments, ideologies and interests. Curiously he fails to mention any of the epistemic methodological criteria that were outlined in Part I. These are standardly appealed to by both scientists and philosophers of science as reasons for adopting one theory rather than another. An account of such reasons for choosing one discourse, or theory, is an important part of scientific methodology that scarcely gets a mention in the entire work of Foucault (an exception being writings just before his death).

Foucault also fails to mention any of our cognitive activities in connection with the invention of new theories (discourses) or the invention of new techniques for solving problems. For example, how did Isaac Newton discover the theory of dynamics for which he is famous? Richard Westfall, an historian of science and a biographer of Newton, tells us: 'In his age of celebrity, Newton was asked how he had discovered the law of universal gravitation. "By thinking on it continually" was the reply.'[3] (Even though the example is from physics, its point about the discovery or invention of theory applies to all sciences, whether physical, biological or social.) It is an important part of our folk psychology, ignored by Foucault and most other sociologists of knowledge, that our thinking processes can be the source of new hypotheses, theories, discourses or whatever. Like several sociologists of knowledge Foucault, as we will see, is in the grip of a single theory of the cause of discourse or theory and fails to recognise what is most obvious, viz. that there may well be a plurality of causes of a discourse and its changes. Social factors might have a role to play, but they need not always have a role, and when they are present they need not be the dominant factor.

These omissions aside, Foucault goes on to mention that he has become increasingly dissatisfied with many of the generally accepted explanations of change:

> the traditional explanations – spirit of the time, technological or social changes, influences of various kinds – struck me for the most part as being more magical than effective. In this work I left the problem of causes to one side; I chose instead to confine myself to describing the transformations themselves, thinking that this would be an indispensable step if, one day, a theory of scientific change and epistemological causality was to be constructed. (*ibid.*, p. xiii; see also p. 50-51).

Foucault makes similar complaints towards the beginning of *The Archaeology of Knowledge* (Foucault (1972), pp. 21-22) when he criticises the stances that prevail within the history of ideas. He says that we must rid ourselves of a mass of notions such as tradition, genius, influence, development, evolution and spirit, since they are not sufficiently explanatory. Moreover they tend to support the idea of underlying continuity of our discourses rather than their discontinuities.

9.2   THE EMERGENCE OF POWER AS *THE* CAUSE

The Nietzschean turn for the post-1969 Foucault is that power is the causal agent of discourse change. In part the turn is indicated in his above-mentioned 1971 paper 'Nietzsche, Genealogy and History' in which he explored Nietzsche's concept of genealogy, and subsequently used the term 'genealogy' for his own theoretical purposes. However one of Foucault's earliest hints about the causal role of power occur in lecture he gave at the Collège de France towards the end of 1970 entitled 'L'Order du Discours' (translated with the English title 'The Discourse on Language').[4] In the lecture Foucault is somewhat tentative about his power hypothesis, in marked contrast with most of his subsequent work in which the causal effect of power on discourse is not qualified or restricted in any way. He says:

> Here then is the hypothesis I wish to advance. ... I am supposing that in every society the production of discourse is at once controlled, selected, organised and redistributed according to a certain number of procedures, whose role is to avert its powers and its dangers, to cope with chance events, to evade its ponderous awesome materiality. (Foucault (1972), p. 216).

As an illustration he mentions exclusion as one kind of broad control and gives three examples of it. The first is prohibition, especially the taboos directed against speaking on certain subjects such as sex and politics. Power gets a mention when Foucault says in this context: 'In appearance, speech may well be of little account, but the prohibitions surrounding it soon reveal its links with desire and power.' (*loc. cit.*) The control that prohibits certain talk about sex foreshadows Foucault's later work on the links between power and sex, just as his second illustration refers to his earlier work on how we treat the insane. This prohibition is the division and rejection embodied in how, in our talk of the sane and the insane, we classify what is reasonable, and what is unreasonable or folly.

The third kind of exclusion, suggested tentatively, is the division between the true and the false. From what Foucault says in the lecture, he does respect the distinction he drew earlier in *The Archaeology of Knowledge* between a proposition and a statement. There is a further allied distinction that is also extremely important to observe, viz. the distinction between (a) what is objectively true or false independently of our discourse constructing-activities, and (b) what we *take* or *believe* to be true or false. Does Foucault intend that the ontological distinction between what in the world makes our statements true or false results from control and exclusion, or from social production? Or does he intend that the epistemological distinction between what we *believe* to be true or *believe* to be false results from control and exclusion? Since Foucault is often understood to be challenging quite primitive and basic philosophical distinctions the above questions need to be put quite sharply.

In the lecture Foucault appears to claim, correctly, that control and exclusion operates only for the latter epistemological distinction when he says:

> It is perhaps a little risky to speak of the opposition between true and false as a third system of exclusion, along with those I have mentioned already. How could one reasonably compare the constraints of truth with those other divisions [e.g., prohibitions on talk of sex and politics or the reason/folly distinction], arbitrary in origin if not developing out of historical contingency – not merely modifiable but in a state of continual flux, supported by a system of institutions imposing and manipulating them, acting not without constraint, nor without an element, at least, of violence?'
> Certainly, as a proposition, the division between true and false is neither arbitrary, nor modifiable, nor institutional, nor violent. Putting the question in different terms, however – asking what has been, what still is, throughout our discourse, this will to truth which has survived throughout so many centuries of our history; or if we ask what is, in its very general form, the kind of division governing our will to knowledge – then we may discern something like a history of exclusion (historical, modifiable, institutionally constraining) in the process of development. (*ibid.*, pp. 217-8)

As an example of this division between truth and falsity Foucault suggests the separation of the poetry of Hesiod from the philosophy of Plato and mentions specifically power as the cause: 'A division emerged between Hesiod and Plato, separating true discourse from false; it was a new division for, henceforth, true discourse was no longer considered precious and desirable, since it had ceased to be a discourse linked to the exercise of power. And so the Sophists were routed.' (*ibid.*, p.218). This is cited merely as an illustration of the way Foucault conveys the impression that the true/false distinction is 'linked to the exercise of power' rather than our *beliefs* about what is true or false. This impression is reinforced by the evocation of Nietzschean phrases such as the 'will to truth' and the 'will to knowledge'. However despite the odd modes of expression it does seem that, after some wavering, Foucault retains, in his lecture, the objectivity of the true/false distinction and its independence from control and exclusion (i.e. power). This is born out in his remark that as far as propositions are concerned, their division into true or false is not arbitrary, modifiable, institutional or violent! (though who would have thought that the truth value of a proposition might be violent!).

As a further illustration, consider his remarks later in the lecture concerning Mendelian genetics:

> People have wondered how on earth nineteenth-century botanists and biologists managed not to see the truth of Mendel's statements. [The translator has used 'statement' rather than 'proposition' here.] But it was precisely because Mendel spoke of objects, employed methods and placed himself within a theoretical perspective totally alien to the biology of his time. ... Mendel, on the other hand, announced that hereditary traits constituted an absolutely new biological object ... . Here was a new object, calling for new conceptual tools, and for fresh theoretical foundations. Mendel spoke the truth, but he was not *dans le vrai* (within the true) of contemporary biological discourse; it simply was not along such lines that objects and biological concepts were formed. A whole change in scale, the deployment of a totally new range of objects in biology was required before Mendel could enter into the true and his propositions [Note: not 'statements'] appear for the most part exact. (Foucault (1972), p. 224).

Here Foucault clearly maintains that it is how the world is that makes Mendel's propositions about inheritance true or false. However, Foucault's odd remark that Mendel was not 'within the true', a phrase he adopts from Canguilhem, needs some explication.

Briefly we can understand 'not being within the true' to mean that, given the prevailing views of inheritance, Mendel's theory was not generally *accepted* as true by his fellow scientists. Using Foucault's terminology we can say that Mendel's discourse on genetics, particularly its enunciative modalities, lacked any authority within the community of biologists and was not taken up. (As a matter of fact this is misleading; it seems that practically no one read Mendel's papers owing to the obscure places in which they appeared.) But this merely tells us *that* Mendel's work was not taken up and not *why* it was neglected for thirty years. Foucault's further remarks that Mendel employed 'a totally new range of objects in biology' and that this put him at some disadvantage with respect to his contemporaries, stems rather from his theory of discourse with its discourse-relative objects and not from any actual investigation of Mendel's theory. In fact Mendel's original papers introduce no 'new objects' and could easily have been understood by any contemporary horticulturist interested in selecting plants and growing them for particular characteristics. The contentiousness of Foucault's account of this episode in the history of science arises from the excesses of his theory of discourses.

The main point to be extracted from these remarks is that the lecture is a transitional work in which Foucault endorses the following. (i) There is a distinction between the truth and falsity of the propositions expressed within our discourse and that this is independent from power. (ii) What we *take* or *believe* to be true or false of our discourse could result from power (exclusion and control). (iii) There is some form of the belief/knowledge distinction.

These three distinctions are largely obliterated in works after the 1970 lecture. In these later works Foucault simply talks of truth and power, or as he puts it power/knowledge, without any suggestion that what makes a discourse true, or what counts as knowledge, could be independent of the nexus of the power relations at work in society. For example, in the first book Foucault published after he had developed his theory of power/knowledge, viz. the 1975 *Surveiller et Punir*, he tells us that power is effervescent throughout society causing a whole range of things, including the very individual persons that exist in society. In coming to understand power we should not always view it negatively as repressive but largely as positively creative:

> We must cease once and for all to describe the effects of power in negative terms: it 'excludes', it 'represses', it 'censors', it 'abstracts', it 'masks', it 'conceals'. In fact, power produces; it produces reality; it produces domains of objects and rituals of truth. (Foucault (1979) p. 194)

The context of this remark is one in which Foucault says that the individual, as well as being an "atom" within society, 'is also a reality fabricated by this specific technology of power that I have called "discipline"'. And after the quotation he wonders: 'Is it not somewhat excessive to derive such power from the petty machinations of discipline?'. Foucault often has sceptical qualms about his excessive remarks, but the qualms rarely last long enough for some qualification to be given. So we are left with the positive, creative, effervescent power that produces, reality, the domains of objects in our discourses, and our 'rituals of truth'. It remains unclear whether rituals of truth are the same as truth; ordinarily one would not think so, but Foucault often fails to separate the two.

Perhaps the following argument lies behind Foucault's claim cited above:

Power causally produces our discourses;

Discourses construct their own domains of objects;

∴ Power causally produces the objects of our discourses (i.e. 'reality', ' rituals of truth').

Let us suppose that the conclusion really does follow from the premises. (Some may wish to claim that there is not even a valid argument here; but transitivity can be restored by treating talk of construction in the second premise as a kind of causal production relation). The second premise is something that Foucault maintains, as discussed in the previous chapter. We have yet to discuss the first premise, viz. that power produces knowledge or discourse. But granting both premises we can see how Foucault makes exaggerated claims on behalf of the creativity of power. But if both premises are false, then there can be no grounds for Foucault's exaggeration. That the second is false is the topic of section 8.4; that the first premise is false will be discussed in sections 9.4 and 9.5.

Has Foucault in his later work abandoned the tentative hypothesis of his 1970 lecture in which claims about "power/knowledge" are severely qualified? Or has he merely forgotten to add qualifications to the "power/knowledge" doctrine as he did in that lecture? My suggestion is the former, given the pervasiveness of the doctrine in his post-1969 work.

9.3   POWER

In his writings of the 1970s and 1980s Foucault accepted Nietzsche's view that society is an all-pervasive nexus of power relations that embrace even knowledge. This latter claim is the topic of subsequent sections. This section explores enough of Foucault's account of power in order to discuss the "power/knowledge" doctrine.

9.3.1   The Counterfactual Account of Power

Foucault endorses what might be called the counterfactual analysis of power, viz. a person A has power over another person B if and only if A gets B to do something that B would not have otherwise done. To illustrate, suppose that B has a range of actions, x, y and z, available to him or her. Then A has power in restricting B's actions to alternatives y and z by removing the possibility x when A would prefer x to y or z, if x were available. B need not necessary be aware that A has so restricted the range of actions; moreover B still has a choice between y and z, and y is something that they do not mind doing. A also has power over B in Foucault's sense if A increases the possibilities available to B by adding w; in the absence of w B would have done x, but if w were to be added to x, y and z then B would choose w. For example a providential uncle might turn up offering enough money to buy a new Alfa Romeo car when the nephew had not even considered this because of lack of money, but was about to purchase one of three available second-hand ten year old cars instead.

To illustrate the counterfactual theory consider the following remarks from Foucault's 'The Subject and Power', and other sources, which are typical of the way Foucault repeatedly expresses his account of power:

it [power] is a way in which certain actions modify others (Foucault (1983), p. 219).

It [power] is a total structure of actions brought to bear on possible actions; it incites, it induces, it seduces, it makes easier or more difficult; in the extreme it constrains or forbids absolutely; it is nevertheless always a way of acting upon an acting subject or acting subjects by virtue of their acting or being capable of action. A set of actions upon other actions. (*ibid.*, p. 220) .

When one defines the exercise of power as a mode of action upon the actions of others ... one includes an important element: freedom. Power is exercised over free subjects, and only in so far as they are free. (*ibid.*, p. 221)

Power is only a certain type of relation between individuals. ... The characteristic feature of power is that some men can more or less entirely determine other men's conduct – but never exhaustively or coercively. A man who is chained up and beaten is subject to force ... not power. But if he can be induced to speak, when his ultimate recourse could have been to hold his tongue ... then he has been caused to behave in a certain way. His freedom has been subject to power. (Kritzman (1988), pp. 83-84)

In a number of other places Foucault emphasises that his notion of the power one might have over another does not exclude the freedom and liberty of the other but, on the contrary, requires it (for example, see Bernauer and Rasmussen (1988), pp. 11-13). That is, in the limit A does not have power over B, but coerces B, when A deprives B of all possibilities of action except what A bids (usually but not always under threat or due to physical force). What Foucault's various remarks have in common is the idea that the person with power is able to affect in some way the actions of another; that is, if A had not exercised power over B, then B would not have acted in the way they did, but in some other way.

The counterfactual condition may be necessary for the power of one person over another but it is not sufficient. If I ask you what is the time and you tell me, then it does not follow that I have power over you in getting you to speak when you would otherwise have remained silent. If you tell me that you are going to have a beer and I ask you to also bring me one, and you do, it does not follow that I have power over you. Something extra is needed alongside the counterfactual condition. The extra condition might have to do with your fear of how I might retaliate if you do not do my bidding; or perhaps it has to do with the way I have implanted in you, independently of your conscious wishes, the desire to please me. More generally, the extra condition might concern whether doing what I ask is against your interests, or whether it runs counter to your desire or your will. In the passages cited, and elsewhere, Foucault does not recognise that some extra condition is needed.

What extra condition(s) need to be added will not be explored at length here. Suffice to note that about the time Foucault began to develop his account of power, Steven Lukes published a book *Power* in which the counterfactual view of power was recognised as insufficient; it was merely the first of several dimensions of power that need to be considered. Lukes' own starting point is: 'A exercises power over B when A affects B in a manner contrary to B's interests' (Lukes (1974), p. 34). The added 'contrary to B's interests' is an important addition to the causal (or counterfactual) clause. Lukes sees the addition as adding an evaluative aspect to the conception of power in which more needs to be said about the normativity of interests in this context. Without a qualification of the sort suggested by Lukes power looms everywhere in our interactions with one another. For Foucault, who believed that power is omni-present, this is to be applauded. For others it is simply more Foucauldian exaggeration to be excised.

To illustrate one excess, consider the case of the relationship between a Socratic educator and an enquirer that Foucault often cites as an instance of power/knowledge. A useful counterexample to Foucault's conception of power is provided by the Socratic educator who gets an enquirer to understand something using the Socratic question/answer method. In Plato's dialogue *Meno*, Socrates gets a slave boy to see what is the answer to a geometrical problem, viz. what is the length of the side of a square double the area of a given square. Socrates gets the boy to see that the answer is the diagonal, not by teaching the boy anything, but merely by putting questions to him; the boy thinks about his problem situation and gives his answer. Just because Socrates' line of questioning gets the slave boy to give answers that he would not otherwise have given without the questioning, it does not follow that Socrates has power over the boy as he thinks his way through to the answers. (The fact that the boy is a slave is irrelevant: it is part of the message of the dialogue that even for a slave boy there is freedom of the intellect.)

In the light of Lukes' conception of power, what one needs to show is that the boy's interaction with Socrates is at least against the boy's current interests (which is not the same as its being in the boy's interests to learn some geometry). But clearly it is not, at least because the boy stays engaged in the line of questioning. The relationship between a Socratic educator and an enquirer does not exemplify power. In fact the Socratic method of acquiring knowledge as exemplified in the *Meno* is a counterexample to the power/knowledge thesis. Note however that the Socratic educator/enquirer relationship is not the same as the teacher/pupil relationship, especially if the latter is understood to exist in some institutional setting in which teachers have conferred on them certain powers of examination, of grading, of curriculum organisation, or punishment, etc, and pupils have certain powers such as whether they complete assignments or not, attend class or not, pay attention, etc. The teacher/pupil relationship can carry with it juridical and other powers determined by rules and statues governing schools. Independently of this social context, at the heart of teaching and learning is a model of educator and enquirer who come together only for the purposes of the enquirer learning about some matter of which the educator is knowledgeable. This can take place outside the rule-governed context of educational institutions. And it can successfully proceed in the absence of any power relations holding between the two, as the episode in the *Meno* reveals.

9.3.2 *Power as War by Other Means: A Relational Ontology of Power*

The above account of power presupposes that power resides in people, or institutions, when they exercise power over one another. While this can be so, it does not represent fully Foucault's conception of power. At best people are, as he puts it, 'vehicles of power' (Gordon (1980), p. 98); they are merely items in a network of independent power relations. Indeed 'the individual is an effect of power' (*loc. cit.*). Even more strongly : 'The individual is not be conceived of as a sort of elementary nucleus ... on which power comes to fasten .... In fact, it is already one of the prime effects of power that certain bodies, certain gestures, certain discourses, certain desires, come to be identified and constituted as individuals' (*loc. cit.*). Thus individuals can have power, and exercise it. But they are also the effects of power; and they are even *constituted* by power relations. Surprisingly, this has strongly behaviourist overtones.

Foucault's notion of social relations of power goes along with what can be called a 'relational ontology' of power. This relational ontology of power does not replace the counterfactual account; rather it is intended to underpin it. What is the relational ontology of power? In the first lecture of 'Two Lectures' (Gordon (1980), pp. 78-92), Foucault sets aside two models of power as primary and adopts a third. He sets aside the juridical model of power in which a monarch can exercise power over the members of a state. And, on the whole, he rejects a second model of power, exemplified in psychoanalytic theory, in which 'power is essentially that which represses' (*ibid.*, pp. 89-90). However he does recognise that power as repression does operate within our society even though it is not the most characteristic kind of power to be found. The first two kinds of power do exist and operate; however there is a third kind of power which is primary and dominant, and is well described by an inversion of von Clausewitz's dictum, viz. 'power is war continued by other means' (*ibid.*, p. 90).

The way in which the third kind of power operates in society is often described using a number of obscure metaphors. What is clear is that Foucault does not always regard power as an attribute of persons or institutions. Rather power is a set of social relations, each relation existing independently of the persons (bodies) that might exemplify it but each still retaining an aim and direction. As he puts the matter:

> Power relations are both intentional and nonsubjective. ... there is no power that is
> exercised without a series of aims and objectives. But this does not mean that it
> results from the choice or decision of an individual subject; let us not look for the
> headquarters that presides over its rationality; neither the cast which governs, nor the
> groups which control the state apparatus, nor those who make the most important

economic decisions ... the rationality of power is characterized by tactics that are
often quite explicit at the restricted level where they are inscribed (the local cynicism
of power), tactics which, becoming connected to one another, attracting and
propagating one another, but finding their base of support and their condition
elsewhere, end by forming comprehensive systems: ... . (Foucault (1981), pp. 94-95).

One of the difficulties in this passage is to explain how Foucault, who
generally adopts an anti-teleological stance, can claim that power is
intentional (i.e., is directed upon some end) yet is non-subjective (i.e., there
is no subject which exhibits the intentionality). Commonly, the entity that
exercises power is a person, class or institution; and they direct power to
some end. But Foucault wishes to remove the illusion that it is these entities
that are primary, and power is merely a property of them. Rather he adopts a
more strongly realist view of power in which it is the relations of power
themselves which are primary and the entities that exemplify them are
secondary and interstitial; at best they occupy the "nodes" of the prevailing
network of power relations. As such, power relations cannot be without any
directionality; they must in some sense have some direction even though
there is no entity such as a person who directs the power. In the light of this
Foucault says that there is still a rationality of power, and that it is exhibited
by the 'tactics' it displays.

Just before this passage, Foucault describes how it is impossible to escape
power within society. He speaks of 'the omnipresence of power' and adds:

Power is everywhere; not because it embraces everything, but because it comes from
everywhere. And "Power", insofar as it is permanent, repetitious, inert, and self-
reproducing, is simply the overall effect that emerges from all these mobilities, the
concatenation that rests on each of them and seeks in turn to arrest their movement.
One needs to be nominalistic, no doubt: power is not an institution, and not a
structure; neither is it a certain strength we are endowed with; it is the name that one
attributes to a complex strategical situation in a particular society. (Foucault (1981),
p. 93).

This also contains obscure metaphors and is very sketchy. But the von
Clausewitz theme of power as war continued by other means reappears as
talk of the tactical or strategic character of omni-present power. And the
omni-presence of power arises from the inadequate view that power is simply
the way our actions, and their complex of interactions, are able to modify the
actions of others.

Martin Kusch has developed a useful analogy to explain how power,
which is not primarily in subjects, can be tactical or strategic. (Kusch (1991),
p. 138ff). He compares power with magnetic attraction that has its effect by
means of the lines of force that Faraday postulated to exist between magnetic
bodies. Just as magnetic attraction is relational and works on bodies *via* the
lines of force that exist between them in a magnetic field, so Foucauldian
power is relational and works on the bodies in social space such as people
and institutions. On this view power is everywhere, just as magnetic

attraction is everywhere (or even better, gravitational fields of attraction). Moreover, just as the amount of magnetic power can vary in different places through three-dimensional space, so power can vary in different places through 'social space'. This is at best an analogy; but Foucault's own realist and relational account of power is itself commonly described metaphorically.

In support of this analogy consider a few more remarks from Foucault:

> Between every point of a social body, between a man and a woman, between the members of a family, between a master and his pupil, between everyone who knows and every one who does not, there exist relations of power .... (Gordon (1980), p. 187)

> Power must be analysed as something which circulates, or rather as something which only functions in the form of a chain. It is never localised here or there, never in anybody's hands, never appropriated as a commodity or piece of wealth. Power is employed and exercised through a net-like organisation. And not only do individuals circulate between its threads; they are always in the position of simultaneously undergoing and exercising this power. ... In other words, individuals are the vehicles of power, not its points of application. (*ibid.*, p. 98)

These remarks also underline Foucault's relational ontology of power: power is a set of relations between items in a social field. In fact Foucault confesses that his use of the word 'power' is usually a shorthand for 'relationships of power' (Bernauer and Rasmussen (1988), p. 11). Power is clearly a relation between two persons as relata, as when one says that, say, a teacher has power over a pupil. But from this talk it does not follow that power itself is something relational between teacher and pupil; commonly we suppose that the source of the power resides in the teacher. However Foucault appears to make this inference. As well as relata such as teacher and pupil, there also exist relations of power between them. Power is not to be ontologically downplayed by treating it as a property dependent on a subject, the teacher who exercises power, and an object, the pupil who receives the effects of power. More strongly, power relations are reified and made ontologically fundamental while their relata, such as teacher and pupil, are themselves the dependent items.

In fact Foucault holds the even stronger view that teacher and pupil are themselves the products of the operation of the independent powers. Individual persons almost drop out of consideration when they are understood to be constituted by the very relations of power that operate on "bodies", as Foucault puts it. Foucault uses the Cartesian terminology of body and soul, but being a materialist the very souls that make us the persons we are, are formed by the action of power. 'It would be wrong to say that the soul is an illusion ... it exists, it has a reality, it is produced permanently around, on, within the body by the functioning of a power ... (Foucault (1979), p. 29). The powers at work here are those in which a "soul" 'is born out of methods of punishment, supervision and constraint' (*loc. cit.*) directed upon the "body". Such a view makes Foucault seem akin to a behaviourist who would claim that the behaviour a "body" exhibits arises out of a history of rewards and punishments. This instructive analogy will not be followed here. But it becomes an important theme in the later Foucault who makes much use of the notion of "governmentality", a neologism which designates 'the way in which the conduct of individuals and groups might be directed; ... To govern, in this sense, is to structure the possible field of action of others' (Foucault (1983) p. 221). This is nothing but the operation of power again as an independent "field" in which one can find dependent relata such as bodies on which a 'technology of power' operates. The upshot of the strongly relational view of power is that material bodies do exist as relata; and for some of these bodies the power relations so act on them that they make a "soul", that is, they make the special characteristics of each person.

Many of the above themes are encapsulated in some remarks made by Foucault makes in a 1984 interview:

> [W]hen I speak of 'relationships of power' ... I mean that in human relations, whatever they are – whether it be a question of communicating verbally, as we are doing right now, or a question of a love relationship, an institutional or economic relationship – power is always present: I mean the relationship in which one wishes to direct the behaviour of another. (Bernauer and Rasmussen (1988), pp. 11)

Here we find the bare counterfactual theory of power, the relational character of power and power's all-pervasiveness. But more as well; in any kind of relationship, such as being interviewed, loving, etc, there is allegedly a central relation of power involved. But what is to be made of the 'always' in 'power is always present'? This could be understood to mean that relations of power are present essentially in these other relations. Loving and being interviewed are not possible unless there is power exercised. If so, then it would follow that there is a regular connection between power and these other relations. A regular connection can arise even when there is no essential, but only a contingent, connection between power and these other relations. In both these cases power would be all-pervasive because of the regular connection. However there need not be any regular connection between power and interviewing or loving. Not only is the connection contingent but it is also intermittent; it might be the case that sometimes when relations, such as interviewing or loving, are present, power relations are also present; at other times they are absent. This third alternative would be inconsistent with Foucault's 'always'.

Consider the case of the multifarious relations that are involved in an interview. It is correct to say that if certain questions were not put, then certain answers would not have been given. But as we have seen this is not sufficient for a power relation to hold between interviewer and interviewee. Suppose that there are power relations, for example the "power ploys" adopted by an interviewer to get certain questions answered; then this does not show that power is a regular accompaniment of all interviews, or that power is essential to the interview. At best it shows that at least on one occasion in the interview power was used; that is, at least on one occasion the interviewer got the interviewee to answer a question he would not have answered, and that so answering was against the desires, or interests, of the interviewee (they were, say, trying to secretive, or "economical with the truth"). Even between those skilled in interviewing and being interviewed, the power relations remain contingent and occasional. Further, one can envisage that in the absence of any "power ploys" a similar interview could still take place.

The same can be said of the relation of love. It is not required that there be either a regular or essential connection between love and some power relations for love to be present. But many love relationships do, as a matter of contingent fact, involve, in varying degrees, occasional exercise of power of various sorts. One consequence of the view that in all human relations power relations are essential or at least regular, is that power is then omni-present. But if power is not omni-present but only intermittently present, then the very idea of power being essentially, or even more weakly regularly, linked to all other human relations, such as love relations or interviewing, etc, cannot be sustained. Further, as was argued in the case of Socratic questioning, the relationship of educator to enquirer need not involve any power relations at any point throughout the enquiry. But if the relationship is given a definite social setting and re-described as a teacher/pupil relationship, then power relations could well be a regular (though not necessarily essential) aspect of the relationship. So one again the connection of power to teachers and pupils may be only contingent and not essential.

That power is omni-present, as Foucault insists, is due to two faulty aspects of his view of power. The first is the weak counterfactual account of power which lacks sufficiency; the second is the view that power is always (regularly or essentially) involved in all other relationships. Given that Foucauldian power is allegedly omni-present, many critics have complained that this entails the emptiness of his concept of power since Foucault gives no criteria for its presence or absence (see Merquior (1985), pp. 113-4). What this complaint turns on is that there are no empirical conditions for testing the presence and absence of specific power relations. As a result, the doctrine of the omni-presence of power is unfalsifiable. Even if one cannot detect power relations in other relations that does not mean they are not there. If one has not found them then one has not looked hard enough for, by the doctrine, they must be there! But this is an arbitrary legislation about the character of power in human relationships.

9.4   POWER/KNOWLEDGE

We now turn to Foucault's 'problematic of pouvoir/savoir', i.e. the doctrine of "power/knowledge". Even if power is to understood in a strongly realist manner as reified relations, there are still the relata to consider which are, if you like, the nodes in the network of power relations. So far we have considered people as one of the main kinds of relata; power relations hold at least between people. Could knowledge be one of these relata? In some way this seems to be a gross category error. It is people who know x (either they know how to x, know that x, etc). And so it is people who are the relata; or, better, that aspect of them that is affected by power relations is the person's knowing x. However Foucault also holds to a doctrine of objective knowledge in which one can speak of what is known independently of any knowing subject. So perhaps the doctrine under examination could be understood in two ways: either, there is knowledge of x that makes reference to some knowing subject, as in 'person P knows x'; or, there is knowledge of x in which no reference to a knower P is made, as in 'x is known'.

Some of these points about knowledge have already been made in Part I in preparation for how we are to understand the "power/knowledge" doctrine. In fact the discussion in Part I makes it very difficult to understand what this doctrine could be. One of the main obstacles is the way the doctrine ignores the normative character of knowledge altogether in supposing that it is a possible relatum in a power relationship. In contrast, and act of belief (but not a belief content, or rational belief) can be a possible relatum because of its lack of normativity. However Foucault hardly mentions belief in this context; all emphasis is on knowledge, and discourse, as we will see later. It would be an empirical matter to determine whether on some occasion power is a cause of our act of believing that p. However what distinguishes knowledge from mere belief is that some of our beliefs display certain epistemic virtues. What makes knowledge is that the knower comes to believe that p on the basis of sufficient evidence, or the knower has reliable belief-forming processes, or whatever it is that one's theories of knowledge set out as the difference between knowledge and belief. As these theories illustrate, relations of power do not make the difference between knowledge and belief. Since these central normative features of knowledge are not acknowledged by Foucault, many philosophers dismiss out of hand talk of "power/knowledge" as an unenlightened slogan.

Perhaps power is to be linked not with knowing *that* but with knowing *how-to*, viz., power produces our skills and abilities. In many cases we acquire our skills socially; however it would be a mistake to conclude that all socialisation involves power. Power *may* sometimes be involved in our learning or acquiring certain kinds of know-how-to, or enabling us to maintain or successfully exercise of such know-how-to; but this is to be determined empirically. Foucault does not mention explicitly know-how-to in his talk of 'power/knowledge'. But often he refers to our changing practices in terms of power/knowledge, for example, our differing practices in, say, what is done to inmates of prisons. Nor does Foucault direct our attention to other uses of know as in 'know why ...', 'know how ...' (both of these concerning explanation), 'know (by acquaintance)', 'know what ...' and so on. Consequently, in what follows we will consider only propositional knowing that ... since this is directly related to discourses and not know-how-to, which often concerns our skills in relation to non-discursive activities (e.g. experimenting, imprisoning, etc).

9.4.1 Francis Bacon on Power and Knowledge

The aphorism 'knowledge is power' is often attributed to the Elizabethan Francis Bacon. What Bacon did say, when discussing science and the fundamental role experiment plays in it, is the following:

> Although the roads to human power and to human knowledge are closely linked together and more or less the same, nevertheless, yet because of the harmful and inveterate habit of dwelling on abstractions it is much safer to begin and raise up sciences from those foundations that relate to the active part, and let that part indicate and determine the contemplative part. (*Novum Organum* Book. II, Aphorism IV).

Here it is the *roads* to human power and human knowledge that are said to be almost the same, from which it does not follow that the ends of the road, viz., human power and knowledge, are the same. Bacon's aphorism also refers to the exercise of our powers of active experimentation (and not contemplation alone) in founding the sciences, and in the discovery of causes (as the previous Aphorism to the one just cited makes clear). Experimental science then becomes the common cause of both knowledge and power. Bacon makes a stronger connection when he says: 'Human knowledge and human power come to the same thing, for where the cause is not known the effect cannot be produced. We can only command Nature by obeying her ...' (*Novum Organum*, Book I, Aphorism III). But the strong talk of 'coming to the same thing' hardly expresses the weaker relationship suggested in the next two clauses cited, which merely express a necessary condition for human power: if human power to be successfully exercised we need to have knowledge of how nature works.

Bacon's aphorisms suggest three different relationships between human knowledge and human power.

(B1) Human powers, as exercised in experimentation, are important for the advancement and discovery of scientific knowledge. This is especially true in experimental sciences. But we can also add to this the human powers of observation, even in those sciences that are not experimental, e.g., astronomy. Also the use of instruments, such as the telescope, in non-experimental sciences, are an enhancement of human powers which assist the advancement of, and discoveries in, such sciences.

(B2) A necessary condition for the enhancement of (most) human powers is growth in scientific knowledge. According to the last Baconian aphorism above, we can only command, and thus have power, over nature if we know how nature works. Power can, in many cases, flow from scientific knowledge to improve the lot of human kind. Bacon did not envisage that science might make life worse. So let us understand him to say merely that scientific knowledge is instrumentally useful for human purposes. Alternatively we can say: 'scientific knowledge makes (certain kinds of) human power possible', or 'if there were no scientific knowledge (of certain sorts) then there would not be human powers (of certain sorts)'. In contrast, Foucault's claim is 'scientific knowledge produces human power'; this is a different way of putting the matter which can be misleading.

(B3) Advocates of a particular line of scientific research or investigation often require funding for their projects. To this end they, or their supporters, might have to exercise a particular kind of power. Further, particular interests, even interests in acquiring power, can spur scientific investigation. To cite an example, not Bacon's, from section 5.3.3, an interest in eugenics was a cause of the growth in late Victorian England of theories of statistics. In fact on one reading of the Strong Programme, social factors can be a cause of the development, or fettering, of the sciences (but not their scientific content, as has been argued). That there needs to be an institutional setting for the development of the science was also envisaged by Bacon in his utopian fable, *The New Atlantis*. He gives an account of an institution called 'Salomon's House' which is specifically organised for the benefit of scientific research, and which, some claim, was inspirational in shaping The Royal Society. Such organisations can themselves be a power, well beyond that of individuals, for the promotion of science.

The difference between (B2) and (B3) is that in (B2) scientific knowledge can be a causal factor in our increased power and control over (mainly) non-scientific matters, while in (B3) power can be a causal factor in the development of the sciences. Also an interest in acquiring power itself may spur certain sorts of scientific inquiry with technological spin-off (e.g. there would be no non-stick fry pans without the power-political interests that dominated the space race between the USA and the USSR). But these are merely contingent, or extrinsic, connections between the social-power context of science and scientific knowledge which need not apply to all cases of scientific investigation or growth in knowledge. Expressed in plausible ways, Bacon's claims are unproblematic and insightful. So how do Foucault's claims about power/knowledge differ from Bacon's, if they do at all? It will be argued that the plausible content of what Foucault says does not go beyond the various Baconian themes; and where it does it becomes quite implausible. As we will see, the pernicious and inveterate habit of dwelling on abstractions, of which Bacon complains, is perpetuated by Foucault when he talks of 'power/knowledge'.

9.4.2 Four Theses About the Connection Between Power and Knowledge

In this section we will examine a number of formulations of the "power/knowledge" doctrine with a view to discovering what it really claims. We have already seen at the beginning of chapter 8 that Foucault rejects the identity interpretation of '/' that so many of his followers adopt when they say 'power is knowledge'. Foucault believes that the '/' relation does stand for something, rather than nothing (in which case the doctrine would have to be rejected). In some places Foucault holds the strong claim 'that power and knowledge directly imply one another' (Foucault (1979), p. 27) and later speaks of 'power relations that make it possible to extract and constitute knowledge' (*ibid.*, p. 185). Talk of mutual implication (logical?), or of constitution, may well have misled some to think that Foucault has in mind a quite strong connection between power and knowledge. But claims that knowledge and power are identical, or intrinsically or necessarily linked, or that they mutually imply one another, or that one constitutes the other, are either wrong or too obscure. So what other linkage does Foucault envisage? At best the link can only be contingent, evidence for which must be found through empirical research and not any *a priori* theory about the causes of "knowledge". Bacon's links are obviously contingent. So, one prime candidate for Foucault's linkage remains to be considered; this is that knowledge and power causally interact with one another.[5]

Though Foucault says different things in the following, clearly causal interdependence of knowledge and power is suggested:

> We should admit rather that power produces knowledge (and not simply by encouraging it because it serves power or by applying it because it is useful); that power and knowledge directly imply one another; that there is no power relation without the correlative constitution of a field of knowledge, nor any knowledge that does not presuppose and constitute at the same time power relations. (Foucault (1979), p. 27).

The parenthetical remarks set Foucault's claim apart from Bacon's claim (B3). He does not intend the first part of his sentence to convey the impression that, for someone who has an interest in acquiring power, they believe that their interest will be best served if they first acquire scientific knowledge or scientific know-how. Rather, it is bare relations of power themselves that are allegedly the causal agent, and not someone's intentional attitude directed towards power.

Three theses can be distinguished in this passage concerning the connection between knowledge and social relations of power. The first concerns the causal primacy of power relations over knowledge as, for example, when Foucault talks of 'power producing knowledge':

Primacy Thesis (A): social power relations cause knowledge.

A weaker Primacy Thesis is indicated by the remark that 'there is no power relation without the correlative constitution of a field of knowledge'. Appropriate abbreviating this, we get a claim of the sort: 'if there is no knowledge then there is no power'; or equivalently 'if there is power then there is knowledge'. We will see later that there is a reading of the conditional that is not assertoric but modal and counterfactual, thereby bring the dependence of knowledge on power close to that of causal dependence. However this is understood, it is much stronger than any claim of Bacon's.

The second thesis is the converse of the Primacy Thesis and concerns the reciprocal effect of knowledge on power as when Foucault says that knowledge 'implies' power. If we take these relations to be causal rather than logical then we have the following thesis:

Reciprocity Thesis: knowledge causally produces relations of power.

This is stronger than Bacon's (B2) in which scientific knowledge does not "causally produce" human power, but *may sometimes* be instrumentally useful in increasing our power. There is also a weaker reading as when Foucault also says: 'nor any knowledge that does not presuppose and constitute at the same time power relations'. Setting aside talk of constituting this, appropriately abbreviated, says that 'there is no knowledge that does not presuppose power' which in turn can be rendered 'if there is knowledge then there is power'. Later we will see that this can be understood alternatively to express the causal dependence of power on knowledge.

These two theses combine to produce a causal interaction thesis as when Foucault says that 'power and knowledge directly imply one another' (taking 'imply' not logically but causally). But this is misleading if such talk is to be understood as mutual causal interaction of the same power and knowledge items. However there can be differential interactions as when Foucault says of the "soul": 'it is the element in which are articulated the effects of certain types of power and the reference of a certain type of knowledge, the machinery by which the power relations give rise to a possible corpus of knowledge, and this knowledge extends and reinforces the effects of this power' (Foucault (1979), p. 29). This appears to say that power produces knowledge, and this knowledge in turn enhances the power effects which produced it. There can also be other kinds of interaction as when some power relation causes some item of knowledge, and this in turn produces a new power relation; that is, items of power and knowledge are linked in a chain, or a spiral. Such an interaction can be expressed as follows:

Interaction Thesis: Some relations of power P_1 causally produce (an item of) knowledge K_1, and this in turn causally produces a different power relation P_2; and so on.

We also need to take into account a claim Foucault's makes elsewhere, viz., that power is also a cause of the *change* of our discourses, or in our bodies of knowledge. Thus there is a corollary of the Primacy Thesis:

Change Thesis: Granted that (social) relations of power produce knowledge, and discourses, then for any change in a set of (social) relations of power there will be a corresponding change in knowledge, or in the discourse, that the power produces.

From here on the Primacy Thesis will be the main focus. We can, by stretching a point, see in the Reciprocity Thesis the more plausible weaker Baconian thesis that our knowledge *can* enhance our human powers. What might be additional to this will become clearer in discussing the Primacy Thesis. This says that knowledge is causally produced by power, but not in the more plausible Baconian way (such as B3).

The Primacy Thesis needs more careful formulation; in fact it will receive four further formulations in an attempt to find a plausible reading of version (A). Foucault holds that all our knowledge is linked in a causal nexus to all-pervasive power and none escapes it; so the thesis applies quite generally. If 'SRP' stands for social relations of power and 'K' for some body of knowledge then the Primacy Thesis can be expressed in a form similar to that of the Causality Tenet of the Strong Programme:

Primacy Thesis (B): For *all* bodies of knowledge K *there exists* a set of social power relations SRP such that SRP cause K.

9.4.3 Does the "Power/Knowledge" Doctrine Apply to All the Sciences?

Can we claim that *all* bodies of knowledge are caught up in the causal nexus? That this is so has been made clear in version (B) of the Primacy Thesis. Some say that Foucault's thesis should be restricted to just the human sciences, and that it does not apply to the other sciences. It is true that Foucault's most extensive use of the Primacy Thesis is with respect to the human sciences. And we also noted in the previous chapter that in *The Archaeology of Knowledge* the notion of a discourse was commonly illustrated in biology, grammar, economics and psychiatry. But of these, biology is not confined to just the biology of humans but to species generally. And we have already noted that in lecture 'The Discourse on Language' (see section 9.2) the case of Mendelian Genetics was not unlinked to aspects of power. Shortly we will see in other formulations of his doctrine (set out in the next sub-section) that Foucault makes no specific restriction to just the human sciences; in the absence of any qualification one can legitimately take him to be talking of all kinds of knowledge in all sciences.

It is clear that Foucault extends his theory to non-scientific areas and to items that are not even discourses, e.g., those about human conduct (e.g. his last two books on sexual conduct in the ancient world) and to discourses *within* prisons, this being distinct from discourses *about* prisons which do contain some human sciences. However in interview in 1978 Foucault was asked specifically about power and the non-human sciences:

> Your analysis of the relations between knowledge and power takes place in the area of the human sciences. It does not concern the exact sciences, does it?
>
> FOUCAULT Oh no, not at all! I would not make such a claim for myself. And anyway, you know, I'm an empiricist: I don't try to advance things without seeing whether they are applicable. Having said that, to reply to your question, I would say this: it has often been stressed that the development of chemistry, for example, could not be understood without the development of industrial needs. ... But what seems to me to be more interesting to analyze is how science, in Europe, has become institutionalized as power. It is not enough to say that science is a set of procedures by which propositions may be falsified, errors demonstrated, myths demystified, etc. Science also exercises power: it is, literally, a power that forces you to say certain things, if you are not to be disqualified not only as being wrong but ... as being a charlatan. Science has become institutionalized as a power through a university system ... (Kritzman (1998) pp. 106-7)

What all of this suggests is that it is an empirical matter for Foucault whether or not any science, including a non-human science such as chemistry, is an instance of the "power/knowledge" doctrine. As the above quotation indicates, when he gave the matter a little thought Foucault does not deny that the non-human sciences fit his "power/knowledge" doctrine. Chemistry is linked to power, even weakly when industrial needs give rise to development in chemistry (see section 4.1.2). According to Foucault, science is not just a set of propositions to which some methodology of test has been applied; this is to focus on the methodological aspects of scientific theories only. Other aspects of the sciences are institutionalised powers. Science is also said to be a power in the sense that it can even force us to say certain things, i.e., it is a power which causes certain enunciative modalities of its discourses to be sanctioned, and others not.

Before the above exchange, the interviewer had said 'One of your theses is that strategies of power actually produce knowledge'. Foucault replies endorsing this, but with a qualification: 'Of course you will always find psychological or sociological theories that are independent of power' (*ibid.*, p.106). That is, some aspects of human sciences are outside the power network. How are we to understand this? In chapter 5 a distinction was drawn between what might be the methodological grounds on which the cognitively contentful propositions of some science are to be accepted, and the power and other social relations in which the non-cognitive aspects of science might be enmeshed. In part Foucault recognises this in his remarks on chemistry; and perhaps this is to carry over to an explanation as to why there is also independence of some of the human sciences from power. It is a science's cognitively contentful propositions that are so independent. But other aspects of science are not so independent. And it is to these non-cognitive aspects of each science that the "power/knowledge" doctrine can apply (but empirical research is needed to show this). This is a division of labour that has already been recognised in chapter 5 in the discussion of the scope of the Strong Programme.

In an interview recorded at the beginning of the year of his death, 1984, Foucault makes a lot clearer what scope his "power/knowledge" doctrine might have when he separates 'relationships of power' from 'games of truth'. He says that medical knowledge has been linked to 'instances and practices of power'. This is an endorsement of the Reciprocity Thesis but not the Primacy Thesis. He goes on to say that though they are linked in this way, there is also a separation, and then makes a comparison with mathematics:

> This fact [the linkage] in no way impairs the scientific validity of the therapeutic efficacy of psychiatry. It does not guarantee it but it does not cancel it our either. Let mathematics, for example, be linked – in an entirely different way from psychiatry – to structures of power; it [the separation] would be equally true, even if it were only in the way it is taught, the manner in which the consensus of mathematicians organises itself, functions in a closed circuit, has its values, determines what is good (true) and evil (false) in mathematics, and so on. That does not at all mean that mathematics is only a game of power but that the game of truth of mathematics is linked, in a certain way and without impairing its validity, to games and institutions of power. ... it is clear that the relationship which can exist between the relations of power and the games of truth in mathematics is entirely different from the one you would have in psychiatry. In any case, one can in no way say that the games of truth are nothing else than games of power. (Bernauer and Rasmussen (1988), p. 16)

What this shows is that Foucault thinks that even a non-human science such as mathematics can, in some way, be subsumed under the "power/knowledge' doctrine. But the passage also shows something else: Foucault has, belatedly, become clearer about what aspects of mathematics (and other sciences such as psychiatry) do, and do not, fall within the scope of "power/knowledge" doctrine. It is quite clear that Foucault wishes to put issues to do with the validity of the claims of psychiatry and mathematics outside the scope of power relations. Strictly speaking validity has to do with arguments. But there is a common (mis)use of the term in which it concerns the truth, or lack of it, of the claims of any science. And it is this, along with methods for determining truth or falsity, that Foucault seems to say are not impaired by power.

Later in the interview Foucault tells us: 'when I say "game" I mean the ensemble of rules for the production of the truth' (*loc. cit.*). On one reading, if they are a collection of rules for producing truths, then games of truth are like rules of inference that have as output truths (given an input of truths as premises). Understood in this way, such rules of validity are outside power relationships. But perhaps Foucault does not intend such inferential rules to be part of 'games of truth'. Whatever the case, 'games of truth' are different from 'games of power', though they are still alleged to be linked in some unspecified way. On this reading there is in Foucault a glimmering of a recognition that, in mathematics and other sciences, matters of power need to be separated from matters to do with validity (methodological test?), and some matters to do with 'games of truth', in mathematics.

The issue that is being addressed here is one that we have met in chapter 5; it concerns the wide scope of the Strong Programme as compared with the narrower scope of its "Weak" Programme rival. Advocates of the Weak Programme, such as Mannheim (see section 4.2), did not want to extend social considerations to matters to do with the validity or truth of scientific claims; but social considerations did legitimately apply to other aspects of the sciences. Advocates of the Strong Programme rejected this and expanded the scope of sociological considerations to all aspects of the sciences, including the very content of the sciences themselves. If he did have such glimmerings of recognition, then advocates of the Strong Programme would see that the final Foucault had ceased to be the radical supporter of their programme that the earlier Foucault has been. He has drifted into the Weak Programme with its separation of matters to do with validity and truth from social matters. This aside, on a certain weak understanding of the power/knowledge doctrine, mathematics and chemistry and other non-human sciences are just as much instances of the doctrine as are the human sciences. Such a weak understanding might leave open the assessment of the propositions of a science independently of any power relations, or social factors.

To sum up. There are two generalisations to be found in the Causality Tenet of the Strong Programme and the Primacy Thesis. One concerns whether all the sciences, from psychiatry to mathematics, have aspects which fall under them. The answer is 'yes'; no science is exempt. So Foucault's "power/knowledge" doctrine applies to all the sciences and not just the human sciences as some have claimed. The other generalisation concerns whether or not the very content of the sciences is to be explained socially, or equally by Foucauldian power relations. Advocates of the Weak Programme say 'no' while advocates of the Strong Programme say 'yes', this being one of the ways in which the programme is strong.

Here a parallel has been drawn with the Strong Programme of chapter 5. The Primacy Thesis bears an exact formal similarity to the Causality Tenet of the Strong Programme. Both theses are to be treated, not as empirical claims within some science, but as schemata that are to be filled out in particular ways through an investigation into the way power, or social factors, causally affect "knowledge". The main difference between the two schemata is that instead of invoking social relations of power, the Causality Tenet speaks of a range of social factors in conjunction with non-social factors as the cause of our "knowledge", or more correctly our beliefs. In this respect the Strong Programme can include within its social factors Foucauldian social power relations, and at the same time gives a fuller picture of what are alleged to be the causes of belief. But both contain an ambiguity about what "object'" is to be causally explained. Is it the *content* of our beliefs? Or our *acts* of believing? Or is it some "object" other than the contents of scientific beliefs, such as the funding of a certain experimental investigation, or they way a certain group of scientists organise themselves? Perhaps it remains unclear whether Foucault supports the Strong or the Weak Programme. But in the light of these similarities, Foucault's "power/knowledge" doctrine inherits all the defects and problems that confront the tenets of the Strong Programme discussed in Part II, unless it is suitably qualified.

9.4.4 The Causal Dependence Version of Power/Knowledge

The Primacy Thesis occurs in many guises in Foucault's writings, especially when he applies it not only to knowledge but also to discourses, to truth and to discourses of truth. To illustrate this consider one extended passage from the second of 'Two Lectures' of 1976 (to which letters in parenthesis have been added):

> My problem is rather this: what rules of right are implemented by [a] the relations of power in the production of discourses of truth? Or alternatively, [b] what type of power is susceptible of producing discourses of truth that in a society such as ours are endowed with such potent effects? What I mean is this: in a society such as ours, but basically in any society, [c] there are manifold relations of power which permeate, characterise and constitute the social body, and these relations of power cannot themselves be established, consolidated nor implemented without the production, accumulation, circulation and functioning of a discourse. [d] There can be no possible exercise of power without a certain economy of discourses of truth which operates through and on the basis of this association. [e] We are subjected to the production of truth through power and we cannot exercise power except through the production of truth. This is the case for every society, but I believe that in ours the relationship between power right and truth is organised in a highly specific fashion. (Gordon (1980) , p. 93)

Versions of the Primacy Thesis occur at various points throughout this quotation. Thus at [a] and [b] Foucault talks of power producing, i.e. causally bringing about, discourses of truth (though [b] also hints at the Reciprocity Thesis in that the truth so produced is said to have potent effects, these effects presumably being more power). At [c] and [d] we are told that in all societies there are no relations of power without a discourse, or a discourse of truth. With appropriate logical transformations this can also be taken to express the Primacy Thesis. Finally [e] contains the Primacy Thesis in the form that power is the cause of truth (and [e] also contains the Reciprocity Thesis).

There are several points to note about this passage. First it claims that power relations that are so productive exist in all societies; there are no societies without them. Second, what do these powers produce? They produce a plethora of different items such as truth, discourses and discourses of truth. At the beginning of this chapter we noted how scrupulous Foucault was in distinguishing the different kinds of genealogy there can be as "objects" of explanation. But here he is quite expansive about the kinds of thing that power is to bring about. Knowledge is not explicitly mentioned among them. However this omission does not mean that knowledge is not a product of power since this is mentioned in many places elsewhere.

Granted these points a further version of the Primacy Thesis can now be expressed as:

Primacy Thesis (C): For any knowledge K (or truth T, or discourse D, or discourse of truth D of T) there are social relations of power, SRP, such that SRP causes K (or, respectively, T, or D, or D of T).

Finally, in [c], [d] and [e] in the above quotation, the power connections are expressed modally. Here we can take some liberties and read the connections as saying: if there were not certain power relations then there would not be certain truths (discourses, discourses of truth, knowledge, etc). In section 5.9 a definition of causation originally given by Hume and recently revived by David Lewis (with modifications; see Lewis (1986), chapter 21)· was outlined. It says: 'A causes B' just means 'if A were not to occur then B would not occur' (where A and B are particular events or states of affairs). This definition of cause (or of causal dependence) in terms of a notion of counterfactual dependence is to be employed in the Primacy Thesis. It captures the notion of particular event causation and has no overtones of determinism, or of causal connections which immediately instantiate laws of nature; these are commonly, but often misleadingly, associated with causation. Given this we will say: 'SRP causes D' just means 'if particular social relations of power SRP were not to prevail then particular discourse D would not obtain'. This is in fact quite close to many of the claims Foucault makes on behalf of the power/knowledge link.

This completes the exposition of Foucault's power/knowledge doctrine. At best it expresses a relation of causal dependence, understood as a relation of counterfactual dependence. Dependence of what on what? In the Primacy Thesis there is a dependent item which Foucault says can variously be truth, a discourse, a discourse of truth, or knowledge (where this again is usually objective knowledge (without a knowing subject), and is usually the contentful knowledge that ..., but can also be know-how ..., etc). And there is a independent item, relations of power (as set out in section 9.3) upon which the above items depend.

9.4.5 An Illustration of the Power/Knowledge Doctrine

So far the "power/knowledge" doctrine has been expressed rather abstractly as a schema with few illustrative examples. The first chapter of *Discipline and Punish* provides an example of how Foucault viewed the inter-relationships of power and knowledge in the particular case of our changing penal practices. The chapter begins with Foucault's graphic description of the public execution of the would-be regicide Damiens in 1757. He then rapidly changes scenarios to eighty years later and gives an outline of quite transformed penal practices in which the day of prisoners is regulated from beginning to end. Foucault expatiates on the differences, one central difference being the shift from bodily physical punishment, both the public exhibition of punishment and the more private carried out within a prison, to the punishment of something else – the soul. Foucault, perhaps in the light of his French Cartesian inheritance, does not speak of the punishment of a person (either physically or mentally) but rather punishment of the body, and then later punishment that is directed upon the soul, as if they were separate items to be punished. Such a difference can be rendered, in a non-question-begging way, as the physical or mental punishment of a person (or human being).

He also carries over the terminology of his theory of discourses to that of practices within penal institutions when he further describes the difference as 'a substitution of objects' (Foucault (1979), p. 17). Here the changed "objects" are the crimes and offences that are judged in a court. Thus judgements are allegedly passed on not just the acts of a person, but also 'the passions, instincts, ..., effects of the environment or heredity, ... drives and desires' (*loc. cit.*). For Foucault 'It [the question] is no longer simply "Who committed it?' But: 'How can we assign the causal process that produced it? Where did it originate in the author himself? Instinct, unconscious, environment, heredity?' (*ibid.*, p. 19)

Foucault is correct in identifying a host of questions that have been regularly asked of any offender over the last 100 years that may not have been asked in the eighteenth century. But he exaggerates. As far as criminal courts are concerned judgements, understood as the verdicts issued by a court, are largely (there are some exceptions) in respect of legal matters; and such judgements culminate in a final verdict about whether or not a person is actually the author of some offence or crime with which they have been charged. These judgements are made by a judge or a magistrate, or more often in English speaking countries by a jury. Another part of the overall penal system will address, if it does, question about the cause of crime, or even motivation, a matter that criminal courts may not even consider when determining their verdicts. Foucault also speaks of the judgements made in sentencing as distinct from judgements of innocence or guilt. Here he is more correct in that matters from psychiatry to forensic medicine do now play a big role, but in earlier centuries may have played little or no role, in sentencing. Foucault is aware of the kind of point just made but rejects it out of hand:

'But, it will be objected, judgement is not actually being passed on them [i.e., the new range of "objects"]; if they are referred to at all it is to explain the actions in question, and to determine to what extent the subject's will was involved in the crime. This is no answer. For it *is* these shadows lurking behind the case itself that are judged and punished' (*ibid.*, p. 17)

This response cannot be wholly correct. It is persons and their acts (not merely bodies or souls, as Foucault might have it) that are judged in a court to determine guilt. It is not the "shadows" behind them that are judged, or even punished (as Foucault oddly says). However as admitted, judgements can occur in sentencing and during prison confinement to assess 'the shadows lurking behind the case'. He also misleads us when he sums up his position, saying that 'a general process has led judges to judge *something other than* crimes' (*ibid.*, p. 22, italics added). On the contrary, they still judge crimes. However others, as well as some judges, might determine other matters to do with crimes, such as what caused a person to so act, or what sentence is appropriate given knowledge of the criminal, or whether the charges be dropped because of criminal insanity. As many others, as well as Foucault, have pointed out, alongside such court procedures there has grown up over the last two hundred years a body of science having to do with human behaviour, including criminal behaviour, that is extensively used in the course of a trial through witnesses who are of medically "expert". The sciences are also used in sentencing, sometimes in attempts at rehabilitation, but not commonly in judging guilt or innocence. To this body of science one could, if one wished, apply Foucault's theory of discourses (though perhaps not in its exaggerated form) as an account of how these human sciences have developed and affected our penal practices.

Here there emerges a connection with the "power/knowledge" doctrine. As Foucault puts it: 'A corpus of knowledge, techniques, 'scientific' discourses is formed and becomes entangled with the practice of the power to punish' (*ibid.*, p. 23). But this remark is an instance only of the more acceptable Reciprocity Thesis in which growing human knowledge can expand our power. It is not an instance of the Primacy Thesis, and so says nothing about power causally producing knowledge. One can readily agree, even though one might reject much else of what Foucault says, that our developing human sciences have vastly influenced the way we deal with those convicted of crimes. But what is not asserted here is that 'the power to punish', or some other power, manages to causally produce knowledge in the human sciences about criminal behaviour. Such 'power to punish' need not even be a spur to discovery in forensic sciences, or to the application of sciences to forensic matters; it need not even goad one on to an understanding of why all those in prisons commit the crimes they do. Later in the chapter

Foucault overlooks this point and argues not only for Reciprocity Thesis but also for the Primacy Thesis on a very slender base of evidence.

More than this, he sees in the changes in the "objects of crime", and in the changes in penal practices, a change in epistemology, the theory of knowledge. This and other unwarranted exaggerations pile up in the following peroration about "power/knowledge":

> Perhaps, too, we should abandon a whole tradition that allows us to imagine that knowledge can exist only where the power relations are suspended and that knowledge can develop only outside its injunctions, its demands, and its interests. Perhaps we should abandon the belief that power makes mad and that, by the same token that the renunciation of power is one of the conditions of knowledge' (*ibid.*, p. 27)

Remarks such as these have led some think that the power/knowledge doctrine is a rival that overthrows theories of knowledge that have dominated our thinking from Descartes, if not before. But this is mere hyperbole bought on by an uncritical acceptance of what Foucault says. Continuing the above quotation, we find Foucault proclaiming, in abstract form, the Primacy Thesis. These remarks have been met before and say: 'We should admit rather that power produces knowledge ... ; that power and knowledge directly imply one another; that there is no power relation without the correlative constitution of a field of knowledge, not any knowledge that does not presuppose and constitute at the same time a power relation' (*loc. cit.*). But nothing previous to this in the chapter has any bearing on the Primacy Thesis expressed in these remarks, viz., the claim that 'power produces knowledge'. The "power/knowledge" doctrine is simply asserted; there is no supporting evidence or illustration.

Even more exaggerated and unwarranted is the further claim: 'it is not the activity of the subject of knowledge that produces the corpus of knowledge, useful or resistant to power, but power-knowledge, the processes and struggles that traverse it and of which it is made up, that determines the forms and possible domains of knowledge.' (*ibid.*, p. 28) This seems to say that it is not people such as psychiatrists who produce some body of psychiatric knowledge. Rather it is power-knowledge itself that allegedly does this. But the reification of power-knowledge does little to explain how it is so efficacious. In fact it makes totally obscure how such psychiatric knowledge might have come about – unless the psychiatric knowledge is contained in a power/knowledge nexus itself, in which case it trivially produces itself.

To sum up, even though the "power/knowledge" doctrine is widely invoked in the first chapter of *Discipline and Punish*, nothing is said there that even supports the central Primacy Thesis that power produces knowledge that ... (in psychiatry or any other field), or even knowledge how to ... (in these fields). It best only half of the "power/knowledge" doctrine holds, viz., the Reciprocity Thesis. But the plausibility of this relies on a Baconian point that few would deny when put in plain language, viz., that our developing knowledge that ... in pure and applied forensic medicine can increase our human powers in enabling us to deal with criminals.

9.5   SIX CRITICISMS OF THE POWER/KNOWLEDGE DOCTRINE

9.5.1   *Power Cannot Cause Truth or Knowledge*

Relations of power are alleged to cause a wide variety of things. Most implausible is the claim that they cause the distinction between the true and the false. Foucault asks himself the following questions:

> Firstly, in what sense is the production and transformation of the true/false division characteristic and decisive for our historicity?   Secondly, in what specific ways has this relation operated in 'Western' societies which produce scientific knowledge whose forms are perpetually changing and whose values are posited as universal?' (Burchell *et. al.* (1991), p. 82).

Here the distinction between truth and falsity is said to be grist to the mill of the 'power/knowledge' doctrine. There is an idealist interpretation of the thesis that is utterly implausible, viz. the very things that make our propositions true are the products of social power relations. That is, not only talk about the roundness and mobility of the Earth but the very roundness and mobility of the Earth itself is the product of social power relations. Similarly not only is talk of mental illness but the very mental illnesses themselves are always caused by social power relations. (Of course the power some people exert can bring on mental illness in others in some cases; but this is not what Foucault intends.) Foucault is strongly realist about his relations of power and the causal effects they allegedly have. But it is a matter of obvious scientific investigation that whatever else relations of power cause, they do not bring about factual states of affairs such as the mobility of the Earth or (on the whole) mental illness in people (e.g. Huntington's Disease which is a genetically based neuro-transmitter disease). Rather, it is these bits of the world that make the *contents* of our beliefs true. In the next sub-section it will be seen that an act/content distinction need to be drawn in formulating the Primacy Thesis. That is, at best power relations might sensibly cause our *act* of believing that the Earth is mobile, or *act* of believing that a person is mentally ill, and not the contained propositional contents.

Let us revisit the argument of section 9.2, slightly reworded, and view it in the light of the Primacy Thesis which now appears as the first premise:

Social relations of power (SRPs) causally produce our discourses;

The discourses we adopt carry with them specific domains of objects;

∴ SRP causally produce the objects that there are (i.e. 'reality', 'truth').

Supposing the argument is valid, the second premise has been argued to be false for Foucault's account of discourses in which they constitute the domains of objects of most sciences whether they be physical, biological or social. The purpose of this section is to show that the first premise is also false.

Do SRPs cause knowledge? This was an issue that arose for the formulation of the tenets of the Strong Programme in section 5.1. There it was decided that social factors could not sensibly be said to cause a person's knowledge that.... Moreover, as was argued in chapter 2, it is the evidence that a person has, or the reliability of their belief-forming processes, which turns belief into knowledge. Since they lack any features of normativity, no SRP can do this. The causes of a person's knowing (as opposed to believing) that p cannot be power relations. Thus we can delete truth and knowledge from the formulations of the Primacy Thesis and replace them by belief which can, on some occasions, be a causal product of power, and which unlike knowledge, is not a normative concept. The fourth version of the Primacy Thesis now says:

Primacy Thesis (D): For all bodies of belief B (respectively discourses D, whether they are discourses of the truth or not) there are SRP such that SRP cause B (respectively D).

9.5.2 Getting the Causal Relations Right

The Primacy Thesis as expressed so far is seriously ill-formed. It concerns causal relations and these standardly hold between events or acts. Do SRPs cause (a) the *propositional contents* of our beliefs or discourse, or do they cause (b) a person's *act* of accepting, adopting, holding, believing or proposing some proposition, or cause a person to engage in some discourse? This was an issue that recurred in the discussion of how to formulate the Causality Tenet of the Strong Programme in section 5.1.2. The verdict was that (a) made no sense, but that (b) was at least a sensible interpretation. Taking this into account, the Primacy Thesis needs to be explicitly set out in terms of the causes of a person's intentional acts of believing, entertaining, etc. Thus in the case of a person's act of belief, or of their engaging in a discourse we have:

Primacy Thesis (E): For all persons x and for all bodies of belief B, or discourses D, there are SRP such that SRP cause either (i) x's *act* of belief in B or (ii) x's *engaging* in discourse D.

This version of the Primacy Thesis is the best that can be done for Foucault's claims about power/knowledge, and it is the one that will be referred to from now on.

This understanding of the Primacy Thesis can be found in Foucault. First, he often asks us to consider the way in which power operates on bodies, which we can take to mean persons, and not just their bodily material. This is set out explicitly as an instance of the Primacy Thesis (E): SRPs are causally active on *person* x in such a way that they come to believe B, or engage in discourse D. The causal links between SRP and discourse are *via* the users of the discourse.

Second, we noted in the previous chapter that discourses were, at least, sets of statements. Importantly statements are akin to speech acts and are to be sharply distinguished from propositions. Now Foucault, as noted in the previous section, gets us to note that validity and 'games of truth' are connected with propositions that are expressed in our sciences; and these, unlike statements, are exempt from power relations. Thus there is a plausible reading of the Primacy Thesis that concerns the power relations allegedly involved in our *acts of statement making*, viz., our speech acts; they do not act on the propositional contents of the speech acts.

Third, speech acts are really a species of event and causal relations are commonly understood to hold between events. So far so good. But discourses are made up of the four rules of formation for objects, enunciative modalities, concepts, and themes; and the discourse has unity in virtue of the regularities that hold between the four rules of formation. It is hard to see how any of these, with the exception of enunciative modalities, can stand in causal relations with SRP. Thus much of what comprises a discourse, viz. objects, concepts, themes and the regularities that hold between them, lie outside the social power causal nexus. However if we modify this and talk instead of x's intentional attitudes such as x's *thinking* that there are certain kinds of objects, or x's *entertaining* certain concepts or x's *adopting* certain themes, then there is a way in which each of these acts might enter into causal relations with SRP.

Enunciative modalities are the only aspects of Foucauldian discourses that could plausibly be regarded as candidates for the causal effects of power. Thus the authoritative discourses of doctors and medical scientists could be viewed as arising out of social power relations that either exist within the community of scientists or arise from outside it. Whether this is so is an empirical matter to be investigated by sociologists of belief and not pronounced upon *a priori* by philosophers.

Can we say that the very regularity that holds between any of these four rules of formation, i.e. that which gives a discursive formation its unity, could itself be the effect of power? In particular, could the special kind of knowledge about the necessary presuppositions that make discourse possible itself be the result of power? For the reasons already given in section 5.1 such knowledge cannot be the result of power. But it might be possible, as indicated above in this section, that acts of belief in such special knowledge could arise from power; but this is an empirical matter to investigate and in now way bears on other version of the Primacy Thesis. Foucault's quest for *savoir* in his technical sense (see sections 8.1.2 and 8.1.3) and its subsumption under the "power/knowledge" doctrine seems to have reached a dead end.

Finally, Foucault's meta-theory of discourses also concerns the way in which objects, concepts, themes and enunciate modalities of a discourse come into existence, are maintained or altered, and then go out of existence in favour of another discourse with different objects, concepts, themes and enunciate modalities. Perhaps power might be a cause of why a person comes to *adopt* a discourse, *persist* in the discourse, *change* aspects of the discourse, and then *reject* the discourse in favour of another. The operation of power on such intentional "objects" is allowed for in formulation (E) of the Primacy Thesis. But what the Thesis now rules out is that items other than power might be the cause of a person adopting, persisting in, etc. a discourse. Such alternative causes might be purely cognitive as when Newton replied to a question about how he had discovered the law of universal gravitation: "By thinking on it continually". Or they might be methodological as when a person adopts, changes or abandons a discourse (or theory) on methodological grounds. Such cognitive and methodological causes are ruled out by Foucault and advocates of the Strong Programme by the very strength of the unqualified theses they propose about the cause of belief.

9.5.3 Is Power the Only Cause of Belief?

The Primacy Thesis mentions one and only one kind of factor, (social) relations of power, as the cause of belief. Unlike the Strong Programme, it fails to mention any biological, psychological or cognitive factor as *a* cause of belief. It also ignores reasons as causes. In particular, it ignores any theory of logic or of scientific method that provide criteria for judging the relative epistemic worth of our beliefs, and in providing, in appropriate cases, rationality explanations of why we believe what we do. In fact Foucault provides no *normative* theory of belief that would tell us which beliefs we *ought* to accept or reject. (Though as we have seen in section 9.4.3, Foucault late in his career came to see that he ought to leave room for such rationality explanations.) Since for most of Foucault's career, 'knowledge' is merely the froth on top of power relations, his theory enables him to point out only the alleged causal effects of power in bringing about certain beliefs and not which beliefs are epistemically worthy. Like Newton we can also think on matters continually and so justify, or change, our beliefs in ways that have nothing to do with power relations. Power might occasionally be a cause of our beliefs but it cannot be the sole cause; it must operate in conjunction with other cognitive factors. In the next section it will be argued that power may not be present at all, or only present in an uninteresting way, in the causation of many bodies of belief. This was argued in chapter 5.9 in the case of the social factors of the Strong Programme; it applies equally to Foucault's power relations.

9.5.4 Is There Evidence For the Primacy Thesis Understood Empirically?

Are there any instances of the Primacy Thesis? Has any person, lay or scientific, had their belief in a scientific theory caused by power? Perhaps this might be true of the lay person who has been taught in a punitive educational régime. But is this true of, say, the scientist who first discovered or invented the theory? To find an instance of the Primacy Thesis one would have, first, to individuate some discourse or body of belief of a person and, next, carefully investigate the full range of possible relations of power that might obtain in the person's immediate community, and outside it, to see what powers, if any, cause the person to so discourse, or to so believe.

More concretely, consider the case of Mendelian genetics that Foucault mentions in his 1970 lecture (see section 9.2). Did Mendel come to believe his two laws of segregation and independent assortment through the causal efficacy of social relations of power, either remote or near? This is an empirical matter for the sociology of belief to discover. The Primacy Thesis itself is not a particular causal claim *within* empirical sociology which sets out what are, say, the particular power relations that surround Mendelian genetics (or theories of grammar, economics, psychiatry, medicine, biology – and even mathematics). The Primacy Thesis, though empirical and contingent, is a higher level meta-thesis about the form any sociological investigation into the cause of belief ought to take. The hard work of finding power-discourse connections is yet to be carried out. Foucault endorses this, in part, when he openly says about the investigation into power and the sciences: 'I'm an empiricist'. (Kritzman (1998), p. 106)

The Primacy Thesis can be understood in two distinct ways. It might be construed (a) as a general directive to look for the causes of discourses/belief, or (b) a general factual thesis about the cause of all discourse/belief.

Consider (a) first. In *Discipline and Punish* Foucault does set out his thesis as a general directive, or a research programme, for the investigation of the causes of discourse in the human sciences. Understood in this way the Primacy Thesis is a directive which says 'for any discourse look for the powers that cause it':

> Instead of treating the history of penal law and the history of the human sciences as two separate series whose overlapping appears to have had one or the other, or perhaps both, a disturbing or useful effect, according to one's point of view, *see* whether or not there is not some common matrix or whether they do not both derive from a single process of 'epistemological-juridical' formation; in short, *make* the technology of power the very principle both of the humanization of the penal system and of the knowledge of man. (Foucault (1979), p. 23, my emphasis).

Here Foucault directs us to investigate whether the discourses/beliefs that accompany our penal practices and the discourses/beliefs within the human sciences have a common source in power. Of course, asking us to look into the matter from this point of view is quite compatible with our discovering some, or failing to discover any, connection between power and beliefs about the penal system, or beliefs within the human sciences.

Let us now consider (b). This is the interpretation of the Primacy Thesis as a general empirical meta-thesis about the cause of all belief/discourse. Causal claims within the social sciences are notoriously difficult to establish, especially causal claims between social factors such as power and belief/discourse. Normally causal claims are tested using, say, Mill's methods or the more sophisticated randomised double-blind statistical tests which are the standard tools-in-trade of medical, some social, and many other investigative sciences. It is unlikely that any such methods of test have been carried out for any alleged power/discourse causal link. So much of what follows is speculative. Owing to the difficulties in testing for social causation, this will be the case for much of what Foucault or sociologists of knowledge say about the aetiology of our discourses/beliefs.

Let us return to the theory of Mendelian genetics. What are the power relations, if any, which either cause Mendel's act of belief in his laws, or cause (aspects of) Mendelian discourse? What were the power relations swirling within and around the monastery in which he carried out his investigations into the morphological features of the sweet peas he grew in the back garden? Perhaps Mendel had to exercise power to get his sweet peas grown or to find time from his ecclesiastical duties to work on his theory. But none of this is causally sensitive to the very theory he invented; power remains an uninteresting background cause. In most accounts of what went on, it was the evidence that Mendel collected concerning sweet peas that was the cause of his scientific beliefs/discourse and not any power relations. Like Newton, Mendel gave the data he found about sweet peas a lot of thought and this provided evidence for his belief in the laws of inheritance.

Without a full empirical investigation into the case of Mendel this is a dogmatic statement about a complex and difficult episode in the history of science. Mendel has been criticised for allegedly cooking his data (thought this is not a well-founded claim) and Fisher has argued that some of his statistical inferences are faulty.[6] Does power have anything to do with either the correct or the dubious aspects of Mendel's work? Hardly, since these objections, if correct, concern the evidential basis for the theory. Perhaps aspects of power are involved in non-discursive aspects of Mendel's science, e.g. his experimentation; but this is not germane to the Primacy Thesis which concerns discourse/belief only. Perhaps Mendelian genetics is not a suitable case to investigate. More interesting would be the possible power relations involved in the IQ debate or in the controversy that still surrounds Cyril Burt's work on twins separated from birth. But that one scientific episode rather than another illustrates the Primacy Thesis is hardly satisfactory since the Thesis is alleged to hold for all belief/discourse.

As we have seen, even Foucault recognises that Mendel's 'propositions appear, for the most part, exact' (Foucault (1972), p. 224). Explanations of why they are true, or can be believed to be true, have nothing to do with power; they have to do with how the world is, or the manner in which evidence bears on hypotheses in the light of some theory of scientific method. The classical account of the aetiology of belief/discourse formation has a lot going for it in a large number of episodes in the history of the human and natural sciences; they constitute a powerful set of counterexamples to the Primacy Thesis. To cast doubt on this claim, the followers of Foucault would have to show that this is wrong for all these alleged counterexamples. They would have to show, for example, that Newton was massively deluded when he said that he arrived at the law of universal gravitation by thinking on it continually, and that he was oblivious to the power relations in which he was engulfed and which allegedly caused his beliefs/discourse.

Since one swallow does not make a summer, so a few seemingly correct instances of the Primacy Thesis do not establish that all discourses are caused by some SRP. The confirmation of the Primacy Thesis cannot be assured without a very large number of case studies in the history of science to support it. So far, the requisite range of case studies are not available. A similar situation prevails for the Strong Programme in the sociology of science, in which it is alleged that quite general social factors, which are broader than Foucauldian social relations of power, cause our scientific beliefs. In fact the Primacy Thesis can be viewed as a special case of the Causality Tenet for the Strong Programme which says that social conditions bring about belief or states of 'knowledge'. In chapter 5 we discussed what the Strong Programme might mean and whether it had any correct instances (see sections 5.6 and 5.7). Along with many of its most fervent critics, such as Laudan ((1981) and (1982)) and Slezak ((1989) and (1991)), it was argued that it has few, if any, instances to its credit. Since Foucault's Primacy Thesis is narrower than the Strong Programme, it has even less chance of having successful instances.

Could much of what Foucault says about psychiatry be understood as a supporting instance of the Primacy Thesis? We have argued in section 9.4.4 that what Foucault says about the development of forensic sciences does not support the Primacy Thesis. One must be careful here in discussing the "power/knowledge" doctrine. Is it the Primacy Thesis or the Reciprocity Thesis that is being supported? We can agree, as Bacon would, that the theory and techniques of psychiatry can yield means to control people; but this has to do only with the Reciprocity Thesis. For psychiatry to support the Primacy Thesis one would have to show that the very theories and techniques of psychiatry are themselves the result of Foucauldian relations of social power. For Foucault this is quite different from showing the Baconian claim that interests in power might have given rise to psychiatric theory. Merely showing that psychiatrists (or others) have an interest in power over other people, and so developed their theories to this end, would not be sufficient to establish psychiatry as an instance of the Primacy Thesis. Foucault wants us to understand the Primacy Thesis to say that, fundamentally, social relations of power are the cause of discourse/belief and not the intentions people have to get power.

Does Hacking's account of the use of statistics in 'How Should We Do the History of Statistics?' (in Burchell *et. al.* (1991), pp.181-96) exemplify the "power/knowledge" doctrine? He describes the way in which bureaucratic power was enhanced in the nineteenth century through knowledge obtained by the collection of data and by the development of statistical theory (e.g., statistics as applied to births, deaths, information from censuses, information pertaining to insurance, etc). However this only provides good evidence for the less disputable Reciprocity Thesis and not the more dubious Primacy Thesis. To establish the latter in the case of statistics, one would have to show what role power played in the emergence of either the very propositions of statistical theory and of probability, or in acts of belief in these. We can admit that power relations might well be involved in the collection of statistical data (such data collection might be backed by laws with penalties for non-compliance). But this is quite different from its involvement in the belief in, or the content of, statistical theory. Though the history of the evolution of our concept of probability is an important matter, in Hacking's work on the history of statistics and probability Foucault's Primacy Thesis is hardly distinguished, and it is certainly not supported by that history. Finally, as was shown in section 5.3.3, Mackenzie's historical work on the growth of statistics in Victorian England does not bear out the claims of the Strong Programme, let alone the Primacy Thesis.

It is quite easy to generate instances of the Primacy Thesis; just set up a university department in which nothing but Foucault is taught in a fundamentalist biblical manner and in which students who believe every word are passed while those who do not are failed by exercise of power of the academic staff. This I take it is an instance of an educational system run along the lines of the chapter called 'The Means of Correct Training' in *Discipline and Punish*. However this hardly supports the more interesting case of the growth of scientific knowledge under discussion.

9.5.5 The Problems of Reflexivity and Relativism

We can ask: what are the causes of belief in Foucault's own theory as expressed in the Primacy Thesis? Does the theory apply to itself or not? Since the Primacy Thesis, PT, is perfectly general we can substitute PT in version (E) for body of belief B, or discourse D, and get the following reflexivity thesis R:

(R) For all persons x and for the belief that PT, or discourse PT, there are
    particular SRP such that SRP cause either (i) x's *act* of belief that PT or
    (ii) x's *engaging* in discourse PT.

Is R, a claim about power causing belief in PT, itself an objectively true or false causal claim? Or is belief in R merely the result of more power play? And so on for this belief in turn? Either R, like any other causal claim is open to test, and is thus objectively true or false; or a regress results. Precisely the same issues arise for the Strong Programme and were discussed in section 6.4 and 6.5.

Foucault has produced for us discourses *about* the causes of discourse concerning penal systems and sexual conduct, and he has produced discourses concerning the causes of belief in the human sciences such as biology, psychology, medicine, and so on. Are Foucault's discourses themselves the result of social power relations? Or are they to be exempt from application to themselves and stand apart from all other discourses? Unlike specific discourses about, say, genetics or economics, are Foucault's own discourses impervious to 'world-creating' relations of power? Do Foucault's discourses have a foundation which all other discourses lack? If so, what is this foundation? Foucault hardly addresses such questions which arise directly from the Primacy Thesis. If there is no foundation to be found for his discourses on power/knowledge", are we to assume that, in university departments where students come to believe the Foucauldian discourses they are taught, the causes of their belief are the power relations prevailing in academia which act on hapless students? The teaching of Foucault's discourses would itself be an exemplification of the technology of power within the University.

Since Foucault adopts a realist view about the causal efficacy of relations of power in producing our discourse/belief, then, to avoid being caught up in the problems concerning relativism, he must accept many of the deeper theoretical claims about power/knowledge in his writings as objective causal claims. Whether these causal claims are true or false, the true/false distinction cannot arise merely as a division caused by social power relations, as is suggested sometimes in his writings. Rather, truth or falsity is determined by the causal structure of the world. Included in this structure are the causes, whatever they be, of our discourses/beliefs. Amongst Foucault's alleged causal claims are all those instances of the Primacy Thesis for each science, from algebra to zoology. The conclusion of the previous section 9.5.4 is that most of these causal claims are objective falsehoods rather than the objective truths Foucault's "power/knowledge" doctrine takes them to be.

The reflexivity paradox that Foucault faces can be expressed in his own words:

> Truth is a thing of this world; it is produced only by virtue of multiple forms of constraint. And it induces regular effects of power. Each society has its regime of truth, its 'general politics' of truth: that is, the types of discourse which it accepts and makes function as true; the mechanisms and instances which enable one to distinguish true and false statements, the means by which each is sanctioned; the techniques and procedures accorded value in the acquisition of truth; the status of those who are charged with saying what counts as true. (Gordon (1980), p. 131)

This is a succinct statement of Foucault's Primacy Thesis concerning discourses about what we take to be true and false. But do the same considerations apply to the Thesis itself? Foucault appears to exempt his own theory of genealogy when he says:

> [T]his is what I would call genealogy, that is, a form of history which can account for the constitution of knowledges, discourses, domains of objects, etc., without having to make reference to a subject which is either transcendental in relation to the field of events or runs in its empty sameness throughout the course of history. (Gordon (1980), p. 117).

But it is precisely this 'transcendental ground' which Foucault seeks when he tries to establish his genealogical theory of the cause of all discourse independently of the nexus of powers with which all other discourses but his own are bound up. If he were to exempt his thesis, could he give a reason for accepting it as objectively true? This question is addressed in the next section.

9.5.6 The Testing of Rival Theories of the Cause of Belief and/or Discourse

Can Foucauldian genealogy be judged better than any other theory of the cause of our beliefs/discourses? Foucault hardly touches on this point, except in one place. He mentions Marxist and psychoanalytic critics who would offer different accounts of the causes of our belief and says in an imaginary reply to them:

> Has there been, from the time when anti-psychiatry or the genealogy of psychiatric institutions were launched ... a single Marxist, or a single psychiatrist, who has gone over the same ground in his own terms and shown that these genealogies that we produced were false, inadequately elaborated, poorly articulated and ill-founded? (Gordon (1980), pp. 86-87).

Here Foucault asks how well his own genealogical theory of the causes of discourse would compare with the quite different Marxist or psychoanalytic theories about the cause of the same discourse. As we noted in section 9.1, Foucault rejects many rival theories of the cause of our discourses such as 'spirit of the time', tradition, genius, evolution, and so on. But as was pointed out he fails to list reasons or evidence as a possible cause of belief.

On what grounds does Foucault claim that theories which rival his own are false or poor in some respect? Is he not appealing to the ways in which we might standardly adjudicate between rival theories concerning some subject matter, in this case causes of belief? Foucault would have to appeal, as many of us do, to principles of scientific method that have nothing to do with power relations, Marxist economic bases, Freudian analyses, or whatever. He would have to use principles of scientific reasoning to show that his own genealogical theory is not inadequate or ill-founded. He would have to use these principles when he claims that his own genealogical theory allegedly offers *better* or *more comprehensive* explanations than psychological or Marxist theories about the causes of discourse.

The theory of scientific rationality gives an account of what it means to say that some theory is ill-founded, better than another, or more comprehensive. To deny that principles of scientific reasoning have a role here is to accept the claim that there are no grounds whatever for rationally evaluating rival theories concerning the source of our beliefs. What would be left? A radical Foucauldian would claim that discourses rise and fall with the varying play of power. Even belief in this is the result of more play of power. But this we have rejected in the previous section as untenable. If Foucault is to claim some superiority for his theory of the cause of our discourse/belief over its rivals then he must admit some criteria for making this claim and state what reason we might have for accepting the criteria. But Foucault is now in territory familiar to those who work on issues to do with scientific method that gets little or no mention in his writings.

Questions that arise immediately at this point are: What is the status of theories of scientific reasoning? What are the grounds for accepting them? Isn't scientific method itself merely more in the way of belief or discourse which arises out of specific power relations? This last question bids us see all forms of reasoning and all theories of scientific method as an instance of the Primacy Thesis. If M is a discourse about logic, principles of reasoning and some theory of scientific method then Primacy Thesis (E) tells us:

For all persons x and for belief M, or discourses M, there are particular SRP such that SRP cause either (i) x's *act* of belief in M or (ii) x's *engaging* in discourse M.

As particular instances of M, the very methods used in testing for causes, such as Mill's or randomised double-blind tests, turn out to be merely more in the way of discourse or belief; so they are grist for the mill of Foucault's genealogy. Here the "power/knowledge" doctrine in the form of the Primacy Thesis has the same grand scope that was claimed for the Strong Programme.

The thesis just expressed makes reason itself the causal product of power relations, thus undercutting any epistemic grounding for principles that could be used to adjudicate between rival theories of the cause of belief. This would be bad news not only for the lovers of reason but also for Foucault who thinks, as was pointed out in 9.1, that his theory of the aetiology of belief is much better than some of the more 'magical' theories that historians of ideas have employed, or that Marxists and psychiatrists have used. In response, some post-modernists, such as Lyotard, might reject the claim that Foucault's theory could in any sense be better when they say that reason is just one narrative amongst others which can be replaced by some different narrative (see for example Lyotard (1984) p. xxiii and p. 7).

Suppose supporters of Foucault were to say: 'Look at the brilliant insights Foucault has given us about the causes of discourse particularly within the human sciences; this at least gives some evidence for his genealogical theory as partly expressed in the Primacy Thesis.' This is an appeal to some allegedly true cases in which power causes certain discourses. The inference from these particular cases to the quite general Primacy Thesis uses standard canons of inductive inference from evidence to hypothesis; in this case induction gives the rational grounds for belief in the Primacy Thesis. Arguing in this way for the Primacy Thesis is quite distinct from hunting out even more power relations that give rise to Foucauldian discourses about power/knowledge. The followers of Foucault are in a bind. On the one hand they like to think that our scientific discourses provide good evidence for belief in Foucault's genealogy. On the other hand they tend to regard theories of induction, confirmation and scientific method as simply more in the way of belief/discourse that arise from power relations. They have given up the good old epistemological stories about what is true or false, or what is well or badly supported by evidence.

What reply can be given? We must be willing to admit that there is a history of the development of theories about principles of reasoning and theories about scientific rationality; it did not arrive God-given. Much work has been done in presenting the history of the development of our forms of reasoning and much more has yet to be done. Philosophers have also tried to find justifications for adopting principles of reasoning and method. One of the more fascinating contemporary debates in the philosophy of science is precisely about these issues. What has Foucault's theory of power to contribute to understanding this history and the modes of justification adopted by philosophers? Power has hardly, if at all, featured in this story. That it might be thought that it should is due to a strong *a priori* imposition of the "power/knowledge" doctrine upon all modes of discourse in the absence of empirical investigation. As noted several times, Foucault simply leaves the whole of epistemology out of consideration; he ignores the role scientific methods play in the renovation of our discourses/beliefs. If, contrary to what many hold, we have to resort in the end to something social in our explanations of our justificatory practices, as many Wittgensteinians suggest, the social element need not be narrowly confined to only relations of power. To argue 'if it is social then it must be a power relation' would be to commit a *non-sequitur*. But as was argued in chapter 7, this route is hardly open even for advocates of the Strong Programme.

9.6 BRIEF COMMENTS ON FOUCAULT'S TALK OF TRUTH

Many of the difficulties with Foucault's views can be traced to his conception of truth. The line taken here has been mildly realist in the sense that what is true or false is so independently of our current epistemic practices and scientific beliefs, especially beliefs about causal relations. This accords with at least Foucault's causal realism concerning power, but is to be set against his discourse relative account of "objects" and the dependent role assigned to knowledge in its power/knowledge nexus. However Foucault does make occasional reference to truths, such as those of Mendelian genetics or mathematics, which are independent of prevailing discourses. Perhaps those more enamoured of epistemologically loaded theories of truth will have rival interpretations to offer of what Foucault says about truth. Here we will consider only two comments on truth that Foucault made in a more considered written reply to a question at the end of an interview. The first says:

> There is a battle 'for truth' or at least 'around truth' – it being understood once again that by truth I do not mean 'the ensemble of truths which are to be discovered and accepted', but rather 'the ensemble of rules according to which the true and the false are separated and specific effects of power are attached to the true.' (Gordon (1980), p. 132)

Foucault's use of the word 'true' is quite non-standard. But it seems clear that he intends to distinguish between objective truth, which can be discovered, from his own idiosyncratic use of the term. But what he says is unclear. He takes truth to be an ensemble of rules. But what do these rules do? We are told that they separate the true from the false. But this makes his definition of truth circular; in the *definiens* he uses the very word 'true' that he is trying to define! It is hard to know how to repair this muddle. Perhaps what Foucault ought to say is that there are rules, such as prohibitions, which determine what we are to believe (as true) and what we are not to believe (take to be false). Thus power allegedly determines what is generally *believed* to be true, but not the truth itself. Is this a truth? Foucault presents it as such. But understood as a universal claim it is simply a version of the Primacy Thesis; and this is false, as has been argued.

It does not take much to see that what might be *believed* to be true as the result of power could be discovered to be false (if it is false). Importantly Foucault needs the distinction between what *is* true and what we *take to be* true if he is to be a critic of our society, unmasking the very power relations he alleges determine our discourses about what we *believe* to be true or false.

In one of his last interviews before his death he said in answer to a question about free speech:

> Nothing is more inconsistent than a political régime which is indifferent to truth; but nothing is more dangerous than a political system which claims to prescribe the truth. The function of 'speaking the truth' should not take the form of law .... The task of speaking the truth is endless: no power can avoid the obligation to respect this task in all its complexity, unless it imposes silence and servitude. (Foucault and Ewald (1984), p. 31).

What this suggests is that it is always possible to get at some discourse-independent truths whatever the system of power/knowledge in which we may be currently causally enmeshed. The very history of discourses that Foucault describes says that we have always been so enmeshed. As all-enveloping as some belief systems/discourses may be, we have been able to cross over their horizon into a new system of belief in which some new truths emerge untrammelled by the previous belief system. If we cannot free ourselves from our current all-enveloping discourse then the task of reaching for at least one new truth not masked by our current discourse becomes hopeless. Foucauldian 'speaking the truth' becomes impossible, i.e. social criticism which exposes the way power causes, or causally maintains, beliefs is impossible.

A second remark of Foucault's says:

> 'Truth' is to be understood as a system of ordered procedures for the production, regulation, distribution, circulation and operation of statements [Note: not propositions].
> 'Truth' is linked in a circular relation with systems of power which produces and sustain it, and to effects of power which induces and extend it. A régime of truth. (Gordon (1980), p. 133)

If we accept the statement/proposition distinction, we can allow that what Foucault says here is simply a reiteration of the Primacy Thesis, along with the Reciprocity Thesis. This highlights one source of confusion for readers of Foucault; an instance of the Primacy Thesis is presented as an account of the dependence of truth on power. However the Thesis, despite the way it gets expressed, says nothing about truth; rather it is about the powers that allegedly produce our discourses, or cause what we *believe* the truth to be. For the Foucault of the 1970 lecture, there were truths independent of our current modes of discourse, such as the truths uncovered by Mendel. Though this notion of truth is submerged in most of his other works of the 1970s it managed to resurface in some of his final thoughts on the matter, e.g. the comments about free speech in which we can move beyond the bounds of current discourse.

Foucault says between the above two quotations from Gordon (1980) concerning truth:

> All this must seem very confused and uncertain. Uncertain indeed, and what I am saying here is above all to be taken as a hypothesis. In order for it to be a little less confused, however, I would like to put forward a few 'propositions' – not firm

assertions, but simply suggestions to be further tested and evaluated. (Gordon (1980), p. 132-33)

Here Foucault is being candid. His views with which we have been wrestling were tentative hypotheses for further investigation! Confused it is – and confusing into the bargain. But uncertain? No, for Foucault's 'power/knowledge" doctrine can be tested. It has failed its test; the central Primacy Thesis in all its generality is false.

To sum up the following can be said:

(1) Foucault's power/knowledge doctrine is obscure; struggling to get out are the more correct Baconian theses. What is meant by 'power', '/' and 'knowledge' are open to a host of different interpretations. The best that can be made for the doctrine are the various versions of the Primacy Thesis and the Reciprocity Thesis.

(2) On some interpretations there may be some true instances of the Primacy Thesis, but this is an empirical matter for sociologists of belief to investigate. Most likely, Foucault's empirical insights are to be attributed more to the Reciprocity Thesis than to the Primacy Thesis of which there are few significant instances.

(3) Foucault's Primacy Thesis leads to nothing of interest in epistemology, and more generally philosophy. Like the Causality Tenet of the Strong Programme it is muddle to be avoided.

(4) Foucault always insists that there must be some connection between power and knowledge and that no knowledge can be free of power. But there might be no connection for some quite central aspects of knowledge, especially those in which methodological principles (mentioned in Part I) are applied to scientific theories. This is something that Foucault does recognise on occasions, but he does not give it the prominence it deserves.

(5) Perhaps we should take Foucault at his word when he says: 'I have to say that I'm not much of a philosopher' (Faubion (1998), p. 249).

NOTES

[1] For an account of explanatory contrasts and the pragmatic aspects of explanation see the collection of essay in Ruben (1993), especially essays VIII, XI and XII·

[2] Sometimes Foucault uses the term 'genealogy' to mean a quite specific kind of knowledge to do, as he says in the first of 'Two Lectures' (Gordon (1980), pp. 78-92), with the 'insurrection of subjugated knowledges', 'local knowledge' and 'historical knowledge of struggle'. In these respects Foucault says 'Let us give the term *genealogy* to the union of erudite knowledge and local memories which allows us to establish a historical knowledge of struggles and to make use of this knowledge tactically today' (*ibid.*, p. 83).

[3] This anecdote comes from Westfall (1980), p. 105. This excellent remark underlines the role that purely cognitive activity can play in the history of thought, something commonly denied in sociological circles. The remark was brought to my attention in Slezak (1991), p. 252.

[4]   The lecture 'The Discourse on Language', is printed at the end of the USA edition of *The Archaeology of Knowledge* published by Pantheon Books New York, pp. 215-237, but is not printed at the end of the English edition. Hence reference is made here to the USA edition of *The Archaeology of Knowledge.*

[5]   Some such connection is suggested but not critically evaluated in Kusch (1991), pp. 163-64. However he calls it an 'internal-essential' relation whereas I will argue that the causal links are, like other causal relations, contingent; they are also extrinsic.

[6]   See R. A. Fisher's commentaries and assessments of Mendel's work in J. H. Bennett (1965).

PART IV

THE GERMAN CONNECTION:

NIETZSCHE

SYNOPSIS OF PART IV

The main chapter of Part IV is devoted to Nietzsche's account of the causes of human belief in either the existence of objects in the world, or in logic or in morality. This fits with the main theme of the book which concerns the non-rational source of our beliefs, since Nietzsche's account falls almost entirely outside the pale of rationality. It begins with Nietzsche's most obscure and troublesome doctrine of the "will to power", since this is alleged to be the main cause of our various bodies of belief. In section 10.1 it is treated as a metaphysical doctrine about the nature of the world, though, as will be seen, there are many other different manifestations of the Protean "will to power'. It also functions as a biological drive in all species as well as a variety of psychological and other drives in humans. Section 10.1 sets out schematically eight theses about the metaphysical "will to power", as well as the related view that events and causal relations are not to be found in the world but are constructions that we project onto the world (this being but one manifestation of the "will to power" in humans to ensure their survival). This leads, in section 10.2, to a consideration of Nietzsche's naturalism, in particular how the metaphysical, biological and psychological aspects of the "will to power" allegedly underlie all investigations in the sciences. Though one can applaud the naturalism, it is argued that the naturalistic basis in "will to power" is flawed. Section 10.3 explores the difficulties in Nietzsche's fictionalism with respect to ordinary substantive objects and their alleged continuing identity, and their elimination in favour of complexes of power. Since, according to Nietzsche, there are no substantive objects in the world but only complexes of power then our conceptual scheme must be false. Our belief in ordinary objects is not based in their real existence; rather they are a further projection we make onto the world in order to make our survival possible. Section 10.4 explores Nietzsche's claim that our belief in the existence of substantive objects, and in logic, depend on false presuppositions about identity. However our belief framework, and our logic, would not exist without the fiction of self-identical objects; they are necessary for our thought and our continued existence even though they are false. So, why do we believe in such fictions? This leads to an application of the "will to power" in the context of the evolution of humans; our inner drives lead us to a construct a world of such fictions which we then believe exists, thereby preserving and enhancing life. Section 10.5 briefly explores Nietzsche's perplexing and contradictory claims about truth. Section 10.6 investigates one aspect of Nietzsche's genealogy of truth and his claim that it is an ascetic ideal. That is, in aiming for, and pursuing, truth we deny ourselves something. What that something might be is also explored in section 10.6.

Section 10.7 investigates Nietzsche's claims about the genealogy of morals, especially Christian slave morality as set out in his book *The Genealogy of Morals*. This involves a further application of the "will to power" as a psychological drive that leads us to adopt, according to the character of the drive, one set of moral beliefs (say, slave morality) as opposed to another (say, master morality). It is argued that Nietzsche's explanation of why we came to adopt the slave morality of Christianity has many dubious hypotheses from psychology, sociology, history and anthropology, as well as his own special hypothesis that it is the "will to power", as exemplified in the psychological drive of *resentment*, that gives rise to slave morality and the overturning of master morality. His genealogical theory has an explanatory lacuna; it remains unclear how those who endorsed master morality became susceptible to slave Christian morality.

There is also a problem for Nietzsche's whole genealogical enterprise. It would appear that in attempting to give a psycho-social account of the original of Christian morality, Nietzsche is indulging in a pursuit of the truth of the matter; and so is committed to the truth asceticism that he criticises. But if he is not to be committed to truth asceticism, it is hard to see what purpose there is in his attempt at a genealogy. The overall thrust of Nietzsche's philosophical project is to show that our various kinds of belief, such as belief in the existence of ordinary objects, in logic and in morality, turn on a number of non-rational drives or forces that fall under the rubric of the "will to power". Viewed in this light, Nietzsche is an important progenitor of postmodernist debunking moves. He has allegedly unmasked the real source of our beliefs; and none of them are to founded in anything rational. However, it will be seen that even Nietzsche pays lip service to the idea that there are methodological principles on which one can adjudicate between rival genealogies of belief in ordinary objects, logic and morals. For he himself proclaims that he is offering a *better* genealogy of these beliefs that anyone else, the term "better' being one from the normative theory of epistemology to which Nietzsche pays little attention. Unless, of course, his aim in producing genealogies lies quite elsewhere in, say, fiction or in promoting a "noble lie". And this brings us back to the problem at the heart of his whole genealogical enterprise. Nietzsche appears to be seeking truths about the origin of a particular type of morality while, at the same time, disparaging all truth seeking as an ascetic ideal. It is argued that this is not the deep problem that Nietzsche has raised for us in philosophy, as some of his commentators claim; rather, it is a muddle within Nietzsche's epistemology that can be readily resolved.

CHAPTER 10

NIETZSCHE'S GENEALOGY OF BELIEF AND
MORALITY

The reception of Nietzsche's writings has had a complete turn-about. During the first half of the twentieth century they were either neglected or suffered from serious misinterpretations by people as different as Bertrand Russell and the Nazis. During the second half Nietzsche apologists have provided a host of rival interpretations, many hagiographical and without critical evaluation. One interpretation of Nietzsche's work is that he himself believed that it could have rival interpretations between which it might be impossible to decide. Of his own enigmatic, obscure, mystifying and even dangerous doctrine of the "will to power" he says: 'Supposing that this also is only interpretation – ... well, so much the better' (*Beyond Good and Evil*, §22). Nietzsche's influence on philosophers in post-World War II France has been very strong, Foucault being no exception. He tells us 'reading Nietzsche was the point of rupture for me' (Kriztman (1988), p. 23). For him: 'It was Nietzsche who specified the power relation as the general focus ... of philosophical discourse .... Nietzsche is the philosopher of power' (Gordon (1980), p. 53). Foucault links his "power/knowledge" doctrine to Nietzsche's "will to power" when he asks about the 'question of truth': 'What is the history of this "will to truth"? What are its effects? How is all this interwoven with relations of power?' (*ibid.*, p. 66). Some scholars might find Foucault's use of Nietzsche suspect, but he has a ready answer: 'I prefer to utilise the writers I like. The only valid tribute to thought such as Nietzsche's is precisely to use it, to deform it, to make it groan and protest' (*ibid.*, pp. 53-4). Groans and protests aside, there are themes in Nietzsche that Foucault has quite rightly discerned and exploited in his work. These include Nietzsche's notion of power, or "will to power", his rejection of the idea that there are substantive objects in favour of an eliminativism that supports the idea of "ersatz" objects, and finally his version of the "power/knowledge" doctrine. But as will be argued, these themes are often as suspect as Foucault's own doctrine of "power/knowledge'.

Mannheim says that even though the sociology of "knowledge" began with Marx, the second important early source was Nietzsche whose 'flashes of insight' provided '... a theory of drives and a theory of knowledge which remind one of pragmatism. He too made sociological imputations, using as his chief categories of "aristocratic" and "democratic" cultures, to each of which he ascribed certain modes of thought' (Mannheim (1936) pp. 278-9). As important as the sociological imputations are, much deeper is the theory of drives, or power, which Nietzsche used to explain matters that other philosophers before him had thought were the domain of rational explanation. For Nietzsche the "will to power" is meant to explain a host of items including our belief in logic, truth, the nature of our conceptual scheme fundamental to which is the notion of an object and, most importantly for him, our moral beliefs, or what we take to be values. According to Mannheim, "drives" do the explaining. But Nietzsche's preferred notion is that of "power", or "will to power" (which, as it is exhibited in humans, can take on the aspect of drives, especially psychological drives).

It will be appropriate to set out first what I take to be the main thesis at work in Nietzsche's "power" doctrine and then modify and apply it to particular cases in what follows. For Nietzsche one of the characteristic features of humans is that we are believers. We form bodies of belief about what we take there to be in the world (including common-sense, mythical, religious and scientific beliefs), and we form beliefs about what we ought to do (morality), how we ought to think (logic), and so on for other norms. For convenience let us label each identifiable body of belief held by some group G over a period of time 'B_G'. These can of course vary from group to group, and over time within the same group. According to Nietzsche, each such body of belief (held over a given time) arises from the operation of the "will to power" in an individual person, or in people taken as a social group. Much interpretative ink has been spilt over Nietzsche's perplexing notion of "will to power"; and a little more will be split subsequently. It is fundamental to an understanding of most of his works, especially those written in the final third of his active life. The "will to power" can arise in a number of ways, from psychological and other drives in persons, to life drives in all species in the biological world, to the more metaphysical doctrine in which the entire world is nothing other than the operation of forces and powers. Exactly what manifestation of the "will to power" is at work is often left quite unspecific and needs careful investigation from case to case to tease out in what way it operates.

The basic "Will to Power" Thesis (WP Thesis) which lies behind much of what Nietzsche says is the following;

WP Thesis: *For all bodies of belief, B_G, held by some group G, there is some manifestation of the "will to power", WP, such that WP gives rise to (or causes, maintains, etc.) G's act of belief in B_G.*

The WP Thesis has the same form as those we have encountered in the Causality Tenet of the Strong Programme and in Foucault's "power/knowledge" doctrine, in particular what has been singled out as the

Primacy Thesis. Of course there are differences. The main difference is Nietzsche's appeal to the operation of the obscure "will to power". However what it operates on is the same in both cases, namely an individual's, or a group's, act of holding some body of belief (though in Nietzsche it operates on a host of other items as well). And the connection between WP and the act of believing B_G is similarly left unclear; but a range of relations can be considered such as those of necessary and/or sufficient condition(s), causation, maintenance, supervenience (see section 3.8, though this relation would not have been available to Nietzsche). Also there is a change thesis as a corollary: changes in the operation of WP lead to the abandonment of some prevailing beliefs B_G and the adoption of other beliefs B^*_G.

Importantly Nietzsche uses the above thesis to debunk rational accounts of why we believe what we do. His task is to show that in many cases that, where we thought reason was in operation and could provide an account of why we believe in logic, truth, morality, the existence of objects, and so on, the "will to power" is really what is at work. To think otherwise is an illusion to be unmasked by the project of Nietzschean analysis, viz., to show that the "will to power" is really in operation. Understood in this way, Nietzsche provides a challenge to all those who might have thought otherwise, one of his special targets being Kant. It also offers another way in which the alleged pretensions of modernism can be undermined, not just through the critique of Kant, but also through an independent critique of logic, truth and morality. The similarity of Nietzsche's underlying thesis to similar theses in the sociology of "knowledge", and in Foucault, is striking. Hence its inclusion here. However the big difference lies in how we are to understand the way in which "will to power" manifests itself, this being what Mannheim referred to as Nietzsche's 'flashes of insight'.

Some scholars make much of the difference between Nietzsche's published texts and the vast amount of material that he left unpublished, advising us to consider the former only and to regard the latter with suspicion.[1] Here the published material will be considered to be primary. However Nietzsche often alludes to issues that we can find more fully discussed in his unpublished notebooks; in that case it is appropriate to refer to this material as well. There are several occasions in which what is in the unpublished material seems to have no counterpart in the published works and goes beyond what is said there, often in an illuminating way; where this material is used it will be clearly indicated. Finally, a note about the mode of reference. Most of Nietzsche's writings are in the form of conveniently numbered paragraphs. So the mode of reference in this chapter will be by means of the title of a book listed in the references (rather than person and date), and a numbered paragraph.

10.1 THE METAPHYSICAL CONCEPTION OF THE "WILL TO POWER".

One common metaphysical, or ontological, question is 'What is the world ultimately made of?'. Nietzsche's position is that of naturalism (see section 3.2), but one that does not adopt the ontology of a competed physics. Nietzsche is one of the great nayayers concerning ontology. He denies the existence of the following: abstract objects such as universals (he favours a strong nominalism); souls, spirits and personal self-identity; Kantian things-in-themselves and any kind of noumenal or other-worldly realm (he lampoons the idea of other-worldly realms advocated from Plato and Christianity to Kant[2]); substances, self-identical objects and atoms (understood as ultimate indivisible continuants); anything in the world that our truths could represent. Since he often says that there are no truths, it is a contentious matter to say what Nietzsche's views on truth are. He maintains that no belief can represent any feature of the world and that the world has no items that could serve as truthmakers for our beliefs. So, what does exist? As will be seen, none of the above exists independently of us. Nietzsche is an extreme projectionist in which it is we that project onto the world constructions of our own making. Nietzsche's projectivism sits happily with much of postmodernism that is equally irrealist and constructivist.

10.1.1 Some Theses of a Metaphysics of "Will to Power".

Nietzsche's ontology is Heraclitean; all that exists is power (*Macht*) or, as is alternatively said, force or energy (*Kraft*). Though aspects of his ontology appears in his published writings, the best account can be found in the final section of his notebooks published in English as *The Will to Power*. The central metaphysical claims can be conveniently summed up in the following list of eight theses. There will be some critical evaluation of these theses; but the main task will be to find a perspective from which to view the main genealogical claims that turn on them.

(M1) Time is infinite in both directions while space is finite 'enclosed by "nothingness"'; there are no gaps of nothingness in what exists in space and time; what exists is continuous, uncreated and has and will exist at all times (*The Will to Power*, §1066-7).[3]

(M2) The world contains a large but finite quantity of force that is conserved (*ibid.*, §1066-7). Talk of conservation suggests that Nietzsche's force is like mechanical force, or perhaps other forces found in physics. But we should not understand his concepts of force or power to be like those of physics. That they are quite different is clear when he says of the concept of force employed in physics that it '... still needs to be completed: an inner will must be ascribed to it, which I designate as "will to power", i.e., as an insatiable desire to manifest power; or as the employment and exercise of power, as a creative drive, etc' (*ibid.*, §619). Though his force is like a vector in that it has direction, as far as theories of physics are concerned it has always been an otiose piece of anthropomorphism to add to our understanding of force an 'inner will to power'; it plays no role in cashing out what is meant by vectorial forces. Perhaps what is intended is that Nietzschean power, like the force of mechanics but unlike power in physics, is also intended to be "vectorial", i.e., Nietzschean power has a direction in that a "will" can be ascribed to it. Thus Nietzschean power is not aimless in its activity but is always directed from one region of space-time towards other regions. This suggests that we should take both force and power as directional, and as primitive and undefined notions in Nietzsche's metaphysics. Perhaps we should refer to the one and only ontological primitive as 'force-power' to indicate directionality due to the operation of a "will". But as long as Nietzschean power is understood to be directional then this is unnecessary.

(M3) There is a large but finite number of regions of space, or centres, at which a number of forces can act, i.e., powers of varying magnitude and direction occur in a given "volume" of space and time (*ibid.*, §1066-7).

(M4) Each centre of power acts in opposition to all other such centres. The "will to power" is not merely reactive but is constantly pro-active and in conflict with other centres of power:

> My idea is that every specific body strives to become master over all space and to extend its force (- its will to power:) and to thrust back all that resists its extension. But it continually encounters similar efforts on the part of other bodies and ends by coming to an arrangement ("union") with those of them that are sufficiently related to it: thus they then conspire together for power. And the process goes on – (*ibid.*, §636).

Nietzsche uses the term 'body' to refer to something which is neither material nor substantival; the term is simply a shorthand for a relatively enduring union of powers. The passage also conveys the idea of power as a ceaselessly active drive and not a disposition that may be activated in only certain conditions. Viewing power anthropomorphically, each centre always has the power to deploy actively whatever forces are at its command (in the style, say, of a military strategist with military forces) in such a way as to maximise the effectiveness of the forces and to resist whatever forces are in opposition.

(M5) There is a constant ebb and flow in the quantity of power, or force, at each of the centres (*ibid.*, §1067). The ebb and flow is chaotic: 'the world is not an organism at all, but chaos' (*ibid.*, §711); 'The total character of the world, however, is in all eternity chaos – in the sense not of a lack of necessity, but of a lack of order, arrangement, form, beauty, wisdom and whatever other names there are for our aesthetic anthropomorphisms' (*The Gay Science*, §109). This suggests that the chaos is only with respect to our aesthetic anthropomorphisms and not because there is a 'lack of necessity' in the world. Not anything goes in Nietzsche's world; it is constrained in some way. Normally such constraints are spoken of in terms of what the laws of nature, or nature's necessities, permit. But as we will see in (M6) and in section 10.1.2, Nietzsche resists the idea that there are laws or natural necessities. So what he might mean here remains unclear.

(M6) Nietzsche's ontology also contains relations that can hold between different centres of power: 'Every centre of force adopts a perspective toward the entire remainder, i.e., its own particular valuation, mode of action, and mode of resistance.' (*The Will to Power*, §567; see also §568). This rather Leibnizian remark suggests that there are relations that every power centre has to every other such centre, but the relations are described anthropomorphically as 'perspectives' or 'modes of valuation, action and resistance'. Nietzsche continues: 'Now there is no other mode of action whatever; and the "world" is only a word for the totality of these actions. Reality consists precisely in this particular action and reaction of every individual part towards the whole' (*ibid.*, §567). This remark serves to underline Nietzsche's view that there is an independent world (of which we are part) and that it comprises only centres of power with the "perspectival" relations that exist between them. In the light of these claims, and (M2), one can see from whence Foucault's talk of power having "intentionality" (i.e., direction or perspective) but without a "subject". These claim also lie behind what was called Foucault's "relational ontology" of power (see section 9.3.2). But what Nietzsche says about power hardly helps to clarify Foucault's often obscure and metaphorical claims about power.

Elsewhere Nietzsche characterises power and its relations less anthropomorphically in supposing that all powers are internally related: 'Supposing that the world had a certain quantum of force at its disposal, then it is obvious that every displacement of power at any point would affect the whole system – thus together with sequential causality there would be a contiguous and concurrent dependence' (*ibid.*, §638). But this follows only on the assumption in (M2) of the conservation of power. Again, referring to powers as dynamic quanta, Nietzsche claims:

> [there are] only dynamic quanta, in a relation of tension to all other dynamic quanta: their essence lies in their relation to other quanta, in their "effect" upon the same. The will to power [is] not a being, not a becoming, but a *pathos* – the most elemental fact from which a becoming and effecting first emerge -' (*ibid.*, §635).

Nietzsche explains how he understands the Greek word 'pathos'; it can also mean an effect, incident or accident that has been experienced or undergone or suffered. Though the language is not particularly clear, Nietzsche is not inviting us to think of the "will to power" as a separate cause behind the effects to which it gives rise; rather it is "in", or identical to, these so-called "effects" as well as the alleged "causes". But there is also the quite strong claim that powers have an 'essence' which is the internal 'relation of tension' each nexus of power has for any other nexus of power. Nietzsche is in other respects anti-essentialist, and could remain so here. Given (M2) about conservation, it is simply a fact about the world in which powers are conserved, that any change in one relation between some dynamic quanta is compensated for elsewhere. To add an essentialist requirement to this is to make a far stronger claim than is needed.

Understanding what is being claimed in these passages is made more difficult by Nietzsche's separate claim that causes and effects do not exist in the world:

> Cause and effect: such a duality probably never exists; in truth we are confronted by a continuum out of which we isolate a couple of pieces .... An intellect that could see cause and effect as a continuum and a flux and not, as we do, in terms of an arbitrary division and dismemberment, would repudiate the concept of cause and effect and deny all conditionality' (*The Gay Science*, §112).

There are no cause and effect relations in reality. They are constructions that we impose on the world; they are arbitrary divisions that we make. As a result the all-pervasive "will to power" is not something that can lurk behind effects as a cause but is a continual flux that constitutes both the supposed cause and effect. Talk of powers as dynamic quanta underlines the view that power is quite fundamental and elemental underlying all creation, destruction, continuing activity and change. It is not something that comes packaged in cause and effect relations. As will be discussed, Nietzsche adopts a constructivist, or projectionist, account not only of cause and effect (see next section 10.1.2) but also of substantive objects, and of truth (sections 10.4 and 10.5).

(M7) Not only are there no causes and effects; there are no substances or subjects either. One consideration Nietzsche uses to establish this arises from some alleged illusions of our language. The sentences of our language are commonly made up of subject or substantival expressions and predicates (verbal or adjectival). But it is wrong to think that reality reflects this grammatical structure; in particular he thinks there are no substantival objects that our subject expressions could refer to. In arguing for this, Nietzsche uses expressions, common in his unpublished notebooks, such as 'quantum of power, or of force'; these expressions surface in an important passage in one of Nietzsche's publications.

> A quantum of force is also a quantum of drive, will, action – in fact, it is nothing more than this driving, willing, acting, and it is only through the seduction of language (and through the fundamental errors of reason petrified in it) – language which understands and misunderstands all action as conditioned by an actor, a 'subject' – that it can appear otherwise. ... the common people distinguish lightning from the flash of light and take the latter as *doing*, as the effect of a subject which is called lightning .... But no such substratum exists; there is no 'being' behind the doing, acting, becoming; 'the doer' is merely a fiction imposed on the doing – the doing itself is everything. Basically, the common people represent the doing twice over, when they make lightning flash – that is doing doubled by another doing: it

posits the same event once as cause and then once again as effect. The natural scientists do not fare any better when they say: 'Force moves, force causes', and the like ... our entire science is still subject to the seduction of language and has not shaken itself free of the monstrous changelings, the 'subjects', foisted upon it (the atom is an example of such a changeling, as is the Kantian 'thing in itself'). (*The Genealogy of Morals*, Part I §13).

This long passage begins with the language of Nietzsche's notebooks; it talks of quanta of force or drive. It then claims that from the grammatical structure of our language in which we use subject expressions to talk about the world which only contains such quanta of force, we should not infer that there are substantive objects corresponding to these subject expressions. From this we allegedly falsely assume, on the basis of our language, that there is a subject lurking behind each effect acting as its cause, i.e., *lightning* which then flashes, *forces* which then move other items, etc. What Nietzsche is denying here is the existence of substantival entities such as objects or persons; and this denial is extended to mechanical forces of science, atoms and Kantian 'things in themselves'. Instead there is just an event , a doing, an acting.[4]

Given the denial of cause and effect relations in (M6), we cannot even say that there is just an effect. Nietzsche has also claimed that our very belief in separate causes and effects is also a product of language, and of divisions we have projected onto the world; but there is no such duality in the world at all. In the light of this, even Nietzsche's talk of 'a quantum of force' is not to be taken strictly as talk of an object. But Nietzsche has overstated his case in assuming that all subject expressions in our sentences should be taken to refer only to substantival entities, or to objects in the sense of things which are substances. We can use the subject expressions of sentences to refer to items which are not substantive objects, such as Nietzsche's own talk of a quantum of force or power, or our scientific talk of force or the movement of heat energy from one region to another (e.g., of a hotter to a colder region of space), or our common-sense (and scientific) talk of events and processes, for example, an increasing breeze (as in yachting). There appears to be a false presupposition that always subject expressions refer to substantive objects; they often do, but not always. So Nietzsche's grammatical argument, while reminding us of an important point, is not generally correct.

(M8) The will to power operates not only from 'centres' but also through persons both in their thought, valuation and action. More commonly when they act through persons they are spoken of as drives. When Nietzsche says '*This world is the will to power – and nothing besides!*' he adds 'And you yourselves are also this will to power – and nothing besides!' (*The Will to Power*, §1067), he emphasises two things. The first is that power is the one and only ontological category. The second is that we ourselves are identical to some cluster of such powers; there is no substantive soul-atom or self but only the complex of driving powers.

There are other aspects of Nietzschean metaphysics that we will pass over. Nietzsche thinks that from the above he can prove that there will be an eternal recurrence of the same history of the universe, given the finitude of force and space (*ibid.*, §1066). It is very doubtful that this conclusion does follow. However if one needs it (it is not needed metaphysically), then it can be added as an independent constraint on how the world evolves over time. Perspectivism is an important idea in Nietzschean metaphysics where it appears as a relation between power centres. It also occurs as a central notion in his theory of knowledge, truth and morality; these are entirely perspectival. However it has been argued that Nietzsche's perspectivism is seriously incoherent;[5] this will not be considered here.

This list of theses is not a complete account of Nietzsche's claims within metaphysics, but it will suffice for our purposes. They set out some of the main features of an otherwise obscure metaphysics of "will to power" that can be attributed to him.[6] But there is a final point worth mentioning. Nietzsche opposed all mechanistic accounts of the world. This is because they often presupposed what Nietzsche denies, viz., the existence of substantial objects and cause-effect relations such as push-pull. Rather he adopted a view of mechanics developed by Boscovich which does away with substantive bodies with mass; these are replaced by forces acting from points, thereby constituting a field of force. Boscovich's foundation for a theory of motion in a field of forces strongly influenced Nietzsche's account of power:

> As for materialistic atomism, it is one of the best refuted theories there are ... thanks chiefly to the Dalmatian Boscovich. ... For while Copernicus has persuaded us to believe, contrary to all the senses, that the Earth does *not* stand fast, Boscovich has taught us to abjure the belief in the last part of the Earth that "stood fast", the belief in "substance", in "matter", in the earth-residuum and particle-atom ... (*Beyond Good and Evil*, §12)

Nietzsche goes on to claim that we must get rid of our "atomistic need", including the idea of the soul-atom. There are other sources for Nietzsche's metaphysics. But Boscovich, as scholars are now claiming,[7] was an early influence upon Nietzsche' view that there were no substances, material things or matter with a continuing and underlying identity. Boscovich's reinterpretation of mechanics showed that such items were unnecessary; so why assume they exist elsewhere, especially in metaphysics.

10.1.2 Nietzsche on the Illusions of Causation and Law.

Even a brief account of Nietzsche's metaphysics, assuming one can be attributed to him, would not be complete without saying something of his view of causation. Because it is not well worked out, it is not without its tensions. As has been mentioned in (M6), Nietzsche understands causality to be a fiction that we anthropomorphically project on the world:

> One should not wrongly reify "cause" and "effect" as the natural scientists do ...; one should use "cause" and "effect" only as pure concepts, that is to say, as conventional

fictions for the purpose of designation and communication – *not* for explanation. In the "in-itself" there is nothing of "causal connections", of "necessity", or of "psychological non-freedom"; there the effect does *not* follow the cause, there is no rule of "law". It is we alone who have devised cause, sequence, ..., law, freedom, motive and purpose; and when we project and mix this symbol world into things as if it existed "in-itself", we act once more as we have always acted – *mythologically*. (*ibid.*, §21).

In this passage Nietzsche does not mean anything Kantian by "in-itself"; rather it refers to whatever there is in the world (power) independently of what we might project on to it. Nietzsche rejection of causation is quite radical. First, as has been seen in (M6), Nietzsche rejects the very idea that there are any relata of cause and effect to be found in nature; they are a division of the continual flux of power that we make. Further, he rejects the idea that there are any relations such as necessary connections to be found in nature. In this respect Nietzsche's position is like Hume's commonly understood position. But it is unlike Hume's in that there is no account of causation in terms of laws or regularities of any sort. Nor are there any Humean associations that we make between the cause and effects that we impose as constructions.

There is an often-cited passage in which Nietzsche says that talk of 'nature's conformity to law' is 'a bad mode of interpretation'; 'it is no matter of fact, no "text"'. Many have argued on the basis of an analogy with human law that there must also be laws of nature: 'Everywhere equality before the law; nature is no different in that respect, no better off than we are' (*ibid.*, §22). Nietzsche claims that the origin of our concept of law of nature is based in an analogy with our own case as makers and subjects of human laws; and from this Nietzsche wishes to claim that there are no laws of nature. There is a genetic fallacy here to avoid. Even if our notion of nature being in conformity with law originally arose from such an analogy, it does not follow either that there are laws in nature, or that there are no such laws; nor does it follow that it is a notion to be debunked. Genetic fallacies are often serious faults in Nietzsche's genealogical considerations based on analogy. That there is a story about the origin of our notion of a law of nature to be told in the sociology of science is one thing;[8] but it is quite another to infer from this that the notion of *law of nature* is thereby suspect. Its origin by analogy to human conformity to law is one thing; it use, development and independent articulation about they way in which nature conforms to natural laws (for which there is much evidence) is quite another.

Following from these remarks, Nietzsche invokes the will to power and the notion of rival interpretations, saying of our talk of 'nature's conformity to law':

But ... that is interpretation, not text; and somebody might come along who, with opposite intentions and modes of interpretation, could read out of the same "nature", and with regard to the same phenomena, rather the tyrannically inconsiderate and relentless enforcement of claims of power – an interpreter who would picture the unexceptional and unconditional aspects of all "will to power" so vividly that almost every word, even the word "tyranny" itself, would eventually seem unsuitable, or a weakening and attenuating metaphor – being too human- but he might, nevertheless, end by asserting that same about this world as you do, namely that it has a "necessary" and "calculable" course *not* because laws obtain in it, but because they are absolutely *lacking*, and every power draws its ultimate consequences at every moment. Supposing that this also is only interpretation – and you will be eager enough to make this objection? – well, so much the better. (*ibid.*, §22)

Here there are two rival interpretations of "nature". The one invokes the idea of nature's conformity to laws (based on an analogy with human conformity to laws); the other rejects the idea that there are laws to be found in nature at all. There is simply the "will to power" understood along the lines of the theses already set out. However the rival interpretations are alleged to agree on something, viz., 'nature has a '"necessary" and "calculable" course'. One might suppose that talk of "necessary" here does not reintroduce the idea of necessities in nature. It is not as if "anything goes" in nature; there are constraints that allow us to calculate what will happen. Nietzsche does not say how we are to make such calculations in the absence of laws. Let us simply suppose that we have some non-law-like "formula" on the basis of which calculation can be made. Granting this, then both the believer in laws and the Nietzschean atheist about laws are able to do the same thing, viz., say something about the future course of nature on the basis of calculations. Finally the quotation ends with Nietzsche's throw-away line that even his own view is just another interpretation. Though Nietzsche does not say it, his claim seems to be that the two interpretations are equivalent as far as 'the same phenomena' exhibited by "nature" are concerned. In more modern terminology we can say that the rival views, viz., the claim that there are laws versus the claim that there are no laws but calculation is still possible, fit the facts equally as well; i.e., both views are underdetermined by what phenomena we could observe.

Much of this fits with a notion of underdetermination employed in analytic philosophy of science. Let us suppose that we have a law of nature that all A are B and write this as 'L(A, B)'; and suppose we can calculate using the formula that all A are B and write this as 'F(A, B)'. Now the formula F(A, B), which we use for calculation says, by hypothesis, no more than that all A are B, which we can write as '(A, B)'. It follows from this that L(A, B) entails F(A, B), which in turn entails (A, B). However the law L(A, B), even though it entails (A, B), goes beyond the empirical content of (A, B) and claims something about the metaphysics of laws of nature, whatever this additional claim may be. Putting matters anachronistically this way, Nietzsche seems to be simply rehearsing the dispute between realists and anti-realists about laws. Empirically they say the same thing, but one goes on to make claims about the metaphysics of laws that the other does not.

However Nietzsche cannot quite put matters this way because his contrast is between a notion of laws of nature based on a human model, and mere calculability. However the same point about underdetermination still holds; so there is alleged to be an empirical equivalence between them. That is, there is no way of deciding between two interpretations of "nature" based on an inspection of "nature" through its associated phenomena. This does seem to be a plausible way of looking at what Nietzsche says. But it now shows that Nietzsche has not left room for a further possibility about laws, viz., one which is not based on a human model and which does make claims about necessities in nature. But that there should be room for such a notion is ruled out elsewhere by his metaphysics of power.

Nietzsche talk of calculability with respect to nature occurs not only in published material but also in unpublished notebooks. Thus he writes of 'formulas' and 'laws' in shudder quotes, saying:

> Let us here dismiss the two popular concepts "necessity" and "law".... "things" do not behave regularly, according to a *rule*; there are no things (- they are fictions invented by us) ...; There is no obedience here: for that something is as it is, as strong or as weak, is not the consequence of an obedience or a rule or a compulsion. The degree of resistance and the degree of superior power – this is the question in every event: if, for our day-to-day calculations, we know how to express this in formulas and "laws", so much the better for us! ... There is no law; every power draws its ultimate consequence at every moment. Calculability exists precisely because things are unable to be other than they are. A quantum of power is designated by the effect it produces and that which it resists' (*The Will to Power*, §634).

Note that the phrase 'every power draws its ultimate consequence at every moment' reappears in the quotation just given above from *Beyond Good and Evil* §22.[9] Elsewhere in his notebooks he also talk of calculation:

> In order for a particular species to maintain itself and increase its power, its conception of reality must comprehend enough of the calculable and constant for it to base a scheme of behaviour on it. ... a species grasps a certain amount of reality in order to become master of it, in order to press it into service. (*ibid.*, §480).
>
> From the fact that something ensues regularly and ensues calculably, it does not follow that it ensures *necessarily*. Necessity is not a fact but an interpretation .... (*ibid.*, §552).

Nietzsche's point that a regularity does not entail necessity is quite correct. But what seems to be admitted in these passages is that there is something akin to regularities from which we can make calculations. In many respects Nietzsche's account of laws and calculability is unsatisfactory in that in denying the former he never spells out what he means by the latter.[10] The best fall-back position might well be that of an anti-realism about laws and causation of the sort commonly attributed to Hume in which there are regularities that can be established but they are not underpinned by any real relations or necessities in nature.

The issues raised here lead nicely on to matters discussed in section 10.3: 'We need "unities" in order to be able to reckon; that does not mean we must suppose that such unities exist' (*ibid.*, §635). If reckoning at least involves calculating, then the claim is that we need "unities" that can be numbered and then used for calculating in some fashion. We can suppose that prime candidates for these "unities" are the complexes of powers that are relatively stable over time. As will be seen in the next section, Nietzsche rejects the idea that the unity of these "unities" is to be cashed out philosophically in terms of substantive objects with continuing identity. It is such "unities" that are said not to exist. But the difficulty in the above quotation is in understanding a need, for the purposes of reckoning and calculating, for "unities" that do not exist. Do we need such "unities" in order to be able to reckon? In physics items such as energy (heat, kinetic, etc), force, fields of force, power, wavelengths, etc, are hardly "unities" yet we can quantify them and employ these quantities in calculation. So the claim that unities are necessary for any reckoning is dubious; reckoning can take place about items that do not have "unity" and which are continuous or discrete in their variability.

Overall, what is to be made of Nietzsche's antirealist projectionist account of the cause-effect relation? There is no good reason to accept it, and plenty as to why we should reject it. Perhaps Nietzsche's projectionist error is to be found in his belief that the continuity he alleges in the "calculable" fluxing of power cannot come naturally packaged as cause-effect relations. But this is wrong. There is no reason to suppose why times slices of power, even time slices at a moment, can not stand in cause-effect relations, as is common in current physics with its talk of the flux of fields and causal chains of energy transmission. In section 5.9 the idea of causation as counterfactual dependence was introduced. There is no reason why we cannot say the following of Nietzschean centres of power: if one centre of power was not acting in a certain way with a given intensity and "will' or direction (at a time), then another centre would not be acting with its intensity and direction (at a time). This is needed if the metaphysical picture of centres of power in opposition to one another is to make sense. But as Hume puts it, and Lewis spells out further, such counterfactual talk is also causal talk. Nietzsche needs some such notion if his account of the operation of the "will to power" in bringing about belief is to have any explanatory force at all (see section 10.8). We have no reason to accept Nietzsche's antirealist projectionism with all its problems about how we can also "calculate" the operations of power in the absence of laws but through the allegedly available "formulae".

10.2 THE "WILL TO POWER" AS THE LEADING HYPOTHESIS OF AN EXPLANATORY AND REDUCTIVE PROGRAMME

Nietzsche used his theory of the "will to power" as the basis of an ambitious programme for the naturalisation of the sciences such as mechanics, biology and psychology, as well as theories of morality and society. He alleges that the explanations and/or interpretations provided by the power hypothesis are superior to all other hypotheses. In discussing exploitation within society Nietzsche harshly declares: '"Exploitation" does not belong to a corrupt or imperfect and primitive society; it belongs to the *essence* of what lives, as a basic organic function; it is a consequence of the will to power which after all is the will to life. If this should be an innovation as a theory – as reality it is the *primordial fact* of all history' (*Beyond Good and Evil*, §259). In Nietzsche's view not only is it a fact that exploitation is an organic function, but also the power hypothesis is supported by the events of history. In addition, the power hypothesis explains our history better than any other hypothesis because it appeals to the very 'essence of what lives as a basic organic function'. Though Nietzsche normally eschews talk of essences, we can assume that explanations which allegedly involve essences are better than those which do not (why 'better' remains methodologically unclear).

The wide scope of the power hypothesis is best illustrated in the following extended passage. Nietzsche begins by supposing that the only reality is our inner drives and speculates whether a more general hypothesis about the "will to power" would suffice for an understanding of even 'the so-called mechanistic (or "material") world':

> In short, one has to risk the hypothesis whether will does not affect will wherever "effects" are recognised – and whether all mechanical occurrences are not, in so far as a force is active in them, will force, effects of will.
>
> Suppose, finally, we succeeded in explaining our entire instinctive life as the development and ramification of *one* basic form of the will – namely, of the will to power, as *my* proposition has it; suppose all organic functions could be traced back to this will to power and one could also find in it the solution of the problem of procreation and nourishment – it is one problem – then one would have gained the right to determine *all* efficient force univocally as – *will to power*. The world viewed from inside, the world defined and determined according to its "intelligible character" – it would be "will to power" and nothing else. (*ibid.*, §36)

The passage begins with a weak supposition of the hypothesis of the "will to power"; but as its theme develops this is strengthened to a supposition of actual explanatory success of a wide range of phenomena. Granting such explanatory success, Nietzsche's overall thesis is then strengthened further. He claims that the events, things and properties postulated in the various sciences are either identical to, or reducible to, or supervenient upon, some aspect of the "will to power" and are not independent of it. One hundred years later we can judge Nietzsche's metaphysical programme to be either a non-starter or a failure. Concerning mechanics and modern physics the "will" has played no role whatever. It would be superficial to see in, say, Einstein's mass-energy equivalence theorem, an instance of a non-material ontology of power. Moreover Nietzsche's programme has provided no impetus to the development of physical theories and is thus a "degenerate" programme (to employ a term of theory appraisal due to Lakatos).

In biology Darwinian evolution and genetics (based on the DNA molecule) leave no room for any notion of the "will to power", including the theory of procreation. Even Richard Dawkins' metaphorically described "selfish gene" theory is not an instance of the will to power. Dawkins uses the anthropomorphic term 'selfish' to characterise the activity of genes, but always insists that this metaphor can be translated back into the language of gene theory that talks about biochemical processes.[11] Turning more generally to the theory of evolution, Nietzsche does hold to a version of the theory of evolution, but one in which an internal will or "oomph" is the driving force. This is not Darwin's theory with its external pressures of natural and sexual selection on a population. Even though his naturalism has a central role for the evolution of human cognition and sociality, Nietzsche constantly railed against Darwin's account of the mechanisms of evolution thereby failing to use it to develop his own evolutionary naturalism. Darwin's theory of natural selection is still a progressive programme in biology compared to which Nietzsche's non-starter power theory fades into non-significance.

One can see in Nietzsche's remarks against Darwinian natural selection Lamarck's view that off-spring could inherit the acquired characteristics of their parents and ancestors, a false view which is bolstered by talk of the "will to power". This is opposed to the Darwinian view of natural selection but it is consistent with the idea that evolution has come about by the inner "oomph" provided by the operation of the "will to power" allegedly in each of us. In a section headed 'Against Darwin' he says: 'The influence of "external circumstances" is overestimated by Darwin to a ridiculous extent: the essential thing in the life process is precisely the tremendous shaping, form-creating force working from within which *utilizes* and *exploits* "external circumstances"' (*The Will to Power*, §647; see also §684-5). Nietzsche recognises here that external circumstances do play a small role; but their role is not that proposed in Darwin's theory of natural and sexual selection.

Nietzsche's adherence to the Lamarckian position is often expressed in his published writings:

> One cannot erase from the soul of a human being what his ancestors liked most to do and did most constantly; whether they were, for example, assiduous savers and appurtenances of a desk and cash box, modest an bourgeois in their desires, modest also in their virtues; or whether they lived accustomed to commanding from dawn to dusk .... It is simply not possible that a human being should *not* have the qualities and preferences of his parents and ancestors in his body, whatever appearances may suggest to the contrary. This is the problem of race (*Beyond Good and Evil*, §264)

But this problem of race and its alleged impossibilities is one of Nietzsche's own invention based on his false Lamarckian views of inheritance in which even assiduous savers of money pass on, genetically rather than behaviourally, their acquired propensity to save to their off-spring. Darwin also holds a strong hereditary principle that would allows that off-spring inherit some of the "qualities" (as Nietzsche says) of their parents. But they would not inherit their "preferences" (as Nietzsche says), say, for modesty in virtue, or being bourgeois. For Darwinian evolution, acquired characteristics cannot be passed on. Though Nietzsche understood what Darwin said he was often prejudiced against it (as the above shows) largely because of his blind adherence to his belief in the "will to power" in each of us as a biological inner "oomph".

In his book on Darwin, Daniel Dennett gets us to think of skyhooks, rather than earth-bound cranes, as a kind of *deus ex machina* emerging from the "sky" to account for the evolutionary process by a magical lifting of one species "into" another. While the metaphor of a skyhook does not capture the idea behind Nietzsche's "inner oomph" evolutionary mechanism, it does capture the idea of totally autonomous force that Nietzsche thinks is at work. In a section called 'Friedrich Nietzsche's Just So Stories', Dennett acknowledges Nietzsche's evolutionary stance. But he points out that Nietzsche lapses into what he calls 'skyhook hunger' when, for example, he complains of '... the prevailing instincts and tastes of the time, which would rather accommodate the absolute arbitrariness, even mechanistic senselessness of all that happens, than the theory of a *will to power* manifesting itself in all things and events' (*The Genealogy of Morals*, Part II §12). Dennett comments that 'Nietzsche's idea of a will to power is one of the stranger incarnations of skyhook hunger, and, fortunately few find it attractive today' (Dennett (1995), p. 466).

The above are brief comments on the lack of success of Nietzsche's naturalism with its underlying appeal to "will to power" as compared with two rival areas of current scientific naturalism in physics and biology. The naturalism can be applauded but not the peculiar account of the "will to power" on which it is based. Some will object to the approach taken here; they claim that Nietzsche was an anti-metaphysician and did not intend to advance metaphysical claims of the sort outlined in this and the previous section. They put emphasis on his throw-away line when discussing his own doctrine of the "will to power" that it is merely one interpretation alongside others. Even if we grant this, it still remains to compare one interpretation with another to see which might be the better. Nietzsche did think that rival interpretations within the genealogy of morals could be better or worse, as we will see;[12] he did not think that rival interpretations cannot be compared on epistemic (and other) grounds. What the above shows is that when comparing the "will to power" with sciences such as physics and biology, the latter come off much better on grounds of correctness and of providing an advancing research programme rather than a degenerating one. Though the program's best known application concerns the genealogy of morals (see section 10.7), we will consider next Nietzsche's attempt to give a naturalistic genealogy of our belief in objects (sections 10.3. and 10.4), and in the very human cognitive ability of believing itself (section 10.5).

10.3 NIETZSCHE'S NATURALISM AND ORDINARY OBJECTS

Given two frameworks of thought with different ontological commitments, such as Nietzsche's power and our common-sense framework that postulates substantive objects, what relationships can there be between them? There are several options including reductionism, eliminativism, supervenience, an anti-realism with respect to both ontologies, and so on. As will be seen, Nietzsche is eliminativist with respect to many of the items not in his preferred ontology. Substances, selves and other-worldly items or realms are said to be fictions just as much as Zeus' thunderbolts, fairy spirits and phlogiston are now fictions. But what of ordinary objects such as individual leaves, drops of water or stones? The smooth pebbles we find on beaches are hardly fictitious and seem to be as real as anything might be, whatever Berkeleyan idealists might say. Nietzsche rejects considerations for scepticism about the existence of substantive objects based on the epistemological grounds of the classical British Empiricists. His reasons for claiming these are fictions are metaphysical and are based on their lack of any identity conditions. Nietzsche does not hold with the Quinean dictum 'no entity without identity'. Rather, since the only things that exist are powers and complexes of power, and these are said to lack identity conditions (they are just fluxes of power), then no object with identity can exist! For Nietzsche the whole idea of identity is a fiction.

Nietzsche's view of the ontological status of common objects like pebbles is unclear. A few of his remarks tend to support the more plausible view that a pebble could be the same as (identical to) a nexus, or a complex, of power which has a great deal of stability over time: 'Duration, identity with itself, being are inherent neither in that which is called subject nor in that which is called object: they are complexes of events apparently durable in comparison with other complexes – e.g., through a different tempo of the event'. (*The Will to Power*, §552c). This remark supports reductionism based on type-type identity; there are pebbles but, appearances to the contrary, they are identical to some relatively stable complex of power. Thus the common-sense view that objects, such as pebbles, do exist is saved, even though there is a reduction to items in Nietzsche's naturalistic ontology of power.

We can employ the Kripke/Putnam theory of reference to show how our words still refer to individual objects and kinds, such as pebbles, even though there is variability about what we believe these items are. As Kripke argues, we can continue to talk of tigers even though we discover them to be reptilian, or to be robots controlled from Mars. On the same grounds we can continue to talk of pebbles even though it turns out that Nietzsche is right and there is no matter, no substance, or no underlying continuants such as atomic ultimates which make up the pebbles and provide identity conditions for them, and that pebbles are really long-lasting, complexes or "unities" of powers. (In what follows let 'substantive objects' refer to objects understood to contain matter, or to be substances, or to be made of unchanging ultimates or atoms, and that in virtue of these, objects have identity conditions and are continuants over time.) On this view we can now take Nietzsche to say that pebbles, and other such items, do exist. So he is a realist about objects. But he is also a reductionist in that such items are reducible to (or are supervenient upon) complexes of power.

However matters are not so simple in the industry of Nietzsche interpretation. By far the largest number of comments in his work suggest that objects such as pebbles are fictional, and that our framework of ordinary objects is to be eliminated (see for example (M7) of section 10.1). Understood this way, Nietzsche is an irrealist about (substantive) objects, so *a fortiori* they cannot be reduced to anything. Whatever else exists, they are not to be counted amongst the existents, nor are they reducible to them. They simply go the way of Zeus, Santa Claus, phlogiston, God, and so on. Nietzsche's writings have a constant ambiguity running through them between a reductive realism about substantive objects and an irrealist eliminative fictionalism about them.

The following is a diagnostic reconstruction of Nietzsche's argument for the rejection of reductive-identity in favour of eliminativism with respect to substantive objects:

(1) All objects are substances, or are bits of substantive matter, with identity conditions;

(2) Anything that is a complex of power has no identity conditions;

(3) Therefore a complex of power cannot be a substance, or substantive matter;

(4) The only things that exist are powers complexes of power (in space-time);

(5) Therefore, there are no substances or bits of substantive matter, with identity conditions.

(6) Therefore, there are no objects.

Also accompanying Nietzsche's irrealism is a doctrine of constructivism: this is the view that substantive objects are fictions we invent, or they are projections we impose on the world, or they are constructions that we make (for what reason will be seen in section 10.4.1). If we accept (4) which expresses Nietzsche's naturalism, the only premises left to examine in this valid argument are (1) and (2).

Nietzsche's argument for the rejection of substantial objects turns on metaphysical matters to do with identity conditions, as indicated in (2). Since Nietzsche's world is in such a state of flux, given the changing kaleidoscope of the complexes of power, nothing has any continuing identity. There is not even the same number of items: 'Continual transition forbids us to speak of "individuals" etc: the number of beings is itself in flux' (*ibid.*, §520). But are these claims about lack of identity correct? We humans are said to change, albeit over a period of about seven years, all the cells in our body; yet we preserve our bodily continuity. Each new part has a historical connection with some other earlier part, and performs a functional role similar to the part it replaces. It is the host of such historical and functional continuities that give us our bodily identity despite total change in its parts. Again there is much discussion in the literature about John Locke's ship *Theseus* in which each plank is replaced by another plank while at sea, thereby raising a question about whether it is the *same* ship that left one port and arrived at another.[13] (Additions can be made to the story; the rejected planks are reassembled raising the question as to whether the same ship *Theseus* is back on the water or another ship is sailing.) Much the same could be said of Nietzschean complexes of power; though the magnitude, direction, and even centre of action of power might change, it does not follow that there is no continuing identity for any such complex of power. There may be sufficient historical continuity and functionality within one complex of power to establish identity over time.

Historical continuity conditions may suffice to establish identity where much stronger conditions would fail to establish any. Thus consider what may be called mereological essentialism:[14] for strong [weak] mereological essentialism, a whole has identity just when all [at least one of] its parts (such as its substance) must have identity and be co-present throughout the life of the whole. Clearly strong mereological essentialism is too strong since it requires a whole to change none of its parts; in contrast the weak version only requires that at least one thing remain the same for the lifetime of the whole. But even the weak version is too strong a requirement in the case of the ship *Theseus,* and in the case of bodily identity if we allegedly change all our cell-parts within a seven year period. And it is too strong a requirement for the identity of complexes of power if at least one direction, or magnitude or centre of some complex of power must remain the same. It would appear that to support premise (2), Nietzsche, the naysayer to all essentialism, says 'Yea' to essentialist identity conditions. It is some such view that underlies that claim that 'continual transition forbids us to speak of "individuals"'. This exposes the over-strong conditions of identity that Nietzsche seems to impose; a more relaxed set of conditions are possible and would relieve him of the need to make the unreasonably strong claim (2).

What of premise (1)? The philosophical doctrine of substance does much work in different contexts. One such piece of work is giving identity to an object that is constituted by some discrete bit of matter which remains unchanging. Another is to ensure that if anything is a substance, then it must come equipped with identity conditions; there can be no substance that lacks identity. But do we think that all our common-sense objects are substances in either of these senses? Though not obviously part of our common sense framework of objects, which includes kinds such as pebbles, the idea that objects are substances has been a constantly accompanying philosophical doctrine that has many advocates from Aristotle to Kant. Locke, Berkeley and Hume raised epistemological worries about our knowledge of substances that Kant attempted to answer. However Nietzsche's challenge does not proceed from their epistemological stance. Rather he seems to accept the Kantian view that objects are substances in accepting (1) – but then so much the worse for our talk of ordinary objects, as the argument shows!

Though (1) seems plausible there are many reasons provided by twentieth-century philosophy of language and metaphysics for rejecting it. As mentioned, the work of Kripke and Putnam shows that even if we were to discover that objects were not substances but were complexes of power, our reference to them would remain invariant despite the falsification of a fairly constantly accompanying philosophical theory of objects as substances. As an example, we have rejected the Aristotelian view that air and fire are pure substances yet we continue to talk of the very same stuff that Aristotle and other Ancient Greeks breathed, or warmed their hands by. Fire, we now think, is quite unlike a substance; rather it is electro-magnetic radiation in the heat and light wave-bands. That there are pebbles is an existence claim that remains unaltered, even if we were to discover that they were not substances but some Nietzschean complex of power (see the discussion of these and related issues in sections 7.7, 8.5 and 8.7). In the light of this, pebbles remain what they are even if we change our metaphysical theory about them so that they are no longer the substantive entities we though they were.

If we reject (1) or (2), we can also reject (5), viz., Nietzsche's eliminativism and fictionalism with respect to substantive objects. Nietzsche's reasons for the elimination of objects and their fictitious character is far from compelling, though it becomes a centre-piece of his metaphysics. He sets aside the more plausible reductionist view in which ordinary objects still exist even if they turn out to be Nietzschean complexes of power. Since the argument above is intended to be a diagnostic reconstruction, it remains to show that (5) is a pervasive feature of Nietzsche's philosophy. It appears in his early writings: 'Logic too depends on presuppositions with which nothing in the real world corresponds, for example on the presupposition that there are identical things, that the same thing is identical at different points in time; but this science came into existence through the opposite belief (that such conditions do obtain in the real world)' (*Human, All Too Human*, Volume I, section 11). And it appears in his late writings when he says that, following Boscovich's refutation of belief in substance, matter and atomism, we need to 'declare war ... against the atomistic need' (*Beyond Good and Evil* , §12) And his unpublished notebooks are replete with such comments.[15] In sum, it appears that even though Nietzsche might be understood to adopt a more plausible reductionist account of what substantive objects are, the weight of his remarks with its talk of fictions supports the more implausible eliminativism with respect to substantive objects along with matter, substance, atomic-ultimates and continuants of any sort.

10.4   THE GENEALOGY OF BELIEF IN SUBSTANTIVE OBJECTS AND IN LOGIC.

If our talk of objects, matter, substance, atoms, selves and the like is really fictional, and thus these items do not exist, how come we talk of them at all? On a reductionist, as opposed to a Nietzschean eliminativist, view a story can be readily told. We bump into and perceive such objects. They are the things which populate the world and with respect to which, in the course of evolution, we have come to develop perceptual apparatus to detect them and cognitive apparatus and language in order to think and talk of them. And this can be the case even if Nietzsche is right and we form false theories about their substantival character and, unbeknownst to us, they are really complexes of power. This is a story that Nietzsche, in his reductionist mood, can tell since he also advocates naturalism and some version of the theory of evolution; he does say that even our organs, including those of perception, have evolved throughout human history and our intercourse with the world.[16] But if substantive objects are really fictional, then scientific naturalism cannot begin to overcome the first hurdle of their non-existence. Perhaps at this point one should downplay the irrealist fictionalism. But since it is a persistent theme, what account does he give of why we believe in such fictional items (and other fictions such as cause-effect relations)? Here the "will to power" plays a crucial role. We will consider in the next sub-section the role of the "will to power" in forming our concept of a substantive object, and thus our common conceptual framework that presupposes the existence of such objects. In sub-section 10.4.2 we will consider the role of the "will to power" in forming our beliefs about logic.

10.4.1 The Role of "Will to Power" in Constructing the Fiction of Objects.

What is to be explained is why we have the conceptual scheme, or framework, that we do with its presupposition that there are substantive objects. However what does the explaining cannot appeal to the existence of such items, as might a realist. Such items are fictions and do not exist. So what can do the explaining?

There are three themes that come together at this point. The first has to do with Nietzsche's theory of evolution and our survival and life enhancement. His explanation appeals to the operation of the "will to power" which, in the course of our evolution, enabled us, as cognitively developing creatures, to create a framework of thought that would be conducive to, and enhance, our survival. The second theme is that the framework need not be true; as will be seen Nietzsche thinks that the entire framework is false. What matters is that, given we evolve into cognitive beings, such a framework is conducive to our survival even if it is quite false. The third theme is this: Nietzsche also holds that this framework (with all its falsity) is necessary; without it there can be no thought at all. In this there is something Kantian: our framework is necessary for thought. But the second theme is very un-Kantian; whereas Kant argued for the synthetic *a priori*, and thus true, character of our conceptual scheme, Nietzsche argues that it is not synthetic *a priori* at all, and is false.[17]

The third theme makes a strong claim: there can be no thought except that which occurs within our (false) conceptual framework. Without a presupposition of substantive objects with continuing identity, thought is alleged to be impossible. But, alas, the second theme insists that thought is always about fictions. The three themes appear in published writings, as will be seen. But the third gets pungent expression in the notebooks: '*Rational thought is interpretation according to a scheme that we cannot throw off*' (*The Will to Power*, §522). And so does the second theme: 'Parmenides said, "one cannot think of what is not"; – we are at the other extreme, and say "what can be thought of must certainly be a fiction"' (*ibid.*, §539). These two remarks tell us that we cannot throw of the conceptual framework in which we think; but, alas, what it presupposes is fictional.

The first theme is that the (false) conceptual scheme we have constructed arises from a drive for survival and enhancement for life. It occurs mixed with the second theme in the following: 'One should not understand this compulsion to construct concepts, species, forms, purposes, laws ("a world of identical cases") as if they enabled us to fix the *real world*; but as a compulsion to arrange a world for ourselves in which our existence is made possible: – we thereby create a world which is calculable, simplified, comprehensible, etc for us' (*ibid.*, §521) That false judgements about a fictitious world constructed by us are necessary for our survival is a common theme Nietzsche's published writings:

> The falseness of a judgement is for us not necessarily an objection to a judgement; in this respect our new language may sound strangest. The question is to what extent it is life-promoting, life-preserving, species-preserving, perhaps even species-cultivating. And we are fundamentally inclined to claim that the falsest judgements (which include the synthetic judgements *a priori*) are the most indispensable for us; that without accepting the fictions of logic, without measuring reality against the purely invented world of the unconditional and self-identical, without a constant falsification of the world by means of numbers, man could not live – that renouncing false judgements would mean renouncing life and a denial of life. To recognise untruth as a condition of life – that certainly means resisting accustomed value feelings in a dangerous way; and a philosophy that risks this would by that token alone place itself beyond good and evil. (*Beyond Good and Evil*, §4; see also *ibid.*, §11 and *The Will to Power*, §511-2)

Here a number of things are quite clearly maintained. First, we have invented a number of fictions including those of self-identical objects that, because of their constructed identity, we can count. Second, such fictions are necessary for us, including not just our thought but also our very existence. Third, such fictions promote and preserve our existence.

What has this to do with the "will to power"? Nietzsche makes the link for us when he says: "Physiologists should think before putting down the instinct of self-preservation as the cardinal instinct of an organic being. A living thing seeks above all to *discharge* its strength - life itself is *will to power*. Self-preservation is only one of the indirect and most frequent *results'* (*Beyond Good and Evil*, §13). The "will to power" has as one of its more indirect but frequent results, our self-preservation. But a much more direct result is its creative power in making us as cognitive beings equipped with a framework of thought in order that we might arrange (partly through calculation), not just for our self-preservation, but for the enhancement of our life. However this framework presupposes nothing real about the world but only fictions. Here we have an instance of the general WP Thesis announced at the beginning of this chapter: for our common conceptual scheme or framework of beliefs which presupposes the existence of substantial objects, there is a manifestation of the "will to power" as a driving force of evolutionary processes which gives rise to and maintains our acts of belief in that framework.

The WP Thesis only tells us *that* the "will to power" is involved here; but it does not tell us *how* or *why* it works in the way it does. But there is a hint as to *how* it works, through the selection of a belief system in terms of Nietzsche's theory of evolution:

> *Origin of knowledge.* – Over immense periods of time the intellect produced nothing but errors. A few of these proved to be useful and helped to preserve the species: those who hit upon or inherited these had better luck in their struggle for themselves and their progeny. Such erroneous articles of faith, which were continually inherited,

until they became almost part of the basic endowment of the species, include the
following: that there are enduring things; that there are equal things; that there are
things, substances, bodies; that a thing is what it appears to be; that our will is free;
that what is good for me is also good in itself. It was only very late that such
propositions were denied and doubted; it was only very late that truth emerged – as
the weakest form of knowledge. It seemed that one was unable to live with it: our
organism was prepared for the opposite; all its higher functions, sense perception and
every kind of sensation worked with those basic errors which had been incorporated
since time immemorial. Indeed, even in the realm of knowledge these propositions
became the norms according to which "true" and "untrue" were determined – down to
the most remote regions of logic.' (*The Gay Science* §110)

Nietzsche's account is evolutionary and almost Darwinian. It starts by
assuming that initially we produced only falsehoods; why only falsehoods
and no truths, or truth-like beliefs, Nietzsche does not say. Some falsehoods
turned out to have survival value in that those who held them are alleged to
have a higher probability of surviving than those who did not. Why some,
rather than other, falsehoods were so efficacious Nietzsche does not say. But
given his resistance to truth talk, he can not say that these beliefs were less
false and more truth-like than others. What are these falsehoods? Amongst
them is the usual suspect, belief in substantival objects. This belief is false
but allegedly conducive to survival.

Nietzsche's position is not without its problems. He does not tell us *why*
the will to power operates in us (across our species) to produce such a
framework of belief. Nor does he tell us *why* it is so successful in preserving
our lives in the absence of the very things the framework postulates, viz.,
substantive objects. It is a puzzle to know why, if objects are fictional, that
belief in such fictions can be life-preserving, and even necessary for such
preservation. We can readily admit that telling the well-chosen lie now and
then can give us an advantage over others. But it is hard to see how persistent
massive untruth can lead to such success. There is an answer to this puzzle
which is available to the realist about substantive objects but is not available
to Nietzsche because of his irrealism, i.e., his fictionalism about objects. It
turns on an instance of inference from the best explanation in which it is the
truth, and not the falsity, of our conceptual framework which best explains
our success in surviving, and in enhancing our life. This argument is also
available to those who are realists and think that there are substantive objects,
even if in the long run they turn out to be complexes of Nietzschean power. It
is not available to irrealists who take objects to be fictions (and who hold that
our conceptual framework is thereby false).[18]

For the realist about objects, our *belief* that there are substantive objects is
successful in preserving and enhancing life because that belief is *true* (or
truth-like). There really are such objects out there (whatever hidden internal
character they might have). And it is the truth of this belief that is a crucial
premise in an explanatory argument to the success we have in holding the

belief. If we failed to believe that there were such things out there in the world when they were really out there, then there would be a marked increase in the difficulty of preserving and enhancing life. Our failure to hold certain beliefs (about, say, a predatory animal lunging towards us) would be detrimental to us. In contrast, if we do believe that there are such substantive objects out there when they are really there, then clearly there is a marked increase in our ability to preserve and enhance life. Our true beliefs are a good guide through life, moreso than those that are false. The case Nietzsche envisages is the one in which we do believe that substantive objects are out there, but they are really not there at all. There might be costs to having such a false belief, but such costs need not always count against the preservation and enhancement of life; if we are in luck there is no cost, but if we are out of luck there could be considerable loss. But it is not easy to see why in this case there could be as massive an increase in the probability of our surviving and leading an enhanced existence as there is in the first case envisaged.

The *truth* in our belief in substantial objects can explain success in survival; the falsity of such beliefs cannot explain our survival (or explain it as well), assuming that we do survive in such conditions of belief.[19] These considerations are available to those who are reductionist and who hold that there are objects but they are complexes of (relatively stable) Nietzschean powers. But they are not available to those who are irrealist and claim that substantive objects are fictions; there are simply not any truths about the existence of objects to explain success in surviving.

That false belief is a necessary condition for life is a theme that also runs through Nietzsche's unpublished notebooks. Thus we are told: 'that a belief, however necessary it may be for the preservation of a species, has nothing to do with truth' (*The Will to Power*, §487). A similar view is expressed in the following, provided the first word 'truth' is understood in the sense of 'what we believe, or hold, to be true': 'Truth is the kind of error without which a certain kind of species of life could not live. The value for *life* is ultimately essential' (*ibid.*, §493).

So far there are two unresolved puzzles about Nietzsche's thesis. We do not know *how* the "will to power" does it work in bring about the requisite false, but necessary, beliefs (unless we appeal to some selection process to do with the truth of our beliefs); it is not even clear that relying on what Mannheim called Nietzsche's 'flashes of insight' can help us here. And granted that the "will to power" does bring such beliefs about, we are at a loss to explain why having such beliefs are so successful in preserving and enhancing life, given that they are false;[20] this remains a mystery. There is a third puzzle that turns on Nietzsche's thesis that can be expressed in two parts of thesis (F) (for falsity). The puzzle is that any reflective Nietzschean must hold two contradictory beliefs.

(F1):	For each member x of a class of belief-entertaining creatures (such as ourselves) there is a set of beliefs B (about, for example, substantive objects) such that x's believing B is a necessary condition for x's continued survival.

(F2):	Beliefs B are false.

Beliefs B are those to which we have given our commitment and which we hold-true. But that B is held-true does not entail that B is true; B could be false. (F2) says exactly this; all the beliefs in B are false. As Nietzsche maintains, B arises from a primitive exercise of the "will to power", not the "will to truth"; if B is held-true a world is thereby constructed in order that we might survive. Can (F1) and (F2) be consistently held together? It might be thought that there is a contradiction between the following two claims: (a) each person holds-true B in order for each to survive – including Nietzsche; (b) Nietzsche (qua philosopher) holds-false B. There is no contradiction between everyone else's holding-true B while Nietzsche, the exception, holds-false B. But there is a contradiction if Nietzsche were to both hold-true B (in order to survive) and also to hold-false B (qua philosopher who exposes illusions about our beliefs). This is a problem for any reflective but irrealist Nietzschean. Qua survivor in the struggle for life they must hold that beliefs B are true; but qua philosopher who exposes the illusory conditions for life they must also hold that beliefs B are false.[21] This is not a problem for a realist since they would claim that (F2) is false since many of the beliefs in B are true.

The above illustrates the workings of Nietzschean genealogy that can bear comparison with the Strong Programme, and Foucault. We have a framework of belief, especially a belief in the existence of substantive objects. Why do we have these beliefs? If they are true, then contrary to the Symmetry Thesis of the Strong Programme, we might appeal, in part, to those very objects in any explanation as to why we believe in their existence (as is done in the "argument for success" above). Or we might, if we are rationalists, appeal to the evidence, if it is available, as a good and sufficient ground for holding these beliefs. But if they are false, then we might have to look for an alternative kind of explanation that does not appeal to the objects themselves or any evidence for them (though note that scientific methodology can explain why we believe in the false as well as the true – see section 6.2.1). This takes us to a domain of investigation that could be the province of the sociology of belief.

For Nietzsche our belief framework is false, so we have to look for an alternative explanation. Nietzsche does not explicitly state anything like the Symmetry Thesis, but his practice is in conformity with it; most of the beliefs he hopes to explain are illusory. However the explanation is not to be sought in social factors, as in the case of the Strong Programme. For Nietzsche, as for Foucault, the explanation is to be found in the operations of the "will to power" as it drives us unrelentingly along. As a consequence of its driving, we have evolved as cognitive beings who must have a conceptual scheme in which we can at least primitively calculate and, later, think about matters well beyond that for which the scheme was originally developed. But allegedly from the beginning the scheme is suffused with error. In order to think at all, including calculating, we need to postulate substantive entities; we cannot think at all with out doing this. Thus we construct a conceptual scheme which is erroneous. But without it we could not survive (at least as cognitive beings). In all of this the "will to power" is ultimately explanatory; it provides the drive for our creative constructing. Nietzsche's genealogy of our overall belief scheme, as false as it is, can be seen as a special case of the WP Thesis mentioned at the beginning. Even though what it postulates as the cause is different, it is a non-rational item, and as such the WP Thesis bears a strong similarity to the Causality Tenet of the Strong Programme.

10.4.2 The Role of "Will to Power" in Constructing the Fictions of Logic

Nietzsche does not criticise any principles of logic. In fact he often employs them, for example when he says that, owing to Boscovich, 'materialistic atomism is one of the best *refuted* theories there are' (*Beyond Good and Evil*, §12, italics added). All refutation involves at least the principle known as *Modus Tollens*. And a little further on he takes himself to have advanced an argument which is, as he says, a *'reductio ad absurdum'* (*ibid.*, §15); and again he says that the concept of a *causa sui* is a 'rape and a perversion of logic' (*ibid.*, §21). If the principles of logic are not abandoned and are used and even praised, then what is Nietzsche's criticism of logic? The attack is more upon what he takes to be the presuppositions, or theory of the status, of logic, and not the principles themselves. Note that Nietzsche was writing at a time when the modern revolution in logic was just underway. He did not notice it, and later he could not have known of its developing ideas. Here we will focus on just two of his central lines of criticism that were advanced early in his career and persisted to the end. The first is, once more, his attack on identity; the second has to do with his account of the evolution of logic and its presuppositions.

The first line of criticism turns on his rejection of identity for individual objects over time and our classification of two or more objects as being of the same kind. Nietzsche seems always to have been a nominalist in that at best there are only individuals and no kinds: kinds are fictions that we construct. However his nominalism, if one can call it that, is even more extreme, given his underlying metaphysics of power. As we have seen there do not really exist individual objects but fluxes of power that are continuous and in some sense not individuable at all. In the light of this he cannot be positively characterised as a nominalist; perhaps 'particularist' would best describe his non-substantive ontology of powers. But at least his nominalism characterises him negatively as a person who denied all abstracta and other-wordly items.

The rejection of logic is tied to his rejection of identity: '*Logic* too depends on presuppositions with which nothing in the real world corresponds, for example on the presupposition that there are identical things, that the same thing is identical at different points of time; but this science came into existence through the opposite belief (that such conditions do obtain in the real world)' (*Human, All too Human*, volume I §11). Talk of 'identical things' could be understood as, say, the identity between the Morning and the Evening Star. But in this context it is best understood to mean that two different things are the *same* in the sense that they are of the same kind, sort or type. Also the idea that an individual item remains the same through time is rejected. If an individual changes its properties over time then, according to Nietzsche, it cannot be the same; no self-identical continuants are to be found in the world at all. (Nietzsche here appears to neglect the possibility that there are essential properties of kinds and continuants that could underpin their identity, since he does appeal to essences for related purposes elsewhere.)

On behalf of the view that there are no kinds in the world, and that kinds are constructs of our own making, Nietzsche tells us in an early unpublished work:

> ... a word becomes a concept insofar as it simultaneously has to fit countless more or less similar cases – which means, pure and simply, cases which are never the equal and thus altogether unequal. Every concept arises from the equation of unequal things. Just as it is certain that one leaf is never totally the same as another, so it is certain that the concept " leaf" is formed by arbitrarily discarding these individual differences and by forgetting the distinguishing aspects ('On Truth and Lies in a Nonmoral Sense', p. 83)

The argument is not a convincing one against the possibility of kinds, though it might say something useful about concept formation. It can be agreed that every leaf is different in some respect from every other leaf, and so none are identical (Nietzsche says "equal"). This is simply one part of Leibniz' principle of identity which says if two objects differ in any respect then they cannot be identical. But this is not an argument against kinds; there can be kinds, yet each instance of the kind meet the principle just mentioned. Even if each instance differs in *some* respect from any other, as Nietzsche says, it does not follow that they must differ in *all* respects. It is still a possibility that they share some common aspects and this may well be sufficient for them to constitute a kind. The same can be said of a continuant, an individual which changes its properties over time; though any two time slices can differ it does not follow that they share no common features that can underlie identity. Nietzsche hardly offers any other considerations against kinds and continuants; but their rejection is of a piece with his extreme nominalism (as is the rejection of all other sorts of abstract entities).

Similar remarks about logic are also scattered throughout his later notebooks, two of which express the view that logic presupposes identity, and that the identities we construct are really a manifestation of the "will to power". In a section given the title 'Origin of Reason and Logic' we find:

> Logic is bound to the condition: assume there are identical cases. In fact, to make possible logical thinking and inferences, this condition must first be treated fictitiously as fulfilled. That is: the will to logical truth can be carried through only after a fundamental *falsification* of all events is assumed. From which it follows that a drive rules here that is capable of employing both means, firstly falsification, then the implementation of its own point of view: logic does *not* spring from will to truth. (*The Will to Power*, §512)
>
> Toward an understanding of logic: *the will to equality is the will to power* ... (*ibid.*, §511)

Note that in this passage Nietzsche talks of what 'makes possible logical thinking and inferences'. He also makes an inference when he says 'it follows that'. So the principles of logic themselves are not under attack. Rather his claim is metaphysical in that individual continuants and kinds are fictions not to be found in the world; they are our own constructions. In order to construct them we have to falsify.

Perhaps what this might mean is suggested by the example of the leaf in which in making the kinds we do we ignore or suppress differences. According to Nietzsche there is a drive at work here in making quite different things equal, more strictly the same, when in reality they are not. In addition there is a will to logical truth that differs from a will to truth. In fact our construction of sameness is the will to make things equal – and this arises from the activity of the "will to power". These considerations provide a clear instance of the WP Thesis in which particular types of act of believing, such as those about logic and its alleged presuppositions of sameness and identity, come about through the operation of the "will to power" in constructing the identities possessed by continuants and kinds. The WP Thesis might need reformulating if what is to be explained is not merely our *act* of belief in kinds and continuants, but the very *content* of the beliefs themselves which give us our concept of the various kinds and continuants there are. That is, the "will to power" not only brings about acts of believing, but also constructs the content of our very concepts. This is a radical extension of the WP Thesis to a new range of entities, in this case concepts and what they mean.

Nietzsche has an evolutionary story to tell about how we evolved as creatures who construct identities:

> *Origin of the logical.* – How did logic come into existence in man's head? Certainly out of illogic, whose realm originally must have been immense. Innumerable beings who made inferences in a way different from our perished; for all that, their ways might have been truer. Those, for example, who did not know how to find often enough what is "equal" as regards both nourishment and hostile animals – those, in

other words, who subsumed things too slowly and cautiously – were favoured with a lesser probability of survival than those who guessed immediately upon encountering similar instances that they must be equal. The dominant tendency, however, to treat as equal what is merely similar – an illogical tendency, for nothing is really equal – is what first created any basis for logic.

In order that the concept of substance could originate – which is indispensable for logic although in the strictest sense nothing real corresponds to it – it was likewise necessary that for a long time one did not see nor perceive the changes in things. The beings that did not see so precisely had an advantage over those that saw everything in "flux". At bottom, every high degree of caution in making inferences and every skeptical tendency constitute a great danger for life. No living beings would have survived if the opposite tendency – ... to err and *make up* things ... – had not been bred to the point where it became extraordinarily strong. (*The Gay Science*, §111).

There is much in this passage that is quite Darwinian and which those who work in philosophy and psychology on the evolution of cognition could applaud. Though Nietzsche does not say it explicitly, he would endorse the claim that those who make invalid inferences have a lesser probability of survival than those who make valid inferences. Rather, his focus is on the concept formation that underpins reasoning, particularly the kinds and substances that we "make up" as he puts it. As well as "making up" concepts of kinds (based on similarity relations) we also have to have sufficient speed in classifying some new item as falling, or not, under the concept. Even a rabbit would have to quickly recognise any new item it came across as falling within the kind cow, or wolf, or moving branches of a tree, etc. The rabbit must judge of some new item that it falls under the first or the third then there is no need to run – but not so in the case of the second. Moreover all three judgements have to be made quickly just in case the new item does in fact fall under the second. Nietzsche also emphasises that there are no such kinds to be found in nature; they are made up by us through ignoring differences. If attention is paid to differences then the classification into kinds can become more refined and vastly extended. But such a complex classification system may take much longer to use without some compensating "speeding up" of our cognitive abilities; simpler ones might do just as well. So there may be no survival value in elaborate systems of classification. In the limit of such elaboration, there would be no kinds but merely one distinct individual after another; a creature who lacked kind classifications would have little chance of survival since no predatory kind could be recognised at all.

Nietzsche does suggest that there might be "truer" ways of making inferences that relates more closely to 'seeing everything in flux' and does not depend on kind classifications or on substances. This possibility can be realised in ordinary quantification logic, something that Nietzsche would not have been aware. In modern logic a sharp distinction is made between the syntax of logic and it semantics. In specifying its semantics a domain of "objects", or a universe of discourse, is specified. Variables range over the "objects" in the domain. And evaluation functions are specified which link syntactical categories to elements of the domain. Thus there are evaluation functions that assign to each name an "object" in the domain, to each predicate expression a set of these "objects" as extension, to each relation a set of ordered pairs as extension, and so on.

What might the "objects" of the domain be? They could be things, or substantive objects, or bits of matter. But they could also be other kinds of entity. The word "object" simply does duty for whatever items there are in the domain. Such items could be kinds or abstracta such as numbers, sets or geometrical objects; they could be possible worlds; they could characters in fiction; they could be events, or processes; they could be discrete sensations, or cross sections of sensations; they could be atoms, or whatever items are postulated in a final physics, including points (or regions) of space-time, masses, quanta of energy, and so on. There is no reason why the domain of "objects' could not be Nietzschean powers, or centres of power, or complexes of power as they prevail at a given time and place. Quantificational logic is ontologically neutral in that it can take any items whatever within its domain and allow its variables to range over them. There is no need to point to· alternative modes of inference that are "truer" than those which limit the domain over which they quantify to only kinds and/or substances. Once this feature of quantification logic is recognised much of Nietzsche's criticism of logic falls by the wayside. In order to have logic there need not be a supposition that logic only works when it applies to substantive objects and to kinds. At best his criticism applies, if it applies at all (and it does not), to our concepts of substance and kinds.

There is a further point that can be raised against Nietzsche's extreme nominalism, if he allows that there can be rules of inference. Consider the rule *Modus Tollens*, which he appears to accept. An instance of it can be expressed as follows: if p then q, and not-q, so not-p. Here there are letters that appear twice. There are two tokens of the type 'p', and two tokens of the type 'q'. These are constants, and indicate that the *same* proposition, or the *same* statement or sentence, q (or p), appears twice in the argument. Here we appeal to items which are said to be the same, i.e., to types or kinds of thing, in this case a proposition, or statement, or sentence. Thus in order to state even our rules of inference, and to indicate that a sentence occurs more than once in an argument, we need to appeal to sameness of type. The very way in which we must express rules of inference is not completely ontologically neutral; we must eschew extreme nominalism and admit types, sorts or kinds.[22] Here Nietzsche might have a point when he claims that logic needs to suppose sameness of some sort. But it is not clear that this is in any way a falsification of the world, or imposes a fiction. It might be construed as a point against the extreme nominalism of Nietzsche that denies that there can be tokens of the same type. After all Nietzsche himself needs such a notion; he needs to say that the same *amount* of force with direction D that acted from centre C at time t is now acting again at a later time t*. That is, there are repeatable samenesses even of amounts of Nietzschean power, the amounts being quantitative and thus "calculable".

What does Nietzsche say about principles of scientific method, rather than principles of deductive inference? Very little. But he does recognise that there are some, and does employ them. Nietzsche makes it clear that rival interpretations of morality (or anything else) can be assessed as better or worse. He claims to give a better account of the genealogy of morals than did Paul Rée in his 1877 book *The Origin of Moral Sensations*, saying that some of his own work at that time was written 'not in order to refute them [Reé's hypotheses] ... but rather, as befits a positive spirit, in order to replace an improbability with something more probable, and occasionally even to replace one error with another' (*The Genealogy of Morals*, Preface §4). Here Nietzsche endorses the time-honoured probabilistic rule that says of two rival hypotheses that we ought to pick that which is more probable with respect to the evidence. What evidence? Later he says of Reé's hypotheses that he wished 'to point a sharp and impartial eye in a better direction, the direction of the real *history of morality* and to warn him off in good time from such English hypothesising *into the blue*' (*ibid.*, §7). For the genealogist it is important to investigate 'what has been documented, what is really ascertainable, what has really existed, in short, the entire long hieroglyphic text, so difficult to decipher, of humanity's moral past' (*loc. cit.*). It is here that Nietzsche spells out where we are to find evidence to test genealogical hypotheses. And we are to reject the English ones because, presumably, they come out of the blue with no evidence. In sum, Nietzsche endorses the common principle that says, given some evidence, chose that hypothesis which is more probable on the evidence. Nietzsche also recognises that genealogical hypotheses can be erroneous, since even he might be merely replacing Reé's errors by errors of his own. With much charity one could read into this something of Popper's notion of verisimilitude; even though he might be replacing one error by another, perhaps Nietzsche's errors are less "erroneous" in that they have greater verisimilitude than Reé's, and so can be adopted on that ground.

Finally Nietzsche could be understood to endorse a coherence criterion when he says about his thought on the genealogy of morals that there has been at work in him 'a *fundamental will* of knowledge'. His older ideas have '... become increasingly inseparable, indeed have even grown into one another and become intertwined'; this is to be contrasted those ideas which '... emerg[e] as isolated, random, and sporadic phenomena'. In the light of this he issues an imperative that says: 'we have no right to any *isolated* act whatever; to make isolated errors and to discover isolated truths are equally forbidden us. Rather, our thoughts ... grow out of us with the same necessity with which a tree bears fruit – all related and connected to one another ...' (*ibid.*, Preface §2). Amongst other things this suggests a methodological prescription not to accept isolated truths and to accept those claims which have a high degree of interconnection and coherence which adds to the support genealogical hypotheses get. As with Foucault (see section 9.5.6), Nietzsche does admit that there are methodological criteria by which rival genealogical hypotheses can be tested.

Nietzsche assails logic because of its alleged dependence on identities that are not present in the world. But as has been seen, this is not a complaint that can be directed against the rules of logic that, on their modern understanding, can be used even to make inferences about Nietzschean powers. At best it is a complaint about the metaphysics of identity. As such his criticisms have no bearing on either deductive or non-deductive methodological principles of reasoning. But what is of interest is Nietzsche's evolutionary account of our construction of identities and samenesses that takes place alongside our cognitive development, and the role the "will to power" plays in this.

10.5   THE GENEALOGY OF BELIEF AND TRUTH.

Nietzsche makes a bewildering number of claims about truth that have exercised commentators. Perhaps no one position satisfactorily accounts for all he says, but some might do a lot better than others. In this section a proposal will be made as to what kind of theory Nietzsche might reasonably be thought to have adopted in his later works. In the next section 10.6, we will look at Nietzsche's account of why we have been so exercised about truth and the genealogy, not so much of truth itself, but our belief in truth.

We have already seen that Nietzsche makes full use of the notions of truth and falsity, especially when he says that falsity is a condition for life. Assuming the law of excluded middle (we have no reason to think that Nietzsche rejects this[23] and that he is some proto-intuitionistic logician), then if some false claim is necessary for life, there must be a truth formed from its negation. Since, as he says, it is our belief in self-identical substances and kinds that are false but necessary for life, then the denial that these exist must be true. What the "true" or "real world" is like (Nietzsche does use such expressions in many of the remarks cited in section 10.4), would be given by his theory of the "will to power" as set out in section 10.1. If this is so, then there are lots of truths to be uttered about Nietzschean powers, even though none of these truths are about substantive objects. Again in this context Nietzsche must admit the following truth, so central to his philosophy. If F is the set of false beliefs necessary for life then the following is a true sentence:

'our believing F is necessary for life'. Room must be made for such 'higher-level' truths about beliefs at a "lower level" which are false.

Nietzsche makes paradoxical remarks about truth, almost to the point of tiresomeness, when a few simple distinctions would remove ambiguity. Thus we find in his notebooks: 'There exists neither "spirit", nor reason, nor thinking, nor consciousness, nor soul, nor will, nor truth: all are fictions that are of no use.' (*The Will to Power*, §480). Most of these Nietzschean ontological rejects we have met before. But it is an exaggeration even for Nietzsche to say that fictions are of no use since he alleges that some of them, though false, are necessary for life. The new reject is truth. But the remark is ambiguous between (a) there are no true propositions at all, and (b) some standardly held *theory* of truth is wrong. More often the latter is what is intended though not explicitly said. If (a) is advanced by Nietzsche, then it is advanced as a truth. Nietzsche, the classical scholar, ought to have been aware of the paradoxes raised by self-referential expressions such as the one uttered by the Cretan Epimenides 'All Cretans are liars'. His own claim 'there is no truth' is subject to the same paradox since Nietzsche advances it as a truth. In contrast to (a), Nietzsche recognised that there are all sorts of truths to countenance, from banal truths such as 'Nietzsche has a big moustache', to the truths of real history on which any genealogical investigation into morals should be based (*The Genealogy of Morals*, Preface §7), and to the deeper truths that genealogists might uncover when, as he says, they 'sacrifice all wishfulness to truth, to *every* truth, even the simple, bitter, ugly, repulsive, unChristian, immoral truth.... For such truths do exist -' (*ibid.*, I §1).

10.5.1 Holding-True as Basic

There are some remarks in the notebooks that do not seem to appear in published writings that are relevant at this point. Here is a selection:

> Believing is the primal beginning even in every sense impression: a kind of affirmation the first intellectual activity! A "holding-true" in the beginning! Therefore it is to be explained: how "holding-true" arose! (*The Will to Power*, §506)
> The *valuation* "I believe that this and that is so" as the *essence* of *"truth"*. ... All our organs of knowledge and our senses are developed only with regard to conditions of preservation and growth. Trust in reason and its categories, in dialectic, therefore the valuation of logic, proves only their usefulness for life, proved by experience – *not* that something is true. ... Therefore, what is needed is that something must be held to be true – *not* that something *is* true. (*ibid.*, §507)

Here talk of 'x's *holding-true* that p' is none other than 'x's *act* of believing that p'. Nietzsche puts great emphasis on our belief-forming capacities that arose at a certain point in our evolution as cognitive beings. He says nothing, and could say nothing, about *how* belief-formation came about; but *that* it came about is undeniable and must be something that is to be fitted within an evolutionary framework. What he does suggest, in line with the views expressed in section 10.4, is that we would have been selected for our ability to form beliefs, to hold-true that p (for some p), where doing so is life-enhancing. Moreover, the belief contents, that p, are false of the world despite being necessary for our existence. Not to hold-true that p would be fatal for life. So for Nietzsche the rock-bottom cognitive act is, for some propositions p, a "holding-true that p". What is allegedly not needed, at rock-bottom, is that p *be* true.

Of course there is a problem here that needs to be resolved. Evolutionary epistemologists might well argue that not all the p can be false; success in evolution depends on getting most of the p right, especially in life-threatening circumstances. They would put strong reliance on the truth of our beliefs as an explanation of why we are, on the whole, successful in the world.[24] In this they are right and Nietzsche is wrong. But with some interpretative wriggle we can try to limit Nietzsche's claims about falsity to those assumptions he thinks are built into our language with its nominative expressions which presuppose substantive objects, or those presuppositions that are built into our conceptual framework about continuants, identities and kinds. In so doing we need not view him as being entirely at odds with the position of most evolutionary epistemologists. Otherwise we are left with the puzzle, mentioned in 10.4.1, of how, with so much falsity, we still manage to be successful.

The upshot of the above, is that holding-true that p is more fundamental than p's being true. Note that the former is independent of the latter; 'x's holding-true that p' entails neither 'p is true', nor 'p is false'. In fact, as Nietzsche requires, p can be false, while it remains true that x holds-true that p. We can also see the workings of the "will to power" in getting x to hold-true that p in the first place, and then continuing to hold-true that p. Like any act of believing, holding-true that p is an instance of the WP Thesis, with some unspecified aspect of the "will to power" bringing this about. Thus it would be possible for Nietzsche to advance an account of belief, and how our beliefs are formed (due to evolutionary processes and "will to power") that need not mention truth at all. However one has to be aware of the following matter. In advancing the WP Thesis, Nietzsche advocates more of what he holds-true, viz., the proposition that some aspect of the "will to power" brings about our various other "holdings-true" (even if these are false).

The last claim about the WP Thesis can be extended to all the metaphysical theses of section 10.1. We can say of these that each is held-true by Nietzsche. But from 'Nietzsche hold-true that p' we should not infer any relativistic truth doctrine of the form: 'p is true relative-to-Nietzsche' (or to Nietzscheans). That x hold-true that p can be a quite objective claim. At this point is would be appropriate to mention, again, the throw-away line at the end of *Beyond Good and Evil* §22 in which Nietzsche says of his own "will to power" doctrine that it, too, might be just another interpretation.

What this might mean, in the light of the above, is that Nietzsche holds-true his "will to power" doctrine; and this is independent of whether or not it is true. But all this comes down to is the claim that Nietzsche believes his "will to power" doctrine (i.e., he believes it to be true). The whole thrust of Part I concerns our rational tradition, with its normative principles of epistemology and methodology which tell us what warrant our beliefs have, Nietzsche's beliefs being no exception. This includes matters such as: what evidence supports our beliefs and how good the support is, what they might explain and how well, how they compare with their rivals and which rival we might chose, and so on. Nietzsche's "will to power" doctrine like any other, is merely a set of alternative hypotheses within philosophy which some believe, and which all can assess for their epistemic worth. There is nothing in Nietzsche, appearances to the contrary, which overturns this aspect of the critical tradition; in fact he adopts some of its methodological principles.

10.5.2 Nietzsche and Contemporary Theories of Truth

It would be a futile enterprise to try to link what Nietzsche says about truth to contemporary theories of truth if there were not something to be gained from it. One of the gains is a further understanding of Nietzsche's rejection of the grammatical structure of our language as a guide to what there is.

Modern theories of truth divide according to whether or not the predicate 'is true' in "'p' is true" picks out a genuine property or not. Those who are commonly called 'inflationists' claim that it does, while those who are 'deflationist' say that it does not. Deflationists add that we are quite wrong to even think that 'is true' functions like a predicate at all. This is a point that Nietzsche would have endorsed.

There are several inflationist theories. The first is a correspondence theory in which the property of being true is to be cashed out in terms of some notion of "correspondence" to facts, states of affairs, or truthmakers. From what we have gleaned from Nietzsche so far, facts, especially facts about substantive objects, are simply not available in his ontology; they are fictions. As he famously says: 'Against positivism, which halts at phenomena – 'There are facts" – I would say: No, facts is precisely what there is not, only interpretations' (*The Will to Power*, §481). But matters might not be this simple. It is possible to propose a proposition of Nietzschean metaphysics which says 'at some place and time a given amount of power P is at work and acts in direction D'. What would make this true, or false? It is rather hard to avoid saying that it is the actual presence, or absence, at that time and place of an amount of power P acting in direction D. And this is what we might standardly regard as a fact, or a truthmaker for the proposition. One of the opening remarks of Wittgenstein's *Tractatus*, 'the world is the totality of facts and not of things', can be adapted to our purposes here. Each elemental fact is simply about what amount of power is acting in what direction at what time and place. Nietzsche's atomic facts are facts about power operating from space-time centres; they are not of simple subject-predicate form but are relational: 'at space-time region R there is an amount of power P acting in direction D'. There is no reason why these cannot be the truthmakers of the statements of Nietzschean metaphysics.

There are two other inflationary theories. On the coherence theory, truth is the property a belief has in virtue of its relation of "cohering" with other beliefs. This theory seems not to have been mentioned by Nietzsche even though, as noted at the end of section 10.4.2, he might be understood to have endorsed a coherence theory of belief and knowledge. The pragmatic theory of truth is sometimes understood to attribute to a belief the property of being useful, or of working. Aspects of a pragmatic theory of truth loom large when beliefs are assessed on the basis of the extent to which they preserve and enhance life.[25] However one draw-back is that when Nietzsche points out that some beliefs are life preserving and enhancing, he also goes to some pains to say that the beliefs are also false. Though this is consistent with our holding the beliefs true for the purposes of our survival, it is hardly sits happily with what is supposed to be a theory of *truth*.

There are a variety of deflationary theories of truth, the most well known being the redundancy view of Frege and Ramsey. The following important equivalence holds, where 's' is some sentence: "'s' is true if and only if s". This holds for all theories of truth (and is in fact and instance of Tarski's T-Schema). On the Frege and Ramsey redundancy theory of truth, this schema is given a particular reading: owing to the equivalence, the expression 'is true' is redundant and can be eliminated from our language. In virtue of the equivalence, simply replace "'s' is true" by 's' wherever the former occurs. Given his rejection of truth and his remarks about language, I am sure Nietzsche would have liked the point about the elimination of 'true'. He might also have liked the disquotational theory of truth in which 'is true' is not a predicate but a function that takes us from a quoted sentence to a not-quoted, or disquoted, sentence. As such 'is true' is merely a device for semantic ascent and descent.

However the theory that might have most attracted him is the performative theory in which 'is true' is not even a predicate picking out a property but merely an indicator of agreement or endorsement. Adding the words 'is true' to the sentences 'Nietzsche had a big moustache', or 'Amount of power P acts here and now in direction D', is not to ascribe a property to either of these sentences of the sort inflationists require. Rather in using the word 'true' the speaker *does* something; the speaker agrees, endorses, asserts, or claims something. The speaker makes a gesture, or nods in agreement in using 'true'. The word 'true' functions like 'Amen' at the end of a prayer. In this respect some versions of the pragmatic theory of truth are akin to the performative theory. They claim that in calling a sentence 'true' one is not adding something further to the statement, such as attributing a property to it; rather, one is doing something such as endorsing, agreeing, etc. (In the case of falsity one is disagreeing, disclaiming etc.)

The thesis that Nietzsche's theory of truth (assuming he has one) has strong affinities to the performative theory of truth has been advanced in Taneseni (1995). But Tanesini extends her thesis to more fully developed considerations based on the prosentential, or anaphoric, theory of truth. In Tanesini's view a deflationary anaphoric theory of truth avoids some of the difficulties that the performative theory has for embedded sentences. We need not enter into the tricky issue of the faults with the original performative theory of truth advanced by Strawson and how anaphoric theories can overcome these problems. One serious problem (recognised by Strawson) is that in order to rule out agreements that are misleading, one has to invoke a notion of *appropriate* agreement. To cash out this notion would require that the conditions specified in the sentence are fulfilled in the world, i.e., a roundabout way of saying 'true'.[26] Thus Nietzsche cannot adopt a purely performative theory of truth without acknowledging some role for the truth-conditions of a sentence.

It is useful to see how Tanesini's account does fit some of Nietzsche's remarks. Nietzsche's notion of (appropriately) holding-true that p is clearly an activity of accepting, or endorsing, or agreeing with the claim that p rather than attributing a property to the proposition that p. Those who hold-true are actively doing something rather than just ascribing a property to a statement. Moreover Nietzsche's claim that 'The *valuation* "I believe that this and that is so" as the *essence* of *"truth"*' explicitly talks of truth as *evaluative*, rather than as descriptive. In holding something true one makes an evaluation, gives one's word, or even stakes one's reputation on it.

Such a view can also explain how the "will to power" can be linked to the notion of truth. For most theories of truth such a claim hardly makes sense. However if truth is akin to evaluating or endorsing, agreeing etc. then a role for the "will to power" does not seem hopelessly implausible. Nietzsche says:

> Will to truth is a making firm, a making true and durable, ... "Truth" is therefore not something there, that might be found or discovered – but something that must be created and that gives a name to a process, or rather to a will to overcome that has no end ... It is a word for the "will to power". (*The Will to Power*, §552)
> The criterion of truth resides in the enhancement of the feeling of power (*ibid.*, §534)

Here the descriptivist aspect of truth is downplayed, and even rejected, while the performative aspect of truth is emphasised in our 'making firm, making true' our beliefs, or giving our say-so. It would be misleading to understand talk of the creation of truth here as merely the making up of truths as when we might, say, fantasise; rather there is creation of truth in the sense of giving our say-so to certain propositions while withholding, or withdrawing, it from others. And in so doing Nietzsche sees a role for the "will to power" in bringing it about that, for some p, we hold-true that p, such as in evolutionary contexts. This yields another instance of the WP Thesis.

Finally, even the criterion of truth, including our ways of testing for truth, are allegedly linked to the way in which our feeling of power is enhanced. In some ways this an over-psychologistic, and even dangerous, criterion for the acceptance of a truth. However as odd as some of these remarks may be, they do fit with a pragmatic-performative view of truth. In assessing this fit, it is not so much whether it is right (see the last footnote for problems with the performative view), but rather just how much of Nietzsche's remarks on truth fit with which of the various theories of truth that are currently available. As unacceptable as are many features of the performative theory, it does offer a good fit with what Nietzsche says.[27]

10.6   THE GENEALOGY OF BELIEF IN TRUTH AND THE ASCETIC IDEAL

As was mentioned in chapter 1, especially in section 1.4 on Kuhn, there are many values that we might wish our hypotheses or theories to realise such as explanatory scope, accuracy of predictions, unity, simplicity, consistency, informativeness, fruitfulness (in leading to new research results), utility, and so on. The list need not include truth (certainly Kuhn puts restrictions on the kinds of truth we should aim for, for example, he allows truth at the level of observation but not at the level of the unobservable). Those with truth phobia will leave truth out; but truth freaks will want it in. And some of the above values might, on closer analysis, depend on truth (e.g., consistency or accuracy of predictions). For Nietzsche one of the big questions was 'why be a truth freak?', 'why pursue truth?', or why, as he put it, the "will to truth"? 'We asked about the *value* of this will [to truth]. Suppose we want truth: *why not rather* untruth? and uncertainty? even ignorance?' (*Beyond Good and Evil*, §1). Despite the value we place on truth and selflessness, Nietzsche wonders whether 'it would still be possible that a higher and more fundamental value for life might have to be ascribed to deception, selfishness, and lust' (*ibid.*, §2). Since Nietzsche's view is that life is best enhanced by beliefs that are false, for him this possibility is also actual. Moreover he claims that truth loses most of its domination over us when we realise that truth claims are not fact-stating, that there are not truths to discover. Though there are said to be no truths (in this last sense), there is still truth-stating which is one of the many activities we perform when we make fundamental endorsements in committing ourselves to holding-true; it is we who create or construct a belief system which is life enhancing and preserving.

So why still the pursuit of truth? Much of the final 'Third Essay' of *The Genealogy of Morals* is devoted to answering this question. The answer lies in Nietzsche's odd belief that truth is an ascetic ideal. That we have such an ideal is part of the genealogy, not so much of truth itself, but of our will to truth. As an example, Nietzsche cites the Christian belief in the rewards of an afterlife which is said to compensate for the suffering and lack of meaning of this life, the more we suffer and deny ourselves the sinful pleasures of this life the greater the afterlife rewards will be (see *The Genealogy of Morals*, III §1 and §28). Here an ascetic ideal turns on holding-true a belief about the afterlife, this being one of the noble lies of Christianity. But this is not so much about truth itself as an ascetic ideal; rather it is about a particular proposition believed to be true and the asceticism required to achieve the state it describes. But Nietzsche's claims are more general than this.

In what follows we will consider what Nietzsche takes asceticism to be and, then, how it relates to science, scepticism and atheism. We will see that he holds to the doctrine of the noble lie as an antidote to truth asceticism. This gives rise to a self-reflective paradox about Nietzsche's own genealogical enterprise and its connection to truth. On the one hand, he wants to claim that the will to truth is an ascetic ideal; and this is in some sense to disparage or debunk the whole idea of truth-seeking. But on the other hand, he also wants to engage in a genealogical investigation into the origin of morality one aim of which is to uncover some truths of history to support his genealogical hypothesis. In fact Nietzsche invokes the notions of truth and falsity quite directly because he does claim that there are truths to be found in this area. But this is more of the ascetic ideal to be debunked. In the long run this paradox is unresolvable; something has to give in Nietzsche's overall point of view. Nor should we understand Nietzsche to be merely advocating the weak claim that we should advocate epistemic values other than truth, as Kuhn largely does (see section 1.4), when he criticises the pursuit of truth as ascetic. This is to ignore the role he gives to the telling of lies as a way of overcoming truth asceticism.

10.6.1 Asceticism and its Alleged Link to Resentment and the "Will to Power"

What does 'ascetic' mean? We commonly use the word in the context of priests, or the religious, who stereotypically are said to deny themselves some of the seeming pleasures of life, such as good food, wine, sex, wealth, a comfortable bed in which to sleep, comfortable clothes (they wear hairshirts), no self inflicted pain (i.e., they flagellate themselves), and so on. They are self-denying or severely abstinent either because they believe it is God's command for the good life, or because of some ideal of purity of body and/or soul, or whatever. In this respect Nietzsche speaks of Wagner's asceticism in 'paying homage to chastity in his old age' (*Genealogy of Morals*, III §2), and in particular the ascetic ideas embodied in his last opera *Parsifal*, to be contrasted with Bizet's *Carmen* (which Nietzsche praised while depreciating Wagner).

The word 'ascetic' comes into English from the Greek 'asketes', monk or hermit. In classical Greek the word 'askeo' meant training in general, such as in athletics or gymnastics, or for the arena, etc. And the word 'askesis' meant a way of life involving training. It can easily be seen how the word was adapted to refer to those who indulged in a training of the self in denial and severe abstinence. Nietzsche's gives full reign to his hostile rhetoric against priests, a type of person which he thinks has appeared at most times in most societies, and not just European:

> The idea at issue in this struggle [with the ideal of the ascetic priest] is the *value* which ascetic priests ascribe to our life: they juxtapose this life ... to a completely different form of existence, which it opposes and excludes .... The ascetic treats life as a wrong track along which one must retrace one's steps ...(*ibid.*, III §11)

In a further passage the existence of this type is linked to the will to power and to resentment:

> It can only be a necessity of the first order which allows this species to grow and flourish in spite of its *hostility to life* – it must somehow be in the *interest of life itself* that such a self-contradictory type does not die out. For an ascetic life is a contradiction in terms: a particular kind of *ressentiment* rules there, that of an unsatisfied instinct and will to power which seeks not to master some isolated aspect of life but rather life itself ...; an attempt is made to use strength to dam up the very sources of strength; .... (*ibid.*, III §11)

Here two aspects of priests are in tension with one another. There is a psychological drive from resentment that sustains denial and extreme abstinence. It grows and becomes triumphant; but it does so at the expense of other instinctual capacities for life. These are diminished and ultimately stunted or crushed. Here Nietzsche names and describes one of the aspects of the will to power that is a psychological drive within us, viz., resentment.

The following claims about the life-denying asceticism of (most) priests can be extracted from Nietzsche's remarks:

(1) resentment plays a leading role in causing the priests to lead an ascetic life;

(2) resentment is an instinctual drive in priests, and is a particular manifestation of the will to power enabling them to master life itself, that is, to resist all the impulses to life and to be hostile to it while continuing to live.

These two claims are well developed in the 'First Essay' of *The Genealogy of Morals*. Nietzsche thinks that one of his great discoveries is the role of resentment in the development of our morality, especially Christian morality (he always uses the French term *'ressentiment'*). Resentment is a bitter, poisonous, reactive (rather than proactive) emotion directed against an Other. It is a characteristic of slave morality that it can be directed against the nobility and their spontaneous and proactive values. It is also a hatred and a seeking of revenge which can be long term in realising its goal of overthrowing the Other, or their values. Because of this Nietzsche says of those who resent: 'A race of such men of *ressentiment* is bound in the end to become *cleverer* than any noble race' (*ibid.*, I §10). Priests are said to be great resenters because they are powerless in comparison with the nobility: 'From powerlessness their hatred grows to take on a monstrous and sinister shape, the most cerebral and most poisonous form. The very greatest haters of world-history have always been priests, as have the most ingenious. ... Human history would be a much too stupid affair were it not for the intelligence introduced by the powerless' (*ibid.*, I §7). In contrast for the noble, cleverness is less important than the cultivation of their proactive instincts. And if they do resent it is expelled in immediate action and does not poison.

This characterisation of resentment presents it as a psychological drive in certain types of person that can be overwhelmingly powerful and in its train cultivates intelligence and cunning in order to overthrow the object of resentment. For Nietzsche, as we will see, the object of resentment is not so much the nobles themselves who are to be dethroned, but rather the moral values they endorse. In this respect Nietzsche talks of 'resentment as an unsatisfied instinct and will to power'. We are now in a position to state one of the main theses of *The Genealogy of Morals* to be discussed in the next two sub-sections. It has the basic form of the WP Thesis:

> For the moral values held by some group, V_G, there is a manifestation of the "will to power" through psychological drives (resentment in the case of advocates of slave morality, proactive vitality in the case of advocates of master morality), which gives rise to, and/or causes or maintains, G to hold the values V_G.

There is also associated with this a change thesis in that the appearance of resentment in a group is said to overthrow one set of values and replace them by another set (e.g., the change from master to slave morality).

But in this section our concern is with truth and the pursuit of truth as an ideal (the case of moral belief is left to the next section). Thus the following conjecture can be proposed, which is also an instance of the WP Thesis. For the (ascetic) ideal of truth adopted or pursued by some group or individual G, T_G, there is a manifestation of the "will to power" through the psychological drive of resentment, which gives rise to, and/or causes or maintains, G to hold, or pursue, T_G (while at the same time preventing G from dropping the ideal for something other than truth).

Is all resentment wrong? This would appear to be the case for Nietzsche. But this is to give resentment a bad name. It could well be an appropriate emotional reaction in some circumstances. Prisoners in a concentration camp can rightly resent their brutal guards and seek redress for their treatment.[28]

This point apart, there appears to be a debunking move now possible. If something arises from resentment, especially our pursuit of truth or our moral values, then this might cast these under a pall of suspicion; they might even be rejected because of their despicable origin. However there is a genetic fallacy at work here. Even if truth, or moral values, had such an origin, it does not follow that they ought to be debunked. What is overlooked is that, even if they arose through resentment, there might now be available an independent story to be told as to why they might be held, or why they might be justified, that is not mere rationalisation. A story about origins is one thing; a story about justification is another. The latter turns mainly on the norms of our critical tradition, despite the (possibly disreputable) origins of what we are attempting to justify. Another way of putting the matter is to invoke some version of the distinction between the context of discovery versus the context of justification, common in the philosophy of science, that is often overlooked or dismissed by those who stand aside from the critical tradition or reject it.

10.6.2 The Ascetic Ideal of Truth and Science

Given truth is an *ascetic* ideal, it remains to be seen from what we are allegedly abstaining or denying ourselves when we pursue truth. Clearly, if we pursue truth, then on pain of contradiction, there will be something that we cannot at the same time pursue, viz., falsity. Nietzsche develops his case as follows. Given the ravages that the ascetic ideal has wrought on some lives, and also the ravages of the Judeo-Christian conception of the world, what might an ideal which opposes it be like? Could science be such an ideal opposed to the ascetic ideal? It at least seems to have no truck with the Judeo-Christian conception of the world, or belief in God. Nietzsche vehemently rejects this suggestion: 'The very opposite of what is being asserted here is the *truth*: ... it constitutes not the opposite of the ascetic ideal but rather *its most recent and most refined form*' (*ibid.*, III §23, italics added for 'truth'). Science is committed to the ascetic ideal of truth. Note well that Nietzsche claims that there is a truth here, viz., that science is the most recent form of the ascetic ideal. If this is a truth, then is it also an instance of the ascetic ideal? It would have to be if *all* truth-seeking is an instance of the ideal. This turns on the rather gross generalisation in which all truth seeking is to be disparaged as a form of asceticism. We could avoid this by rejecting the claim that all cases of truth-seeking involve an ascetic ideal. This would rescue Nietzsche's own particular truth-claim just cited, and more generally his genealogical hypotheses, from the taint of truth asceticism. But there is no clear criterion for exempting some, but not other, truths from asceticism.

Dwelling at length on the reasons Nietzsche gives for his claims that science is the latest manifestation of truth asceticism is unhelpful. In part

Nietzsche mounts an attack on the character of the people he alleges work in science (scientists and scholars are all nerds who deny life). It also remains quite unclear what science (as opposed to scientists – these are quite different) denies or abstains from in being committed to the truth-asceticism. Certainly the Mertonian ethos of science (section 4.3) bid scientists to abstain from fraud, deceit and falsification of results. Yet, we will see that this is the kind of denial that Nietzsche thinks is part of the asceticism of science. Is Nietzsche even right about science and truth seeking? No. As mentioned at the beginning of this section there are those who bid us adopt a number of values for science that do not include a strong role for truth (though truth is hard to ignore for some values that do not invoke it directly). Astronomy up to and including Copernicus did not always invoke theoretical truth as one of its goals. Both instrumentalists and realists want to "save the phenomena"; but instrumentalists will stop at this while realists will want theoretical truth as well. Nineteenth century physicists and chemists during Nietzsche's time mainly adopted an instrumentalist stance about the existence of atoms.

There are some scientists who have openly embraced the postulation of what they know to be false, something that Nietzsche will recommend to truth ascetics, as will be seen later. Only one such case will be mentioned: Descartes. In investigating a range of rival hypotheses, he tells us: 'I shall even assume here some [hypotheses] which it is certain are false' (Descartes (1983), §45, p. 105). The example he gives of a hypothesis, which as he says, is known to be true is that God created all creatures. The hypothesis that he toys with, while knowing that it is false, is that of evolution. And prior to that he says: 'even though these things [hypotheses] may be thought to be false, I shall consider that I have achieved a great deal if all the things that are deduced from them are entirely in conformity with the phenomena; for, if this comes about, my hypothesis will be as useful to life as if it were true' (loc. cit.). What Descartes explores here is the idea of underdetermination in which there can be two rival theories which fit the observational evidence equally as well. On extra-scientific grounds, e.g., the Bible, we can determine that one theory is true while some other is false. However this should not preclude our using the false theory. So, according to Descartes in a slightly Nietzschean mood, the deliberate adoption of the false can have its role in science, and even in life. But this is not a Nietzschean noble lie, as Descartes does not deceive us; he tells us that he is adopting a falsity.

10.6.3 The Ascetic Ideal in Truth, Scepticism and the Arts

Others opposed to the ascetic ideal might be philosophical idealists, by which Nietzsche sometimes appears to mean philosophical sceptics. Seekers after knowledge with a sceptical bent, Nietzsche says, are supposed to be free spirits who do not take on any old belief (with or with out passion). Rather they sift through beliefs very carefully – with an aim to what? Uncovering the truth! On this Nietzsche comments: 'These men are far from *free* spirits: *for they still believe in the truth*!"... it is in their belief in truth that they are more inflexible and absolute than anyone else" (*The Genealogy of Morals*, III §24). He then goes on to ask: 'But what *compels* these men to this absolute will to truth, albeit as its unconscious imperative, is the *belief in the ascetic ideal itself* ... it is the belief in a *metaphysical* value, the value of *truth in itself*.' (*loc. cit.*) Is Nietzsche saying that the pursuit of particular truth claims is an ascetic ideal, regardless of what theory of truth one adopts? Or is he claiming that some particular *theory* of 'truth in itself' commits us to the ascetic ideal while the uncovering of particular truths (as in science) does not?

We get a clue in support of the second view when he cites a passage from an earlier book to illustrate his position:

> The truthful man, in the bold and ultimate sense presupposed by the belief in science *affirms in the process another world* from that of life, nature and history; and in so far as he affirms this "other world", what? Must he not then in the process – deny its counterpart, this world, *our* world? The belief upon which our science rests remains a *metaphysical belief*. (*ibid.*, III §24)

What Nietzsche seems to suppose is that we have adopted a strongly metaphysical theory of truth about other worldly items; and it is this theory of truth that is entrapped in the ascetic ideal. But the passage involves us in two errors: a fallacious inference based on false premises. The truthful man of science need not affirm another world at all. Scientists take themselves to be investigating this world of life, nature and history and not some other. Atoms, molecules, genes, tectonic plates, laws of nature, etc. are all of this world and no other. And now the bad inference. Suppose they do affirm some other world as well. It does not follow that they must deny this one. They could remain dualists about worlds and claim that both exist, viz., this one and some other. But what this passage does show is that even if Nietzsche was trying to throw off the influence of Kant, by assuming some crude two worlds interpretation of him, it does not follow that Nietzsche ought to saddle science and scientists with such a view of the world. Actually Nietzsche goes on to talk of the two worlds view of Plato. But few scientists from the scientific revolution onwards took a Platonic two-worlds view of science. It is all this-world, even though it might postulate unobservable entities.

In the same section Nietzsche takes another tack. He claims that the ascetic ideal involves an unquestioning attitude to truth because it does not allow us to ask for a justification for our belief in the truth, or to question it in any way: 'The will to truth itself requires critique ... the value of truth must for once, by way of experiment, be *called into* question' (*ibid.*, III §24). But philosophers have not been so blinded by the ascetic ideal of truth that they shun all attempts at justifying belief in truth. They have questioned the role of truth, but perhaps not as often as, or in the way, Nietzsche might have liked.

The qualification 'by way of experiment' might be taken to mean that we are to conduct a thought experiment, say along the lines of Cartesian doubt, but applied to truth. This in itself does not support the idea that if one does not carry out the experiment one is thereby committed to truth-asceticism. Finally, Nietzsche proposes a quite weak criticism of truth asceticism if we are to merely question our truth seeking 'by way of experiment'. As already noted, there are epistemic values other than truth that can be properly invoked either along side truth or in contrast with it.

In the next section of *The Genealogy* Nietzsche sets aside science and sceptical idealism as hopelessly implicated in the ascetic ideal and considers art and artists. He tells us that 'art, in which the *lie* is sanctified and the *will to deceive* has good conscience on its side, is much more fundamentally opposed to the ascetic ideal than science'. He continues: 'The artist in the service of the ascetic ideal is therefore the most essential *corruption* of the artist possible' (*ibid.*, III §25). Nietzsche allows that artists can be deceptive and tell lies, both big and small, without feeling bad about it. However we need to distinguish between artists who write fictions and who make it clear that they are doing so, from artists that tell lies. The artist of fictions indulges in no subterfuge or deception about their art and its contents; but the artist of lies must indulge in some subterfuge or deception, otherwise they would not be lying. The issue of the noble lie will emerge more fully in the next sub-section.

Finally, there is a further odd claim that needs examining: 'Even when examined from the point of view of physiology, science rests on the same foundation as the ascetic ideal' (*loc. cit.*). All of science is to be examined in relation to the ascetic ideal. How is this examination to be carried out? Surely not using one of the sciences! But that is exactly what Nietzsche proposes when he says that a physiological examination is to reveal something of science. Nietzsche makes a big appeal to the explanatory role of physiology as one of the ways in which the genealogy of morals is to be examined. This is evident given the number of indexed references there are to it. But does Nietzsche mean the bodily science of physiology? Sometimes. This is very evident in the important note at the end of the 'First Essay' of *The Genealogy of Morals* in which medical science, including physiology, is asked to take a big role in the investigation into values. In particular it is to investigate whether, say, some value best promotes the welfare of the many as opposed to the welfare of the few (*ibid.*, I §17). This is clearly an empirical matter; but it is doubtful whether physiology is the science to answer it.

Nietzsche often takes his genealogical project to be an investigation into cultural pathology. Just as the physical body may exhibit symptoms that the medical science of physiology would investigate, so our cultural or political body might be investigated for its pathology. But this is merely a metaphorical extension of the word 'physiology'. What is needed is an account of what the study of the alleged pathology of culture might be like; there is no well-developed cultural pathology as there is bodily pathology in medicine. Perhaps this is what Nietzsche hopes his investigations in *The Genealogy of Morals* will provide, viz., hypotheses about cultural pathology. One aspect of this is the topic of this section, viz., what role our epistemic notions might play, especially the notion of truth. What Nietzsche hopes to provide are genealogical hypotheses not only about morality but also about our will to truth, or our seeking of truth, something which talk of it as an ascetic ideal is meant to cast some light. For by calling truth an ascetic ideal Nietzsche hopes to uncover one bit of our pathology, viz., our will to truth. But this is paradoxical in that it involves more truth-seeking, the very thing that the debunking of the ascetic ideal is to achieve (a matter considered further in section 10.6.5).

10.6.4 The Ascetic Ideal, Atheism and the Noble Lie

Here we come to the culmination of Nietzsche's investigation into the ascetic ideal and what truth-seeking denies us. It turns out not to be the advocacy of values other than truth (coherence, simplicity, informativeness, etc) but something we might disvalue, namely, the big noble lie. The final possible opponent to the ascetic ideal Nietzsche considers is atheism. Since atheism denies God's existence, is it not thereby free of all asceticism? No, says Nietzsche because atheism is also dominated by the 'will to truth':

> But this will [to truth of atheism], this *remnant* of the ideal, is, if one is willing to believe me, the strictest, most spiritual formulation of the ideal itself, absolutely esoteric, stripped of all outworks – not so much its remnant, then, as its *core'* (*ibid.*, III §27)

So atheism also falls victim to the ascetic ideal of truth – if we are willing to believe him (which we need not!). And now to a following remark which takes us to the culmination of Nietzsche's analysis of truth-asceticism. He says of atheism: 'it is an awe-inspiring *catastrophe*, the outcome of a two-thousand year training in truthfulness, which finally forbids itself *the lie in the belief in God*' (GM II§27).

Note the phrase *'forbids the lie in the belief in God'*. What is the ascetic denial or abstinence of atheists? They deny themselves the lie of the belief in God! One must assume that to overcome asceticism then atheists, like artists just mentioned, must allow themselves to deceive and to lie (now and then? or always over all matters? – this is unclear). The core idea behind lying is that one tells another the opposite of the truth-value one thinks obtains. Instead of always exercising the will to truth they must exercise the will to deceive – and deliberately so; this Nietzsche says they can do with a good conscience and not feel bad about it. In general, all truth seekers are ascetic because they deny themselves lies. They do not merely deny themselves

falsehoods, because, in the course of truth-seeking, one might think one has
hit on the truth when instead on has hit on the false. Nietzsche appears to
offer us stark alternatives: either be committed to the asceticism of the will to
truth (even if we hit on the false), or allow ourselves to lie and to deceive –
the more world-embracing and world-shattering the lie, the better. For some
of Nietzsche's readers the gripping style and rhetoric of the 'Third Essay' of
The Genealogy of Morals about truth-asceticism and the will to truth ends in
a damp squib about bidding us to try the alternative to truth-telling – and lie!
And this despite the fact that, in writings within the same year, we are told
elsewhere with the same urgency that truth is to be sought at all costs.[29]

The moral question of whether or not we should tell lies is outside the
scope of this chapter. But there are cases where Nietzsche claims not merely
that we have adopted falsehoods that are necessary for life. Rather, some
have deliberately got others to successfully adopt something they take to be
false. One example is Christian morality with its doctrines of Christian love
and selflessness. This Nietzsche argues the Ancient Jews managed to foist
onto the rest of the world while not believing in these values themselves. In
fact they did it out of an extremely spiteful resentment, thereby exhibiting the
extent to which the "will to power" operated in them. This is something
Nietzsche found praiseworthy in the ancient Jews, not least because they
were not dominated by the denials imposed by truth-asceticism. Further
discussion of such heady claims must wait until the section 10.7.4.

10.6.5 Nietzsche's Genealogy – a Lie or an Example of the Ascetic Ideal?

Let us now turn to Nietzsche's own genealogical enterprise, and in particular
his own view about truth being an ascetic ideal. Does calling something an
ascetic ideal disparage it in some way? And are all truths to be characterised
as part of the ascetic ideal? Nietzsche does not spell out an answer to the
second question. So far no reason has been given for accepting that any
pursuit of truth in any field commits us to the ideal, even pursuit of the truth
about whether Nietzsche had a big moustache. Thus if I am a truth seeker
about matters to do with either God's existence, or Nietzsche's facial
hairiness, then I am in pursuit of the ascetic ideal – truth. But perhaps the
following remarks from *The Anti-Christ* might indicate a qualification. 'This
time I should like to pose the decisive question: is there any difference
whatever between a lie and a conviction?' We can pass over Nietzsche's
comments on this and note his two concluding remarks: 'For in order to lie
one would have to be able to decide *what* is true here' (*The Anti-Christ*, §55).
Does this mean that even our noble artists of the lie are caught up in the
ascetic ideal of the truth? Do they have to detect the truth in order to lie? Not
necessarily. One can lie to a person in the belief that you are misleading them
as to the truth, but in fact be telling them the truth. Secondly Nietzsche says:

'Ultimately the point is to what *end* a lie is told' (*ibid.*, §56). Given this remark little lies would not be permissible; but world-shattering noble lies are a different matter.

Whatever the scope of Nietzsche's doctrine of truth-asceticism, it is pretty clear that for Nietzsche the ascetic ideal is to be overcome. And he fully recognises a consequence of this: 'A depreciation of the ascetic ideal inevitably entails a depreciation of science' (*The Genealogy of Morals*, III §25) We are entitled to extract the antecedent here and say that truth as an ascetic ideal is to be depreciated. In connection with such an overcoming, we find Nietzsche saying of truth-asceticism: 'All I have been concerned to indicate here is this: in the most spiritual sphere, too, the ascetic ideal has at present only *one* kind of real enemy capable of *harming* it: the comedians of this ideal – for they arouse mistrust of it' (ibid., §27).[30] Nietzsche takes himself to be the comedian of this ideal and his role is to arouse our mistrust. Now we might distrust something, investigate it, and find that, with or without some renovation, it can be given our credence, in much the same way as Descartes proceeded about our belief about the external world. But does Nietzsche go beyond this and invite not merely distrust but rejection of truth altogether? It would seem so; and this invites paradox.

In the 'Third Essay' of *The Genealogy of Morals* Nietzsche gives us a version of a genealogical investigation into truth and truth seeking in which it is alleged that it is one of our ascetic ideals. As such it is to be depreciated. But then any genealogical investigation that tries to advance hypothesis H_x about the genealogy of x (where 'x' could be morality, truth, or whatever), is to be depreciated, if one of the epistemic goals of the enterprise is to get at the truth of H_x. So it follows that if Nietzsche advances his own genealogical hypotheses about the origins of morality, then they too partake of an ascetic ideal and are to be depreciated. Now Nietzsche does tell us that he is advancing his genealogical hypotheses as truths; so his genealogical hypotheses are to be depreciated. The argument can be set out as follows:

(1) All ascetic ideas are to be depreciated;
(2) Any truth (which results from truth-seeking in any area, or just in genealogies) is an ascetic ideal;
(3) So, truths (which result from truth-seeking in any area, or just in genealogies) are to be depreciated;
(4) If genealogical hypotheses H_x are advanced as truths (which result from truth-seeking about the genealogy of x) then H_x is to depreciated;
(5) Conclusion: Nietzsche's own genealogical hypotheses (about the origin of morality, or truth-seeking, etc) are to be depreciated (by Nietzsche in particular).

There are ways out of this that involve rejecting dubious premises, such as (2), or by introducing qualifications. But these are not ways out available to the comedian of the ascetic ideal. Instead, in rejecting the ascetic ideal of truth we appear to be left with the alternative of telling deliberate lies. Now lies might be one way out of extreme life-threatening situations. But it does not seem that the matter of hypotheses about the genealogy of some x would be so life-threatening that one had to lie about what hypotheses one should or should not adopt. The canons of how we assess any hypothesis in any science, including the hypotheses of the genealogy of x, must come into play. Otherwise one is not carrying out any kind of genealogical investigation. One might be merely speculating, fantasizing or brainstorming. But would deliberate lies be countenanced?

In his book on Nietzsche, Nehamas tackles the above problem of how we are to understand the various tensions in Nietzsche position. He puts the problem differently from that above: 'His (Nietzsche's) genealogical account of morality may raise the question of its own status, but it cannot answer it – not, at least, without another genealogy of Nietzsche's own practice, which would itself raise the same question again' (Nehamas, (1985), p. 132). He calls this 'the circle that animates the very structure of the *Genealogy*' (*ibid.*, p. 133). There are differences between the account in the above paragraph and Nehemas' account of Nietzsche's predicament that can be passed over here. But the consequences to be drawn from the predicament do differ. One view is that Nietzsche's problems are based on a muddle of bad reasoning that he could have easily avoided, combined with a lack of appreciation of what a fallibilist epistemology is like in which hypotheses are to be assessed. As Berkeley said 'we raise the dust and then complain that we cannot see' (wisely said not by the English type of genealogist but by an Irishman of another type). We can grant that some truth-freaks have pursued truth to the detriment of all other epistemic values. But in claiming this there is no further reason to jettison truth seeking all together.

In contrast Nehamas thinks that the problems that he rightly locates in Nietzsche can be resolved by pursuing Nietzsche's overall views even further. He broadens the whole matter by saying:

> Nietzsche's problem is that he wants to attack the [philosophical] tradition to which he belongs and also escape it. An explicit attack ... would perpetuate that tradition. A complete escape from it directly into art ... would simply change the subject but leave that tradition intact. ... His unparalleled solution to this problem is to try consciously to fashion a literary character out of himself and a literary work out of his life (*ibid.*, p. 137).

But insofar as it is not merely a life that is literary but also the texts that the life produces, then we can ask whether the texts are merely fictions (not the same as deliberate lies), or do they sometimes express the results of truth-seeking, or do they depreciate truth-seeking by telling deliberate lies? The problem that Nietzsche faces with respect to truth is not easily avoided.

Perhaps we might say that Nietzsche is a writer of fictions but with a rhetorical purpose. The purpose is not to tell yet another alleged truth about the world, or about the genealogy of some x, but to counteract all such attempts with a deliberate fiction which is designed to antagonise those who hold as true some prevailing conception of the world but which, unbeknownst to them, is equally fictitious. Nietzsche did believe that the whole Christian story and its account of morals was deeply false and defective. So, one might suppose, he wrote a seemingly plausible alternative fiction which attempts to undermine it. Such is the view of Smith (amongst others) in the notes to his translation of *The Genealogy of Morals* (see pp. xxiv-xxv). He doubts that Nietzsche's genealogy does document, or explain, the origin of Judeo-Christian morality (a matter we will look into in the next section). Rather Nietzsche's purpose is to be rhetorical in the original sense of persuading the reader of some desired conclusion which may have little to do with the truth. The contrast he makes is between the patient historical work that needs to go into genealogy carried out with the intention of uncovering truths (something Nietzsche speaks of (*ibid.*, Preface §7)), with the violence[31] of an act of interpretation which attempts simply impose a view on the reader, quite independently of its truth, and even with the admission of the interpreter that this might also be another falsehood. As Nietzsche says, perhaps of his own genealogical interpretation: 'violating, adapting, abridging, omitting, padding out, spinning out, re-falsifying, and whatever else belongs to the *essence* of interpretation' (*ibid.*, III §24).

There is something to be said for Smith's view. But if we take this on board we seem to have a case of Nietzschean schizophrenia with which to deal. On the one hand there is the attempt to genuinely say something about the genealogy of morals and of truth (even if there is no guarantee that the account is correct). However both genealogies are to be off-set against the asceticism of truth-seeking. There is also an ambiguity to be resolved between the quite different matters of truth seeking which might still result in the false, the writing of fictions and, finally, the telling of lies. Some might see in all of this difficulties for the whole prospect of our philosophical tradition and the need for its renovation. This is the view of apologists who see in Nietzsche's writings the end of metaphysics, or philosophy, or whatever. Others will view it less charitably as a misapplication of what is best in our philosophical tradition that perforce leaves that tradition untouched. The latter take Nietzsche at his word when to *The Genealogy of Morals* he adds the subtitle 'A Polemic'.

10.7   THE GENEALOGY OF MORALS: PSYCHO-SOCIAL HISTORY AS FICTION OR REALITY?

Nietzsche tells us in his 'Preface' to *The Genealogy of Morals* that his book concerns 'my thoughts on the *origin* [Herkunft] [32] of our moral prejudices' (*ibid.*, §2). These had led him to reject the theological idea that God is the origin of good and evil, and even reject the idea that the Kantian categorical imperative could play a role here. He gives up 'looking for the origin [Ursprung, source, beginning] of evil *behind* the world' (*ibid.*, §3), or in rational thought, and instead looks for its origin *in* the world itself. His approach is clearly empirical and sociological when he asks: 'under what conditions did man invent the value-judgements good and evil?' (*loc. cit.*). This is one kind of genealogical origin that can be investigated, viz., the actual historical adoption of a morality not so much by an individual but by a social group or a people. But another kind of origin [Herkunft] that genealogy can investigate concerns evaluations and asks why we attach the value we do to our values: *'And what value do they themselves possess?'* (*loc. cit.*). Nietzsche's answers are cast in consequentialist terms when he asks: 'Have they helped or hindered the progress of mankind? Are they a sign of indigence, of impoverishment, of the degeneration of life? Or do they rather reveal the plenitude, the strength, the will of life ...' (*loc. cit.*) According to Nietzsche Schopenhauer had praised the values of compassion, self-abnegation and self-sacrifice and these 'became for him "values in themselves", on the basis of which he *said no* to life' (*ibid.*, §5). For Nietzsche Schopenhauer's values are those of slave morality that need to be re-evaluated.

Concerning the evaluative investigation Nietzsche says: 'we stand in need of a *critique* of moral values, *the value of these values itself should first of all be called into question*. This requires a knowledge of the conditions and circumstances of their growth, development, and displacement ...; knowledge the like of which has never before existed nor even been desired' (*ibid.*, §6). And how is this knowledge to be obtained? He says by investigating 'the real *history of morality*' rather than indulging in 'English hypothesising *into the blue*' (*ibid.*, §7). There were English genealogists of morality, such as the Darwinians, who had influenced the genealogical work of Paul Rée. But none of these had taken on what is 'a hundred times more important to a genealogist of morals', namely to investigate 'what has been documented, what is really ascertainable, what has really existed, in short, the whole long hieroglyphic text, so difficult to decipher of humanity's moral past' (*loc. cit.*). All of these remarks indicate that Nietzsche is interested in the real events of the history of morality and an accounting of them on the basis of what evidence can be found. This is important given what was said in the previous section about the ascetic ideal of truth-seeking. Nietzsche project as outlined in the 'Preface' is truth-seeking about real history. Nothing is said about ascetic ideals, or fictions or the noble lie.

This section concerns Nietzsche's explanation of the historical origin of morality and the evaluations that have been made. Section 10.7.2 concerns the use of etymology and physiology to provide evidence about past moral evaluations. Section 10.7.3 concerns Nietzsche's explanation of how Christian slave morality came to prevail over master morality using the typology of slave and master morality, psychological theses about resentment and the "will to power", and claims about social organisation. This combination of explanatory factors is Nietzsche's psycho-social history of morality. It will be shown that at best it is a very incomplete and inconclusive explanation. Finally section 10.7.4 considers Nietzsche's conspiratorial belief, based on his genealogy of morals, that the Ancient Jews foisted on to the world moral values that they did not hold. But before these, a brief account of master and slave morality is needed.

10.7.1 Master and Slave Morality

Though the distinction between master or aristocratic morality and slave or herd morality is made in earlier works, it emerges fully in his 1886 book *Beyond Good and Evil* . The distinction indicates a typology, and an order of raking, of moral values. But it should not be assumed that the pairs of labels are descriptive of the actual historical moralities of masters and slaves. Rather they are typologies or "ideal types" of moral values which appear to cluster together and which might do some explanatory work concerning morality. Nor should we assume that each can be found exclusively in one social group or one individual: there are 'attempts at mediation between these two moralities ... and at times they occur directly alongside each other – even in the same human being, within a *single* soul' (*Beyond Good and Evil*, §260) (but presumably not at the same time).

Typically in master morality there is a clustering of values such as strength, power, nobility, honour, courage, gratitude, being life-affirming, and being hard and having a severe heart (especially with respect to oneself). In contrast in slave morality there is a clustering of values such as sympathy, pity, self-sacrifice, humility, acting for others, narrow utility, and a pessimism about the human condition.[33] Typically each morality disvalues what the other values. The good/bad distinction is one made from the stance of master morality in praise of master morality and in condemnation of slave morality. The good/evil distinction is one made from the stance of slave morality in praise of slave morality and in condemnation of master morality. The two sets of values are neither exclusive nor exhaustive of all values, (some having an ambiguous place within the bi-partite typology).

In the 1887 book *The Genealogy of Morals* talk of master and slave morality almost disappears. Also Nietzsche links his moral typology with social classes, saying that the two groups in society that first exhibited them were not masters and slaves but a ruling class allegedly divided into a knightly-aristocratic or warrior sub-class and a priestly sub-class. While the first class enjoyed 'a rich, burgeoning, even overflowing health' maintained by 'war, adventure, hunting, dancing' and other 'free, high-spirited activity' (*The Genealogy of Morals*, I §7) and were not consumed by jealousy, the priestly class were the antithesis of this. Nietzsche assumes without much evidence that, because the priestly class were impotent and could not act like the knightly warrior class, they became consumed with jealousy, and then by a deep and poisonous resentment. The "will to power" is also said to operate in both sub-classes. In the case of the warrior class the "will to power", as expressed in their drives of exuberance and vitality, sustain the values of their master morality. In the priestly class the "will to power" takes the form of the psychological drives leading to resentment. One of Nietzsche's central claims is that, through resentment at the group level, the priestly class dethrones the warrior class, or at least the values they endorse. Here mixed together are the ingredients of Nietzsche's psycho-social history with its moral typologies, a theory of psychology about motivation which includes the drive of resentment, and a theory of classes (with an upper class divided into priests and warriors). One would need a social and/or anthropological examination of societies (not carried out by Nietzsche) to test many of these claims.

The particular case that Nietzsche discusses is Ancient Jewish society in which its class structure is linked to the moral typology. Despite dropping talk of slave morality, Nietzsche does speak of the Jews as having bought about 'the slave revolt in morals' (*loc. cit.*). Though there had been a history of enslavement of the Jews, it was not as slaves that they bought about the revolt in morality by getting others to adopt Christian values. Nor was it necessarily the priests who carried out the revolt. Nietzsche attributes priestly resentment to the Jewish people as a whole when he says; 'The Jews were the priestly people of *ressentiment par excellence*' (*ibid.*, I §16). Once again there is a heady mixture of the sociology and history of classes in Ancient Jewish society that needs to be investigated in respect of the morality they adopted, or got others to adopt.

There are also rather lurid passages that may be ambiguously interpreted, as when Nietzsche speaks of 'Jewish hatred – the deepest and most sublime form of hatred' (*ibid.*, I §8) But in this context it is hatred and resentment that are allegedly creative in making new values. There is also a lurid account of the behaviour of the aristocratic warrior class who, while said to be friendly and respectful and even tender to one another, are to outsiders 'not much better than predators on the rampage' who are 'a throwback to that inextinguishable horror with which, for hundreds of years, Europe regarded the raging blond Germanic Beast' (*ibid.*, I §11). Given this characterisation of the behaviour and values of the aristocratic-knightly class, resentment by those not in that class might be quite appropriate. It is characterisations such as these, along with hypotheses about resentment as the psychological drive of the "will to power", that are part of the psycho-social explanation of why some values get dropped and replaced by others.

10.7.2 Etymology as Evidentiary Source

Nietzsche begins his investigation into the genealogy of morals with an excursion into the etymology of our central value words such as 'good' and 'evil' (and others) and their equivalents in German, Latin and Ancient Greek (*ibid.*, I, §2-§5). Etymology he thinks, supports an original master/slave morality:

> What pointed me in the *right* direction was actually the question of what the designations of 'good' coined in various languages meant from the etymological perspective. I found that they all led back to the *same transformation of concepts* – that 'refined' or 'noble' in the sense of social standing is everywhere the fundamental concept, from which 'good' in the sense of 'having a refined soul', 'noble' in the sense of 'superior in soul', 'privileged in soul' necessarily developed. This development always ran parallel with that other one by means of which 'common' or 'plebeian' or 'low' ultimately slide over into the concept 'bad'. (*ibid.*, I, §4)

Nietzsche is aware that in some cases he might be straining matters in claiming, say, that the Latin 'bonus' might be traced back not to a good (man) but to a man of war or warrior ; or that the German 'gut' is traceable back to 'the godly' as in 'den Göttlichen' (*ibid.*, I §5). But nonetheless he persists with his thesis, which comes in two parts: (a) etymology supports an original master/slave typology of morality in early societies; (b) the factual sociological thesis that it was those with power, the masters, who got to label what they valued as good and bad, remnants of which can be uncovered etymologically.

Some commentators disagree with him. The only example Nietzsche cites for which there is any agreed, good evidence is the connection between 'schlecht' [bad] and 'schlicht' [simple, the simple common man]; this arose as recently as the first half of the seventeen century. For the rest, as his translator Smith says, 'Nietzsche's etymologies are often speculative and tendentious' (*ibid.*, p. 140, footnote 14).

Given the above, it is surprising to discover that elsewhere Nietzsche casts doubt on the possibility of an etymological investigation revealing anything about the genealogy of morals. We have already seen that he advises us to be suspicious of language (section 10.1.1, thesis (M6)), especially in inferring from grammatical structure to anything about reality. The same suspicions can be raised concerning meaning, and thus etymology. Nietzsche takes this even further when he points out that in any genealogical investigation into the origins of some Y, one should be aware that in the replacement of an earlier X by a later Y, Y tends to obliterate aspects of X. In particular, meanings associated with the linguistic practices concerning X get obliterated when replaced by those of Y. The issue here is akin to recent talk of rival conceptual schemes which are incommensurable because there is no way in which one can define or express the concepts of the one in the other. The strong way in which Nietzsche expresses this claim makes etymological investigations in the service of genealogy a nearly impossible enterprise.

Nietzsche makes comments along these lines in his discussion of punishment (see *ibid.*, II §12 -13). We currently assume in the case of punishment that it has certain aims (such as deterrence, isolation, revenge, forfeiture, and so on), and that the word 'punishment' has certain meanings. But we cannot assume these current aims and meanings were present at the origin of punishment practices. Nor can we assume that once some punishment practice came into existence, for whatever reason, that its continued existence maintains the original meaning and aims. As Nietzsche puts it:

> that anything which exists, once it has somehow come into being, can be reinterpreted in the service of new intentions, repossessed, repeatedly modified to a new use by a power superior to it; that everything that happens in the organic world is part of a process of *overpowering*, *mastering*, and that, in turn, all overpowering and mastering is a reinterpretation, a manipulation, in the course of which the previous 'meaning' and 'aim' must necessarily be obscured or completely effaced. ...
> But all aims, all uses are merely *signs* indicating that a will to power has mastered something less powerful than itself and impressed the meaning of a function upon it in accordance with its own interests.' (*ibid.*, II §12).

The earlier meanings we attach to terms are said to be 'obscured or completely effaced' once later and different meanings emerge. And this process of changing aims and meanings, Nietzsche claims, is just more of what we have constructed through the workings of the "will to power". In the same vein Nietzsche says sententiously of 'the fluid aspect of its [punishment's] "meaning"': 'all concepts in which a whole process is summarised in signs escapes definition; only that which is without history can be defined' (*ibid.*, II §13). In the case of the term 'punishment' it now has so many meanings that it defies definition. And it has a history that, in the light of the previous remarks, will have effaced some of the earlier meanings.

Given that the meanings of words used in earlier and different practices are effaced in later practices, it is hard to see that etymology could tell us anything about the meanings of moral terms embedded in earlier practices. These meaning are not recoverable. So either etymology cannot be used to tell us anything about the original use of moral vocabulary, or the doctrine of effacement just outlined need to be modified so that the meanings of some words can be traced even though they are embedded in different practices.

The latter seems most likely since doctrines of incommensurability, though seemingly exciting, are often overstated.

Nietzsche is aware that his ruminations on etymology might be inconclusive when he appends to the end of the 'First Essay' of *The Genealogy* a 'Note' suggesting

> ... that some philosophical faculty or another might render outstanding service to the promotion of the *historical* study *of morality* through offering a series of academic prizes' in which the question to be answered by philosophers, philologists and historians is: *'What indications for the direction of further research does linguistics, and in particular the study of etymology, provide for the history of the development of moral concepts'* (*ibid.*, I, §17).

This suggests a gratifyingly empirical approach to the matter.

The 'Note' continues arguing that before psychology is employed to investigate morality, we need to use physiology and the medical sciences generally to investigate the value for healthy life of every 'thou shalt' that has been promulgated in history as a code of ethics. As he puts it:

> ... all tables of commandments ... certainly require *physiological* investigation and interpretation prior to psychological examination. Equally, all await a critique from the medial sciences. The question: what is the *value* of this or that table of commandments and 'morality'? should be examined from the most varied perspectives: in particular the question of its value *to what end?* cannot be examined too closely.' (*loc. cit.*)

It is hard to see what current physiology and medical science could contribute. However it is clear that Nietzsche is calling for an empirical investigation into the extent to which a particular value might contribute to the 'survival capacity of a race (or for increasing its powers of adaptation to a particular climate or for the preservation of the greatest number)', and whether this value would also contribute to producing a 'stronger type' (*loc. cit.*). Though he says, uncharacteristically, that he does not wish to pronounce *a priori* on the matter, values which promote the well-being of the many might be in opposition to those which promote the well-being of the few. For Nietzsche the genealogist, the sciences are to be pressed into service to determine empirically which value best promotes what end. But there still remains a purely value question for philosophers to answer; 'the *problem of value*, the determination of the *hierarchy of values*' (*loc. cit.*).

10.7.3 Nietzsche's Incomplete and Unsatisfactory Explanatory Sketch

One of the main questions to be raised in the 'First Essay' of *The Genealogy* (§6 – §9) is: Why did Christian morality, with its espousal of slave values, come to prevail (amongst certain groups of people)? Nietzsche draws on a number of hypotheses to give a psycho-social historical explanation. What will be examined here is the extent to which the explanation is successful; in fact it does not get much beyond an explanation sketch (Hempel (1965), p. 238). Also, even if it is a sketch, it will be argued that it is incomplete in the sense that there is a gap in the explanation which is not even sketched in, and it is hard to see how it might be without introducing dubious hypotheses. The explanation sketch will be set out as a number of hypotheses each of which, when filled out, would be premises in an explanatory argument.

Hypothesis (1). This concerns the two moral typologies of slave/herd morality and master/aristocratic morality (see section 10.7.1).

Hypothesis (2). This concerns the WP Thesis in relation to morality, and says: For a given group of people G, and some body of moral belief or values that they hold, V_G, there is a specific action of the will to power on those people over that time, WP_G, such that WP_G gives rise to, or causes, maintains, etc, (most of) the group's holding moral beliefs V_G.

Hypothesis (3). This concerns the role of resentment. As we have noted, Nietzsche tells us: 'The slave revolt in morals begins when *ressentiment* itself becomes creative and ordains values' (*ibid.*, I §10). The creative act arises when the resenters say 'no' to the values of master morality and overturn them by establishing new values. This claim gives us two further theses that fill out the WP Thesis just listed as Hypothesis (2):

(3a) Resentment is an instance of the operation of the will to power in some people as a psychological drive, or as motivational;

(3b) Ressentiment in group G (or a collection of individuals) causes G's belief (or assertion, or advocacy) that the values of slave morality are good and ought to be adopted, and those of master morality are evil and are to be rejected.

In expressing (3b) we need to pay attention to the intentional attitude specified towards the claim that some morality is good/bad/etc. In (3b) 'believe', 'assert', 'advocate' have been used with the latter two being understood not to entail the first. The difference becomes important because, as will be seen, Nietzsche held that the Ancient Jews advocated the values of slave morality while not themselves holding these values to be of value; and they were successful in their advocacy.

If we grant (3a), then much weight falls on the empirical claim (3b), viz., that resentment is at the root of Christian love, or altruism, etc. Central to Nietzsche's view is the claim: 'from the trunk of that tree of revenge and hatred, Jewish hatred,... a hatred the like of which has never been seen on earth – from this tree grew forth something equally incomparable, a *new love*, the deepest and most sublime of all the kinds of love' (*ibid.*, I §8). Nietzsche rejects what might be taken as a commonly held origin for Christian love as arising in opposition to hatred and revenge: 'But let no one think that it [love] somehow grew up as the genuine negation of that thirst for revenge, as the antithesis of Jewish hatred. No, the opposite is the case!' (*loc. cit.*) Nietzsche even claims that so subterranean and cunningly Machiavellian is Jewish revenge and resentment that the Jews even crucified the advocate of such love, Jesus Christ, 'so that the "whole world", all the opponents of Israel, might unthinkingly bite on just this bait' (*loc. cit.*). The bait here is not just Christian love but also the whole doctrine of redemption after life as symbolised by the crucifixion. A gap in Nietzsche's explanation begins to emerge here, namely, just how did those who resented get the Other to 'bite the bait' and abandon the values of their master morality and adopt slave values. This will be commented upon again at the end of this section.

But what of the particular claim of (3b)? Resentment might sometimes give rise to some values of slave morality, but not all, one prime counterexample being Christian love. Writing in 1912 Max Scheler argued that amongst the discoveries about the origin of our moral values 'Nietzsche's discovery that *ressentiment* can be the source of such value judgements is the most profound' (Scheler (1972), p. 43). Despite maintaining the profundity of the discovery and claiming that Christian values 'can very easily be perverted into *ressentiment* values' (*ibid.*, p. 82), Scheler also says that 'I consider his theory to be completely mistaken' (*ibid.*, p. 84), the mistake being due to 'his misjudgement of the *essence* of Christian morality' (*ibid.*, p. 105). There are several parts to Scheler's alternative view. One is that, in the ancient view best exemplified in Greek philosophy, love is always directed from the lower to the higher, to that which is more noble, more perfect and often other-worldly. In contrast Christian love is more embracing and can also proceed from the higher to the lower, from the noble to the vulgar, and be extended to the ugly as well as the beautiful, to the sick as well as the healthy, to the poor as well as the rich, etc.

Also important for Scheler is the motivation with which one loves. The wrong motivation is at work when one attempts to flee from oneself and to busy oneself in the affairs of others in order to lose oneself; or when one acts out of concern for one's self or interests or self-preservation. In contrast, the right motivation is that love which is 'motivated by a powerful feeling of security, strength and inner salvation' and 'the clear awareness that one is rich enough to share one's being and possessions' (*ibid.*, pp. 88-9). The structure of motivation for Christian love is what fully distinguishes it from other kinds of love. As Scheler expresses it, Christian love is not without a motivational structure that has some aspects of master morality in that the love arises quite freely and 'spring[s] form a spontaneous overflow of force' (*loc. cit.*). But even then it does not sit well with the rest of Nietzsche's moral typology. On the basis of these and other considerations Scheler's objections have point. Nietzsche has simply overlooked the way in which Christian love can be free of *resentment*, and wrongly assumes that all cases of Christian love must arise through it.

It is also hard to find a trace of resentment in Aristotle's definition of his third kind of *philia*, the love involved in what he calls 'complete friendship':

'But complete friendship is the friendship of good people similar in virtue; for they wish goods in the same way to each other in so far as they are good, and they are good in themselves. Hence they wish goods to each other for each other's own sake.' (*Nicomachean Ethics*, 1156b 6-10). Here the motivation is simply the mutual wish of well-being without either wishing to serve an interest of theirs, this being the case in the first two kinds of *philia*. And such a motivation clearly rules out the kind in which hatred, revenge or resentment would be involved.

In sum, the claims of hypothesis (3), though central to Nietzsche's position, are highly dubious.

The hypotheses listed so far concern Nietzsche's moral typology and the workings of the "will to power" through the psychological drive of resentment. These can be collectively called his 'psycho-moral' hypotheses. These hypotheses are quite general and are not anchored in any actual hypotheses and facts about groups and societies, particularly their class structure; nor are they anchored in the particularities of a given society. But they need to be if they are to tell us something about the very matter that Nietzsche wants to explain, viz., how Christian slave morality came to prevail, or be adopted, in Ancient Jewish and other societies. These socio-historical hypotheses, when combined with the other hypotheses, give us what can be called Nietzsche's 'psycho-social' explanation.

Hypothesis (4). This concerns the class structure of a given group. Talk of master/slave morality might indicate that the class structure is a simple master/slave division. But this is not the case in *The Genealogy* where Nietzsche clearly envisages an upper class of rulers that is further divided into two subclasses, the aristocratic knights and the priests.

Hypothesis (5) This concerns the link between class and morality. It is not to be assumed that the members of the ruling class all adopt master morality. It is an empirical matter to discover what moral typology prevails in what class and the variability of that adoption throughout the class. In *The Genealogy* it is only the knightly sub-class that is said to adhere to master morality while the other members of the aristocracy, the priestly caste or class, adopt slave morality. Nietzsche attempts an explanation of why one subclass of the ruling class should adopt a different morality. He refers to the tendency of priests to 'combine brooding with emotional volatility' and to an 'almost unavoidable intestinal sickness and neurasthenia which afflicts priests at all times' (*ibid.,* I §6). But these medical claims are stereotypical and have little empirical basis. They are to be founded more in Nietzsche's dislike of the priests he encountered in Europe than in any sociological or anthropological investigation into priests, or their counterparts, in other cultures at other times (for example, the role and nature of priests in Polynesia at the time of first European contact). In the end Nietzsche simply tells us: 'by now it will be clear how easily the priestly mode of evaluation may diverge from the knightly-aristocratic mode and then develop into its opposite' (ibid., I §7). The unresolved problem for Nietzsche is to explain how, within the ruling class in which master morality is assumed to prevail, a division of moralities could arise. And to this he has to appeal to medical and other hypotheses as well as aspects of hypotheses (2) and (3). What all of this shows is that, for Nietzsche, it is difficult to say what is the source of his supposed split in the master class into its warrior and priestly sub-classes, and to establish any clear morality-class links.

Hypothesis (6). This brings us to the actual history of a particular social group or nation, Nietzsche's example being Ancient Jewish society. Here a host of historical facts can be invoked, including those that relate to the sociological hypotheses (5) and (6), that is Ancient Jewish class structure and what class-morality links might hold. However it turns out that class considerations are irrelevant. With one swoop Nietzsche abandons all his talk of a split in the aristocratic class and replaces it with talk of the Jews, *considered as a whole*, as a *'priestly people'* (ibid., I §7) thereby postulating a psychology of an entire people, and not some sub-class, a characterisation which is adopted in the rest of *The Genealogy*. Consequently the Other whose values they are said to resent and overthrow is not any class within Ancient Jewish society but their outward enemies and conquerors, whoever these may be at various times. It then follows that the Jews as a whole are the bearers of resentment.

These hypotheses are the core of the explanatory sketch that Nietzsche gives. What is to be explained is why the Other adopted the values of Judeo-Christian slave morality (the Other presumably including members of the Roman Empire who became Christians, and who were generally not Jewish, St. Paul being an exception). Since Nietzsche's account of the hypotheses he employs hardly goes beyond what is set out above, his explanation is rightly called an 'explanatory sketch'. This is a common kind of explanation that Hempel finds in history: 'a more or less vague indication of the laws and initial conditions considered as relevant, and it needs "filling out" in order to turn into a fully fledged explanation. This filling-out requires further empirical research, for which the sketch suggests the direction' (Hempel (1965), p. 238). Of course the further empirical research might show that the claims employed in the *explanans* are false, in which case the explanation fails. This has already been suggested for some of Nietzsche's hypotheses.

We can summarise Nietzsche's explanatory sketch as follows, with its moral, psychological and historical claims:

(1) There is an earlier phase in which the Jewish people did not adopt slave morality;

(2) Historical circumstance altered so that the Jewish people were either about to be overwhelmed by their enemies or were actually conquered;

(3) In order to escape from or alleviate their situation, the Jews developed their latent power of revenge or of resentment;

(4) Resentment in the Jews caused the Jews to believe/assert/advocate that slave morality is good.

(5) So, The Jews believed/asserted/advocated that slave morality is good.

(6) So, the enemies, or the conquerors, (or simply the Other) also believed that slave morality is good.

It is now clear that while the explanation might get us to (5) about what the Jews advocated, it does not get us to (6) about the change in morals of the Other. It does not explain how the enemies or conquerors of the Jews also came to accept the view that slave morality is good. There is an explanatory gap here that must account for the transformation of the Other's presumed rejection of their earlier master morality and their later acceptance of, or belief in, slave morality. Resentment is not a psychological drive at work in the Other but only in those who overthrow the values of the Other. What is lacking is a further account of the susceptibility of the Other to abandon their master morality and adopt slave morality.

It is hard to see how the most cunning hatred of the priests could bring off the transformation of values of the aristocratic-knightly Other who, as Nietzsche characterises them, did not suffer from any psychological disabilities, and had 'burgeoning, ever overflowing health'. They appear to lack any psychological 'Achilles heel' whereby the vengeful priests can work their poison. Nietzsche does resort to the metaphor of slave morality spreading like a poison when he says: 'The progress of this poison through the entire body of mankind seems inexorable' (*ibid.* I §9). But this analogy merely highlights the fact that Nietzsche's explanation has a serious gap; it leaves out of account the susceptibility of the rest of humanity to the 'bait' of this Jewish poison. Given the overflowing health and vigour of the warrior class depicted by Nietzsche, and his further characterisation of them as the blond Germanic beast regarded with horror by the rest of Europe, and as 'predators on the rampage' who 'walk away without qualms from a horrific succession of murder, arson, violence and torture, as if it were nothing more than a student prank' (*ibid.*, I §11), it is hard to see how such a "master" class could succumb to slave morality. There is something seriously amiss in Nietzsche's psychological story.

At the very point at which Nietzsche introduces the master/slave morality distinction he adds: 'at times they occur directly alongside each other – even in the same human being, within a *single* soul' (*Beyond Good and Evil*, §260). This suggests that Nietzsche perhaps overdraws the sources and character of master morality and that a much more complex story is to be told than one which depends on a simple bi-partite moral typology. Both moral typologies can be instanced within the one person, presumably exerting their influence at different times and not the same time (i.e., one does not both praise and condemn the same act of pity or compassion at the same time). One might be inclined to extend this claim to the moral typologies existing within the one class. But Nietzsche never says this. In fact he goes on to talk of a ruling and a ruled group and 'the ruling group determines what is "good ..."' (*loc. cit.*). There is no suggestion that the ruling group might be divided. But it might be helpful if he did, for he needs an account of how the ruling group divides into two, and how resentment arises in one sub-group and not the other. Given that people are not to be slotted into a simple bi-partite moral typology, there *might* be a little room to explain not just how a class divides over their moral values, but also how a class becomes susceptible to the morality of the resenters. But this is a speculative attempt at a rescue and not anything conclusive.

Perhaps the fault lies in Nietzsche's quite restricted appeal, in his explanations, to psychological factors only, such as resentment, while ignoring other quite different kinds of factors, such as intellectual maters. Thus why did the early, rather brutal warrior Anglo-Saxons, not unlike Nietzsche's characterisation of the warrior class at least in their war-like activities, succumb to Christianity? Some suggest that it was, in part, an intellectual matter. The Anglo-Saxons believed strongly that each person was fated to the life that befell them and they could do nothing about it. The Old English word for 'fate' is 'wyrd', literally 'what will be'. Though the early Christian missionaries adapted their doctrines to fit with this conception of fate, they also challenged it through the Christian doctrine of redemption and salvation, something that the Anglo-Saxons found to be an attractive alternative to their rather bleak view of a fully determined human existence. In explaining why the Anglo-Saxons adopted Christianity such intellectual, non-psychological matters cannot be ignored.[34] Nietzsche's tendency to put emphasis on psychological explanations due to the "will to power" ignores plausible alternatives. Again a quite different non-psychological story would have to be told as to why, say, Polynesians throughout the Pacific abandoned their rather brutal way of life when they became Christian in the nineteenth century. Again their susceptibility to conversion is not to be explained psychologistically along Nietzschean lines (even if we agree with Nietzsche about the dubious claim that their missionary converters were also resenters).

Nietzsche's theory is just one of several theories of the genealogy of morals that are available to us now. Nietzsche's genealogy turns on psychological hypotheses about the drive of resentment and the operation of the "will to power". But not all "genealogies" of morality need be psychological in this sense. One rival genealogy arises from the very theory that Nietzsche always denied, viz., Darwin's theory of evolution in its recent incarnation as evolutionary psychology as applied in the case of our morals. There are also alternative genealogies about particular moral stances to be found in Marx that are not psychological in character; these offer rival claims about the origin and sustaining of particular moral values of capitalism such as thrift, or individual effort. All of these, Nietzsche's included, attempt to provide truth-oriented explanations rather than a tale about a noble lie. And each can be compared with the other, as any hypotheses in science can be compared, to assess which is the most successful (a task not carried out here).

10.7.4 Nietzsche's Jewish Conspiracy Theory

Need those who are resentful, and foist new values on others while overturning their old values, also believe and adopt the values they foist? It would appear not, according to Nietzsche. And here lies a rather odd view that Nietzsche held about the Ancient Jews: they conspired against the world in an act of tremendous exercise of the "will to power" to foist a new value system on the world but without believing those values. And for this Nietzsche praises, rather than condemns, them. This view arises directly out of his theory of the genealogy of Christian morality. If this theory is dubious (as it is) then Nietzsche can be viewed as holding a curious conspiracy theory about the Ancient Jews. That Nietzsche held a Jewish conspiracy theory all of his own is argued for in Maccoby (1999). Though he praised the Jews for the powers they exhibited, Nietzsche did provide ammunition for anti-Semites that flowed directly from his own genealogical theory of Christian morality.

Nietzsche was no simple anti-Semite like his sister Elizabeth Förster and her circle. In fact he heartily disliked such people and their views. Many passages from his works can be cited to this effect. The following is just one brief comment, not to be noted for its liberal approach to the matter: 'it might be useful and fair to expel the anti-Semitic screamers from the country' (*Beyond Good and Evil*, §251). But Nietzsche is not beyond sharing in the conspiratorial theories that abounded in the nineteenth century about the Jews. Of modern Jews Nietzsche says: 'The Jews, however, are beyond any doubt the strongest, toughest, and purest race now living in Europe: they know how to prevail even under the worst of conditions (even better than under favourable conditions), by means of virtues that today one would like to mark as vices ...' (*loc. cit.*). Nietzsche's praise of the Jews is cast in the language of 'race' and 'purity' which, after the events of the twentieth century, can only be distasteful. He also uses the same language to talk of the Russians in the same vein.

More significantly he continues: 'That the Jews, if they wanted it – or if they were forced into it, which is what the anti-Semites want- *could* even now have preponderance, indeed quite literally mastery over Europe, that is certain: that they are *not* working and planning for that is equally certain (*loc. cit.*).[35] Again this can be taken as praise for the Jews, even when Nietzsche goes on to say that what they also seek is assimilation into Europe and to cease being the 'Wandering Jew'. But there are other undertones here as well. Nietzsche makes two claims, both with a high degree of certainty; (a) the Jews now (i.e., at the time of his writing, the 1880s) have the power or ability to quite literally have mastery over Europe; (b) they are not now working and planning for such mastery (that is they are not currently conspiring). Now Nietzsche and the Anti-Semite differ over (b); the anti-Semite thinks the Jews are currently conspiring in working and planning for such mastery, perhaps with equal certainty as Nietzsche believes the opposite. However both Nietzsche and the anti-Semite believe, perhaps with equal certainty, that the Jews right now have the power to have mastery over Europe. Some might find the vague phrase 'quite literally mastery over Europe' troubling (mastery over what? – Football? Mathematics? Making money? Political life?) But what is clear is that he attributes sufficient power or ability to the Jews to have, at that time, mastery over whatever in Europe; it is not as if this mastery is to emerge well into the future. And this is something to which most contemporary anti-Semites would have also subscribed.

Nietzsche gives no evidence for claim (a) despite his alleged certainty of it. There is no evidence for attributing such power of mastery to the Jews. In fact there is much evidence to the contrary. Witness the powerlessness of the Jews in nearly all 19th and 20th century Eastern European pogroms, and their subsequent powerlessness to obtain any mastery at all in the face of the Nazi onslaught against them. In this respect the Jews are no different from any other groups one might care to investigate (such as Gypsies, etc). The Jews as a whole, despite the prominence of a few families like the Rothschilds, were relatively powerless in this respect, and their influence was fairly negligible. Rather it was the non-Jews who saw every little improvement of the conditions for Jews as a threat to them and as evidence for the growth in Jewish power and their conspiring for mastery in Europe. Nietzsche simply exaggerates the powers of Jews as a group. But that they allegedly have such a power of mastery is an important part of what counts as having a conspiracy theory of the Jews. The pervasiveness of the 'will to power" in Nietzsche's thinking misleads him here.

Nietzsche loved the Jews of the Old Testament but hated the New Testament (see *The Genealogy of Morals*, III §22). He also attributed an extraordinary power to the Ancient Jews. In *The Anti-Christ* Nietzsche discusses the Ancient Jews and reviews what he said in *The Genealogy of Morals* about *'ressentiment* morality', which is none other than Judeo-Christian morality, and then adds that amongst the Ancient Jews:

...the instinct of *ressentiment* here become genius had to invent *another* world from which that *life-affirmation* would appear evil, reprehensible as such. Considered psychologically, the Jewish nation is a nation of the toughest vital energy which, placed in impossible circumstances, voluntarily, from the profoundest shrewdness in self-preservation, took the side of all *décadence* instincts – *not* as being dominated by them but because it divined in them a power by means of which one can prevail against 'the world'. The Jews are the counterparts of *décadents*; they have been compelled to *act* as *décadents* to the point of illusion, they have known, with a *non plus ultra* of histrionic genius, how to place themselves at the head of all *décadence* movements (as the Christianity of Paul) so as to make of them something stronger than any party *affirmative* of life. For the kind of man who desires to attain power through Judaism and Christianity, the *priestly* kind, *décadence* is only a *means*; this kind of man has a life interest in making mankind *sick* and in inverting the concepts 'good' and 'evil', 'true' and 'false' in a mortally dangerous and world-calumniating sense. (*The Anti-Christ*, §24)

Much of this simply reiterates themes concerning the genealogy not just of the good/evil distinction, but also of the true/false distinction (which should be the hold-true/hold-false distinction). But what is quite explicit here is the claim that Jews do not really hold-true the morality they have got much of the rest of the world to believe in order to escape their 'impossible' historical situation. They are not *décadents* themselves in that they hold with the decadent (in Nietzsche's view) morality they have got others to accept. Rather they are the *counterparts* of *décadents* and act out the role of *décadents* to the point of illusion, creating the impression that they also hold with the morals that they have got others to accept.

This is to attribute and extraordinary power to the Ancient Jews. Not only are the urges of the "will to power" so strong in them that they get others to accept the values of slave morality, they also are also powerful enough to create the illusion that they also believe slave morality. Nietzsche praises the Jews for this display of power. But it is also a conspiracy theory of Nietzsche's own unique making about an alleged power of the Ancient Jews. Add to this the claim that it makes the whole world 'sick', there is justification for saying that Nietzsche 'handed on concepts that were utilized by the anti-Semites he despised' (Maccoby (1999), p. 15). And all of this flows quite naturally from Nietzsche's account of the genealogy of Christian, or Judeo-Christian, morality from resentment and his doctrine of the "will to power".

10.8   ADDENDUM ON NIETZSCHE'S GENEALOGICAL PROJECT

Finally there remain three issues to discuss. The first concerns the logical status of the WP Thesis; the second, linked with first, concerns the debunking move; the third concerns Nietzsche's rejection of explanation and causation in terms of which the WP Thesis has been formulated.

Nietzsche tells us: 'To impose upon becoming the character of being – that is the supreme will to power' (*The Will to Power*, §617). The character of being that we have imposed on the Nietzschean world of fluxes of power includes a number of items that involve some notion of identity such as substantial objects, matter, ultimate atoms of being, continuants and kinds, none of which really exist. We have also imposed a logic, a conceptual scheme presupposing the existence of these fictitious items, and a language in which we express "truths" about these fictions. And finally we have imposed a morality upon ourselves. One way of expressing these impositions is in the form of the WP Thesis which says: For all groups G, and all bodies of belief they hold, B_G, there is some manifestation of the "will to power", WP, such that WP gives rise to (or causes, maintains) G's act of belief in B_G. This is quite schematic and needs to be filled in with reference to specific groups and their beliefs in some area, and specific manifestations of the will to power. In some cases this has been done, for example moral evaluations such as those of master and slave morality, in which the operation of the "will to power" has been expressed more specifically as the drive of resentment. As such this instance of the WP Thesis is open to test. And it can be applied in specific socio-historical circumstances to explain the origin and maintenance of moral belief. (As argued, the prospects of passing its tests are dim.) Another instance of the WP concerns belief in kinds and continuing identities in which the "will to power" manifests itself as an evolutionary force; life is preserved and enhanced by holding-true such beliefs. However Nietzsche links his theory of evolution to the "will to power" understood as an inner "oomph" driving all organic and cognitive development. This can be replaced by a Darwinian theory of selection, but at the cost of dropping talk of the "will to power" from the WP Thesis – in which case one would loose the very feature that makes it a distinctive thesis.

For other kinds of belief no specific aspect of the "will to power" is mentioned that could yield an empirically testable claim. If no such aspect is fleshed out (as in the case of resentment), then one is simply left with vacuous talk of the "will to power" for which there are no empirical test conditions at all. The thesis then becomes one which is metaphysical in the pejorative sense of the positivists. Less pejorative is Popper's view in which such untestable (and irrefutable) claims as the WP Thesis become a piece of influential metaphysics which can be used to solve problems when applied in some context (Popper (1963) chapter 8, Part 2). With respect to such problem-solving contexts it is still possible to compare irrefutable claims like the WP Thesis with other such claims according as they solve more or less problems, do it more or less simply, do it fruitfully, and so on. In this respect the WP Thesis is a much less successful solver of problems than its rivals, as was argued in section 10.2.

Also important for the WP Thesis is the scope of 'all' as in 'for *all* groups and *all* beliefs they hold-true'. This matter also arose in connection with the

Causality Tenet of the Strong Programme, and it arose for Foucault's "power/knowledge" doctrine. Nietzsche intends his doctrine imperialistically; it applies in all cases of (group) belief. As such it contrasts with the "weak" programme in the sociology of knowledge advocated by Mannheim. Within the weak programme there is room for models of explanation that do not appeal to causes such as social factors or the workings of the "will to power", but rational ones instead. It is here that the debunking moves are made possible. Where one thought that there might have been a rational explanation available, instead an explanation in terms of power, or "will to power" or social factors, steps in replacing the rational model of explanation. The so-called illusions of reason within modernity exposed by postmodernists under the influence of Nietzsche and others, often turns on the possibility of finding a replacement Nietzschean explanation which does not appeal to rationality and at the same time undermines any rational explanation. But as has been suggested, many of the replacements commit the genetic fallacy; how a belief system came to be adopted in the first place, and how it might be maintained is one thing; what justification it might have is a quite independent matter. There are still the autonomous (but naturalisable – see sections 3.7 and 3.8) norms of reason that can explain why we have some of the beliefs we do.

Finally, in section 10.1.2 we saw that Nietzsche claims that there are no causes, or effects in nature, and perforce no causal relations between them. At best there is a continuum of the flow of power that we divide up for our purposes; we construct both the cause and the effect events and the relation between them. The WP Thesis has been expressed in causal terms; to make it consistent with Nietzsche's overall view, either we need to abandon all talk of causation, or, more weakly, adopt an anti-realist account of causation. Moreover Nietzsche tends to downplay explanation in favour of description only: '"Explanation" is what we call it, but it is "description" that distinguishes us from older stages of knowledge and science. Our descriptions are better – we do not explain any better than our predecessors' (*The Gay Science*, §112). According to Nietzsche, our attempts at explanation yield nothing more than more description. Nietzsche follows the above remark by claiming that the items in terms of which we would give explanations do not exist such as 'lines, planes, bodies, atoms,…' (loc. cit.). To this list could he could add laws of nature and causal claims. But, as was noted in section 10.1.2, there must be formula that we can use for calculation. However we cannot explain using these formula; rather, we simply describe what is going on.

But Nietzsche does have a few comments to make on the nature of explanation that suggests that it might not be inappropriate to keep the realist interpretation. The problem is, says Nietzsche, that the formula 'do not involve any *comprehension'* (*loc. cit.*). So, if we are to get beyond mere description to explanation, then we require the missing ingredient of *comprehension*. Little is said about what this might be in the published works. According to Poellner, who discusses the difference, explanation 'would involve a comprehension of the "quality" in the cause which "compels"' (Poellner (1995), pp. 39-40). Such talk might be construed to fit the doctrine of the ever obscure "will to power" in which one of the qualities of the "will to power" is its capacity to compel. If this is the case, then the WP Thesis can be taken to be explanatory, rather than offering more in the way of description, because it makes an appeal to qualities that compel. This can be taken to suggest that Nietzsche does have in mind an account of explanation of the sort that can be found in the philosophical tradition, viz., an appeal to the intrinsic (but not necessarily essential) natures of things, or qualities, though Nietzsche would not like to put the matter this way.[36] But all of this requires a much more substantial account of cause and causal explanation than arises out of Nietzsche's philosophical theory which allows for very little other than our own projections on the world. And Nietzsche does help himself to a real, and so non-projected, item when he appeals to some obscure "quality" of the "will to power" which is to give his explanations some bite that takes them beyond mere descriptions.

Finally, what is Nietzsche's view about the place of the norms of reason within naturalism? We have seen that Nietzsche is a naturalist, but one who proposes, not the sciences as the basis for naturalism, but rather the "will to power" (see section 10.2). And he is an extreme kind of nominalist or particularist, not with respect to objects but with respect to the one and only item in his ontology, centres of power. In being a nominalist/particularist, he is in the company of advocates of the Strong Programme and Foucault. What of norms of reason within his naturalism? We saw in section 10.4 that even though Nietzsche did not reject principles of reasoning he thought them to be fictions that depended upon other fictions, viz., objects with continuing identity. And we have also seen that Nietzsche endorses some quite standard principles of methodology (see end of section 10.4.2). But we have also noted that Nietzsche thinks that truth, or truth seeking, is an exemplification of the ascetic ideal to be avoided. Though Nietzsche does not appear to have discussed the matter, it would not be too far wrong to assume that he also thinks of our norms of reason as more of what we project onto the world in order to facilitate our survival. And when it comes to explaining why we believe what we do, even when principles of reason are employed, this is simply more in the way of the operation of the obscure "will to power" and not to be explained in terms of any independent, but naturalisable, norms.

NOTES

[1]   There is even a mounting literature on this issue. For a recent sensible account see L. Williams (2001), pp. 63-77.

[2]   See Nietzsche's *Twilight of the Idols*, section IV entitled 'How the "Real World" at Last Became a Myth: History of an Error'.

[3]   The book *The Will to Power* is a compilation of notebook entries that were unpublished in Nietzsche's lifetime. The current English translation does reflect what Nietzsche wrote, but with provided headings and organisation. It gives the most extended account of his obscure "will to power".

[4]   Much the same is said in *Twilight of the Idols*, Section III, "'Reason' in Philosophy" §5 where it is claimed that the metaphysics of our language gives rise to a 'rude fetishism', and leads to the belief 'in the ego as substance, and which *projects* its belief in the ego-substance on to all things – only thus does it create the concept "thing"'. The passage ends with the well-known remark about language: 'I fear we are not getting rid of God because we still believe in grammar'.

[5]   The doctrine of perspectivism may well render his metaphysical system incoherent. This is argued in chapter 6 §2 of Poellner (1995), which is in turn discussed in Hales and Welshon (2000) pp. 81-4.

[6]   Some commentators doubt that Nietzsche does have such a metaphysics. Whether or not he does would be part of a critical and textual evaluation not attempted here. My view is that he does have a metaphysics the central feature of which is an eschewal of most of the traditional categories of metaphysics in favour of the processes of power; there might be local teleology for individual power centres but no overall teleology. In some respects Nietzsche's position has affinities with Leibniz's metaphysics with monads understood not as substances but as centres of force or power with their own perspectives.

[7]   Nietzsche's acquaintance with the views of Boscovich seem to date from the 1870s when he was teaching at the University of Basel. See Whitlock, ' Roger J. Boscovich and Friedrich Nietzsche: A Re-Examination', in Babich and Cohen (eds.), pp. 187-201; and Poellner (1995), pp. 48-57.

[8]   See Weinert (1995) which contains a number of papers on what are laws of nature. The paper by Giere argues for a conception of physical sciences without laws. In this respect only, the position, adopted by others, is also similar to Nietzsche's. The papers by Ruby and Steinle discuss the history and sociology of the concept of laws of nature, including the claim that they are of anthropomorphic origin. But again this has no bearing on whether there are, or are not, laws of nature.

[9]   The German is also the same in the passage from *Beyond Good and Evil* §22 and in the extract from the notebooks. This is an interesting case in which a remark about calculability in the absence of laws, and the operation of the "will to power", occurs first in a work published in 1886 (though there may be earlier notebook sources than this) and then in a notebook entry in the spring of 1888. (See 14[79], p. 50, lines 4-5 of G. Colli and M. Montinari (eds.) *Nietzsche Werke*, volume VIII3 Berlin, Walter de Gruyter.) It betokens a continuity of thought about such matters that might be downplayed by those who give less credence to notebook entries than is given here, however sketchy and incompletely thought through the notebook entries may be.

[10]   Some other commentators feel an unresolved tension in Nietzsche's views about real causation and laws of nature on the one hand, and on the other, claims about calculability; see Hales and Welshon (2000) chapter 4, especially pp. 94-5. What Nietzsche says about laws in his published works would have difficulty in passing muster these days, given the advances that have been made in our understanding of laws and causality. But this is not to say that all

problems have been solved; rather it is to say that there is clarity now where in Nietzsche there remains unclarity.

[11]   See Dawkins (1976) p. 48 and p. 95. However in Dawkins (1983) there are some remarks about genes that do have a Nietzschean flavour, especially when he talks of genes exerting power: 'We are going to use the metaphor of power. An active replicator is a chunk of a genome that, when compared to its alleles, exerts phenotypic power over its world, such that its frequency increases of decreases relative to that of its alleles' (p. 91). Elsewhere Dawkins speaks of his having built up an '...image of a turmoil of selfish replecators, battling for their own survival at the expense of their alleles ...' (p. 250). However note that Dawkins talks of the "metaphor of power". Presumably he would want to cash out the talk of power in terms of the underlying chemistry. And Nietzsche would resist Dawkins' talk of selfishness; rather powers are intrinsically directed at, and opposed to, one another.

[12]   In *The Genealogy of Morals* Preface §4, Nietzsche compares Paul Reé's account of the genealogy of morals negatively with his own genealogy and hopes 'to replace an improbability with something more probable'. This section shows that Nietzsche clearly thought that often one interpretation could be judged better than another, and that his throw-away line at the end of *Beyond Good and Evil* §22 was not to be understood to entail that all interpretations were on a par. See the end of section 10.4.2.

[13]   The topic of the identity of objects is an old one dating from Ancient Greek discussions of the identity of the ship *Theseus* cited by Plutarch and discussed by Hobbes. Locke (of whom Nietzsche cites approvingly Schelling's remark 'je méprise [despise] Locke' (*Beyond Good and Evil*, §252)) has a better discussion of issues concerning object and personal identity in *An Essay Concerning Human Understanding* Book II, Chapter 27, than any found in Nietzsche (who tended to make pronouncements rather than give arguments).

[14]   On the different kinds of mereological essentialism see Peter Simons (1987), especially index under 'essentialism, mereological'; see also Wiggins (1980) chapters 4 and 6.

[15]   See especially *The Will To Power*, Book Three, I, sections 3 and 5.

[16]   See *The Will to Power* §640-58 which has been given the title 'The Organic Process'. Nietzsche extends evolution to consciousness saying that 'it is our relation with the "outer world" that has evolved it' (*ibid.*, §524).

[17]   The utility and necessity of fictions in Nietzsche is stunningly presented by an early interpreter of Nietzsche, Hans Vaihinger, in his philosophy of "as if" in a book of that title.

[18]   That Nietzsche can claim both that we have massively false beliefs, and that these are necessary for thought and for survival, is also challenged in Poellner (1995) pp. 140-51. He calls it Nietzsche's 'argument from utility' and judges it to be a failure. Like many evolutionary epistemologists, Nietzsche recognises that our beliefs have massive utility for life; but unlike them Nietzsche also insist that the beliefs are false. He does not see that from their utility an argument can be mounted to the effects that our beliefs are more likely true than false, or that they have high verisimilitude.

[19]   Expressed more carefully, this is not intended to rule out cases where false beliefs are not also, in special circumstances, conducive to the preservation of life. Birds do not see large panes of glass, especially when the glass reflects bush opposite to it. Birds then fly into the glass, often killing themselves, while believing that the reflection of bush in the glass indicates bush ahead. To prevent this, silhouettes of large flying birds are often placed on the glass. Other birds avoid what they believe to be large birds, and in so doing avoid the fatal glass. Through being tricked by a false belief, their life is preserved when it might not have been. But such avoidance behaviour depends on a prior behaviour of generally avoiding larger birds in the first place; and this can have survival value, even in the case of a false positive such as this.

[20]   For other passages expressing a similar view, see *The Will to Power* §483, §492, §497, §532, §584 and §609.

[21] The above contradiction for the reflective Nietzschean seems to follow directly from what Nietzsche says. He does not note this problem, nor does it appear that his commentators do. It is not unresolvable; but some work needs to be done to remove it. It is this work that is missing in the literature. One suggestion would be to distinguish, as Carnap does (see Carnap (1956) 'Empiricism, Semantics and Ontology'), between questions *within* a framework of belief and questions *about* the framework. Alternatively one could adopt Mackie's notion of an "error theory" (Mackie (1977), chapter 1) concerning our belief in substantive objects.

[22] The point that nominalists who do not admit sameness of type have difficulty in even expressing the rules of inference is made by Putnam in section II of 'Philosophy of Logic" reprinted in Putnam (1979) chapter 20, pp. 323-57.

[23] Nietzsche does not discuss the Law of Excluded Middle explicitly, but he does discuss the Law of Non-Contradiction in *The Will to Power*, §516. He does not seem to reject it. Rather he wrongly psychologises it and downgrades it from a 'necessity' to a psychological inability on our part to affirm and deny the same thing at the same time. Also there is a reiteration of issues to do with self-identicals that are not strictly relevant. In line with his fictionalism Nietzsche says that the laws of logic 'contain no *criterion of truth*, but an *imperative* concerning that which *should* count as true'. This reiterates his claim that the laws of logic and their alleged assumptions are things we construct and they are necessary for life.

[24] This matter also arose in section 10.4.1 concerning the truth of our beliefs being a much better explainer of our success in surviving than the falsity of our beliefs, and the more careful statement of this in terms of probability of success.

[25] This is a view adopted in an earlier paper of mine (see Nola, (1987), especially section III). It took its cue from remarks like: 'But what is truth? Perhaps a kind of belief that has become a condition for life?' *(The Will to Power*, §532) 'Truth is the kind of error without which a certain species of life could not exist. The value for life is ultimately decisive' (*ibid.*, §493).

[26] For some remedies for this problem see Tanesini (1995) and Kirkham (1992), pp. 29-30 and chapter 10. Kirkham also discusses the above difficulty for the performative theory (*ibid.*, pp. 309-10), namely that one can signal agreement dishonestly. Also any agreement must be "appropriate" in the sense that certain conditions in the world must obtain before one can *properly* signal agreement. Thus it is only *appropriate* to signal agreement to the sentence 'Nietzsche has a big moustache' if it is the case that Nietzsche has a big moustache. But a serious lack in the performative theory is an account of the link between a sentence and the fulfilment, in the world, of the semantic conditions specified by the sentence. For those who do not accept the performative theory as the final world about truth, this is where the real issues concerning any theory of truth must lie.

[27] It might also fit with the later Foucault's forays into the notion of "truth-telling" (or parrhesia) as his commitment to the 'power/knowledge" doctrine began to subside.

[28] This may need qualification on empirical grounds. It may be psychologically inaccurate about the responses people can have when confined in terror, since they often come to dislike their fellow inmates while coming to terms with life with their guards.

[29] 'Truth has to be fought for every step of the way, almost everything else dear to our hearts, on which our love and our trust in life depend, has had to be sacrificed to it. Greatness of soul is needed for it; the service of truth is the hardest service. For what does it mean to be *honest* in intellectual things? That one is stern towards one's heart ...' (*The Anti-Christ*, §50). This occurs in a section in which Nietzsche rejects belief in the sense of a conviction about, say, the existence of God, in which the potency of the belief is taken as an indication of truth. He also rejects feelings of pleasure as a criterion of truth; hence his comment about being stern about what one's heart tells one. If truth involves asceticism then Nietzsche himself is deeply ascetic here.

[30]   Here the translation by Kaufmann is given because it renders the German more accurately. He translates "Komödianten' as 'comedians', whereas the Smith translation, which has been followed so far, has the less graphic 'play-actors'.

[31]   Here Smith alludes to the aphorism from *Thus Spake Zarathustra* with which Nietzsche prefaces the 'Third Essay' of *The Genealogy*.

[32]   The German for 'origin' is 'Herkunft'. This is an origin which can also be concerned with descent, extraction and pedigree, or lack of it.

[33]   Some of the contrasting values of master and slave morality can be found in *Human, All Too Human* §44-50 and *Beyond Good and Evil* §260-1.

[34]   On the matter of the early Anglo-Saxons, their conversion to Christianity and the problem of wyrd (fate), see the section entitled 'Destiny' pp. 302-3 in Crossley-Holland (ed.) (1984). See also the 'Introduction' to Heaney (1999) in which the seemingly Christian world of the author of the poem Beowulf is contrasted with the poem's preoccupation with pagan wyrd.

[35]   Similar more extended remarks can be found in *Daybreak,* §205.

[36]   This account of explanation adopted by Nietzsche leaves out some ingredients, including his insistence that we only explain the unfamiliar by means of the familiar. This is taken to mean, in the context of talk of powers and forces, that we make an analogy based on the inner compulsions we feel, using once more ourselves as a model: 'It will do to consider science as an attempt to humanize things as faithfully as possible: as we describe things and their one-after-another, we learn how to describe ourselves more and more precisely' (*The Gay Science,* §112). Luckily not all modern scientific explanation need adopt this stricture.

CHAPTER 11

EPILOGUE

The causes of our beliefs are multifarious. For some beliefs held by some people at some time, there may well be good reasons. This is a quite weak claim; anything stronger would require an epidemiological investigation into the occurrence of rational belief, or knowledge. However, in explaining why some scientist came to hold the beliefs they did, we can rightly appeal to principles of logic, epistemology and methodology (the structure of these explanations can be found in sections 5.4 and 6.1). However it would be wrong to make the strong claim that for all explanations of all scientific beliefs held by all people at all time, some rationality model is applicable. On the contrary, there are multifarious ways in which people can come to accept, or believe or entertain some proposition. As has been argued, claims about knowledge must be intrinsically rationally based, otherwise they cannot be knowledge. Knowledge is a normative critical notion. However it must be allowed that what one person holds as an item of knowledge, another will hold simply as a belief, often on grounds that have little to do with normativity.

As has been noted, sociologists of "knowledge", and others, ignore the normative, critical dimension of knowledge. Moreover they wish to make claims about the causes of all beliefs held by all people at all times, without admitting that there might be exceptions to these quite general claims (Mannheim, and perhaps Marx, do attempt to restrict their claims). The social cause models they advocate have little to do with rational models of explanation. Or if they do admit that there might be something "rational" about the causes of belief, especially in science, then the kind rationality is not autonomous but is alleged to turn on social causes once again (for example, the sociologists' appeal to a communitarian solution to Wittgenstein's doctrine of rule following).

For some specified body of belief, Marx and Marxists would trace its connection to something social, the forces and relations of production that constitute the materialist basis for history. (The links are often quite unspecific and are variously said to cause, condition, determine, or be functionally appropriate). Mannheim would trace its connection to what he calls *Seinsverbundenheit des Wissens*, the existential determination of knowledge being understood to be implicitly social. Again, contemporary advocates of the sociology of knowledge, especially in the guise of the Strong

Programme, will point to a wide typology of social factors that are omnipresent in all cases of the cause of scientific belief (in conjunction with non-social factors that cannot by themselves explain the variation in belief). For Foucault's doctrine of "power/knowledge", there is a causal inter-linking of power with knowledge, and of knowledge with power. (He never questioned his assumption that there must be some such linkages, but their precise nature often eluded him.) Finally, Nietzsche held that all our different bodies of belief were linked in some way to some aspect of the workings of the "will to power", whether it be a purely metaphysically understood aspect of the "will to power", or biological aspects of power, or power as expressed in various human psychological drives.

All of these theories (and many others not mentioned here) have a common character. They appeal to what are clearly non-rational factors to explain belief. As such, the causal explanatory models they must adopt remain quite at variance with rational explanation models. Of course, the various theoreticians disagree amongst themselves as to what non-rational factors are causally efficacious, from Marxist forces and relations of production, to social and power relations of various sorts. But this is merely an in-house difference that should not mask the large difference that sets all of them apart from those who advocate the use of rationality models of explanation at least some of the time. All their doctrines are universal in character, admitting no role, however small, for principles of rationality as explainers. Or if they do, ultimately a social story is to be told about the status and authority of the principles themselves.

If the non-rational models of the explanation of scientific belief are alleged to be successful, then this puts the rational models of explanation in jeopardy. For not both kinds of radically different explanations can be correct. The rational models can then be unmasked as mere rationalisations at best, or, at worst, they are exposed as ideological foils covering up what are the real causes of belief. Such is the unmasking that many postmodernists provide for what they would regard as enlightenment excesses in over-promoting the role of rationality in our beliefs. And in this they are sometimes correct. Some have seen in Marx or Nietzsche, or some other unmasker of the foibles of modernity, an important lesson to be learned about the excesses of rationality. But often this ploy is simply overdone, the case of scientific beliefs and knowledge being a good example.

As has been shown, all the writers discussed in Parts II, II and IV are naturalists of one sort or another. And they appeal to purely naturalistic items in their explanations of belief. But their naturalism is usually bought at the expense of any account of the normativity of rationality, or the justificatory methods of knowledge or science, even though they make covert appeal to such normativity. As a result they are inclined to jettison the normative notion of rationality and the critical role it can play. Instead they offer us a display of the workings of non-rational items such as power, or the will to power, or whatever. And this is thought to be unmasking enough. One consequence is that normative and critical notions such as knowledge, or the norms of method, are totally lost. In particular, there is no notion of knowledge available at all. On the models of belief formation they adopt, there is no room for the idea that our beliefs even remotely track the state of affairs they are about; their causal origins lie in social or power relations quite unconnected with what the belief is about. Nor is there any account of the justificatory reasons for the belief, or of the reliability for the truth of the processes that form beliefs. Their naturalising project is an unnatural one.

Given that there may be non-rational social causal, or other such models for the explanation of belief, the stage is set for the debunking of rational explanations that appeal to methodological and epistemic principles of reason. In the case of the Strong Programme the debunking of rational explanations is simply proclaimed through the Symmetry Tenet; it bids us adopt the same kind of social cause model for all beliefs regardless of their epistemic status. And this is taken to exclude any role for rationality explanations. What little argument for this is explored in section 6.3 and found wanting. In the case of Foucault there is little recognition of the role of rationality explanations given the prominence of his "power/knowledge" doctrine (though belatedly towards the end of his life he did come to recognise that such considerations might have a place, but says little more than this).

The case against Nietzschean debunking is slightly different. Take, for example his claim to have shown that the values of slave morality, such as pity or altruism, arise from resentment. Add to this central claim of his genealogy of morals the further claims that resentment is a "bad" thing (for whatever Nietzschean reason), and that anything that arises from such a negative motivation must itself also be highly questionable. From these premises the conclusion is drawn that the values of slave morality, such as pity or altruism, are to be debunked. Is this the unmasking of our prevailing morality that casts all its evaluations into oblivion? There are several objects to this.

The first turns on how good an explanation the operation of the "will to power" as resentment offers for why we came to be, say, altruistic (supposing that we are to some large extent). Here the question is not so much the value of altruism, that is, whether it is a good or a bad thing. Rather the matter is just how good an explanation resentment can offer of how altruism came about, and why it still prevails as a value. And this is not a matter that has immediately to do with values; rather, it has to do with an empirically based explanation of why we adopt altruism, or believe in altruism. Nietzsche's genealogical explanation is psycho-social in its appeal to the workings of resentment in a culture. However it has a strong explanatory rival in accounts based in evolutionary psychology (sociobiology) that cover not only animal altruistic behaviour but human altruistic behaviour as well. Just how good are these two rival hypotheses in explaining at least the prevalence of altruism? Though the case cannot be argued here, the explanation based in evolutionary psychology shows Nietzsche's psycho-social explanation to be rather distant unsuccessful competitor.

The second objection turns on whether the negative connotations of resentment are passed on to what it allegedly gives rise to, viz., altruism. But it can be asked whether resentment is always a bad thing? The answer must be 'no'. There may well be cases where resentment would be appropriate if it were to arise, for example, amongst the inmates of a concentration camp towards their guards. So we cannot infer in all cases that resentment is necessarily a "bad" thing. But, thirdly, assuming that it is, does it follow that what it gives rise to is thereby "bad"? There is a genetic fallacy here. Even if resentment were to be a motive for adopting altruism as a value, it does not follow that there is no justification for altruism on quite other grounds. And philosophers have been at pains to show that there is something worthy about altruism on other grounds. Thus the Nietzschean debunking move against what he dubs the slave morality of altruism is based on both dubious premises and argument. There are better rival explanations of the emergence of altruism, and viable philosophical theories of why it is often a good thing.

Similar kinds of considerations can be extended to other debunking moves. For demaskers of modernity to get their debunking moves about our beliefs under way, two models for the explanation of our beliefs need to be in contention. One is a rationality model and the other is some form of non-rational model such as the social cause model. Some considerations are then given for adopting the second kind of model and rejecting the first. And in some cases the debunkers can be right. Our illusions are exposed in the course of accepting the debunking move. In some respects Foucault and Nietzsche are masters of the debunking move. But they are too cleaver by half a good deal of the time. They often rely on the very models of explanation that they try to discredit when they propose their debunking move. After all why do we think that, in some cases, the debunkers have a good explanation, and so are right? Or why do we think that they have an unsatisfactory explanation? Either way, reason has to come to the rescue.

REFERENCES

Adorno, T. (*et. al.*) (1976) *The Positivist Dispute in German Sociology*, trans, by G. Adey and D. Frisby, London, Heinemann.

Althusser, L. (1969) *For Marx*, Harmondsworth, Penguin.

Anderson, A. and Belnap, N. (1975) *Entailment: The Logic of Relevance and Necessity, Volume I*, Princeton NJ, Princeton University Press.

Anstey, P. (2000) *The Philosophy of Robert Boyle*, London, Routledge.

Armstrong, D. (1968) *A Materialist Theory of Mind*, London, Routledge and Kegan Paul.

Armstrong, D. (1973) *Belief, Truth and Knowledge*, Cambridge, Cambridge University Press.

Armstrong, D. (1978) *Nominalism & Realism: Universals & Scientific Realism Volume I*, Cambridge, Cambridge University Press.

Armstrong, D. (1981), 'Naturalism, Materialism and First Philosophy', in *The Nature of Mind*, Brighton, Harvester Press, pp. 149-65.

Austin, J. (1962) *How to Do Things with Words*, Oxford, Clarendon.

Babich, B. and Cohen, R. (eds.) (1999), *Nietzsche, Epistemology and Philosophy of Science: Nietzsche and the Sciences II*, Dordrecht, Kluwer Academic Publishers.

Bacon, F. (1973) *The Advancement of Learning*, London, J. M. Dent (Everyman Library).

Bacon, J. (1986) 'Supervenience, Necessary Coextension, and Reducibility', *Philosophical Studies* **49**, 163-76.

Baker, G. and Hacker, P. (1984) *Scepticism, Rules and Language*, Oxford, Backwell.

Baker, G. and Hacker, P. (1985) *Wittgenstein: Rules, Grammar and Necessity: Volume 2 of an Analytical Commentary on the Philosophical Investigations,* Oxford, Backwell.

Barnes, B. (1982) *T. S. Kuhn and Social Science*, London, Macmillan.

Barnes, B., Bloor, D. and Henry, J. (1996) *Scientific Knowledge: A Sociological Analysis,* Chicago, The University of Chicago Press.

Barnes, B. and Dolby, R. (1970) 'The Scientific Ethos: A Deviant Viewpoint', *Archives Europeenes de Sociologie* **11**, 3-25.

Barnes, J. (1979), *The Presocratic Philosophers: Volume 1: Thales to Zeno*, London, Routledge and Kegan Paul.

Ben-David, J. (1991) *Scientific Growth: Essays on the Social Organisation and Ethos of Science*, Berkeley, University of California Press.

Bennett, J. (ed.) (1965), *Experiments in Plant Hybrydisation: Gregor Mendel*, Edinburgh and London, Oliver and Boyd.

Bernauer, J. and Rasmussen, D. (eds.) (1988), *The Final Foucault*, Cambridge MA, MIT Press.

Blackburn, S. (1984) *Spreading the Word*, Oxford, Clarendon.

Blackburn, S. (1993) *Essays in Quasi-Realism*, New York, Oxford University Press.

Bloor, D. (1981) 'The Strengths of the Strong Programme', *Philosophy of the Social Sciences* **11**, 199-213.

Bloor, D. (1982) 'Durkheim and Mauss Revisited: Classification and the Sociology of Knowledge', *Studies in History and Philosophy of Science* **13**, 267-304.

Bloor, D. (1983) *Wittgenstein: A Social Theory of Knowledge,* London, Macmillan.

Bloor, D. (1991) *Knowledge and Social Imagery*, Chicago, The University of Chicago Press (second edition; first edition 1976, London Routledge and Kegan Paul).

Bloor, D. (1991a) 'Ordinary Human Inference as Material for the Sociology of Knowledge', *Social Studies of Science* **21**, 129-39.

Bloor, D. (1997) *Wittgenstein, Rules and Institutions*, London, Routledge.

Bloor, D (1997a) 'The Conservative Constructivist', *History of the Human Sciences* **10** (#1), 123-5.

Bloor, D. (1999) 'Anti-Latour' *Studies in History and Philosophy of Science* **30A**, 81-112.

Bluck, R. (1961) *Plato's Meno*, Cambridge, Cambridge University Press.

Boghossian, P. (1989) 'The Rule-Following Considerations', *Mind* **98**, pp. 507-49.

Borradori, G., (1994) *The American Philosopher: Conversations with Quine, Davidson, Putnam, Nozick, Danto, Rorty, Cavell, MacIntyre and Kuhn*, The University of Chicago Press, Chicago.

Braddon-Mitchell, D. and Jacksdon, F. (1996) *Philosophy of Mind and Cognition*, Oxford, Blackwell.

Brandom, R. (2000) *Articulating Reasons: An Introduction to Inferentialism*, Cambridge MA, Harvard University Press.

Brante, T., Fuller, S. and Lynch W. (eds.) (1993) *Controversial Science: From Content to Contention*, Albany NY, State University of New York Press.

Brentano, F. (1973), *Psychology from an Empirical Standpoint*, London, Routledge and Kegan Paul; first German publication in 1874.

Brown, J. (1989) *The Rational and the Social*, London, Routledge.

Brown, J. (2001) *Who Rules in Science?: An Opinionated Guide to the Science Wars*, Cambridge MA, Harvard University Press.

Bulmer, R. (1967) 'Why is the Cassowary Not a Bird? A Problem of Zoological Taxonomy Among the Karam of the New Guinea Highlands', *Man* (New Series) **2**, 5-25.

Bunge, M. (1991) 'A Critical Examination of the New Sociology of Science Part 1', *Philosophy of the Social Sciences* **21**, 524-60.

Bunge, M. (1992) 'A Critical Examination of the New Sociology of Science Part 2', *Philosophy of the Social Sciences* **22**, 46-76.

Burchell, G., Gordon, C. and Miller, P. (eds.), *The Foucault Effect: Studies in Governmentality*, Chicago, University of Chicago Press.

Burnyeat, M. (1976), 'Protagoras and Self-refutation in Plato's *Theaetetus*', *Philosophical Review* **85**, 172-95.

Cahoone, L. (ed.) (1996) *From Modernism to Postmodernism: An Anthology*, Oxford, Blackwell.

Carnap, R. (1956) *Meaning and Necessity; A Study in Semantics and Modal Logic*, Chicago, University of Chicago Press.

Cartwright, N. (1983) *How the Laws of Physics Lie*, Oxford, Clarendon.

Cartwright, N. (1999) *The Dappled World: A Study of the Boundaries of Science*, Cambridge, Cambridge University Press.

Coady, C. (1992) *Testimony: A Philosophical Study*, Oxford, Clarendon.

Cohen, G. (1978) *Karl Marx's Theory of History: A Defence*, Oxford, Clarendon.

Cohen, L. (1992) *An Essay on Belief and Acceptance*, Oxford, Clarendon.

Collingwood, R. (1972) *Essay on Metaphysics*, Chicago, Gateway.

Collingwood, R. (1994) *The Idea of History* Oxford, Oxford University Press (revised edition).

Collins, H. (1981) 'What is TRASP?: The Radical Programme as a Methodological Imperative' *Philosophy of the Social Sciences* **11**, 215-224.

Collins. H. (1985) *Changing Order: Replication and Induction in Scientific Practice*, London, Sage.

Crossley-Holland, K. (ed.) (1984) *The Anglo-Saxon World: An Anthology*, Oxford, Oxford University Press.

Currie, G. (1980) 'The Role of Normative Assumptions in Historical Explanation', *Philosophy of Science* **47**, 456-73.

D'Amico, R. (1999) *Contemporary Continental Philosophy*, Boulder, Westview.

Day, J. (ed.) (1994) *Plato's Meno in Focus*, London, Routledge.

Davidson, D. (1980) *Essays on Actions and Events*, Oxford, Clarendon.

Dawkins, R. (1976) *The Selfish Gene*, London, Paladin.

Dawkins, R. (1983) *The Extended Phenotype*, Oxford, Oxford University Press.

Dennett, D. (1995), *Darwin's Dangerous Idea: Evolution and the Meanings of Life*, London, Penguin.

Descartes, R. (1983), *Principles of Philosophy*, Dordrecht, Reidel.

Devitt, M. (1991) *Realism and Truth*, Princeton NJ, Princeton University Press (second edition).

Donavan, A., Laudan, L. and Laudan, R., (eds.) (1992) *Scrutinizing Science*, Baltimore, The Johns Hopkins Press, second edition.

Dretske, F. (1969) *Seeing and Knowing*, London, Routledge and Kegan Paul.

Dretske, F. (1981) *Knowledge and the Flow of Information*, Oxford, Blackwell.

Dreyfus H. and Rabinow, P. (1983) *Michel Foucault: Beyond Structuralism and Hermeneutics*, Chicago, The University of Chicago Press (second edition).

Dupré, J. (1993) *The Disorder of Things*, Cambridge MA, Harvard University Press.

Durkheim, E. and Mauss, M. (1963) *Primitive Classification*, London, Cohen and West.

Earman, J. (1992) *Bayes or Bust? A Critical Examination of Bayesian Confirmation Theory*, Cambridge MA, MIT Press.

Eddington, A. (1929) *The Nature of the Physical World*, Cambridge, Cambridge University Press.

Editors of *Lingua Franca* (eds.) (2000) *The Sokal Hoax: The Sham that Shook the Academy*, Lincoln, University of Nebraska Press.

Ellis, B. (1996) 'Natural Kinds and Natural Kind Reasoning', in P. Riggs (ed.) *Natural Kinds, Laws of Nature and Scientific Methodology*, Dordrecht, Kluwer Academic Publishers, pp. 11-28.

Ellis, B. (2001) *Scientific Essentialism*, Cambridge, Cambridge University Press.

Elster, J. (1985) *Making Sense of Marx*, Cambridge, Cambridge University Press.

Etchemendy, J. (1990) *The Concept of Logical Consequence*, Cambridge MA, Harvard University Press.

Faubion, J. (ed.) (1998) *Michel Foucault: Aesthetics, Methodology and Epistemology: Essential Works of Foucault 1954-84: Volume 2*, New York, The New Press.

Faubion, J. (ed.) (2000) Michel Foucault: Power: Essential Works of Foucault 1954-84: Volume 3, New York, The New Press.

Feyerabend, P. (1962), 'Explanation, Reduction and Empiricism', in Herbert Feigl and Grover Maxwell (eds.), *Minnesota Studies in the Philosophy of Science: Volume III: Scientific Explanation, Space and Time*, Minneapolis, University of Minnesota Press, pp. 28-97.

Feyerabend, P. (1975) *Against Method*, London, NLB (first edition).

Feyerabend, P. (1978) *Science in a Free Society,* London, NLB.

Feyerabend, P. (1981) *Realism, Rationalism and Method: Philosophical Papers Volume I*, Cambridge, Cambridge University Press.

Feyerabend, P. (1987) *Farewell to Reason*, London, Verso.

Feyerabend, P. (1993) *Against Method*, London, NLB (third edition).

Feyerabend, P. (1995) *Killing Time*, Chicago, The University of Chicago Press.

Field, H. (1972) 'Tarski's Theory of Truth', *Journal of Philosophy* **69**, 347-375.

Fisher, R. (1926) *The Design of Experiments*, Edinburgh, Oliver and Boyd.

Fodor, J. (1983) *The Modularity of Mind*, Cambridge MA, The MIT Press.

Fodor, J. (1984) 'Observation Reconsidered', *Philosophy of Science* **51**, 23-43.

Fogelin, R. (1994) *Pyrrhonian Reflections on Knowledge and Justification*, New York, Oxford University Press.

Foley, R. (1994) 'Quine and Naturalized Epistemology', *Midwest Studies in Philosophy: Philosophical Naturalism* **XIX**, 243-60.

Forman, P. 1971 'Weimar Culture, Causality, and Quantum Theory 1918-27: Adaptation by German Physicists and Mathematicians to a Hostile Intellectual Environment', in R. McCormmach (ed) *Historical Studies in the Physical Sciences* **3** Philadelphia, University of Pennsylvania Press, pp. 1-116.

Foucault, M. (1970), *The Order of Things: An Archaeology of the Human Sciences*, London, Tavistock.

Foucault, M. (1972) *The Archaeology of Knowledge and The Discourse on Language*, New York, Pantheon.

Foucault, M. (1979), *Discipline and Punish: The Birth of the Prison*, Harmondsworth, Penguin.

Foucault, M. (1981) *The History of Sexuality: Volume 1*, Harmondsworth, Penguin Books.

Foucault, M. (1983), 'The Subject and Power', in Dreyfus and Rabinow (1983), pp. 208-26..

Foucault, M. and Ewald, F. (1984), 'The Regard for Truth'; an interview translated by Paul Patton, *Art and Text* **16**, p. 31.

Fox, J. (1988) 'It's All in the Day's Work: A Study of the Ethnomethodology of Science', in Nola (ed.) (1988).

Frege, G. (1977) 'Thoughts', in *Logical Investigations* P. T. Geach (ed.) Oxford, Blackwell, pp. 1-30.

Freundlieb, D. (1994) 'Foucault's Theory of Discourse and Human Agency' in C. Jones and R. Porter (eds.), *Reassessing Foucault: Power, Medicine and the Body,* London, Routledge.

Friedman, M. (1998) 'On the Sociology of Scientific Knowledge and its Philosophical Agenda', *Studies in History and Philosophy of Science* **29A**, 239-271.

Gallop, D. (1990) *Aristotle on Sleep and Dreams*, Peterborough, Broadview Press.

Garfinkel, H. (1967) *Studies in Ethnomethodology*, Englewood Cliffs NJ, Prentice-Hall.

Garfinkel, H., Lynch, M., and Livingstone, E. (1981) 'The Work of a Discovering Science Constructed with Materials from the Optically Discovered Pulsar', *Philosophy of Social Science* **11**, 131-158.

Garver, N. (1996) 'Philosophy as Grammar', *The Cambridge Companion to Wittgenstein*, H. Sluga and D. Stern (eds.) Cambridge, Cambridge University Press, pp. 139-70.

Gettier, E. (1963) 'Is Justified True Belief Knowledge?', *Analysis* **23**, 121-3.

Ghins, M. (2003, forthcoming) 'Thomas Kuhn on the Existence of the World', *International Studies in the Philosophy of Science*.

Gibbard, A. (1990) *Wise Choices, Apt Feelings*, Cambridge MA, Harvard University Press.

Giere, R. (1988) *Explaining Science: A Cognitive Approach*, Chicago, The University of Chicago Press.

Gilbert, M. (1989) *On Social Facts*, London, Routledge.

Gingerich, O. (1975) '"Crisis" versus Aesthetic in the Copernican Revolution' in A. Beer and K. Strand (eds) *Copernicus Yesterday and Today:Vistas in Astronomy,* volume 17, Oxford, Pergamon Press, pp. 85-95.

Glymour, C. (1980) *Theory and Evidence*, Princeton NJ, Princeton University Press.

Glymour, C., Scheines, R., Spirtes, P., and Kelly, K. (1987) *Discovering Causal Structure,* New York, Academic Press.

Goldfarb, W. (1985) 'Kripke on Wittgenstein on Rules', *The Journal of Philosophy* **82**, 471-88.

Goldman, A. (1986) *Epistemology and Cognition*, Cambridge MA, Harvard University Press.

Goldman, A. (1992), *Liasons: Philosophy Meets the Cognitve and Social Sciences*, Cambridge MA, The MIT Press Bradford Book.

Goldman, A. (1999) *Knowledge in a Social World*, Oxford, Clarendon.

Goodman, N, (1955) *Fact, Fiction and Forecast,* Cambridge MA, Harvard University Press.

Gordon, C. (ed.) (1980) *Michel Foucault: Power/Knowledge: Selected Interviews and Other Writings*, Brighton, Harvester Press.

Greenberg, D. (1967) *The Politics of Pure Science*, New York, The New American Library (Plume Book).

Gross, P. and Levitt N. (1994) *Higher Superstition: The Academic left and Its Quarrels with Science*, Baltimore, The Johns Hopkins University Press.

Guthrie, W. (1955) *Plato: Protagoras and Meno*, translator, Harmondsworth, Penguin.

Gutting, G. (ed.), (1980) *Paradigms and Revolutions*, Notre Dame IN, University of Notre Dame Press.

Haack, S. (1978) *Philosophy of Logics*, Cambridge, Cambridge University Press.

Hacker, P. (1972) *Insight and Illusion* Oxford, Oxford University Press.

Hackforth, R. (1952), *Plato's Phaedrus*, Cambridge, Cambridge University Press.

Hacking, I. (1999) *The Social Construction of What?* Cambridge MA, Harvard University Press.

Hahn, L. and Schilpp, A. (eds) (1986) *The Philosophy of W. V. Quine*, La Salle IL, Open Court.

Hales, S. and Welshon, R. (2000), *Nietzsche's Perspectivism*, Urbana and Chicago, University of Illinois Press.

Hampton, J. (1998) *The Authority of Reason*, Cambridge, Cambridge University Press.

Harman, G. (1968) 'Knowledge, Inference and Explanation', *American Philosophical Quarterly* **5**, 164-73.

Harper, P. (ed.) (1991), *Huntington's Disease,* London, W. B. Saunders.

Hawthorn, G. (1991) *Plausible Worlds: Possiblity and Understanding in History and the Social Sciences*, Cambridge, Cambridge University Press.

Hayden, M. (1981), *Huntington's Chorea,* Berlin, Springer-Verlag.

Heaney, S. (1999) *Beowulf*, London, Faber and Faber.

Hempel, C. (1965) *Aspects of Scientific Explanation and Other Essays in the Philosophy of Science*, New York, The Free Press.

Hempel, C. (1983) 'Valuation and Objectivity in Science', in R. S. Cohen and L. Laudan (eds.) *Physics, Philosophy and Psychoanalysis,* Dordrecht, Reidel, pp. 73-100.

Hendry, J. (1980) 'Weimar Culture and Quantum Causality', *History of Science* **18**, 155-180.

Hesse, M. (1980) 'The Strong Thesis of Sociology of Science', in *Revolutions and Reconstructions in the Philosophy of Science,* Brighton, The Harvester Press, pp. 29-60.

Hessen, B. (1931) 'The Social and Economic Roots of Newton's *Principia*', in N. I. Bukharin *et. al.* (eds) *Science at the Cross Roads*, London, Frank Cass (second edition 1971), pp. 149-212.

Hintikka, J. (1962) *Knowledge and Belief*, Ithaca NY, Cornell University Press.

Hintikka, M. and Hintikka, J. (1986) *Investigating Wittgenstein*, Oxford, Blackwell.

Hobson, J. A. (1988) *The Dreaming Brain* New York, Basic Books.

Hollis, M. and Lukes, S. (eds.) (1982) *Rationality and Relativism* Oxford, Blackwell.

Horwich, P. (ed.) (1993) *World Changes: Thomas Kuhn and the Nature of Science*, Cambridge MA, The MIT Press Bradford Book.

Howson, C. (1976) *Method and Appraisal in the Physical Sciences*, Cambridge, Cambridge University Press.

Howson, C. and Urbach, P. (1993) *Scientific Reasoning: The Bayesian Approach*, La Salle IL., Open Court (second edition).

Hunter, M. (1990) 'Science and Heterodoxy: An Early Modern Problem Considered', in D. Lindberg and R. Westman (eds) *Reappraisals of the Scientific Revolution*, Cambridge, Cambridge University Press, pp. 437-60.

Hunter, M. (ed.) (1994) *Robert Boyle Reconsidered*, Cambridge, Cambridge University Press.

Hume, D. (1985) *Essays Moral, Political and Literary*, Liberty Press, Indianapolis.

Hyde, D. and Priest, G. (eds.) (2000) *Sociative Logics and Their Applications: Essays by the Late Richard Sylvan*, Aldershot, Ashgate.

Irzik, G. and Grünberg, T. (1995) 'Carnap and Kuhn: Arch Enemies or Close Allies', *The British Journal for the Philosophy of Science* **46**, 285-307.

Irzik, G. and Grünberg, T. (1998) 'Whorfian Variations on Kantian Themes: Kuhn's Linguistic Turn', *Studies in History and Philosophy of Science* **29**, 207-221.

Jackson, F. (1998) *From Metaphysics to Ethics: A Defence of Conceptual Analysis*, Oxford, Oxford University Press.

Jacob, J. R. (1972) 'The Ideological Origins of Robert Boyle's Natural Philosophy' *Journal of European Studies* **2**, 1-21.

Jacob, M. C. (1976) *The Newtonians and the English Revolution 1689-1720*, Hassocks, The Harvester Press.

Jacob, M. C. (1995) 'Reflections on the Ideological Meanings of Western Science from Boyle and Newton to the Postmodernists', *History of Science* **33**, 333-57.

Jeffrey, R. (1983) *The Logic of Decision*, New York McGraw-Hill (second edition).

Kahneman, D. Slovic, P. and Tversky, A. (eds.) (1982) *Judgement Under Uncertainty: Heuristics and Biases*, Cambridge, Cambridge University Press.

Kim, J. (1993) *Supervenience and Mind*, Cambridge, Cambridge University Press.

Kirkham, R. (1992) *Theories of Truth: A Critical Introduction*, Cambridge MA, MIT Press.

Koertge, N. (ed.) (1998) *A House Built on Sand: Exposing Postmodernist Myths About Science*, New York, Oxford University Press.

Kornblith, H. (1993), *Inductive Inference and its Natural Ground*, Cambridge MA, The MIT Press.

Kornblith, H. (1994) *Natualizing Epistemology*, Cambridge MA, MIT Press, second edition.

Kripke, S. (1980) *Naming and Necessity*, Oxford, Blackwell.

Kripke, S. (1982) *Wittgenstein on Rules and Private Language*, Oxford, Blackwell.

Kritzman, L. (ed.) (1998), *Michel Foucault: Politics Philosophy Culture: Interviews and other Writings 1977-1984,* London, Routledge.

Kroon, F. and Nola, R. (2001). 'Ramsification, Reference Fixing and Incommensurability' in P. Hoyningen-Huene and H. Sankey (eds.) *Incommensurability and Related Matters*, Dordrecht, Kluwer, pp. 91-121.

Kuhn, T. (1970) *The Structure of Scientific Revolutions*, Chicago, The University of Chicago Press (second edition with 'Postscript'; first edition 1962).

Kuhn, T, (1970a) 'Reflections on My Critics' in Lakatos and Musgrave (eds) 1970.

Kuhn, T. (1977) *The Essential Tension* Chicago, The University of Chicago Press.

Kuhn, T. (1983) 'Rationality and Theory Choice', *The Journal of Philosophy* **80**, 563-70.

Kuhn, T. (1991) 'The Road Since Structure', in Fine, A., Forbes, M. and Wessels, l. (eds) *PSA 1990, Volume Two*, East Lansing MI, Philosophy of Science Association, 3-13.

Kuhn, T. (1992) *The Trouble With the Historical Philosophy of Science; Robert and Maurine Rothschild Distinguished Lecture,* Cambridge MA: Department of the History of Science, Harvard University, 3-20.

Kuhn, T. (2000) *The Road Since Structure*, edited by J. Conant and J. Haugeland, Chicago, The University of Chicago Press.

Kuipers, T. (2000) *From Constructivism to Constructive Realism*, Dordrecht, Kluwer Academic Press.

Kukla, A. (2000) *Social Constructivism and the Philosophy of Science*, London, Routledge.

Kusch, M. (1991) *Foucault's Strata and Fields: An Investigation into Archaeological and Genealogical Science Studies*, Dordrecht, Kluwer Academic Publishers.

Kusch, M. (1996) 'Sociophilosophy and the Sociology of Philosophical Knowledge', in S. Knuuttila and I. Niiniluoto (eds.) *Methods of Philosophy and the History of Philosophy; Acta Philosophica Fennica*, **61** (Helsinki, Societas Philosophica Fennica), pp. 83-98.

Kusch, M. (2001) '"A General Theory of Social Knowledge?" Aspirations and Shortcomings of Alvin Goldman's Social Epistemology', *Studies in History and Philosophy of Science* **32A**, 183-92.

Lakatos, I. (1978) *The Methodlogy of Scientific Research Programmes: Philosophical Papers Volume I*, Cambridge, Cambridge University Press.

Lakatos, I. and Musgrave, A. (eds.) (1970) *Criticism and the Growth of Knowledge*, Cambridge, Cambridge University Press.

Latour, B. (1987) *Science in Action*, Cambridge MA, Harvard University Press.

Latour, B. (1988) *The Pasteurizatrion of France*, Cambridge MA, Harvard University Press.

Latour, B. and Woolgar, S. (1986) *Laboratory Life: The Construction of Scientific Facts*, Princeton NJ, Princeton University Press (second edition).

Laudan, L. (1977) *Progress and its Problems* London, Routledge and Kegan Paul.

Laudan, L. (1981) 'The Pseudo-Science of Science?', *Philosophy of Social Science* **11**, 173-98; reprinted in Laudan (1996), pp. 183-205.

Laudan, L. (1982) 'More on Bloor' *Philosophy of Social Science* **12**, 71-4; reprinted in Laudan (1996), pp. 205-9.

Laudan, L. (1984) *Science and Values*, Berkeley, University of California Press.

Laudan, L. (1990) *Science and Relativism: Some Key Controversies in the Philosophy of Science,* Chicago, The University of Chicago Press.

Laudan, L. (1996) *Beyond Positivsm and Relativism: Theory, Method and Evidence*, Boulder, Westview.

Laudan, L., Donovan, A., Laudan, R., Barker, P., Brown, H., Thagard, P. and Wykstra, S. (1986) 'Scientific Change: Philosophical Models and Historical Research', *Synthese* **69**, 141-22.

Leakey, R. (1981) *The Making of Mankind*, New York, Elsevier-Dutton.

Lennon, K. (1990) *Explaining Human Action*, Oxford, Blackwell.

Lewis, C. (1912) 'Implication and the Algebra of Logic', *Mind* **21**, 522-31.

Lewis, C. and Langford, C. (1959) *Symbolic Logic*, New York, Dover (second edition; first edition 1932).

Lewis, D. (1983) *Philosophical Papers Volume I*, New York, Oxford University Press.

Lewis, D. (1986) *Philosophical Papers Volume II*, New York, Oxford University Press.

Lewis, D. (1986a) *On the Plurality of Worlds,* Oxford, Blackwell.

Lewis, D. (1999) *Papers in Metaphysics and Epistemology*, Cambridge, Cambridge University Press.

Lewis, D. (2000) 'Causation as Influence', *The Journal of Philosophy* **XCVII**, 182-97.

Lewis, P. (2001) 'Why the Pessimistic Induction is a Fallacy', *Synthese* **129**, 371-80.

Locke, J. (1975) *An Essay Concerning Human Understanding*, Oxford, Clarendon.

Lukács, G. (1978) *Marx's Basic Ontological Principles*, London, The Merlin Press.

Lukes, S. (1974) *Power: A Radical View*, London, Macmillan

Lynch, M. (1993) *Scientific Practice and Ordinary Action*, Cambridge, Cambridge University Press.

Lyotard, J-F. (1984) *The Postmodern Condition: A Report on Knowledge*, Minneapolis, University of Minnesota Press.

Maccoby, H. (1999) 'Nietzsche's love-hate affair: Are life-affirming Jews nearer to Superman than decadent Christians?', *The Times Literary Supplement*, June 25, No. 5021, pp. 14-5.

Machlup, F. (1980) *Knowledge: Its Creation, Distribution, and Economic Significance. Volume I: Knowledge and Knowledge Production*, Princeton NJ, Princeton University Press.

MacKenzie, D. (1981) *Statistics in Britain: 1865-1930: The Social Construction of Knowledge*, Edinburgh, Edinburgh University Press.

Mackie, J. (1977) *Ethics: Inventing Right and Wrong*, Harmondsworth, Penguin.

Mannheim, K. (1936) *Ideology and Utopia*, London, Routledge and Kegan Paul.

Marion, M. (1998) *Wittgenstein, Finitism, and the Foundations of Mathematics*, Clarendon, Oxford.

Marx, K. (1970) *A Contribution to the Critique of Political Economy*, Moscow, Progress Publishers.

Marx, K. and and Engels. F. (1976) *Collected Works, Volume 5: The German Ideology*, London, Lawrence and Wishart.

Merquior, J. (1985) *Foucault*, London, Fontana.

Merton, R. (1973) *The Sociology of Science*, Chicago, The University of Chicago Press.

Merton, R. (1993) *On the Shoulders of Giants*, Chicago, The University of Chicago Press.

Musgrave, A. (1980) 'Kuhn's Second Thoughts', reprinted in G. Gutting (ed.) (1980).

Musgrave, A. (1999) *Essays on Realism and Rationality*, Amsterdam - Atlanta GA, Rodopi.

Nehamas, A. (1985) *Nietzsche: Life As Literature*, Cambridge MA, Harvard University Press.

Newton-Smith, W. (1985) 'The Role of Interests in Science', in A. Phillips Griffiths (ed.) *Philosophy and Practice: Royal Institute of Philosophy Lecture Series 18*, Cambridge, Cambridge University Press, pp. 59-73.

Nietzsche, F (1966) *Beyond Good and Evil*, translated by W. Kaufmann, New York, Vintage.

Nietzsche, F. (1968) *Twilight of the Idols* and *The Anti-Christ*, translated by R. Hollingdale, Harmondsworth, Penguin.

Nietzsche, F (1968) *The Will to Power*, edited by W. Kaufmann, New York, Vintage.

Nietzsche, F. (1969) *The Genealogy of Morals* and *Ecce Homo*, New York, Vintage.

Nietzsche, F (1974) *The Gay Science*, translated by W. Kaufmann, New York, Vintage.

Nietzsche, F (1979) 'On Truth and Lies in a Nonmoral Sense', in D. Breazeale (ed.) *Philosophy and Truth: Selections from Nietzsche's Notebooks of the Early 1870s*, Atlantic Highlands NJ, Humanities Press.

Nietzsche, F. (1982) *Daybreak: Thoughts on the Prejudices of Morality*, translated by R. Hollingdale, Cambridge, Cambridge University Press.

Nietzsche, F. (1986) *Human, All Too Human*, translated by R. Hollingdale, Cambridge, Cambridge University Press.

Nietzsche, F (1996) *The Genealogy of Morals*, translated by D. Smith, Oxford, Oxford University Press.

Niiniluoto, I. (1999) *Critical Scientific Realism*, Oxford, Clarendon.

Nola, R. (1987) 'Nietzsche's Theory of Truth and Belief', *Philosophy and Phenomenological Research* **47**, 525-62.

Nola, R. (ed.) (1988) *Relativism and Realism in Science*, Dordrecht, Kluwer.

Nola, R. (1991) 'Ordinary Human Inference as Refutation of the Strong Programme', *Social Studies of Science* **21**, 107-29.

Nola, R. (1999) 'On the Possibility of a Scientific Theory of Scientific Method', *Science & Education* **8**, 427-39.

Nola, R. and Sankey, H. (eds.) (2000) *After Popper, Kuhn and Feyerabend: Recent Issues in Theories of Scientific Method*, Dordrecht, Kluwer.

Nola, R. and Irzik, G. (2003, forthcoming) 'Incredulity Toward Lyotard: A Critique of a Postmodernist Account of Science and Knowledge', *Studies in History and Philosophy of Science*.

Nozick, R. (1981) *Philosophical Explanations*, Cambridge MA, Harvard University Press.

Okasha, S. (2000) 'The Underdetermination of Theory by Data and the "Strong Programme" in the Sociology of Knowledge, *International Studies in the Philosophy of Science* **14**, 283-97.

Pais, A.(1982) *'Subtle is the Lord ...': The Science and the Life of Albert Einstein*, Oxford, Oxford University Press.

Papineau, D. (1993) *Philosophcal Naturalism*, Oxford, Blackwell.

Pearl, J. (2000) *Causality: Models, Reasoning and Inference*, Cambridge, Cambridge University Press.

Pickering, A. (ed.) (1992) *Science as Practice and Culture*, Chicago, The University of Chicago Press.

Poellner, P. (1995) *Nietzsche and Metaphysics*, Oxford, Clarendon.

Popper, K. (1959) *The Logic of Scientific Discovery* London, Hutchinson (first published 1934).

Popper, K. (1962) *The Open Society and Its Enemies: Volume II: The High Tide of Prophecy: Hegel, Marx, and the Aftermath* London, Routledge and Kegan Paul.

Popper, K. (1963) *Conjectures and Refutations* London, Routledge and Kegan Paul.

Popper, K. (1972) *Objective Knowledge,* Oxford, Clarendon.

Popper, K. (1974) 'Replies to my Critics: II The Problem of Demarcation', in P. Schilpp (ed.) (1994) pp. 976-1013.

Preston, J. (1997) *Feyerabend: Philosophy, Science and Society*, Cambridge, Polity Press.

Preston, J. (2000) 'Science as Supermarket: 'Postmodern' Themes in Paul Feyerabend's Later Philosophy of Science' in Preston *et. al.* (2000).

Preston, J., Munévar, G. and Lamb, D. (eds.) (2000) *The Worst Enemy of Science? Essays in Memory of Paul Feyerabend*, Oxford, Oxford University Press.

Prior, A. (1960) 'The Autonomy of Ethics', *The Australasian Journal of Philosophy* **38**, 199-206.

Psillos, S. (1999) *Scientific Realism: How Science Tracks the Truth*, London, Routledge.

Putnam, H. (1975) *Mind, Language and Reality: Philosophical Papers Volume 2*, Cambridge, Cambridge University Press.

Putnam, H. (1978) *Meaning and the Moral Sciences*, London, Routledge and Kegan Paul.

Putnam, H. (1979) *Mathematics Matter and Method: Philosophical Papers Volume 1,* Cambridge, Cambridge University Press, second edition.

Putnam, H. (1990) *Realism With a Human Face*, Cambridge MA, Harvard University Press.

Quine, W. V. (1969) 'Epistemology Naturalized' in *Ontological Relativity and other Essays*, Cambridge MA, Harvard University Press, pp. 69-90.

Quine, W. (1992) *Pursuit of Truth*, Cambridge MA, Harvard University Press (revised edition).

Quine, W. and Ullian, J. (1978) *The Web of Belief*, New York, Random House.

Ramsey, F. P. (1990) *Philosophical Papers*, D. H. Mellor (ed.), Cambridge, Cambridge University Press.

Raz, J. (1975) *Practical Reason and Norms*, London, Hutchinson.

Reichenbach, H. (1949) *The Theory of Probability*, Berkeley, University of California Press.

Rosenau, P. M. (1992) *Post-Modernism and the Social Sciences*, Princeton NJ, Princeton University Press.

Ross, A. (ed.) (1996) *Science Wars*, Durham NC, Duke University Press.

Rouse, J. (1987) *Knowledge and Power: Toward a Political Philosophy of Science*, Ithaca, Cornell University Press.

Ruben, D-H. (ed.) (1993) *Explanation*, Oxford, Oxford University Press.

Russell, B. (1918) 'The Philosophy of Logical Atomism', reprinted in R. C. Marsh (ed) (1956) *Logic and Knowledge: Essays 1901-50*, London, George Allen & Unwin Ltd, pp. 175-281.

Salmon, W. (1967) *The Foundations of Scientific Inference*, Pittsburgh, University of Pittsburgh Press.

Salmon, W. (1990) *Four Decades of Scientific Explanation*, Minneapolis, University of Minnesota Press.

Salmon, W. (1990a) 'Rationality and Objectivity in Science, *or* Tom Bayes meets Tom Kuhn', in C. Wade Savage (ed.), *Scientific Theories: Minnesota Studies in the Philosophy of Science volume XIV*, University of Minnesota Press, Minneapolis.

Salmon, W. (1998) *Causality and Explanation*, New York, Oxford.

Samuels, R., Stich, S. and Tremoulet, P. (1999) 'Rethinking Rationality: From Bleak Implications to Darwinian Models', in E. Lepore and Z. Pylyshyn (eds) *What is Cognitive Science?* Oxford, Blackwell, pp. 74-120.

Sargent, R-M. (1995) *The Diffident Naturalist: Robert Boyle and the Philosophy of Experiment*, Chicago, The University of Chicago Press.

Sayer-McCord, G. (ed.) (1998) *Moral Realism*, Ithaca NY, Cornell University Press.

Scheler, M. (1972), *Ressentiment*, New York, Schocken Books (first published in German in 1912, and in an expanded edition of 1915).

Schilpp, P (ed.) (1974) *The Philosophy of Karl Popper*, La Salle IL, Open Court.

Schmitt, F. (ed.) (1994) *Socializing Epistemology: The Social Dimensions of Knowledge*, Lanham MD, Rowman and Littlefield.

Schurz, G. (1997) *The Is-Ought Problem*, Dordrecht, Kluwer Academic Publishers.

Searle, J. (1995) *The Construction of Social Reality*, London, Allen Lane Penguin Press.

Segerstråle, U. (1993) 'Bringing the Scientist Back In: The Need for an Alternative Sociology of Scientific Knowledge', in Brante, Fuller and Lynch (eds.) (1993), 57-82.

Sellars, W. (1963) *Science, Perception and Reality*, London, Routledge and Kegan Paul.

Shapere, D. (1982) 'The Concept of Observation in Science and Philosophy', *Philosophy of Science* **49**, 485 -585.

Shapin, S. (1994*) A Social History of Truth*, Chicago, The University of Chicago Press.

Shapin, S. and Schaffer, S. (1985) *Leviathan and the Air-Pump,* Princeton NJ, Princeton University Press.

Shope, R. (1983) *The Analysis of Knowing*, Princeton, Princeton University Press.

Sidgwick, H. (1884) *The Methods of Ethics*, London, Macmillan (third edition).

Simonds, A. (1978) *Karl Mannheim's Sociology of Knowledge*, Oxford, Clarendon.

Simons, P. (1987) *Parts: A Study in Ontology*, Oxford, Clarendon.

Slezak, P. (1989), 'Scientific Discovery by Computer as Empirical Refutation of the Strong Programme', *Social Studies of Science* **19**, 563-600.

Slezak, P. (1991), 'Bloor's Bluff: Behaviourism and the Strong Programme', *International Studies in the Philosophy of Science* **5**, 241-56.

Sober, E. (1998) *Reconstructing the Past: Parsimony, Evolution and Inference*, Cambridge MA, MIT Press.

Sokal, A. (1996a) 'Transgressing the Boundaries — Toward a Transformative Hermeneutics of Quantum Gravity', *Social Text* **14** (#1 &#2), 217-252; reprinted in Sokal and Bricmont (1998), pp. 199-240.

Sokal, A. (1996b) 'A Physicist Experiments with Cultural Studies' *Lingua Franca* (May-June), 62-64.

Sokal, A. and Bricmont, J. (1998) *Intellectual Impostures: Postmodernist Philosophers' Abuse of Science*, London, Profile Books.

Sosa, E. (1991) *Knowledge in Perspective: Selected Essays in Epistemology*, Cambridge, Cambridge University Press.

Sperber, D. (1996) *Explaining Culture: A Naturalistic Approach*, Oxford, Blackwell.

Stove, D. (1991) *The Plato Cult and Other Philosophical Follies*, Oxford, Blackwell.

Strawson, P. (1952) *Introduction to Logical Theory*, Methuen, London.

Stroud, B. (1966) 'Wittgenstein and Logical Necessity', in G. Pitcher (ed.) *Wittgenstein*, New York, Anchor.

Tanesini, A. (1995) 'Nietzsche's Theory of Truth', *The Australasian Journal of Philosophy* **73**, 548-59.

Tymoczko, T. (1979) 'The Four Colour Problem and its Philosophical Significance', *The Journal of Philosophy* **76**, 57-83.

van Fraassen, B. (1980) *The Scientific Image*, Oxford, Clarendon.

van Fraassen, B. (1989) *Laws and Symmetry*, Oxford, Clarendon.

von Wright, G. H. (1982) *Wittgenstein*, Oxford, Blackwell.

von Wright, G. H. (1983) *Practical Reason: Philosophical Papers Volume 1*, Oxford, Blackwell.

Weinberg, S. (1993) *Dreams of a Final Theory*, London, Hutchison Radius.

Weinberg, S. (2001) *Facing Up: Science and Its Cultural Adversaries*, Cambridge MA, Harvard University Press.

Weinert, F. (ed.) (1995), *Laws of Nature: Essays on the Philosophical, Scientific and Historical Dimensions*, Berlin, Walter de Gruyter.

Westfall, R. (1980), *Never at Rest: A Biography of Isaac Newton*, Cambridge, Cambridge University Press.

Wiggins, D. (1980), *Sameness and Substance*, Cambridge MA, Harvard University Press.

Will, C. (1993) *Was Einstein Right? Putting General Relativity to the Test*, Oxford, Oxford University Press (second edition).

Williams, B. (1978) *Descartes: The Project of Pure Enquiry*, Harmondsworth, Penguin.

Williams, L. (2001) *Nietzsche's Mirror: The World as Will to Power*, Lanham, Rowman & Littlefield.

Williams, M. (2001) *Problems of Knowledge: A Critical Introduction to Epistemology*, Oxford, Oxford University Press.

Windschuttle, K. (1996), *The Killing of History*, Sydney, Macleay Press, revised edition.

Wittgenstein, L. (1967) *Philosophical Investigations*, Blackwell, Oxford (third edition).

Wittgenstein, L. (1978) *Remarks on the Foundations of Mathematics*, Blackwell, Oxford (third edition).

Wittgenstein, L. (1980) *Culture and Value*, Chicago, The University of Chicago Press.

Wolterstorff, N. (1996) *John Locke and the Ethics of Belief*, New York, Cambridge University Press.

Wolff, K. (ed.) (1971) *From Karl Mannheim*, New York, Oxford University Press.

Wolpert, L. (1993) *The Unnatural Nature of Science*, Cambridge MA, Harvard University Press.

Yearly, S. (1982) 'The Relationship Between Epistemological and Sociological Cognitive Interests: Some Ambiguities Underlying the Use of Interest Theory in the Study of Scientific Knowledge', *Studies in the History and Philosophy of Science* **13**, 353-388.

NAME INDEX

Boston Studies in the Philosophy of Science

Editor: Robert S. Cohen, *Boston University*

Boston Studies in the Philosophy of Science

18. P. Mittelstaedt: *Philosophical Problems of Modern Physics.* Translated from the revised 4th German edition by W. Riemer and edited by R.S. Cohen. [Synthese Library 95] 1976
ISBN 90-277-0285-3; Pb 90-277-0506-2

19. H. Mehlberg: *Time, Causality, and the Quantum Theory.* Studies in the Philosophy of Science. Vol. I: *Essay on the Causal Theory of Time.* Vol. II: *Time in a Quantized Universe.* Translated from French. Edited by R.S. Cohen. 1980   Vol. I: ISBN 90-277-0721-9; Pb 90-277-1074-0
Vol. II: ISBN 90-277-1075-9; Pb 90-277-1076-7

20. K.F. Schaffner and R.S. Cohen (eds.): *PSA 1972.* Proceedings of the 3rd Biennial Meeting of the Philosophy of Science Association (Lansing, Michigan, Fall 1972). [Synthese Library 64] 1974
ISBN 90-277-0408-2; Pb 90-277-0409-0

21. R.S. Cohen and J.J. Stachel (eds.): *Selected Papers of Léon Rosenfeld.* [Synthese Library 100] 1979
ISBN 90-277-0651-4; Pb 90-277-0652-2

22. M. Čapek (ed.): *The Concepts of Space and Time.* Their Structure and Their Development. [Synthese Library 74] 1976
ISBN 90-277-0355-8; Pb 90-277-0375-2

23. M. Grene: *The Understanding of Nature.* Essays in the Philosophy of Biology. [Synthese Library 66] 1974
ISBN 90-277-0462-7; Pb 90-277-0463-5

24. D. Ihde: *Technics and Praxis.* A Philosophy of Technology. [Synthese Library 130] 1979
ISBN 90-277-0953-X; Pb 90-277-0954-8

25. J. Hintikka and U. Remes: *The Method of Analysis.* Its Geometrical Origin and Its General Significance. [Synthese Library 75] 1974
ISBN 90-277-0532-1; Pb 90-277-0543-7

26. J.E. Murdoch and E.D. Sylla (eds.): *The Cultural Context of Medieval Learning.* Proceedings of the First International Colloquium on Philosophy, Science, and Theology in the Middle Ages, 1973. [Synthese Library 76] 1975
ISBN 90-277-0560-7; Pb 90-277-0587-9

27. M. Grene and E. Mendelsohn (eds.): *Topics in the Philosophy of Biology.* [Synthese Library 84] 1976
ISBN 90-277-0595-X; Pb 90-277-0596-8

28. J. Agassi: *Science in Flux.* [Synthese Library 80] 1975
ISBN 90-277-0584-4; Pb 90-277-0612-3

29. J.J. Wiatr (ed.): *Polish Essays in the Methodology of the Social Sciences.* [Synthese Library 131] 1979
ISBN 90-277-0723-5; Pb 90-277-0956-4

30. P. Janich: *Protophysics of Time.* Constructive Foundation and History of Time Measurement. Translated from German. 1985
ISBN 90-277-0724-3

31. R.S. Cohen and M.W. Wartofsky (eds.): *Language, Logic, and Method.* 1983
ISBN 90-277-0725-1

32. R.S. Cohen, C.A. Hooker, A.C. Michalos and J.W. van Evra (eds.): *PSA 1974.* Proceedings of the 4th Biennial Meeting of the Philosophy of Science Association. [Synthese Library 101] 1976
ISBN 90-277-0647-6; Pb 90-277-0648-4

33. G. Holton and W.A. Blanpied (eds.): *Science and Its Public.* The Changing Relationship. [Synthese Library 96] 1976
ISBN 90-277-0657-3; Pb 90-277-0658-1

34. M.D. Grmek, R.S. Cohen and G. Cimino (eds.): *On Scientific Discovery.* The 1977 Erice Lectures. 1981
ISBN 90-277-1122-4; Pb 90-277-1123-2

35. S. Amsterdamski: *Between Experience and Metaphysics.* Philosophical Problems of the Evolution of Science. Translated from Polish. [Synthese Library 77] 1975
ISBN 90-277-0568-2; Pb 90-277-0580-1

36. M. Marković and G. Petrović (eds.): *Praxis.* Yugoslav Essays in the Philosophy and Methodology of the Social Sciences. [Synthese Library 134] 1979
ISBN 90-277-0727-8; Pb 90-277-0968-8

Boston Studies in the Philosophy of Science

37. H. von Helmholtz: *Epistemological Writings*. The Paul Hertz / Moritz Schlick Centenary Edition of 1921. Translated from German by M.F. Lowe. Edited with an Introduction and Bibliography by R.S. Cohen and Y. Elkana. [Synthese Library 79] 1977
ISBN 90-277-0290-X; Pb 90-277-0582-8

38. R.M. Martin: *Pragmatics, Truth and Language*. 1979
ISBN 90-277-0992-0; Pb 90-277-0993-9

39. R.S. Cohen, P.K. Feyerabend and M.W. Wartofsky (eds.): *Essays in Memory of Imre Lakatos*. [Synthese Library 99] 1976
ISBN 90-277-0654-9; Pb 90-277-0655-7

40. Not published.

41. Not published.

42. H.R. Maturana and F.J. Varela: *Autopoiesis and Cognition*. The Realization of the Living. With a Preface to "Autopoiesis' by S. Beer. 1980
ISBN 90-277-1015-5; Pb 90-277-1016-3

43. A. Kasher (ed.): *Language in Focus: Foundations, Methods and Systems*. Essays in Memory of Yehoshua Bar-Hillel. [Synthese Library 89] 1976
ISBN 90-277-0644-1; Pb 90-277-0645-X

44. T.D. Thao: *Investigations into the Origin of Language and Consciousness*. 1984
ISBN 90-277-0827-4

45. F.G.-I. Nagasaka (ed.): *Japanese Studies in the Philosophy of Science*. 1997
ISBN 0-7923-4781-1

46. P.L. Kapitza: *Experiment, Theory, Practice*. Articles and Addresses. Edited by R.S. Cohen. 1980
ISBN 90-277-1061-9; Pb 90-277-1062-7

47. M.L. Dalla Chiara (ed.): *Italian Studies in the Philosophy of Science*. 1981
ISBN 90-277-0735-9; Pb 90-277-1073-2

48. M.W. Wartofsky: *Models*. Representation and the Scientific Understanding. [Synthese Library 129] 1979
ISBN 90-277-0736-7; Pb 90-277-0947-5

49. T.D. Thao: *Phenomenology and Dialectical Materialism*. Edited by R.S. Cohen. 1986
ISBN 90-277-0737-5

50. Y. Fried and J. Agassi: *Paranoia*. A Study in Diagnosis. [Synthese Library 102] 1976
ISBN 90-277-0704-9; Pb 90-277-0705-7

51. K.H. Wolff: *Surrender and Cath*. Experience and Inquiry Today. [Synthese Library 105] 1976
ISBN 90-277-0758-8; Pb 90-277-0765-0

52. K. Kosík: *Dialectics of the Concrete*. A Study on Problems of Man and World. 1976
ISBN 90-277-0761-8; Pb 90-277-0764-2

53. N. Goodman: *The Structure of Appearance*. [Synthese Library 107] 1977
ISBN 90-277-0773-1; Pb 90-277-0774-X

54. H.A. Simon: *Models of Discovery* and Other Topics in the Methods of Science. [Synthese Library 114] 1977
ISBN 90-277-0812-6; Pb 90-277-0858-4

55. M. Lazerowitz: *The Language of Philosophy*. Freud and Wittgenstein. [Synthese Library 117] 1977
ISBN 90-277-0826-6; Pb 90-277-0862-2

56. T. Nickles (ed.): *Scientific Discovery, Logic, and Rationality*. 1980
ISBN 90-277-1069-4; Pb 90-277-1070-8

57. J. Margolis: *Persons and Mind*. The Prospects of Nonreductive Materialism. [Synthese Library 121] 1978
ISBN 90-277-0854-1; Pb 90-277-0863-0

58. G. Radnitzky and G. Andersson (eds.): *Progress and Rationality in Science*. [Synthese Library 125] 1978
ISBN 90-277-0921-1; Pb 90-277-0922-X

59. G. Radnitzky and G. Andersson (eds.): *The Structure and Development of Science*. [Synthese Library 136] 1979
ISBN 90-277-0994-7; Pb 90-277-0995-5

Boston Studies in the Philosophy of Science

Boston Studies in the Philosophy of Science

83. É. Meyerson: *The Relativistic Deduction.* Epistemological Implications of the Theory of Relativity. Translated from French. With a Review by Albert Einstein and an Introduction by Milič Čapek. 1985   ISBN 90-277-1699-4
84. R.S. Cohen and M.W. Wartofsky (eds.): *Methodology, Metaphysics and the History of Science.* In Memory of Benjamin Nelson. 1984   ISBN 90-277-1711-7
85. G. Tamás: *The Logic of Categories.* Translated from Hungarian. Edited by R.S. Cohen. 1986
                                                                                ISBN 90-277-1742-7
86. S.L. de C. Fernandes: *Foundations of Objective Knowledge.* The Relations of Popper's Theory of Knowledge to That of Kant. 1985   ISBN 90-277-1809-1
87. R.S. Cohen and T. Schnelle (eds.): *Cognition and Fact.* Materials on Ludwik Fleck. 1986
                                                                                ISBN 90-277-1902-0
88. G. Freudenthal: *Atom and Individual in the Age of Newton.* On the Genesis of the Mechanistic World View. Translated from German. 1986   ISBN 90-277-1905-5
89. A. Donagan, A.N. Perovich Jr and M.V. Wedin (eds.): *Human Nature and Natural Knowledge.* Essays presented to Marjorie Grene on the Occasion of Her 75th Birthday. 1986
                                                                                ISBN 90-277-1974-8
90. C. Mitcham and A. Hunning (eds.): *Philosophy and Technology II.* Information Technology and Computers in Theory and Practice. [*Also* Philosophy and Technology Series, Vol. 2] 1986
                                                                                ISBN 90-277-1975-6
91. M. Grene and D. Nails (eds.): *Spinoza and the Sciences.* 1986   ISBN 90-277-1976-4
92. S.P. Turner: *The Search for a Methodology of Social Science.* Durkheim, Weber, and the 19th-Century Problem of Cause, Probability, and Action. 1986.   ISBN 90-277-2067-3
93. I.C. Jarvie: *Thinking about Society.* Theory and Practice. 1986   ISBN 90-277-2068-1
94. E. Ullmann-Margalit (ed.): *The Kaleidoscope of Science.* The Israel Colloquium: Studies in History, Philosophy, and Sociology of Science, Vol. 1. 1986
                                                      ISBN 90-277-2158-0; Pb 90-277-2159-9
95. E. Ullmann-Margalit (ed.): *The Prism of Science.* The Israel Colloquium: Studies in History, Philosophy, and Sociology of Science, Vol. 2. 1986
                                                      ISBN 90-277-2160-2; Pb 90-277-2161-0
96. G. Márkus: *Language and Production.* A Critique of the Paradigms. Translated from French. 1986   ISBN 90-277-2169-6
97. F. Amrine, F.J. Zucker and H. Wheeler (eds.): *Goethe and the Sciences: A Reappraisal.* 1987
                                                      ISBN 90-277-2265-X; Pb 90-277-2400-8
98. J.C. Pitt and M. Pera (eds.): *Rational Changes in Science.* Essays on Scientific Reasoning. Translated from Italian. 1987   ISBN 90-277-2417-2
99. O. Costa de Beauregard: *Time, the Physical Magnitude.* 1987   ISBN 90-277-2444-X
100. A. Shimony and D. Nails (eds.): *Naturalistic Epistemology.* A Symposium of Two Decades. 1987   ISBN 90-277-2337-0
101. N. Rotenstreich: *Time and Meaning in History.* 1987   ISBN 90-277-2467-9
102. D.B. Zilberman: *The Birth of Meaning in Hindu Thought.* Edited by R.S. Cohen. 1988
                                                                                ISBN 90-277-2497-0
103. T.F. Glick (ed.): *The Comparative Reception of Relativity.* 1987   ISBN 90-277-2498-9
104. Z. Harris, M. Gottfried, T. Ryckman, P. Mattick Jr, A. Daladier, T.N. Harris and S. Harris: *The Form of Information in Science.* Analysis of an Immunology Sublanguage. With a Preface by Hilary Putnam. 1989   ISBN 90-277-2516-0
105. F. Burwick (ed.): *Approaches to Organic Form.* Permutations in Science and Culture. 1987
                                                                                ISBN 90-277-2541-1

Boston Studies in the Philosophy of Science

106. M. Almási: *The Philosophy of Appearances.* Translated from Hungarian. 1989
ISBN 90-277-2150-5

107. S. Hook, W.L. O'Neill and R. O'Toole (eds.): *Philosophy, History and Social Action.* Essays in Honor of Lewis Feuer. With an Autobiographical Essay by L. Feuer. 1988
ISBN 90-277-2644-2

108. I. Hronszky, M. Fehér and B. Dajka: *Scientific Knowledge Socialized.* Selected Proceedings of the 5th Joint International Conference on the History and Philosophy of Science organized by the IUHPS (Veszprém, Hungary, 1984). 1988 ISBN 90-277-2284-6

109. P. Tillers and E.D. Green (eds.): *Probability and Inference in the Law of Evidence.* The Uses and Limits of Bayesianism. 1988 ISBN 90-277-2689-2

110. E. Ullmann-Margalit (ed.): *Science in Reflection.* The Israel Colloquium: Studies in History, Philosophy, and Sociology of Science, Vol. 3. 1988
ISBN 90-277-2712-0; Pb 90-277-2713-9

111. K. Gavroglu, Y. Goudaroulis and P. Nicolacopoulos (eds.): *Imre Lakatos and Theories of Scientific Change.* 1989 ISBN 90-277-2766-X

112. B. Glassner and J.D. Moreno (eds.): *The Qualitative-Quantitative Distinction in the Social Sciences.* 1989 ISBN 90-277-2829-1

113. K. Arens: *Structures of Knowing.* Psychologies of the 19th Century. 1989
ISBN 0-7923-0009-2

114. A. Janik: *Style, Politics and the Future of Philosophy.* 1989 ISBN 0-7923-0056-4

115. F. Amrine (ed.): *Literature and Science as Modes of Expression.* With an Introduction by S. Weininger. 1989 ISBN 0-7923-0133-1

116. J.R. Brown and J. Mittelstrass (eds.): *An Intimate Relation.* Studies in the History and Philosophy of Science. Presented to Robert E. Butts on His 60th Birthday. 1989
ISBN 0-7923-0169-2

117. F. D'Agostino and I.C. Jarvie (eds.): *Freedom and Rationality.* Essays in Honor of John Watkins. 1989 ISBN 0-7923-0264-8

118. D. Zolo: *Reflexive Epistemology.* The Philosophical Legacy of Otto Neurath. 1989
ISBN 0-7923-0320-2

119. M. Kearn, B.S. Philips and R.S. Cohen (eds.): *Georg Simmel and Contemporary Sociology.* 1989 ISBN 0-7923-0407-1

120. T.H. Levere and W.R. Shea (eds.): *Nature, Experiment and the Science.* Essays on Galileo and the Nature of Science. In Honour of Stillman Drake. 1989 ISBN 0-7923-0420-9

121. P. Nicolacopoulos (ed.): *Greek Studies in the Philosophy and History of Science.* 1990
ISBN 0-7923-0717-8

122. R. Cooke and D. Costantini (eds.): *Statistics in Science.* The Foundations of Statistical Methods in Biology, Physics and Economics. 1990 ISBN 0-7923-0797-6

123. P. Duhem: *The Origins of Statics.* Translated from French by G.F. Leneaux, V.N. Vagliente and G.H. Wagner. With an Introduction by S.L. Jaki. 1991 ISBN 0-7923-0898-0

124. H. Kamerlingh Onnes: *Through Measurement to Knowledge.* The Selected Papers, 1853-1926. Edited and with an Introduction by K. Gavroglu and Y. Goudaroulis. 1991
ISBN 0-7923-0825-5

125. M. Čapek: *The New Aspects of Time: Its Continuity and Novelties.* Selected Papers in the Philosophy of Science. 1991 ISBN 0-7923-0911-1

126. S. Unguru (ed.): *Physics, Cosmology and Astronomy, 1300–1700.* Tension and Accommodation. 1991 ISBN 0-7923-1022-5

Boston Studies in the Philosophy of Science

127. Z. Bechler: *Newton's Physics on the Conceptual Structure of the Scientific Revolution.* 1991
ISBN 0-7923-1054-3
128. É. Meyerson: *Explanation in the Sciences.* Translated from French by M-A. Siple and D.A. Siple. 1991
ISBN 0-7923-1129-9
129. A.I. Tauber (ed.): *Organism and the Origins of Self.* 1991   ISBN 0-7923-1185-X
130. F.J. Varela and J-P. Dupuy (eds.): *Understanding Origins.* Contemporary Views on the Origin of Life, Mind and Society. 1992   ISBN 0-7923-1251-1
131. G.L. Pandit: *Methodological Variance.* Essays in Epistemological Ontology and the Methodology of Science. 1991   ISBN 0-7923-1263-5
132. G. Munévar (ed.): *Beyond Reason.* Essays on the Philosophy of Paul Feyerabend. 1991
ISBN 0-7923-1272-4
133. T.E. Uebel (ed.): *Rediscovering the Forgotten Vienna Circle.* Austrian Studies on Otto Neurath and the Vienna Circle. Partly translated from German. 1991   ISBN 0-7923-1276-7
134. W.R. Woodward and R.S. Cohen (eds.): *World Views and Scientific Discipline Formation.* Science Studies in the [former] German Democratic Republic. Partly translated from German by W.R. Woodward. 1991   ISBN 0-7923-1286-4
135. P. Zambelli: *The Speculum Astronomiae and Its Enigma.* Astrology, Theology and Science in Albertus Magnus and His Contemporaries. 1992   ISBN 0-7923-1380-1
136. P. Petitjean, C. Jami and A.M. Moulin (eds.): *Science and Empires.* Historical Studies about Scientific Development and European Expansion.   ISBN 0-7923-1518-9
137. W.A. Wallace: *Galileo's Logic of Discovery and Proof.* The Background, Content, and Use of His Appropriated Treatises on Aristotle's *Posterior Analytics.* 1992   ISBN 0-7923-1577-4
138. W.A. Wallace: *Galileo's Logical Treatises.* A Translation, with Notes and Commentary, of His Appropriated Latin Questions on Aristotle's *Posterior Analytics.* 1992   ISBN 0-7923-1578-2
Set (137 + 138) ISBN 0-7923-1579-0
139. M.J. Nye, J.L. Richards and R.H. Stuewer (eds.): *The Invention of Physical Science.* Intersections of Mathematics, Theology and Natural Philosophy since the Seventeenth Century. Essays in Honor of Erwin N. Hiebert. 1992   ISBN 0-7923-1753-X
140. G. Corsi, M.L. dalla Chiara and G.C. Ghirardi (eds.): *Bridging the Gap: Philosophy, Mathematics and Physics.* Lectures on the Foundations of Science. 1992   ISBN 0-7923-1761-0
141. C.-H. Lin and D. Fu (eds.): *Philosophy and Conceptual History of Science in Taiwan.* 1992
ISBN 0-7923-1766-1
142. S. Sarkar (ed.): *The Founders of Evolutionary Genetics.* A Centenary Reappraisal. 1992
ISBN 0-7923-1777-7
143. J. Blackmore (ed.): *Ernst Mach – A Deeper Look.* Documents and New Perspectives. 1992
ISBN 0-7923-1853-6
144. P. Kroes and M. Bakker (eds.): *Technological Development and Science in the Industrial Age.* New Perspectives on the Science–Technology Relationship. 1992   ISBN 0-7923-1898-6
145. S. Amsterdamski: *Between History and Method.* Disputes about the Rationality of Science. 1992   ISBN 0-7923-1941-9
146. E. Ullmann-Margalit (ed.): *The Scientific Enterprise.* The Bar-Hillel Colloquium: Studies in History, Philosophy, and Sociology of Science, Volume 4. 1992   ISBN 0-7923-1992-3
147. L. Embree (ed.): *Metaarchaeology.* Reflections by Archaeologists and Philosophers. 1992
ISBN 0-7923-2023-9
148. S. French and H. Kamminga (eds.): *Correspondence, Invariance and Heuristics.* Essays in Honour of Heinz Post. 1993   ISBN 0-7923-2085-9
149. M. Bunzl: *The Context of Explanation.* 1993   ISBN 0-7923-2153-7

Boston Studies in the Philosophy of Science

Boston Studies in the Philosophy of Science

171. M.A. Grodin (ed.): *Meta Medical Ethics*: The Philosophical Foundations of Bioethics. 1995
ISBN 0-7923-3344-6

172. S. Ramirez and R.S. Cohen (eds.): *Mexican Studies in the History and Philosophy of Science.* 1995
ISBN 0-7923-3462-0

173. C. Dilworth: *The Metaphysics of Science.* An Account of Modern Science in Terms of Principles, Laws and Theories. 1995
ISBN 0-7923-3693-3

174. J. Blackmore: *Ludwig Boltzmann, His Later Life and Philosophy, 1900–1906* Book Two: The Philosopher. 1995
ISBN 0-7923-3464-7

175. P. Damerow: *Abstraction and Representation.* Essays on the Cultural Evolution of Thinking. 1996
ISBN 0-7923-3816-2

176. M.S. Macrakis: *Scarcity's Ways: The Origins of Capital.* A Critical Essay on Thermodynamics, Statistical Mechanics and Economics. 1997
ISBN 0-7923-4760-9

177. M. Marion and R.S. Cohen (eds.): *Québec Studies in the Philosophy of Science.* Part I: Logic, Mathematics, Physics and History of Science. Essays in Honor of Hugues Leblanc. 1995
ISBN 0-7923-3559-7

178. M. Marion and R.S. Cohen (eds.): *Québec Studies in the Philosophy of Science.* Part II: Biology, Psychology, Cognitive Science and Economics. Essays in Honor of Hugues Leblanc. 1996
ISBN 0-7923-3560-0
Set (177–178) ISBN 0-7923-3561-9

179. Fan Dainian and R.S. Cohen (eds.): *Chinese Studies in the History and Philosophy of Science and Technology.* 1996
ISBN 0-7923-3463-9

180. P. Forman and J.M. Sánchez-Ron (eds.): *National Military Establishments and the Advancement of Science and Technology.* Studies in 20th Century History. 1996
ISBN 0-7923-3541-4

181. E.J. Post: *Quantum Reprogramming.* Ensembles and Single Systems: A Two-Tier Approach to Quantum Mechanics. 1995
ISBN 0-7923-3565-1

182. A.I. Tauber (ed.): *The Elusive Synthesis: Aesthetics and Science.* 1996   ISBN 0-7923-3904-5

183. S. Sarkar (ed.): *The Philosophy and History of Molecular Biology: New Perspectives.* 1996
ISBN 0-7923-3947-9

184. J.T. Cushing, A. Fine and S. Goldstein (eds.): *Bohmian Mechanics and Quantum Theory: An Appraisal.* 1996
ISBN 0-7923-4028-0

185. K. Michalski: *Logic and Time.* An Essay on Husserl's Theory of Meaning. 1996
ISBN 0-7923-4082-5

186. G. Munévar (ed.): *Spanish Studies in the Philosophy of Science.* 1996   ISBN 0-7923-4147-3

187. G. Schubring (ed.): *Hermann Günther Graßmann (1809–1877): Visionary Mathematician, Scientist and Neohumanist Scholar.* Papers from a Sesquicentennial Conference. 1996
ISBN 0-7923-4261-5

188. M. Bitbol: *Schrödinger's Philosophy of Quantum Mechanics.* 1996   ISBN 0-7923-4266-6

189. J. Faye, U. Scheffler and M. Urchs (eds.): *Perspectives on Time.* 1997   ISBN 0-7923-4330-1

190. K. Lehrer and J.C. Marek (eds.): *Austrian Philosophy Past and Present.* Essays in Honor of Rudolf Haller. 1996
ISBN 0-7923-4347-6

191. J.L. Lagrange: *Analytical Mechanics.* Translated and edited by Auguste Boissonade and Victor N. Vagliente. Translated from the *Mécanique Analytique, novelle édition* of 1811. 1997
ISBN 0-7923-4349-2

192. D. Ginev and R.S. Cohen (eds.): *Issues and Images in the Philosophy of Science.* Scientific and Philosophical Essays in Honour of Azarya Polikarov. 1997   ISBN 0-7923-4444-8

Boston Studies in the Philosophy of Science

193. R.S. Cohen, M. Horne and J. Stachel (eds.): *Experimental Metaphysics*. Quantum Mechanical Studies for Abner Shimony, Volume One. 1997   ISBN 0-7923-4452-9
194. R.S. Cohen, M. Horne and J. Stachel (eds.): *Potentiality, Entanglement and Passion-at-a-Distance*. Quantum Mechanical Studies for Abner Shimony, Volume Two. 1997
                                        ISBN 0-7923-4453-7; Set 0-7923-4454-5
195. R.S. Cohen and A.I. Tauber (eds.): *Philosophies of Nature: The Human Dimension*. 1997
                                                        ISBN 0-7923-4579-7
196. M. Otte and M. Panza (eds.): *Analysis and Synthesis in Mathematics*. History and Philosophy. 1997   ISBN 0-7923-4570-3
197. A. Denkel: *The Natural Background of Meaning*. 1999   ISBN 0-7923-5331-5
198. D. Baird, R.I.G. Hughes and A. Nordmann (eds.): *Heinrich Hertz: Classical Physicist, Modern Philosopher*. 1999   ISBN 0-7923-4653-X
199. A. Franklin: *Can That be Right?* Essays on Experiment, Evidence, and Science. 1999
                                                        ISBN 0-7923-5464-8
200. D. Raven, W. Krohn and R.S. Cohen (eds.): *The Social Origins of Modern Science*. 2000
                                                        ISBN 0-7923-6457-0
201. Reserved
202. Reserved
203. B. Babich and R.S. Cohen (eds.): *Nietzsche, Theories of Knowledge, and Critical Theory*. Nietzsche and the Sciences I. 1999   ISBN 0-7923-5742-6
204. B. Babich and R.S. Cohen (eds.): *Nietzsche, Epistemology, and Philosophy of Science*. Nietzsche and the Science II. 1999   ISBN 0-7923-5743-4
205. R. Hooykaas: *Fact, Faith and Fiction in the Development of Science*. The Gifford Lectures given in the University of St Andrews 1976. 1999   ISBN 0-7923-5774-4
206. M. Fehér, O. Kiss and L. Ropolyi (eds.): *Hermeneutics and Science*. 1999 ISBN 0-7923-5798-1
207. R.M. MacLeod (ed.): *Science and the Pacific War*. Science and Survival in the Pacific, 1939-1945. 1999   ISBN 0-7923-5851-1
208. I. Hanzel: *The Concept of Scientific Law in the Philosophy of Science and Epistemology*. A Study of Theoretical Reason. 1999   ISBN 0-7923-5852-X
209. G. Helm; R.J. Deltete (ed./transl.): *The Historical Development of Energetics*. 1999
                                                        ISBN 0-7923-5874-0
210. A. Orenstein and P. Kotatko (eds.): *Knowledge, Language and Logic*. Questions for Quine. 1999   ISBN 0-7923-5986-0
211. R.S. Cohen and H. Levine (eds.): *Maimonides and the Sciences*. 2000   ISBN 0-7923-6053-2
212. H. Gourko, D.I. Williamson and A.I. Tauber (eds.): *The Evolutionary Biology Papers of Elie Metchnikoff*. 2000   ISBN 0-7923-6067-2
213. S. D'Agostino: *A History of the Ideas of Theoretical Physics*. Essays on the Nineteenth and Twentieth Century Physics. 2000   ISBN 0-7923-6094-X
214. S. Lelas: *Science and Modernity*. Toward An Integral Theory of Science. 2000
                                                        ISBN 0-7923-6303-5
215. E. Agazzi and M. Pauri (eds.): *The Reality of the Unobservable*. Observability, Unobservability and Their Impact on the Issue of Scientific Realism. 2000   ISBN 0-7923-6311-6
216. P. Hoyningen-Huene and H. Sankey (eds.): *Incommensurability and Related Matters*. 2001 ISBN 0-7923-6989-0
217. A. Nieto-Galan: *Colouring Textiles*. A History of Natural Dyestuffs in Industrial Europe. 2001
                                                        ISBN 0-7923-7022-8

Boston Studies in the Philosophy of Science

Also of interest:

R.S. Cohen and M.W. Wartofsky (eds.): *A Portrait of Twenty-Five Years Boston Colloquia for the Philosophy of Science, 1960-1985*. 1985   ISBN Pb 90-277-1971-3

Previous volumes are still available.

KLUWER ACADEMIC PUBLISHERS – DORDRECHT / BOSTON / LONDON